Advances in Intelligent Systems and Computing

Volume 965

The series "Advances in Intelligent Systems and Computing" contains publications on theory, applications, and design methods of Intelligent Systems and Intelligent Computing. Virtually all disciplines such as engineering, natural sciences, computer and information science, ICT, economics, business, e-commerce, environment, healthcare, life science are covered. The list of topics spans all the areas of modern intelligent systems and computing such as: computational intelligence, soft computing including neural networks, fuzzy systems, evolutionary computing and the fusion of these paradigms, social intelligence, ambient intelligence, computational neuroscience, artificial life, virtual worlds and society, cognitive science and systems, Perception and Vision, DNA and immune based systems, self-organizing and adaptive systems, e-Learning and teaching, human-centered and human-centric computing, recommender systems, intelligent control, robotics and mechatronics including human-machine teaming, knowledge-based paradigms, learning paradigms, machine ethics, intelligent data analysis, knowledge management, intelligent agents, intelligent decision making and support, intelligent network security, trust management, interactive entertainment, Web intelligence and multimedia.

The publications within "Advances in Intelligent Systems and Computing" are primarily proceedings of important conferences, symposia and congresses. They cover significant recent developments in the field, both of a foundational and applicable character. An important characteristic feature of the series is the short publication time and world-wide distribution. This permits a rapid and broad dissemination of research results.

** Indexing: The books of this series are submitted to ISI Proceedings, EI-Compendex, DBLP, SCOPUS, Google Scholar and Springerlink **

More information about this series at http://www.springer.com/series/11156

Tareq Ahram
Editor

Advances in Artificial Intelligence, Software and Systems Engineering

Proceedings of the AHFE 2019 International
Conference on Human Factors in Artificial Intelligence
and Social Computing, the AHFE International
Conference on Human Factors, Software, Service
and Systems Engineering, and the AHFE International
Conference of Human Factors in Energy,
July 24–28, 2019, Washington D.C., USA

 Springer

Editor
Tareq Ahram
Institute for Advanced Systems Engineering
University of Central Florida
Orlando, FL, USA

ISSN 2194-5357   ISSN 2194-5365   (electronic)
Advances in Intelligent Systems and Computing
ISBN 978-3-030-20453-2   ISBN 978-3-030-20454-9   (eBook)
https://doi.org/10.1007/978-3-030-20454-9

This Springer imprint is published by the registered company Springer Nature Switzerland AG
The registered company address is: Gewerbestrasse 11, 6330 Cham, Switzerland

Advances in Human Factors and Ergonomics 2019

AHFE 2019 Series Editors

Tareq Ahram, Florida, USA
Waldemar Karwowski, Florida, USA

10th International Conference on Applied Human Factors and Ergonomics and the Affiliated Conferences

Proceedings of the AHFE 2019 International Conference on Human Factors in Artificial Intelligence and Social Computing, the AHFE International Conference on Human Factors, Software, Service and Systems Engineering, and the AHFE International Conference of Human Factors in Energy, held on July 24–28, 2019, in Washington D.C., USA

Advances in Affective and Pleasurable Design	Shuichi Fukuda
Advances in Neuroergonomics and Cognitive Engineering	Hasan Ayaz
Advances in Design for Inclusion	Giuseppe Di Bucchianico
Advances in Ergonomics in Design	Francisco Rebelo and Marcelo M. Soares
Advances in Human Error, Reliability, Resilience, and Performance	Ronald L. Boring
Advances in Human Factors and Ergonomics in Healthcare and Medical Devices	Nancy J. Lightner and Jay Kalra
Advances in Human Factors and Simulation	Daniel N. Cassenti
Advances in Human Factors and Systems Interaction	Isabel L. Nunes
Advances in Human Factors in Cybersecurity	Tareq Ahram and Waldemar Karwowski
Advances in Human Factors, Business Management and Leadership	Jussi Ilari Kantola and Salman Nazir
Advances in Human Factors in Robots and Unmanned Systems	Jessie Chen
Advances in Human Factors in Training, Education, and Learning Sciences	Waldemar Karwowski, Tareq Ahram and Salman Nazir
Advances in Human Factors of Transportation	Neville Stanton

(continued)

(continued)

Advances in Artificial Intelligence, Software and Systems Engineering	Tareq Ahram
Advances in Human Factors in Architecture, Sustainable Urban Planning and Infrastructure	Jerzy Charytonowicz and Christianne Falcão
Advances in Physical Ergonomics and Human Factors	Ravindra S. Goonetilleke and Waldemar Karwowski
Advances in Interdisciplinary Practice in Industrial Design	Cliff Sungsoo Shin
Advances in Safety Management and Human Factors	Pedro M. Arezes
Advances in Social and Occupational Ergonomics	Richard H. M. Goossens and Atsuo Murata
Advances in Manufacturing, Production Management and Process Control	Waldemar Karwowski, Stefan Trzcielinski and Beata Mrugalska
Advances in Usability and User Experience	Tareq Ahram and Christianne Falcão
Advances in Human Factors in Wearable Technologies and Game Design	Tareq Ahram
Advances in Human Factors in Communication of Design	Amic G. Ho
Advances in Additive Manufacturing, Modeling Systems and 3D Prototyping	Massimo Di Nicolantonio, Emilio Rossi and Thomas Alexander

Preface

Researchers and business leaders are called to address important challenges caused by the increasing presence of artificial intelligence and social computing in the workplace environment and daily lives. Roles that have traditionally required a high level of cognitive abilities, decision making, and training (human intelligence) are now being automated. The *AHFE International Conference on Human Factors in Artificial Intelligence and Social Computing* promotes the exchange of ideas and technology enabling humans to communicate and interact with machines in almost every area and for different purposes. The recent increase in machine and systems intelligence has led to a shift from the classical human–computer interaction to a much more complex, cooperative human–system work environment requiring a multidisciplinary approach. The first part of this book deals with those new challenges and presents contributions on different aspects of artificial intelligence, social computing, and social network modeling taking into account those modern, multifaceted challenges.

The *AHFE International Conference on Human Factors, Software, and Systems Engineering* provides a platform for addressing challenges in human factors, software, and systems engineering pushing the boundaries of current research. In the second part of the book, researchers, professional software and systems engineers, human factors, and human systems integration experts from around the world discuss next-generation systems to address societal challenges. The book covers cutting-edge software and systems engineering applications, systems and service design, and user-centered design. Topics span from analysis of evolutionary and complex systems, to issues in human systems integration and applications in smart grid, infrastructure, training, education, defense, and aerospace.

The third part of the book reports on the *AHFE International Conference of Human Factors in Energy*, addressing oil, gas, nuclear, and electric power industries. It covers human factors/systems engineering research for process control and discusses new energy business models.

All in all, this book reports on one of the most informative systems engineering event of the year.

The research papers included here have been reviewed by members of the International Editorial Board, whom our sincere thanks and appreciation goes to. They are listed below:

Software, and Systems Engineering/Artificial Intelligence and Social Computing

Human Factors in Energy

Lauren Reinerman-Jones, USA
Kristiina Söderholm, Finland

We hope that this book, which reports on the international state of the art in human factors research and applications, will be a valuable source of knowledge enabling human-centered design of a variety of products, services, and systems for global markets.

July 2019 Tareq Ahram

Contents

Software, Service and Systems Engineering

Artificial Intelligence and Social Computing

Using Information Processing Strategies to Predict Message Level Contagion in Social Media

Sara Levens[(⊠)], Omar ElTayeby, Tiffany Gallicano,
Michael Brunswick, and Samira Shaikh

University of North Carolina at Charlotte, 9201 University City Blvd,
Charlotte, NC 28223-0001, USA
{slevens, oeltayeb, tgallica, mbrunswi,
sshaikh2}@uncc.edu

Abstract. Social media content can have extensive online influence [1], but assessing offline influence using online behavior is challenging. Cognitive information processing strategies offer a potential way to code online behavior that may be more predictive of offline preferences, beliefs, and behavior than counting retweets or likes. In this study, we employ information processing strategies, particularly depth of processing, to assess message-level influence. Tweets from the Charlottesville protest in August 2017 were extracted with favorite count, retweet count, quote count, and reply count for each tweet. We present methods and formulae that incorporate favorite counts, retweet counts, quote counts, and reply counts in accordance with depth of processing theory to assess message-level contagion. Tests assessing the association between our message-level depth of processing estimates and user-level influence indicate that our formula are significantly associated with user level influence, while traditional methods using likes and retweet counts are less so.

Keywords: Depth of processing · Emotion contagion · Influence ·
Social media

1 Motivation

Social media has become a major distributor of news and social commentary. As the influence of social media on society increases, so too does the need to measure that influence on the behavior of individuals and groups. One measure of reach is a high follower to friend ratio, referred to as the *golden Twitter ratio*; it is indicative of influence as a function of a profile's social network size [2]. Although the golden ratio may be a good proxy for a user's reach, it does not provide a framework for assessing the potential influence or contagion of individual messages or posts. Furthermore, there are not many ways to assess the potential offline impact of an online message. A common approach to evaluating a post's impact is to tabulate the number of shares it receives [3], but we believe that this approach is limited and likely has a low capacity for assessing the offline impact of an online message. In the present paper, we present

© Springer Nature Switzerland AG 2020
T. Ahram (Ed.): AHFE 2019, AISC 965, pp. 3–13, 2020.
https://doi.org/10.1007/978-3-030-20454-9_1

an alternative method for measuring the offline contagion of online information, which is guided by an information processing theory from the field of cognition.

Information processing strategies offer a potential way to analyze online behavior that may be more predictive of offline preferences, beliefs, and behavior than 'like' or 'retweet' counts. In this paper, we propose formulas that are based on information processing theory, particularly depth of processing, to assess message-level influence on social media.

1.1   Assessing Message Level Influence Using Depth of Processing

While there are a number of factors that influence depth of processing, one of the most powerful factors is based on whether someone created the message themselves, which is known as the *generation effect* [4]. Extensive research has shown that information constructed by individuals themselves tends to be better remembered than information created by others [4, 5]. This is because individuals construct content based on their experiences, knowledge, and interests, which leads to an increase in brain connectivity and synaptic development. Consequently, the generation effect is often measured through learning and memory-based tasks that require individuals to recall content. For example, the number of words accurately recalled and the speed of recall have been used to measure depth of processing in learning and memory research [4–7]. However, these methods are primarily laboratory-based and as such are not feasible to use when measuring content generation on social media. New methods need to be developed for indexing learning and the depth of processing that occurs on social media in the naturalistic environment in which it occurs.

In an attempt to reconcile research on the generation effect that has been conducted offline with the context of online social interaction, we present a novel model of message-level contagion. We developed our model to explain cognitive processing in an SNS (social networking service) context based on the degree of content generation. Our goal is to develop a method that incorporates depth of processing principles to predict message influence. We test our method by first examining the association between weighted variable formulas of Twitter response and user-level influence as indexed by the golden ratio, and second by examining the association between weighted variable formulas of Twitter response and contagion potential as measured by emotional intensity of the tweet content. We assume that each distinct reaction towards the tweets corresponds to a different cognitive level of processing.

We contend that a favorite is the simplest interaction in response to a tweet. Retweeting requires greater processing than favoriting because after reading the content and pushing a button to retweet, the user is required to confirm the desire to retweet and is invited to add a comment before pressing a button to confirm the retweet. We also expect a tweet receives more processing by people who share it because they are likely to be more selective of what they share with others than what they simply choose to favorite. Quoted tweets require even more processing than retweets do because a user must think of and write original text to accompany the message. Replies generally require greater cognitive effort than quoted tweets do because they involve directing the communication at one or more people in particular. They are also prominently displayed in the "tweets & replies" section of a user's profile. Original tweets require the greatest depth of processing because they tend to require the greatest effort (Fig. 1).

Fig. 1. Depth of processing based on cognitive and physical effort

2   Data Collection

To assess message-level influence on Twitter, datasets must contain counts of all response options for each post; however, our initial data pull excluded some of the original tweets that inspired replies and quoted tweets because the original tweets did not match our search criteria. To create as complete a picture as we could for a given event, we conducted a second round of data collection using the Twitter API and collected the missing tweets. This was accomplished by simply querying the Twitter API with the ID of each original tweet that was contained within the quoted or replied-to tweet. In our dataset there are 42,273 quoted tweets in total, the most quoted tweet was quoted 425 times in our dataset. There are 126,331 replied to tweets in total,

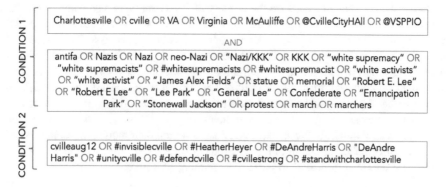

Fig. 2. List of hashtags and keywords used to collect our data corpus for Charlottesville protest event. The hashtags were split into two Conditions. In Condition 1, there are two sets of keywords and hashtags and the search criteria is that the tweet should match at least one item from each set. Condition 2 is a set of key words, where the search criteria is to match at least one item from the set.

the most replied to tweet was replied to 932 times. In the next section, we describe in more depth how we obtained a complete dataset with favorite, retweet, quoted tweet, and reply counts for each original tweet.

2.1   Obtaining the Dataset

We have outlined the stages below that describe our data collection:

1. Downloaded tweets matching our rules through GNIP's Historical PowerTrack.
2. Used Twitter's API to collect the missing tweets that were referenced in the quoted tweets in our original dataset.
3. Applied web scraping to gather the number of replies for those quoted tweets.

First, we used GNIP's Historical PowerTrack to download all tweets matching the keyword and timespan rules that we defined. We started with a collection of 807,954 tweets that were posted within the period of February 7 and October 10, 2017. The total number of unique users in this corpus was 335,183. The corpus was collected using a diverse set of keywords rules for querying the tweets as listed below. The tweets contained 807,954 original tweets and 4718,329 retweets of English and non-English languages. We used 2 conditions to pull the data. Condition 1 is a combination of two sets of words that must be combined to identify a relevant tweet. For Condition 2, the tweets should match one of the words in that set.

More depth how we obtained a complete dataset with favorite, retweet, quoted tweet, and reply counts for each original tweet (Fig. 2).

Next, we identified those tweets that were most often quoted in our corpus. Then, we again queried Twitter to download these tweets, since these were not in our original corpus. We used Twitter GNIP API to pull them.

We first sorted the quoted tweets by the frequency of which they were quoted to prioritize the order in which they were collected. Because the API has a rate limit, we first experimented with collecting the top 500 quoted tweets (comprising the initial dataset), and then gathered the top 5,000 tweets (comprising the expanded dataset). We collected the top quoted tweets on the January 2019. There were also some tweets that we failed to collect because they were deleted before we started the collection process, which resulted in a final expanded dataset of 4,122 tweets (instead of 5000 tweets). In accordance with Twitter's policies, these tweets are prohibited from inclusion in research anyway. In addition, it is important to consider that these tweets might not have contained the keyword rules that we specified in the first stage when we used GNIP. The reason is that those original quoted tweets are related to the topic of the Charlottesville protest, and they are an important piece in measuring the influence of the followers on a message level. Finally, for these collected tweets, we find their number of replies to have complete information about all of the reactions they received. We used the well-known HTML parser BeautifulSoup[1] to parse their number of replies.

---

[1] https://www.crummy.com/software/BeautifulSoup/.

2.2 Finite Options for How People Can Interact with Tweets

It is also important to keep in mind that it is impossible for a user on Twitter to quote and reply at the same time, because quoting is like retweeting with a comment and retweeting is categorized as a separate action by itself from replying. However, it is possible to find tweets that have been quoted and replied to at the same time, so we parsed the number of replies directly through the link of the quoted tweet using the HTML parser. Our findings are organized by presenting an initial dataset of 500 tweets, followed by an expanded dataset, in which we applied the same techniques to obtain a full dataset of 4,122 tweets to more fully test our directed generation models.

2.3 Data Sample

In Table 1 we present a sample of the collected tweets that have complete information about their number of favorites and retweets, as well as the number of times they have been quoted and replied to.

Table 1. Five representative examples of the tweets collected.

Display name	Tweet body	#Fav	#RT	#Quote	#Rep
George Wallace	That little baby grew up believing in goodness and justice and standing up to hate. Heather Heyer. Say her name. https://t.co/82noPD35uU	133632	48099	425	1139
Donald J. Trump	We ALL must be united & condemn all that hate stands for. There is no place for this kind of violence in America. Lets come together as one!	181635	53815	386	63022
Donald J. Trump	Condolences to the family of the young woman killed today, and best regards to all of those injured, in Charlottesville, Virginia. So sad!	103739	21196	257	59971
NowThis	This journalist was on the ground at Charlottesville and everyone needs to hear her story https://t.co/iaExx9pEVf	42518	16072	113	612
NBC News	Heather Heyer's mother concludes: "I'd rather have my child, but by golly, if I gotta give her up, we're gonna make it count https://t.co/yxmqPhodSR	14655	18572	106	3764

3   Method

For the purpose of measuring the extent to which tweets were contagiousness, we developed three simple formulas, where each formula is represented by the sum of the number of reactions (favorite, retweet, quote and reply) and is weighted with different values. Based on the generation effect phenomenon [4], we used the number of each type of reaction that a tweet received. We used the number of each type of reaction (referred to later in this study as the raw metric) as input into a formula to create a metric of the contagion level of the message. Our goal is to know the total representative picture of each tweet without leaving out any indicator that would be relevant for assessing the contagion level. Thus, in these developed formulas we included the raw metric for each type of interaction to get a broad picture of the contagion level rather than using one-dimensional measures, such as favorites. Each raw metric indicates a different level of cognition and depth of processing. Excluding one of the raw metrics from the formula is equivalent to missing a signal that represents one of the cognitive levels. Our formula uses the combination of the raw metrics to capture nuanced insight about how deeply people engaged with the tweet.

The scheme of the formulas is structured as the sum of the raw metrics, where each metric is multiplied by a different weight. Specifically, each raw metric is multiplied by a weight that corresponds to a different depth of processing level. Specifically, the weighting of each raw metric depends on the depth of processing of the information in the message to which the follower reacted. This depth of processing is presented by the efforts made by the follower in response to the message. As stated in the introduction, when followers respond to a message, their response (e.g., a like or reply) indicates how much time or effort they expended to process the message. We consider that cognition levels range from favoriting (requiring the least amount of effort) to replying (requiring the most effort). Therefore, the favorite is weighted with the lowest factor, and replies are weighted with the highest factor. As the ideal weight of each variable in predicting depth of processing is unknown, we test three initial formulae that differentially weigh each raw metric:

$$W\_1234 = (\#Favorites * 1) + (\#Retweets * 2) + (\#Quotes * 3) + (\#Replies * 4)$$

(1)

$$W\_1244 = (\#Favorites * 1) + (\#Retweets * 2) + (\#Quotes * 4) + (\#Replies * 4)$$

(2)

$$W\_1246 = (\#Favorites * 1) + (\#Retweets * 2) + (\#Quotes * 4) + (\#Replies * 6)$$

(3)

In the subscript area following the W (indicating the weight), the weight of each interaction is listed in ascending order, and the multiplication of these numbers produces the desired weighting for measuring contagion. Each equation is an attempt towards testing the underlying representation of the ratios among types of interactions and the differences in cognition. Equation W_1234 is the straightforward formula that tests the linearity among the cognition levels; the weights are linearly incremented as we go up the cognition levels from 1 to 4. In equation W_1244, we test whether quoted tweets and replies should be weighted equally to reflect a similar depth of processing, and they are both weighted exponentially higher than favorites and retweets because they both involve some degree of original content generation. Finally, in equation W_1246, we continue to test the more significant weighting on original content while significantly weighting original content more heavily than a quoted tweet.

4 Evaluation and Results

In order to evaluate our depth of processing theory and test our formulas, we ran correlations among our weighted formulas, our raw metrics, and a measure that has been associated with greater influence and contagion at the user level (the golden ratio) as well as with a measure of emotion content. Emotional content was measured by the famous Linguistic Inquiry and Word Count (LIWC) engine by calculating the percentages of different emotions expressed in the tweets [8]. In Table 2, we show the Pearson correlations between our formulas and the golden ratio (user level influence), and in Table 3, we show the Pearson correlations between our variables and the emotional intensity of the tweets, which signals the potential of message contagion potential.

4.1 Correlation with User-Level Influence

Table 2 illustrates the association between the raw metrics, our formulas, and the golden ratio. The white cells reflect the Pearson correlation for the initial 500 dataset, and the grey-shaded cells reflect the Pearson correlations for the expanded 4,122 dataset. As illustrated in Table 2, the golden ratio has a significant association in the 500 dataset with quotes and replies but not with likes or retweets. Interestingly, all three of our formulas show a significant association with the golden ratio, validating our basic theory that synergizing all raw metrics and weighting them according to depth of processing confers value. In the expanded 4,122 tweet dataset, the favorite, retweet, and reply metrics all correlate with the golden ratio, but replies are much more predictive than likes and retweets—again validating our belief that responses are indicative of a comparatively greater depth of processing. Furthermore, replicating the same basic patterns from the initial dataset in the expanded dataset confirms that all of the formulas predict the golden ratio. Moreover, the formulas are a stronger predictor than merely examining likes and retweets, which are common metrics examined in the context of viral and contagious content.

Finally, in our examination and comparison of the three formulas, we concluded that one of the formulas is consistently a stronger predictor than the other two based on having higher Pearson correlation values. Specifically, formula W_1246 has stronger correlations than formulas W_1244 and W_1234, suggesting that an exponential weighting schema that has replies as exponentially (versus linearly) higher in weighting than quotes, retweets, and likes respectively is a better predictor of user-level influence than linear weighting schemas that merely increase the multiplication number by one for each increase in the level of response.

Table 2. Correlations among Twitter follower, friends, the golden ratio variable, and raw message metrics (e.g., favorite counts) vs. our weighted variables for the initial dataset of 500 tweets (coded as the white cells) and the expanded dataset of 4,122 (coded as the grey cells).

	GR	Fav	RT	Qt	Rep	W_1234	W_1244	W_1246
GR	-	0.071	0.033	.408*	.764*	.113*	.113*	.140*
Fav	.053**	-	.978*	.139*	.484*	.995*	.995*	.993*
RT	.032*	.962**	-	.149*	.447*	.989*	.989*	.986*
Qt	-.001	.165**	.180**	-	.498*	.177*	.178*	.193*
Rep	.327**	.425**	.363**	-.001	-	.530*	.530*	.557*
W_1234	.087**	.989**	.975**	.205**	.515**	-	1.000*	.999*
W_1244	.087**	.989**	.975**	.205**	.515**	1.000**	-	.999*
W_1246	.106**	.981**	.964**	.219**	.563**	.998**	.998**	-

Note: * = p<.01, ** = p<.001

4.2   Correlation with Emotions

To further validate our approach in another domain associated with message contagion, we examine associations between our raw metric variables, formula values and emotion intensity metrics. Emotional content is associated with contagion [9–11], and multiple studies have demonstrated that emotional content has unique viral properties [12–14]. Emotional content is also associated with higher depth of processing [15]. Accordingly, higher emotional content should be associated with greater depth of processing—and as our formulas are designed to reflect depth of processing on social media, our formulas should also be associated with the level of the emotional content. To test this, we used LIWC. Then, we conducted correlations between our raw metrics, formulas, and the affective metrics from LIWC (positive emotion [posemo], negative emotion [negemo], anxiety, anger, sadness). Results are in accordance with our theory because our formulas are better at predicting emotion than raw metrics, and they combine all metrics into one variable.

Table 3. Correlations among Twitter favorites, retweets, quoted tweets, and replies in the 4,100 tweet dataset and emotion variables, as depicted by LIWC.

	Fav	RT	Qt	Rep	W_1234	W_1244	W_1246
Posemo	.028t	.024	−.009	.066**	.034*	.034*	.037*
Negemo	.027t	.022	−.010	.034*	.028t	.028t	.029t
Anxiety	.001	.003	−.004	.017	.004	.004	.005
Anger	.038*	.037*	−.007	.009	.037*	.037*	.036*
Sadness	−.002	−.007	−.003	.009	−.001	−.001	.000

Note: * = p < .01, ** = p < .001

5 Discussion

The purpose of this paper was to discover an effective method for predicting depth of processing and, by extension, an estimated likelihood of offline behavior. We pursued this goal by testing three formulas' abilities to predict the golden mean at a significant level. All three formulas performed at a significant level, and we identified the formula that functioned the best for this task, which turned out to be the one involving the most exponential amount of weighting among the levels of effort involved in reacting to Twitter messages. We used the golden ratio as the test for the formulas because it is a measure that indicates influence that is independent of the measures we used. The golden ratio also allows us to examine the link between message level impact and user level influence. Users who participate in online Twitter conversations on events of significance are more likely to develop a twitter following if there messages are impactful and impactful and emotional messages are more likely to be processed at greater depth.

This study makes a significant contribution in several ways. It adopts an innovative approach to predicting offline behavior by utilizing an information processing strategy, specifically, depth of processing. In doing so, the study provides empirical data supporting the position that there are vast differences in the extent of processing that occurs between simplistic Twitter responses (e.g., favoriting a tweet, retweeting a tweet) and Twitter responses that involve writing original content in some way, whether that content accompanies a quoted tweet or is a reply tweet. This study is significant in its testing of the influence each type of Twitter response has on depth of processing by exploring the extent to which three formulae predicted the golden mean. The study's results provide further validation to the significance of the generation effect in influencing depth of processing based on the relatively heavier weighting of tweets involving original content generation. And finally, the method of this study can be helpful to others who confront the problem of having quoted tweets and reply tweets in a dataset without the originating tweets that inspired them.

The favoriting action occurred substantially more often than the quoted tweets, and quoted tweets vastly outnumbered replies in the dataset. This uneven distribution supports our assignment of heavier weights to responses that require a greater amount of work. The uneven distribution also supports conventional power laws in which a small number of people perform most of the work, sometimes referred to as the 80/20 rule in which 20% of users perform 80% of the work [16]. In pure power laws, the jump between the first and second position is substantially larger than the jump between the second and third position and so forth [16].

With consideration to the inverse of the formulas tested in this paper, influencers have earned their following through the ability to inspire their followers to engage with their content. In this paper, we found a significant correlation between depth of processing and influencers, who were identified through their golden ratio of having a significant number of followers despite not following many people comparatively. This study offers general validation to communication strategies that integrate influencers as effective message sources.

6 Conclusions and Future Work

The purpose of this paper was to discover an effective method for predicting depth of processing on social media and, by extension, an estimated likelihood of offline behavior. We pursued this goal by testing three formulas' abilities to predict the golden mean at a significant level. All three formulas performed at a significant level, and we identified the formula that functioned the best for this task, which turned out to be the one involving the most exponential amount of weighting among the levels of effort involved in reacting to Twitter messages. We used the golden mean as the test for the formulas because it is a measure that indicates influence that is independent of the measures we used. To strengthen the internal validity of this research, a study could run the formulas in the context of a random group of users who are not united by tweeting about the same event.

Future research can explore the hidden audience that reads tweets but does not even favorite them. Although the data from this study suggests that the audience that does not even favorite a tweet has the least amount of processing, we expect that other reasons also exist and that this is a heterogenous group. Specifically, we believe some might not leave a mark because they don't like the tweet and do not want to be confrontational by posting a reply or quoted tweet, whereas others might engage in a deep level of processing but desire anonymity for social reasons, professional reasons, or both categories.

Other future research can consider other factors that influence the level of processing information receives. For example, highly emotional material, regardless of valence, receives greater depth of processing and is recalled more easily [15]. Accordingly, emotional content has a greater capacity for influencing our behavior. Depth of processing has also been found to be strongly impacted by repetition. When individuals readily attend to material, repetition greatly improves learning, thus indicating a greater depth of processing [17]. Repeated exposures can even promote learning when individuals are disengaged [18]. Accordingly, information encountered repeatedly on social media has a greater capacity to influence behavior. Thus, the strongest equation in this study can be expanded to incorporate additional measures.

Next steps also include the use of machine learning to determine the optimum weighting formula to best predict message level influence. By working with industry partners or conducting mixed method experiments that combine online and offline behavior, we can construct a dataset to assess the association between online behavior (or its absence in hidden audiences) and offline behavior. We can use a data driven and machine-learning approach to determine the optimal weights for our algorithm to incorporate the full range of social media response behaviors to detect message contagion. Ideally this algorithm would be flexible enough to function as a test of encoding and learning in an online environment.

Also, future research is needed to test the formulas' prediction of a new kind of influencer, known as a nanoinfluencer [19]. Nanoinfluencers have the golden Twitter ratio on a small scale—they have no more than 5,000 followers and typically are highly specialized in a niche area. Testing the formulas used in this paper can help communication strategists assess how nanoinfluencers compare to major influencers with regard to the depth of processing scores they achieve, controlling for the number of followers each type of influencer has.

In addition, this work can be extended into a corporate context by assessing the strength of ties that exist between companies and the people who follow them on social media. By collecting longitudinal data, researchers can analyze changes in depth of processing over time as a measure of a company's communications function.

Acknowledgements. This work was supported in part by funding from the Charlotte Research Institute Targeted Research Internal Seed Program.

References

1. Tufecki, Z., Wilson, C.: Social media and the decision to participate in political protest: observations from Tahrir Square. J. Commun. **62**, 363–379 (2012)
2. Parsons, J.: What is The Ideal Twitter Follower to Following Ratio? Web blog post, 1 April 2017. Follows.com
3. Kelly, N.: 4 ways to measure social media and its impact on your brand. Web blog post, 15 June 2010. Socialmediaexaminer.com
4. Bertsch, S., et al.: The generation effect: a meta-analytic review. Mem. Cogn. **35**(2), 201–210 (2007)
5. Winstanley, P., Ligon Bjork, E.: Processing instructions and the generation effect: a test of the multifactor transfer-appropriate processing theory. Memory **5**(3), 401–422 (1997)
6. Symons, C.S., Johnson, B.T.: The self-reference effect in memory: a meta-analysis. Psychol. Bull. **121**(3), 371 (1997)
7. Stahl, S.A., Fairbanks, M.M.: The effects of vocabulary instruction: a model-based meta-analysis. Rev. Educ. Res. **56**(1), 72–110 (1986)
8. Pennebaker, J.W., Francis, M.E., Booth, R.J.: Linguistic Inquiry and Word Count. LIWC 2001, vol. 71. Lawrence Erlbaum Associates, Mahway (2001)
9. Ahmed, S., Jaidka, K., Cho, J.: The 2014 Indian elections on Twitter: a comparison of campaign strategies of political parties. Telematics Inform. **33**(4), 1071–1087 (2016)
10. Fowler, J.H., Christakis, N.A.: Dynamic spread of happiness in a large social network: longitudinal analysis over 20 years in the Framingham Heart Study. BMJ **337**, a2338 (2008)
11. Kramer, A.D.I., Guillory, J.E., Hancock, J.T.: Experimental evidence of massive-scale emotional contagion through social networks. Proc. Natl. Acad. Sci. **111**(24), 8788–8790 (2014): 201320040
12. Fan, R., Xu, K., Zhao, J.: Easier contagion and weaker ties make anger spread faster than joy in social media. arXiv preprint arXiv:1608.03656 (2016)
13. Brady, W.J., et al.: Emotion shapes the diffusion of moralized content in social networks. Proc. Natl. Acad. Sci. **114**(28), 7313–7318 (2017)
14. Berger, J., Milkman, K.L.: What makes online content viral? J. Mark. Res. **49**(2), 192–205 (2012)
15. Xu, X., et al.: Effects of level of processing on emotional memory: gist and details. Cogn. Emot. **25**(1), 53–72 (2011)
16. Shirky, C.: Here Comes Everybody: The Power of Organizing Without Organizations. Penguin, New York (2008)
17. Sawyer, A.G.: The effects of repetition of refutational and supportive advertising appeals. J. Mark. Res. **10**(1), 23–33 (1973)
18. Krugman, H.E.: The impact of television advertising: learning without involvement. Public Opin. Quart. **29**(3), 349–356 (1965)
19. Karwowski, M.: 5 Reasons Your Brand Needs Nanoinfluencers in 2019. Web blog post, 28 January 2019. Campaign US

EmoVis – An Interactive Visualization Tool to Track Emotional Trends During Crisis Events

Samira Shaikh[✉], Karthik Ravi, Tiffany Gallicano,
Michael Brunswick, Bradley Aleshire, Omar El-Tayeby,
and Sara Levens

University of North Carolina at Charlotte, Charlotte, NC 28262, USA
{samirashaikh, kravil, tgallica, mbrunswi, baleshil,
oeltayeb, slevens}@uncc.edu

Abstract. The goal of this research is to develop a novel tool that can aid social science researchers in inferring emotional trends over large-scale cultural stressors. We demonstrate the usefulness of the tool in describing the emotional timeline of a major crisis event – the 2017 Charlottesville protests. The tool facilitates understanding of how large-scale cultural stressors yield changes in emotional responses. The timeline tool describes the modulation of emotional intensity with respect to how the Charlottesville event unfolded on Twitter. We have developed multiple features associated with the tool that tailor the presentation of the data, including the ability to focus on single or multiple emotions (e.g., anger and anxiety) and also delineate the timeline based on events that precede crises events, in this case, the Charlottesville protests. By doing so, we can begin to identify potential antecedents to various protest phenomena and their accompanying emotional responses.

Keywords: Emotion analysis · Natural language processing · Crisis events · Interactive tools · Social media analysis

1 Motivation

Communication about large-scale social events has primarily shifted to social media in recent years [1]. Concomitantly, content from traditional media platforms such as news outlets is guided by perspectives that are detached from actual events [2]. Social media, on the other hand, can be used to disseminate information as an event unfolds. Individuals witnessing events firsthand can post content as events change and evolve. As a result, local events can rapidly grow to become the national or international sociocultural issues that involve individuals who otherwise would not have been emotionally involved [3].

An example of a local event that entered the national consciousness due to social media was the 2017 Charlottesville protest that occurred in Virginia, USA. Despite an abundance of news coverage, it remains unclear how specific emotions changed and unfolded over the course of this protest event. In addition, we know very little about

© Springer Nature Switzerland AG 2020
T. Ahram (Ed.): AHFE 2019, AISC 965, pp. 14–24, 2020.
https://doi.org/10.1007/978-3-030-20454-9_2

how information influences people's emotional experience over the course of a long-term dynamic series of related events, such as those that characterize the Charlottesville protests.

Understanding how information modifies emotions expressed over the course of many months is critical for elucidating the role of emotions in dynamic large-scale cultural events. Emotions have been shown to engender certain actions tendencies; for example, anger can increase activism [4] and sadness can increase donations [5], suggesting that emotion may be critical for catalyzing social movements. Accordingly, case study research on emotion responses on social media is critical for building theory on emotionally influenced actions, enhancing the ability to predict emotional responses to various types of situations and to contribute to professionals' ability to develop public responses to emotionally salient and meaningful events.

Our goal is to build and elucidate the emotional timelines of crises events that encompass longer timescales, spanning not only weeks but months ahead of the crisis event to understand how large-scale cultural stressors yield changes in emotional responding. To situate our methodological approach, we selected an event of national and international significance that involved a large time range of social media activity and intense emotional responses from individuals on social media. Listed below are the main contributions of this study:

1. We make available a *large corpus* of approximately 8 million tweets relating to the Charlottesville event from February to October 2017, as well as the real-world event details that preceded and followed the Charlottesville protest.
2. We make available an *interactive visualization tool* using RShiny with a simple web interface that can be adapted to different crisis events and timelines. This tool can aggregate, normalize, and plot the changes in emotion for the timeline of the protest event. In this paper, we showcase five emotions for the purpose of analysis: anger, anxiety, sadness, positive emotion, and negative emotion.
3. We describe our approach through sample analyses and present *critical inferences* from the Charlottesville protest emotion timeline.

Our interactive visualization tool allows us to make inferences and describe the modulation of emotional intensity with respect to the crisis event. This tool can be used to describe factors and mechanisms that may cause events to yield different emotional responses. In the next section of this paper, we explain our contributions in the context of prior work.

2   Related Work

The creation of the emotional timeline tool is an interdisciplinary effort; as such, we draw upon literature pertinent to emotion analysis from natural language text and the design of visualization tools to aid in the analysis of large-scale heterogeneous data. We describe related work in both areas in the sections below.

2.1   Emotion Analysis from Social Media Text

There is a great body of work from psychology that theorizes about emotions [6, 7]. For instance, Ekman [6] identified the six basic emotions as anger, disgust, fear, happiness, sadness, and surprise. Prior research in automated analysis of emotions has focused on training machine-learning classifiers that automatically discover the emotions in tweets. EmoTex [8] applies supervised learning methods to detect emotions. Recently, researchers in the natural language processing community [9, 10] have introduced shared tasks of detecting the intensity of emotion felt by the producer of a social media message. State-of-the-art systems in these shared tasks [11, 12] use approaches of different ensemble models and applied feature vectors, including word embeddings, semantic characteristics, and syntactic features to represent text. The current approaches towards automated detection of emotions in text rely heavily on advances in neural networks, specifically recurrent neural networks and convolutional neural networks to accurately predict emotions, in not only English but also other languages [13, 14]. However, neural networks do not facilitate interpretable results, which traditional, bag-of-words approaches offer. Accordingly, we use a traditional dictionary-based approach to automatically label emotions in social media messages.

2.2   Visualization Tools for Analysis of Timeline Data

Emotion analysis can be an insightful way to understand people's reactions to major events, such as how emotions spread during the course of 9/11 and Hurricane Katrina [15, 16]. Digital media has ushered in the age of mass amateurization in which anyone can be a writer, publisher, and photographer [17]. Rather than a traditional one-to-many model of communication, digital media foster an environment of many-to-many dialogue [17]. The many-to-many dialogue can, however, create challenges when analysts and researchers are trying to distill overall messages and topic trends.

Researchers have focused on combining visualization and automated analytical approaches to improve the ability to approach timeline data with sophisticated analysis techniques and conduct particularly high-level and complex tasks. For example, in the medical field, scientists have focused on enhancing the understanding of qualitative and quantitative data through a timeline visualization in the context of patients with pulmonary embolism [18]. Urban studies researchers have used timeline-based visual analytics to explore data and develop pattern interpretations to guide the development of smart cities [19].

Time is one of the most important properties of emotional trend analysis. The state-of-the-art approaches in natural language processing of emotions from social media text and their analysis [20] lay the important groundwork in extraction of emotion, but presenting this data in an accessible form that allows users to draw individual conclusions and observations is difficult. Timeline analysis has been used with visualization tools to demonstrate the effectiveness of achieving various tasks related to temporal sentiment and affective analysis of tweets [21]. Early work in this area probed highly significant events in our history, namely 9/11, to examine how emotions changed in response to specific incidents that occurred on that day [15]. These researchers examined language content in pager texts to evaluate emotional trends over time [15]. Another prior work developed an interactive analytic tool to support multi-dimensional emotion analysis from social media text to automatically detect an individual's emotions expressed at different times and summarize those emotions to reveal their emotional style. The disadvantage of this tool is that it is limited to only one person [22]. Taking into account a larger number of people can aid in understanding the emotional dynamics that give rise to social and political movements, as well as large scale reactions to crisis events [20].

3   Method

The goal of the present paper is to present a tool that meets the following key objectives: *Objective (1)* present large amounts of data in an organized and meaningful fashion that facilitates the observation of data trends of emotions; *Objective (2)* allow the user to tailor the presentation of that data based on their needs or preferences; *Objective (3)* organize the presentation of the data as a function of time to enable examination of unfolding of dynamic events; *Objective (4)* be potentially modifiable to present a variety of text features.

3.1   Data Collection

To achieve key Objective 1, we identified a significant emotional event that unfolded over a long period of time: the 2017 Charlottesville protest. The 2017 Charlottesville protest was a local event that entered the national consciousness due to social media. In Charlottesville, Virginia, white nationalists and counter-protestors clashed violently following the decision to remove a Robert E. Lee statue, placing a national focus on white supremacy and creating a backlash that heavily condemned racism and hate [23, 24]. Despite an abundance of news coverage, it remained unclear how specific emotions changed and unfolded over the course of Charlottesville.

To obtain a comprehensive database of tweets relating to the Charlottesville protest, we synthesized ~ 8 million tweets that contained at least one of the keywords or hashtags shown in Fig. 1. We used 2 conditions to pull the data. Condition 1 is a combination of two sets of words that must be combined to identify a relevant tweet. For Condition 2, the tweets should match one of the words in that set.

We obtained the data using the Twitter GNIP Historical PowerTrack API [25]. After the process of filtering duplicates, our final dataset consisted of 807,954 tweets that were posted between February 7 and October 11, 2017, from 335,183 unique Twitter IDs. We extracted the text of the tweet, the timestamp it was posted on, and the information of the user who made the tweet from this resulting data for further pre-processing. In the text of the tweet, we replaced all instances of URLs, numbers, hashtags, and user mentions with <url>, <number>, <hashtag>, and <user> respectively. We also ignored the occurrence of emojis in tweets for the present work, in keeping with our focus on language analysis. We plan to include emoji analysis as part of our future work.

Fig. 1. List of hashtags and keywords used to collect our data corpus for Charlottesville protest event. The hashtags were split into two Conditions. In Condition 1, there are two sets of keywords and hashtags and the search criteria is that the tweet should match at least one item from each set. Condition 2 is a set of keywords, where the search criteria is to match at least one item from the set.

3.2   Automated Analysis of Emotion

The Linguistics Inquiry and Word Count (LIWC) tool is a widely used and validated software for identifying text associated with themes and thought processes, feelings, personality, and motivation [26]. While we present a focus on emotional content in this article, in accordance with *Objective 4*, the tool can easily be modified to present other LIWC features of interest, enabling a broader range of social and psychological insights.

Using the pre-processed text data, we used the LIWC [26] tool to compute the proportion of words in each tweet that relate to (a) sadness (e.g., *crying, grief*), (b) anxiety (e.g., *worried, fearful*), (c) anger (e.g., *hate, kill, annoyed*), and the higher level emotional categories of (d) affect (e.g., *happy, cried*), (e) positive emotion (e.g., *love, nice, sweet*), and (f) negative emotion (e.g., *hurt, ugly, nasty*). The output of LIWC is a numeric integer that measures the presence of each emotion (or feature) in the text. We then aggregate the individually computed emotion values for tweets posted per minute. Doing this makes the data less computationally expensive for future operations of visualization and interactive options in the tool. This process also smoothens the distribution of data by reducing the fluctuations of individual emotion over the range of 0 to 60 s. In addition, for displaying the entire timeline from February 7 to October 11, 2017, we aggregate the emotions per tweets posted in 15-minute intervals. This process gives us the timeline data per minute and per fifteen minute intervals for all six emotions (listed above) in our analysis.

3.3   Tool Implementation and Design

We developed the tool using dygraphs, RShiny, and deployed it on the Shiny application server. RShiny is a web framework for building web applications using R programming language. Shiny helps us to turn our analysis into an interactive web application running HTML, CSS, and Javascript. We used dygraphs to explore and interpret dense datasets, which is the aggregated emotion values in our case. The tool is available to the research community at the following link: https://proudme.shinyapps. io/emotion/. The tool was designed and implemented with our key *Objectives 1–4* as guiding principles. We describe the tool next and also describe how the design fulfills our key objectives.

4   EmoVis – Emotional Trends Visualization Tool

In this section, we describe the tool we developed and its features. Figure 2 shows a screenshot of the tool with annotations for key features (a–f), explained below.

(a)   *Main Window:* The main window is the central part of the tool where the user can interact with the data. It displays the trends of emotions that are selected and displayed for the chosen time range. In the current screenshot (Fig. 2a), the rate of anger and sadness emotions is displayed along with the tweet count over our entire data collection period on the x-axis (Feb–Oct 2017). On the right y-axis is the tweet count for each time point, ranging from 0 to 100,000. On the left y-axis are the values of the anger and sadness emotions, ranging from 0 to 5. Hovering over the graph and the timelines allows the user to see the actual values of each point on the graph. Under the x-axis is a smaller graph showing the number of tweets per day, which is also interactive and can be used to select the date range. For example, sliding the bars on either end of the bottom graph inward will narrow the time range, while sliding them outward will widen the time range. The bottom snapshot also allows us to view the data in its entirety. Based on this view, we can easily see that the number of tweets remain relatively even throughout the collection period, with a substantive increase (seen as a hump) in August.

(b)   *Options Window:* The options window allows users to customize their options for the displayed graph. The *Information Displayed by Day* and *Information Displayed by Month* options allow the user to either focus on a particular day or on a particular month respectively. These options can be used to answer questions such as "Are the trends of emotions in the month of February different from those in March?" and "Do the trends of emotions differ depending upon the day of the week when people are tweeting?" Another option that is supported is the ability to add one's own plot title using the *Input Plot Title* function. This allows the user to create a title in the main window graph for purposes such as taking a screenshot for further use. The next option available is the ability to *Add Event Annotations*, *Select LIWC Emotion Dimension*, and select *Normalization Technique* (explained in detail in parts d, e and f below).

(c)   *Multiple Tabs:* The multiple tabs allow the user to navigate easily between the complete timeframe of data collection (Feb–Oct 2017) and an event-specific time window (*Objective 3*). In this instance, the event-specific time window is August, in which the largest Charlottesville protest occurred. This feature allows researchers to conduct a more focused exploration of the Twitter activity that occurred surrounding the Charlottesville protest (*Objective 2*).

(d)   *Options for Normalization Technique:* We have 4 normalization technique options for tailoring the data presentation in the tool (*Objective 3*). The data presented in the tool is averaged by intensity or amplified by intensity (respectively dividing or multiplying the feature integer with the tweet count/word

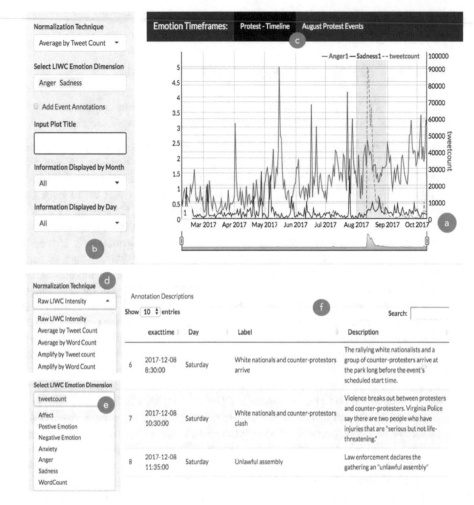

Fig. 2. Screenshots of the EmoVis tool and its various functions: (a) shows the main window where the timeline trends are displayed; (b) shows the Options window where the user can select various options to update the display in the main window; (c) shows the two tabs, one of which displays the time range of the analysis during the entire data collection period (Feb–Oct 2017) and one of which solely focuses on August to obtain depth; (d and e) show the options that are available for normalization techniques and emotion dimensions respectively; (f) shows the annotations of offline events that can be placed on the timeline to better situate the emotion trends.

count). Accordingly, there are 4 presentation options: (1) Average by Tweet Count, (2) Average by Word Count, (3) Amplify by Tweet Count, or (4) Amplify by Word Count.

(e) *Options for Emotion Dimensions:* Users have the option to select up to 6 emotions from the list: (Sadness, Anxiety, Angry, Affect, Positive Emotion, and Negative Emotion).

(f) *Annotations Descriptions:* The event annotations allow the user to contextualize the Twitter activity and emotion intensity with respect to specific events that occurred surrounding and leading up to the Charlottesville protest (Objective 3). In Fig. 2f, three events occurred at different times during the day on August 8, 2017 (as shown in the column exactTime). We also include a short label and a textual description of the event to better situate any changes in emotion due to events that occur offline.

Having discussed the functions present in the tool in this section, in the next section we illustrate the utility of the tool by making inferences that would otherwise have been difficult to make in the absence of the tool.

5   Discussion

We describe the utility of the tool via three illustrative examples. Through each example, we show inferences that are made possible when viewing the data through the tool that would have been difficult to make in the absence of the EmoVis tool.

Example 1: Figure 3 illustrates how the presence of Negative Emotion and Anger evolves with respect to Tweet Count on August 12 and 13, the day of and the day after the protest. We show event annotations on the timelines, starting with 6 and leading through event 17. As shown in Fig. 2f, event 6 is the event indicating that White Nationalist and counter protestors arrived at the site of the protests in Charlottesville. We can also see how the Negative Emotion and Anger levels increased over time during the day, along with the tweet count. While Negative Emotion and Anger show similar patterns, Negative Emotion is higher than Anger. This snapshot of the timeline in August reflected the immediate and profound role social media has in disseminating information on a large socio-cultural scale. As the Unite the Right rally began, occurred, and resulted in violent deaths, tweet count soared alongside negative emotion. Emotion, especially, anger, remained high while tweet count slowly fell in the days after the event.

Example 2: As seen in Fig. 2a, throughout the overall timeline from February to October, anger continually grows alongside the tweet count. Interestingly the tool allowed us to see that at the onset of the timeline, there was an emotional lag such that emotions peaked 2 or 3 days after early critical events. As early key events and decision points occurred, and small-scale local reactions began to occur, the emotional lag shortened and emotions peaked 1 or 2 days after the key events that preceded the protest. Once the protest occurred and the event reached national prominence, the emotion lag disappeared, and emotion peaks were more immediate. This emotional lag behavior that we could discover using the tool allows us to hypothesize that the degree of emotional lag can be modeled to predict when emerging events might reach the emotional mass needed to catalyze larger scale movements or national media attention.

Example 3: In Fig. 2a, we also observe that the tool allowed us to infer that more distinct peaks occurred in the months more closely preceding August largely due to the high-profile nature of the events (e.g., there was ample media attention), resulting in

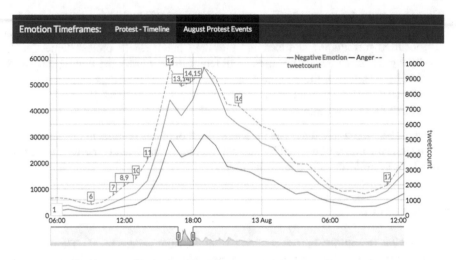

Fig. 3. Screenshots of the EmoVis tool show trends of Negative Emotion and Anger, along with Tweet Count during the Charlottesville protest. We focus on the protest and counter-protest that took place on August 12, 2017, the events leading up to it, and what followed after. Boxes with numbers in one graph (e.g., 6, 7, 8) indicate offline events that took place during that time.

widespread emotional reactions. The events that garnered more media coverage and national presence displayed higher levels of negative emotion from higher tweet counts than localized events.

6   Conclusion and Future Work

In this paper, we have presented a visual analytics tool that we have named EmoVis, which helps users conduct multidimensional emotional analysis of text. EmoVis automatically analyzes the social media messages related to a social event by inferring the emotional content in the tweet text over discrete time frames. It then visually summarizes the complex emotion analysis results in a timeline-based tool. In addition, EmoVis allows a user to interact with the timelines to further investigate the distribution of each emotion across multiple events happening over the event timeline.

While the current functionality supports a wide range of analysis options, several avenues of improvement suggest themselves. As part of our future work, we plan to develop the EmoVis tool by integrating real-time data. Achieving this functionality can help facilitate the analysis of events as they evolve in the offline world. In addition, in the present version of the tool, we have leveraged the widely accepted LIWC software in extracting the emotion values associated with the text. We are exploring other methods that could augment the LIWC dimensions. Another area of improvement would be the addition of the social network of users in the tool. This information could be used in addition to the emotion timelines to make inferences that include, for example, the amount of influence users have and the corresponding weight that could be assigned to emotions in a given moment based on their influence. This measure can help in understanding the users that have the ability to emotionally influence others.

Acknowledgements. This work was supported in part by funding from the Charlotte Research Institute Targeted Research Internal Seed Program.

References

1. Perrin, A.: Social media usage. Pew Research Center, 52–68 (2015)
2. Flaxman, S., Goel, S., Rao, J.M.: Filter bubbles, echo chambers, and online news consumption. Public Opin. Q. **80**(S1), 298–320 (2016)
3. Wahl-Jorgensen, K.: The emotional architecture of social media. In: A Networked Self and Platforms, Stories, Connections, pp. 93–109. Routledge, Abingdon (2018)
4. Power, M., Dalgleish, T.: Cognition and Emotion. From Order to Disorder. Psychology Press, Hove (1997)
5. Västfjäll, D., Slovic, P., Mayorga, M.: Whoever saves one life saves the world: confronting the challenge of pseudoinefficacy. Submitted for publication (2014)
6. Ekman, P.: Are there basic emotions? Psychol. Rev. **99**, 550–553 (1992)
7. Plutchik, R.: Emotions and psychotherapy: a psychoevolutionary perspective. In: Emotion, Psychopathology, and Psychotherapy, pp. 3–41. Academic Press, San Diego (1990)
8. Hasan, M., Rundensteiner, E., Agu, E.: Emotex: detecting emotions in twitter messages (2014)
9. Mohammad, S.M., Bravo-Marquez, F.: WASSA-2017 shared task on emotion intensity. arXiv preprint arXiv:1708.03700 (2017)
10. Mohammad, S., et al.: Semeval-2018 task 1: affect in tweets. In: Proceedings of the 12th International Workshop on Semantic Evaluation (2018)
11. Goel, P., et al.: Prayas at emoint 2017: an ensemble of deep neural architectures for emotion intensity prediction in tweets. In: Proceedings of the 8th Workshop on Computational Approaches to Subjectivity, Sentiment and Social Media Analysis (2017)
12. Duppada, V., Jain, R., Hiray, S.: SeerNet at SemEval-2018 task 1: domain adaptation for affect in tweets. arXiv preprint arXiv:1804.06137 (2018)
13. Abdullah, M., Hadzikadic, M., Shaikh, S.: SEDAT: sentiment and emotion detection in Arabic text using CNN-LSTM deep learning. In: 2018 17th IEEE International Conference on Machine Learning and Applications (ICMLA). IEEE (2018)
14. Abdullah, M., Shaikh, S.: TeamUNCC at SemEval-2018 task 1: emotion detection in English and Arabic tweets using deep learning. In: Proceedings of the 12th International Workshop on Semantic Evaluation (2018)
15. Back, M.D., Küfner, A.C.P., Egloff, B.: The emotional timeline of September 11, 2001. Psychol. Sci. **21**(10), 1417–1419 (2010)
16. Atkeson, L.R., Maestas, C.D.: Catastrophic Politics: How Extraordinary Events Redefine Perceptions of Government. Cambridge University Press, Cambridge (2012)
17. Shirky, C.: Here Comes Everybody: The Power of Organizing Without Organizations. Penguin, London (2008)
18. Bade, R., Schlechtweg, S., Miksch, S.: Connecting time-oriented data and information to a coherent interactive visualization. In: Proceedings of the SIGCHI Conference on Human Factors in Computing Systems. ACM (2004)
19. Zheng, Y., et al.: Visual analytics in urban computing: an overview. IEEE Trans. Big Data **2**(3), 276–296 (2016)

20. Zollo, F., et al.: Emotional dynamics in the age of misinformation. PLoS ONE **10**(9), e0138740 (2015)
21. Wang, F.Y., et al.: SentiCompass: interactive visualization for exploring and comparing the sentiments of time-varying twitter data. In: 2015 IEEE Pacific Visualization Symposium (PacificVis). IEEE (2015)
22. Zhao, J., et al.: PEARL: an interactive visual analytic tool for understanding personal emotion style derived from social media. In: 2014 IEEE Conference on Visual Analytics Science and Technology (VAST). IEEE (2014)
23. Peters, M.A., Besley, T.: White supremacism: the tragedy of Charlottesville. Educ. Psychol. Theor. **49**(14), 1309–1312 (2017)
24. Principe, M.: When white supremacists march: the message of charlottesville. Against Curr. **32**(4), 4 (2017)
25. http://support.gnip.com/gnip2.0/
26. Pennebaker, J.W., Francis, M.E., Booth, R.J.: Linguistic Inquiry and Word Count. LIWC 2001, vol. 71. Lawrence Erlbaum Associates, Mahway (2001)

Social Convos: A New Approach to Modeling Information Diffusion in Social Media

Gregorios Katsios$^{(\boxtimes)}$, Ning Sa, and Tomek Strzalkowski

ILS Institute, University at Albany, Albany, NY, USA
{gkatsios,nsa,tomek}@albany.edu

Abstract. A common approach, adopted by most current research, represents users of a social media platform as nodes in a network, connected by various types of links indicating the different kinds of inter-user relationships and interactions. However, social media dynamics and the observed behavioral phenomena do not conform to this user-node-centric view, partly because it ignores the behavioral impact of connected user collectives. GitHub is unique in the social media setting in this respect: it is organized into "repositories", which along with the users who contribute to them, form highly-interactive task-oriented "social collectives". In this paper, we recast our understanding of all social media as a landscape of collectives, or "convos": sets of users connected by a common interest in an (possibly evolving) information artifact, such as a repository in GitHub, a subreddit in Reddit or a group of hashtags in Twitter. We describe a computational approach to classifying convos at different stages of their "lifespan" into distinct collective behavioral classes. We then train a Multi-layer Perceptron (MLP) to learn transition probabilities between behavioral classes to predict, with high-degree of accuracy, future behavior and activity levels of these convos.

Keywords: Socio-behavioral computing · Social media ·
Information diffusion · Simulation

1 Introduction

A common view of social media (Twitter, Facebook, Reddit, GitHub, etc.) is that of a network of independent participants and the information that flows between them. Typically, users in this network are viewed as nodes, connected by various types of temporary and semi-permanent links that include group membership, friendship, followership, and similar links. Information travels though these links, allowing users to broadcast messages to their connections and to comment on and re-broadcast messages received from others. Most research on social media analytics assumes this view, which lends itself to formal analytics using graph and information propagation theories.

In reality, this user-node-centric view of social behavior is not sufficient. The behavioral impact of connected user collectives, which we shall call "convos", which often act like hybrid organisms, makes it difficult to view individual users as independent agents. It is our hypothesis that the collective-centric representation can adequately capture a collective's social dynamics, which is an important factor in predictions of

© Springer Nature Switzerland AG 2020
T. Ahram (Ed.): AHFE 2019, AISC 965, pp. 25–36, 2020.
https://doi.org/10.1007/978-3-030-20454-9_3

future behavior, including such phenomena as information cascades. To test our hypothesis, we define a convo as a set of users connected by a common interest in an (possibly evolving) information artifact. A convo is part conversation and part task-oriented collaboration, and its character varies across social media types and subject matter. In GitHub, the convos form around the repositories; in Reddit, they arise in subreddits where topics of interest to the group are discussed; in Twitter, convos are circumscribed by groups of hashtags, or just by a message that continues to be retweeted. We should note that convos are by their nature transient phenomena, different than communities (which are defined by connectivity) and interest groups (which are defined by topics); convos transcend both and are more readily compared to conversations, gatherings/movements, or task-oriented groups.

Social behavior in a convo is exhibited by how much its various participants control the direction of the ongoing interaction, how much they are involved in the activities, and how they relate to each other. Convos are highly distributed; the participants are not always present, they do not necessarily know one another, and they may be involved in multiple convos at the same time. In a convo, however, everyone can "hear" everyone else, just like in a real conversation. In a convo-based representation, information spread, by definition, is confined to a convo and is equivalent to its growth or decay. Since convos are group interactions, they also exhibit collective behaviors, such as sociability, cohesion, task focus, etc. [1] that give us better insight into the convo's future: will it grow or shrink, and how fast.

Our objective is to identify a convo's behavioral indicators and related observable features that allow for an accurate characterization of its internal workings and evolution over time, in terms of activity levels. Our approach is built around a general hypothesis that current or past collective behavior of the convo's participants, including members and visitors, has a measurable impact on its future activity levels. Our goals are twofold:

1. Given a set of features that describe a convo's activities within an observation interval (e.g., a week or a month), we wish to classify the resulting convo-slices (i.e., the week-long or month-long snapshots of events and users) into a number of distinctive classes within a multi-dimensional feature space. Such classes, or clusters, can be identified by computing separable densely populated regions where many data points (convo-slices) converge within close proximity of each other. One characteristic of interest is the convo's potential for popularity and activity changes (growth, decline, steady) in the next interval; however, the actual time is not a factor in our analysis, and we consider the convo's behavioral characteristics more broadly, as is explained in detail below.
2. Once the convo-slices are classified, we use a Multi-Layer Perceptron (MLP) to learn the probability of a convo transitioning from one class to another within a future time interval. In the first instantiation of this method, we trained an MLP that can predict the correct state transition for the immediately following time interval with 77% accuracy.

In this paper, we focus our experiments primarily on convos in GitHub (repositories) where the common artifact (the repository) is clearly defined, and where most user interactions are conducted by directly manipulating the repository. In a recent

U.S. Government evaluation, we demonstrated that a collective-centric representation of GitHub can lead to highly accurate predictions of future activities, including occurrence of information cascades. In the following section, we discuss the features currently considered in repository classification and explain the rationale behind them. We then briefly describe how this phenomenon can also be observed in Twitter and Reddit.

2 Motivation

We are exploring a hypothesis that a repository in GitHub is a form of collaborative task-oriented interaction, and thus may display similar socio-behavioral characteristics as other collaborative settings. In particular, social behaviors such as topic control, involvement, influence, etc. [1] can be observed from the communications between the users. In GitHub, most communications occur through a limited set of commands, called "events" that users post to request changes, make contributions, raise issues, etc. [2], which is nonetheless rich enough to convey key behavioral indicators, such as new topic introduction, subsequent mentions, disagreement, etc., and to a lesser degree, sentiment. We further ask: how does the repository's socio-behavioral structure inform and predict this repository's success as measured by an increased level of activity and outside interest? For example, other studies [3] showed that a more equitable collaboration leads to more successful outcomes in task-oriented groups. Can a similar effect be observed in GitHub?

We performed a series of preliminary investigations to verify if the above correlation might hold in GitHub repositories. To do so, we selected a random subset of public repositories that experienced a substantial (at least 20%) growth in either activity (number of events per time unit) or popularity (number of users) over a period of 2 to 3 months anytime within a 2-year span (2015–2016). We observed two types of change: (1) spikes, generally short-lived increases in popularity and/or activity spanning no more than a single month after which the repository returns to its "regular" activity level; and (2) shifts, more sustained increases (or decreases) where the new level of activity and popularity is maintained over at least 2 months. In both cases, certain regularities in the repository's users' collective behavior were noted: these included an influx of new users, and a significant diversification of activities among the repository's core users (members). More sustained shifts occurred when the new users stayed with the repository becoming steady contributors.

The above consideration, also signaled in our earlier paper [2], led to the initial conceptualization of the convo phenomenon. If GitHub is like any other social media, as we argued, would the convos exist in these media as well? In some cases, the answer is pretty straightforward: in Reddit, a sub-reddit consisting of posts, each with its own set of comments, naturally forms a convo. In Twitter, convos form around hashtags, or frequently co-occurring groups of hashtags. In both cases, the interaction is mostly verbal, and the language is not constrained; the information artifact, unlike the GitHub repository, is highly distributed: a sum of the original posts and the various comments attached to them.

3 Related Work

The approach presented here draws on sociolinguistic analysis of group interactions, particularly [1], [4] and [5] that exploited language features to model social dynamics in a group. This included identifying leaders, influencers, powerful players, as well as estimating group cohesion. Further research stemming from this line of work, [3] demonstrated that groups displaying a certain balance of social behavior, for example, power equitability, tend to be more successful in tasks than groups with strong, dominating personalities. Since successful projects tend to attract attention, we stipulated that the collective behavioral makeup may provide a key to solving the puzzle of "unpredictable cascades" in social networks.

Further relevant work is a series of publications by Hofman and others [6–9] that exploited the notion of influence as a factor in predicting information cascades in social networks. The overall conclusions from these works was that cascades can only be predicted with up to about 50% accuracy, and that the most influential people are not the sources of most cascades. This sobering result has nonetheless revealed the inherent limitations of these studies. For one, the notion of influence was largely limited to just one factor, known as network centrality [1] and further confined to the physical connectivity, i.e., discernible "follow" links, but not considering the patterns of information flow. Specifically, one can be influential without necessarily being directly connected to others. This limitation leads to another key issue: actual information content flow in the network was not considered. Messages were simply seen as sealed packages that people sent to one another. In our work presented here, we (partly) unseal these information packages to gain more insight into group social behavior, specifically topic control behavior [10]. Related GitHub specific work include [11] and [12].

4 Simulating Future Behavior

Our goal is to predict the behavioral trend a convo will take in the near future, based on its socio-behavioral makeup. We explain this process in detail using GitHub repositories; the process for other media is analogous.

Using the feature values generated by a repository at time t, we attempt to predict the behavioral trend that the repository will adopt at time $t + 1$, where the temporal unit is selected so that a sufficient amount of activity can be observed in most repositories. The intuition is to use the known feature-vector at a previous time interval to predict the class of the unknown feature-vector at a future time interval. Once the class is selected, we can project the repository's future behavior based on the mean feature vector for that class (alternatively, selecting a feature vector from the class probability distribution).

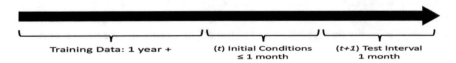

Fig. 1. Simulation set up.

The test is constructed as shown in Fig. 1; predict each repository's behavior in a future interval, given an initial condition immediately preceding the test interval.

4.1 Types of Events

Most activities in GitHub are conducted through a relatively small set of 14 types of events that are posted by users in appropriate repositories. We further divide these 14 types of events into three disjoint categories; *Popularity Events, Collaborative Events* and *Contributive Events*. The Popularity category contains WatchEvent and ForkEvent. These are the events by which external users show interest in the repository but make no contribution to it. The Collaborative category contains IssuesEvent, IssueCommentEvent, CommitCommentEvent, PullRequestReviewComment and GollumEvent. These are the events by which both external and internal users can communicate various problems, suggestions, agreements, and disagreements with any aspect of the repository, including any previously posted events. Finally, the Contributive category contains PushEvent, PullRequestEvent, CreateEvent, DeleteEvent, MemberEvent, PublicEvent and ReleaseEvent. These are the events that can be performed almost exclusively by the repository's members (except for one variant of the PullRequestEvent) and deliver (or attempt to deliver) specific changes to the repository's content.

The reason we perform this separation is based on the observation that many bursts of activity in a repository are triggered and sustained by Issue and Comment events performed by GitHub users that are not members of the repository.

In other social media the set of event types is more constrained, and typically consists of just two types: a posting and a comment. In Twitter a re-tweet simply forwards the original message, while a quote and a reply may also add new content. Due to the hierarchical structure of these convos, the original post/tweet is naturally introducing a new topic (not unlike the GitHub Create command). The subsequent comments/retweets carry this original topic but also can introduce new topics, thus giving specific posters a degree of topic control, and consequently influence. This explicit post-reply branching structure in Twitter and Reddit allows for cost effective topic tracking in even very large datasets (as a contrast, consider a chatroom conversation where everyone posts into the moving stream of messages).

4.2 Features

Some of the features described in Tables 1, 2 and 3 are simplifications of more detailed socio-behavioral metrics that require a significant amount of time to compute. For example, the *Top Group GINI* feature is computed based on an estimated equitability of contributions among the users based on the counts of events posted (a rough equivalent of the involvement metric) rather than content contributions (a topic control metric). Nonetheless, this approximation allows for rapid (in minutes) testing of our approach using thousands of repositories.

The feature set is divided in three logical groups, the first containing the *event-centric* features, the second the *user-centric* features and third the *topic-centric* features. Event-centric features use statistics collected though all event types and are listed in Table 1, which contributes 16 values (2 event count variables, and 14 event distribution percentages).

To calculate the user-centric features, we define *top group* as the set of users that jointly generate 90% of events during time *t*, while each person generates at least 10%. This is the set of users whose joint behavior determines the repository's socio-behavioral profile at a given time. User-centric features, listed in Table 2, use the statistics collected through contributive and collaborative events occurring in the repository at time *t*. As a result, we construct two sub-sets of values; one sub-set reflects the contributive behavior, occupying 19 values in the feature-vector, and the other subset reflects the collaborative behavior, occupying another 19 values.

Topic-centric features are calculated by examining the event log of a repository and recognize events that form topically related chains e.g., an Issue being raised and then responded to by others, or a code change requested (PullRequestEvent) that is then discussed and approved or disallowed. This contributes another 18 values to our feature set. Thus, concatenating all subsets of features together, each repository-slice (a repository snapshot at time interval *t*) will have a feature-vector consisting of 72 values.

Table 1. List of Event-centric features.

Name	Explanation
Total Event Count	Total number of events recorded in a repository during time *t*
Event Rate of Change	Rate of change in the number of events from time *t-1* to time *t*
Total Event Distribution	The percentage of activity per event type generated inside the repository at time *t*. This series of variables captures the nature of activity within a repo

Table 2. List of User-centric features.

Name	Explanation
Top Group Size	The number of users that belong to the *top group* at time *t*. This value corresponds most directly to the level of activity in a repo
Out of Top Group Size	The number of people that generate events for the repository at time *t* but are not members of the *top group*. This is the set of potential top group members that may predict a future growth
Top Group Size Difference	The difference in the number of new users in *top group* from time *t-1* to time *t*
Top Group Event Count	Number of events by the members of the *top group* at time *t*. Correlates with involvement distribution among the top users
Top Group Mean Events	Average number of events by the *top group* users at time *t*
Top Group Stdev Events	Standard deviation of the number of events by *top group* at *t*
Top Group Event Rate of Change	Rate of change in the number of events made by *the top group* from time *t-1* to time *t*. This may reflect change to the group size or an increased workload

(continued)

Table 2. (*continued*)

Name	Explanation
Top Group Size Rate of Change	Rate of change in the size of the top group from time *t-1* to time *t*. This is a companion variable to the above
Top Group GINI	The GINI coefficient of the distribution of events generated by the *top group* at *t*. A measure of equitability of involvement
Top-10 User Event Distribution	The percentage of events that the top-10 members of the top group have generated at time *t*

Table 3. List of Topic-centric features.

Name	Explanation
Number of Topics	Total number of topics within time *t*. A topic is defined as any sequence of events of length 2 or more that share the same service assigned id. Equivalent to local topics in conversation
Number of Topics (L2)	Number of topics which have at least 2 comments. These are topics of minimum length of 3 or 4 (open, 2+ comments, close)
Number of Topics (U2)	Number of topics which have at least 2 different commenters. Corresponds to the citation measure in [1]
Topic Length	Maximum topic length within time *t*
Topic Followers	Maximum number of users posting comments on a topic
Topics Created	Number of topics created within the time *t*
Topics Closed	Number of topics closed within the time *t*
Number of Users	Total number of users involved in all topics
Top-10 Topic Control Distribution	The topic control distribution among the top-10 members of the repository at time *t*. Equivalent to the socio-behavioral measure of Topic Control [1]

Convo-Slice Vector Construction. For each convo-slice (in GitHub, a repository snapshot at time *t*) we calculate the vector containing the features listed above, if the convo-slice has a sufficient number of events in any of the time intervals under consideration, and more than one member. This threshold is adjusted dynamically but we generally require that the top group members contribute on average at least 3 distinct events.

4.3   Clustering

Clustering can help us determine the number of behavioral trends within the training data and the distribution of the rates by which repositories change from time *t* to *t + 1*. After we have performed feature extraction from the training data, we attempt to group similar time-repository vectors into clusters.

We first apply Principal Component Analysis (PCA) dimensionality reduction [13], reducing the vectors dimensions from 72 to 5. Then, we apply Hierarchical Density Based Spatial Clustering of Applications with Noise (HDBSCAN) [14]. This algorithm looks for densely packed observations and makes no assumption about the number or the shape of the clusters. Additionally, this clustering algorithm marks observations that lie in low-density regions as outliers.

Behavioral Trends. Using the clusters found by HDBSCAN, we can identify the various behavioral trends that exist in our training data. Each behavioral trend consists of the 'average' statistics generated by using the feature-vectors within the cluster, such as mean number of events, mean top group size, mean rates of change, etc. Some features, such as the Total Event Distribution, Top-10 User Event Distribution or the Top-10 Topic Control Distribution cannot be summarized as averages across vectors. In these cases, we calculate the individual mixture distributions by assuming each feature represents a generalized Bernoulli distribution [15] and [16].

4.4   Training Set for Machine Learning

The main goal of this module is to use the known feature-vector at the previous time stamp to predict the class of the unknown feature-vector at their next time stamp. The assumption is that the evolution of repositories from time t to $t + 1$ can be projected based on the evolution patterns observed in the past. In order to achieve this, we represent the evolution pattern in the training data by assigning labels representative of the convo's next time interval to each current convo-slice.

After clustering, let's consider each cluster to be a class $(C_1, C_2, \ldots, C_K)$. Then, all feature-vectors in cluster 1 will be assigned class C_1, etc. We construct the input to the learning algorithm as follows: For feature-vector v_i, with time t-1, we retrieve v'_i with time t. Let v'_i belong to some class C_p. Then, we construct an instance of training data such that: $x = v_i$ and $y = C_p$.

Classifier: Multi-layer Perceptron. The machine learning algorithm chosen is the Multi-layer Perceptron (MLP) artificial neural network [17]. The activation function of our MLP is the Rectified Linear Unit (ReLU) [18], which results in better gradient propagation with fewer vanishing gradient problems compared to sigmodal activation functions [19]. Multi-class classification is performed using a Softmax layer at the output.

5   Predicting Collective Behavior

After training, we apply the classifier as follows: given a feature-vector generated by a convo from the initial condition, predict the behavioral trend that the convo will adopt in its next time interval. Having this prediction, we can estimate the convo's user activity levels for that interval. We would like to note that we do not perform any updates on our system based on the initial condition; we merely use it to generate a new feature-vector.

5.1 Discovered Behavioral Trends

HDBSCAN clustering can produce different results depending on what value the *minClusterSize* parameter is set to. This parameter determines the smallest number of points that can be considered a cluster. In our experiments, we have examined *minClusterSize* values that range from 2 to 3000, which in turn produce a number of clusters in the range of 3 to 623. For the remainder of the discussion, we will focus on clustering results that yield 15 behavioral trends.

We set this empirical threshold for two reasons. First, due to the observation that clustering results that produce a larger number of clusters lead towards lesser quality behavioral trends. Second, since the total number of vectors does not change, dividing them into many smaller clusters culminates in a harder classification task since each class will have less reference points to be learned from. From the discovered behavioral trends, we have identified 4 fundamental behaviors in GitHub:

1. Collaborative work, when multiple users contribute content and comments inside a repository.
2. Individual work, when individual users work by adding content with little or no interaction.
3. External interest without involvement, when external users show interest in a repository (Watch, Fork) but don't get involved otherwise.
4. External interest with response, when external users are engaged by repository members.

We used the presence of these 4 types of behaviors in the clusters as a general gauge of the quality of the clustering. It is interesting to note that the same classes exist in the other media, although one may prefer to describe them a bit differently. For example, the "collaborative work" class of GitHub corresponds to multiple posts or tweets that form a debate between the contributors. Similarly, instead of "individual work" in Reddit and Twitter, we often observe "single contributor with no response". "External interest without involvement" in GitHub best corresponds to a one-time post/tweet that gets commented on/retweeted once by others without any further involvement from the original posters, which is also quite common. Finally, the "external interest with response" maps on situations where there are branching commentaries that may draw the original contributors into the interaction. This apparent correlation of behavioral types discovered across different media platforms provides a partial validation of the convo approach described here.

5.2 Evaluation and Results

In order to train our system, we used GitHub data collected during the period of 1/1/2015 to 8/30/2017. We focused our experiments on convos (repositories) of three particular domains of interest: Cryptocurrencies (Crypto), Cybersecurity (Cyber) and Common Vulnerabilities & Exposures (CVE).

Moreover, we measure how well our classifiers can predict the next behavioral trend that the convo will follow. We perform a 5-fold cross validation using our training data and calculate the accuracy and F1-score metrics. The results, listed in

Table 4, make explicit the strong correlation between the amount of training data and the quality of the predictions; to train a classifier for GitHub, we require at least 500K convo-slices (feature vectors). In our earlier experiments with 2 years-worth of GitHub data we recorded 85% accuracy over 8 million convo-slices – a relatively minor gain for significantly more data.

Table 4. Training data vs. classifier performance.

Domain	Extracted convo-slices	Accuracy	F1-score
Crypto	49837	35%	19%
Cyber	587786	77%	23%
CVE	221134	51%	26%

Experimental Set Up. To evaluate our approach, we use Normalized Root Mean Squared Error (NRMSE), which is a measure of accuracy between different models [20]. We compare our model against a naïve, but reasonable baseline which replicates the activity levels observed in the initial condition. This baseline assumes there was no change in any of the repositories while transitioning from one time-slice to the next.

To determine whether our system can capture and predict collective behavior trajectories, we have prepared a series of 7 experiments for each domain. We train our system following the architectural pipeline discussed in the previous sections by utilizing GitHub data gathered from 2015 and 2016, keeping the 2017 data aside for testing. The first experiment will use January 2017 as initial condition and make activity estimations for February 2017. Then, for the second experiment we add January 2017 to the training set, use February 2017 as initial condition and carry out estimations for March 2017. We repeat this process, for the remainder of the test set, until August 2017.

Results. Using the NRMSE metric and baseline discussed above, we evaluate the performance of our system. In addition to GitHub, we train our system on Reddit and Twitter data following similar methods. This is done in order to demonstrate the applicability of our approach in different social media platforms.

Table 5. NRMSE evaluation results (B: baseline, P: predictions).

	GitHub						Reddit						Twitter					
	Crypto		CVE		Cyber		Crypto		CVE		Cyber		Crypto		CVE		Cyber	
	B	P	B	P	B	P	B	P	B	P	B	P	B	P	B	P	B	P
Feb-17	1.81	1.67	4.43	3.59	5.21	4.31	3.17	2.47	3.92	2.32	3.28	1.92	5.4	5.96	2.9	2	6.29	8.39
Mar-17	2.63	2.59	3.40	3.11	3.68	3.61	3.60	2.63	2.41	1.90	3.60	2.05	2.6	3.98	2.9	2.1	4.11	5.21
Apr-17	5.28	2.79	3.14	2.75	3.56	3.43	5.42	2.05	2.98	2.40	3.81	1.95	16	16	3	2.1	9.26	9
May-17	2.27	2.02	3.27	3.51	5.43	4.74	5.02	3.32	5.64	1.52	2.40	1.63	1.8	4.66	0.9	1.5	6.87	9.94
Jun-17	2.73	2.63	3.16	2.91	7.30	4.82	2.73	2.65	4.33	2.37	2.99	2.11	4.2	5.23	3.2	1.9	16	8.5
Jul-17	1.85	1.78	3.56	3.57	11.96	6.08	3.09	2.62	2.72	2.21	2.70	1.79	4.6	5.26	0.8	1.1	4.99	7.36
Aug-17	2.51	2.45	4.04	3.40	11.93	11.01	6.15	3.52	2.72	1.89	2.91	1.75	5.8	6.62	2.7	2.6	14.7	3.31

Table 5 shows our system's performance vs. the naïve baseline in all three platforms and across the three scenarios. As already seen, the GitHub models that were trained using more data perform better. Reddit and Twitter data available for these scenarios were significantly smaller, thus yielding less accurate models.

6   Future Work

Future work will focus on a large-scale experimental validation of the convo design and ways of maximizing its benefits in tasks such as cascade prediction in social media both within and across social media platforms. The latter may involve solving the problem of cross-convo communication, with the convos viewed as hybrid entities rather than sets of individuals. Given the generally different levels of focus in these 3 media, even when dealing with the same topic (such as cybersecurity issues, or Bitcoin's price), is there an activation threshold, possibly measured in terms of information intensity or topic persistence coming out of one convo, that would cause a reaction in another convo on a parallel platform? Currently such cross-platform correlations are poorly understood, and we plan to focus initially on cases where they were observed, e.g., in crypto currency price manipulation campaigns.

7   Conclusion

In this paper we presented a novel approach to analyzing and predicting collective behavior in social media, starting with GitHub, and then mapping our approach onto Twitter and Reddit. The approach is centered at the notion of the convo: a collective user activity around an information artifact (a repository, a subreddit, a hashtag). We postulate that the activity within a social media platform is best represented by a collection of convos, which form, persist, and disappear dynamically over time. Each convo is viewed as a basic unit of analysis: the users, the artifact they handle, and the observable trace of their activity. This approach allows for direct incorporation of socio-behavioral elements into the analysis that have been largely missing in previous studies. While the work presented here is preliminary, we believe it would lead to new capabilities of understanding and predicting behavior in social networks, including information cascades.

Acknowledgments. This work was supported by the Defense Advanced Research Projects Agency (DARPA) under Contract No FA8650-18-C-7824. All statements of fact, opinion or conclusions contained herein are those of the authors and should not be construed as representing the official views or policies of AFRL, DARPA, or the U.S. Government.

References

1. Broadwell, G.A., Stromer-Galley, J., Strzalkowski, T., Shaikh, S., Taylor, S., Liu, T., Boz, U., Elia, A., Jiao, L., Webb, N.: Modeling sociocultural phenomena in discourse. Nat. Lang. Eng. **19**, 213–257 (2013)
2. Strzalkowski, T., Harrison, T., Sa, N., Katsios, G., Khoja, E.: GitHub as a social network. In: International Conference on Applied Human Factors and Ergonomics, pp. 379–390. Springer (2018)

3. Oliveira, A.W., Boz, U., Broadwell, G.A., Sadler, T.D.: Student leadership in small group science inquiry. Res. Sci. Technol. Educ. **32**, 281–297 (2014)
4. Beebe, S.A., Masterson, J.T.: Communicating in Small Groups: Principles and Practices. Scott, Foresman, Glenview (1986)
5. Huffaker, D.: Dimensions of leadership and social influence in online communities. Hum. Commun. Res. **36**, 593–617 (2010)
6. Bakshy, E., Hofman, J.M., Mason, W.A., Watts, D.J.: Everyone's an influencer: quantifying influence on Twitter. In: Proceedings of the 4th ACM International Conference on Web Search and Data Mining, pp. 65–74. ACM (2011)
7. Cheng, J., Adamic, L., Dow, P.A., Kleinberg, J.M., Leskovec, J.: Can cascades be predicted? In: Proceedings of the 23rd International Conference on World Wide Web, pp. 925–936. ACM (2014)
8. Hofman, J.M., Sharma, A., Watts, D.J.: Prediction and explanation in social systems. Science **355**, 486–488 (2017)
9. Martin, T., Hofman, J.M., Sharma, A., Anderson, A., Watts, D.J.: Exploring limits to prediction in complex social systems. In: Proceedings of the 25th International Conference on World Wide Web, pp. 683–694. International World Wide Web Conferences Steering Committee (2016)
10. Strzalkowski, T., Broadwell, G., Stromer-Galley, J., Shaikh, S., Taylor, S.: Modeling leadership and influence in online multi-party discourse. In: COLING Conference, pp. 596–617. Mumbai, India (2012)
11. Tsay, J., Dabbish, L., Herbsleb, J.: Influence of social and technical factors for evaluating contribution in GitHub. In: Proceedings of the 36th International Conference on Software Engineering, pp. 356–366. ACM (2014)
12. Casalnuovo, C., Vasilescu, B., Devanbu, P., Filkov, V.: Developer onboarding in GitHub: the role of prior social links and language experience. In: Proceedings of the 10th Joint Meeting on Foundations of Software Engineering, pp. 817–828. ACM (2015)
13. Pearson, K.: Liii. On lines and planes of closest fit to systems of points in space. The Lond. Edinb. Dublin Phil. Mag. J. Sci. **2**, 559–572 (1901)
14. McInnes, L., Healy, J.: Accelerated hierarchical density based clustering. In: Data Mining Workshops (ICDMW), IEEE International Conference, pp. 33–42. IEEE (2016)
15. Johnson, N.L., Kotz, S., Balakrishnan, N.: Discrete Multivariate Distributions. Wiley, New York (1997)
16. Agresti, A.: An Introduction to Categorical Data Analysis. Wiley, Hoboken (2018)
17. Rosenblatt, F.: Principles of neurodynamics. Perceptrons and the theory of brain mechanisms. Technical report, CORNELL Aeronautical Lab Inc Buffalo, NY (1961)
18. Hahnloser, R.H., Sarpeshkar, R., Mahowald, M.A., Douglas, R.J., Seung, H.S.: Digital selection and analogue amplification coexist in a cortex-inspired silicon circuit. Nature **405**, 947 (2000)
19. Glorot, X., Bordes, A., Bengio, Y.: Deep sparse rectifier neural networks. In: Proceedings of the 14th International Conference on Artificial Intelligence and Statistics, pp. 315–323 (2011)
20. Hyndman, R.J., Koehler, A.B.: Another look at measures of forecast accuracy. Int. J. Forecast. **22**, 679–688 (2006)

Manipulated Information Dissemination and Risk-Adjusted Momentum Return in the Chinese Stock Market

Hung-Wen Lin[1], Jing-Bo Huang[2], Kun-Ben Lin[3],
and Shu-Heng Chen[4(✉)]

[1] Nanfang College of Sun Yat-sen University, Guangzhou, China
[2] Sun Yat-sen University, Guangzhou, China
[3] Macau University of Science and Technology, Macau, China
[4] National Chengchi University, Taipei, Taiwan
chen.shuheng@gmail.com

Abstract. We study the manipulated information dissemination and risk-adjusted momentum return in the Chinese stock market. In this paper, we employ excess media coverage as a proxy for manipulated information dissemination. The raw momentum returns are negative across all degrees of manipulated information dissemination, but turn into significantly positive after controlling for risks. These outcomes hint that the manipulations of information dissemination contribute to price instabilities, so raw momentum returns are negative but turn into positive owing to risk adjustments. Moreover, we also discover that the stocks with high manipulated information dissemination exhibit big size characteristic and resist market risk well.

Keywords: Manipulated information dissemination · Risk adjustments · Momentum

1 Introduction

As is known to all, the stock price movements actually stem from the changes in information. Information dissemination has crucial impacts on the stock markets [1, 2]. In details, information dissemination improves the incorporation of news into prices [3]. Similarly, some studies hold that information dissemination enhances stock price efficiency [4]. Hence, information dissemination really deserves our detections.

The Chinese stock market exhibits significantly negative momentum returns no matter how long the formation and holding periods are [5]. If a market has significantly positive momentum returns, there is a clear pattern of continuation in stock prices. Stock prices are likely to maintain their past trends and show stable characteristic under positive momentum returns. On the contrary, stock prices easily reverse and present volatile patterns if a market has significantly negative momentum returns, suggesting that price reversal and contrarian are prevalent in this market. The stock prices are greatly volatile in China. Momentum is a result of gradual information dissemination and momentum returns are larger within the stocks accompanied with slow information

© Springer Nature Switzerland AG 2020
T. Ahram (Ed.): AHFE 2019, AISC 965, pp. 37–45, 2020.
https://doi.org/10.1007/978-3-030-20454-9_4

dissemination [6, 7], which suggests that stock price stability is negatively related to the information dissemination.

Some literatures have studied information dissemination in the Chinese stock market. For instance, price changes are driven by information diffusion and no price reversal will be perceived in the short run [8]. Besides, information dissemination can explain the increases in idiosyncratic volatility in China [9]. These discoveries again imply information dissemination is related to stock price stability. We are interested in the artificially manipulated information dissemination, which is measured as excess media coverage in this paper. Therefore, we shed light on the relationship between manipulated information dissemination and momentum return by portfolio approach and risk adjustment analyses in the Chinese stock market.

Media coverage can be regarded as a proxy for information dissemination and it affects stock prices and returns [10]. In the Chinese stock market, the promotions in media coverage positively affect stocks [11]. That is, the stock returns increase with media coverage. This finding suggests that media coverage enhances stock price stability.

The credible news from media significantly influences an IPO corporation's stock returns and the uncertain tone from news may do harm to the corporation [12]. In addition, the trends of stock prices are positively correlated with the tones from media reports [13]. From the perspective of news tone, media coverage is also able to stabilize stock prices. In the above veins, it seems that the discussions on relationship between information dissemination and stock prices do not reach very consistent outcomes. Therefore, this contradiction in the literature also stimulates our motivation to carry out this research.

The remainder of this paper is organized as follows. We introduce the calculations of momentum, excess media coverage, construction procedures of different portfolios and data in Sect. 2. Section 3 contains the empirical findings. Section 4 concludes this paper.

2   Empirical Setup and Data

In this section, we describe the computations of excess media coverage, momentum return and the construction details of different portfolios. Moreover, momentum return computations include two parts.

First, raw momentum returns computations is based on rolling procedure [14]. Second, three factor models including CAPM, three-factor model [15] and four-factor model [16] are used to compute the risk-adjusted momentum returns.

The momentum portfolios consist of the portfolio from independent classifications and the portfolios from the dependent classifications. In this regard, we are able to dissect the relationship between manipulated information dissemination and momentum return completely.

2.1 Excess Media Coverage Computations

The proxy for manipulated information dissemination is excess media coverage and it is calculated by the logics as follows [17].

$$ln(1 + no.art.)_i = \alpha + \sum_{n=1}^{4} \beta_n Exp_{n,i} + \varepsilon_{i,media} \qquad (1)$$

where $ln(1 + no.art.)_i$ is the natural log of the number of articles of stock i, $\sum_{n=1}^{4} \beta_n Exp_{n,i}$ contains the explanatory variables and their coefficients. They are the natural log of market capitalization, a dummy is defined as 1 when a stock is indexed in $CSI\,300$ and 0 otherwise, a dummy is defined as 1 when a stock is listed in Shenzhen stock exchange and 0 otherwise and the natural log of the number of analysts.

2.2 Raw Momentum Return Computations

In each month t, we rank the stocks into z groups by stock returns during the past t-6 to t-1 months. The $\frac{1}{z}$ stocks with highest past returns are winner portfolio, while the $\frac{1}{z}$ stocks with lowest past returns are loser portfolio. We create a long position for winner portfolio and a short position for loser portfolio to make momentum portfolio in each month t.

Within the future $t + 1$ to $t + 6$ months, we maintain the positions produced in month t and calculate the average return of winner portfolio, $R_W^{t+1,t+6}$ and loser portfolio, $R_L^{t+1,t+6}$. Finally, a time series of $R_W^{t+1,t+6} - R_L^{t+1,t+6}$ is generated by the rolling procedure.

2.3 Capital Asset Pricing Model

$$ER_{i,t} = \alpha_i + \beta_{mkt,i} EMKT_t + e_{i,t} \qquad (2)$$

where $ER_{i,t}$ is the stock return in excess of risk-free rate, $EMKT_t$ is the market return in excess of risk-free rate.

2.4 Three-Factor Model

$$ER_{i,t} = \alpha_i + \beta_{mkt,i} EMKT_t + \sum_{p=1}^{2} \beta_{p,i} Factor_{p,t} + e_{i,t} \qquad (3)$$

where $\sum_{p=1}^{2} \beta_{p,i} Factor_{p,t}$ denotes the size, book-to-market factor and related coefficients. They are SMB_t (return differences between small and big stocks), and HML_t (return differences between high book-to-market and low book-to-market stocks).

2.5   Four-Factor Model

$$ER_{i,t} = \alpha_i + \beta_{mkt,i}EMKT_t + \sum_{p=1}^{2} \beta_{p,i}Factor_{p,t} + \beta_{w,i}WML_t + e_{i,t} \qquad (4)$$

where WML_t represents the momentum factor (return differences between the stocks with highest past returns and those with lowest past returns).

2.6   Momentum Portfolios with Various Classifications

These various classifications will make relatively complete analyses for the effect of manipulated information dissemination (excess media coverage) on risk-adjusted momentum returns. Interactive momentum portfolio is generated from the intersections of independent classifications for the stocks by excess media coverage and stock returns. This kind of momentum portfolio is noted as *Inter (R, C)*.

Break-down momentum portfolio first classifies the stocks by returns and second classifies the stocks by excess media coverage, which is noted as *BD (R, C)*. We first classify the stocks by excess media coverage and second classify the stocks by returns to construct the conditional momentum portfolio. *Con (C, R)* represents this portfolio.

2.7   Data Descriptions

The monthly data is collected from China Infobank and China Stock Markets and Accounting Research (CSMAR). In details, the data period ranges from June 2005 to September 2016.

We find that the Chinese stock market has a significantly negative momentum return of -0.026 (p-value $= 0.005$) with 6-month formation and holding period, suggesting this market exhibits contrarian. The total mean of excess media coverage is insignificant of 0.004 (p-value $= 0.917$). The mean of residual is assumed to be zero in OLS regression, which indicates that excess media coverage is normal during our sample period.

3   Empirical Findings

The risk-adjusted momentum returns are regression intercepts of CAPM (1F), three-factor model (3F) and four-factor model (4F) [18, 19]. We want to observe whether momentum returns survive after risk adjustments [20]. We also show the factor coefficients of factor models. Owing to the length of this paper, we do not show the results of raw momentum returns but they are available if required.

3.1   Risk-Adjusted Momentum Returns

This section shows the risk-adjusted momentum returns by the regression intercepts of our factor models. We also offer p-values of the intercepts to confirm the significance of the risk-adjusted momentum returns.

Displayed in Table 1 are the risk-adjusted momentum returns (*MR*). The momentum portfolios are *Inter (R, C)*, *BD (R, C)*, and *Con (C, R)*. The levels of excess media coverage contain three levels: high excess media coverage (*HC*), medium excess media coverage (*MC*) and low excess media coverage (*LC*).

Table 1. Risk-adjusted momentum returns

	1F	*p*-value	3F	*p*-value	4F	*p*-value
Panel A: Risk-adjusted MR in Inter (R, C)						
HC	0.095	0.000	0.106	0.000	0.106	0.000
MC	0.113	0.000	0.109	0.001	0.110	0.001
LC	0.109	0.000	0.108	0.000	0.111	0.000
Panel B: Risk-adjusted MR in BD (R, C)						
HC	0.109	0.000	0.122	0.000	0.120	0.000
MC	0.102	0.000	0.102	0.000	0.105	0.000
LC	0.126	0.000	0.121	0.000	0.123	0.000
Panel C: Risk-adjusted MR in Con (C, R)						
HC	0.111	0.000	0.125	0.000	0.124	0.000
MC	0.118	0.000	0.112	0.000	0.113	0.000
LC	0.113	0.000	0.114	0.000	0.117	0.000

The raw momentum returns that we do not show here are all negative across all levels of excess media coverage. When we control for risks by factor models, all the risk-adjusted momentum returns are significantly positive. These results indicate that the contrarian of the Chinese stock market is not robust if we adjust risk. In other words, the contrarian of Chinese stock market may come from the manipulation of information dissemination, so it disappears after controlling for risk.

3.2 Factor Sensitiveness of Factor Models

We provide the factor coefficients of factor models to detect the factor sensitiveness in this section. The *p*-values are also reported to confirm the significance of factor sensitiveness.

Table 2. Factor sensitiveness of CAPM

	b	*p*-value
Panel A: Coefficients in Inter (R, C)		
HC	0.053	0.725
MC	0.946	0.000
LC	0.388	0.005
Panel B: Coefficients in BD (R, C)		
HC	0.183	0.202
MC	0.298	0.017

(*continued*)

Table 2. (*continued*)

	b	*p*-value
LC	0.866	0.000
Panel C: Coefficients in Con (C, R)		
HC	0.228	0.142
MC	0.840	0.000
LC	0.315	0.033

Displayed in Table 2 are the factor coefficients of CAPM. *b* represents the coefficient of *EMKT*. The momentum portfolios are *Inter (R, C)*, *BD (R, C)*, and *Con (C, R)*. The levels of excess media coverage contain three levels: high excess media coverage (*HC*), medium excess media coverage (*MC*) and low excess media coverage (*LC*).

An interesting outcome occurs in this table. Among *Inter (R, C)*, *BD (R, C)*, and *Con (C, R)*, the factor coefficients of *EMKT* are all insignificant (*p*-values = 0.725, 0.202 and 0.142, respectively) with high excess media coverage. This result implies that the stocks whose information is manipulated to disseminate widely are not affected by the market risk. By contrast, the stocks with medium excess media coverage or low excess media coverage are easily affected by the market risk. For instance, in *Inter (R, C)*, the coefficient is significantly positive of 0.946 (*p*-value = 0) with medium excess media coverage.

Table 3. Factor sensitiveness of three-factor model

	b	*p*-value	*s*	*p*-value	*h*	*p*-value
Panel A: Coefficients in Inter (R, C)						
HC	−0.062	0.677	−0.565	0.092	0.984	0.047
MC	0.959	0.000	0.261	0.619	0.082	0.916
LC	0.390	0.007	0.015	0.963	−0.008	0.987
Panel B: Coefficients in BD (R, C)						
HC	0.074	0.598	−0.792	0.014	0.689	0.141
MC	0.297	0.023	0.025	0.931	0.037	0.931
LC	0.932	0.000	0.209	0.671	−0.687	0.343
Panel C: Coefficients in Con (C, R)						
HC	0.124	0.419	−0.827	0.019	0.587	0.251
MC	0.891	0.000	0.337	0.503	−0.360	0.627
LC	0.317	0.039	−0.057	0.869	−0.083	0.870

Displayed in Table 3 are the factor coefficients of Three-Factor Model. *b* represents the coefficient of *EMKT*, *s* is the coefficient of *SMB* and *h* is the coefficient of *HML*. The momentum portfolios are *Inter (R, C)*, *BD (R, C)*, and *Con (C, R)*. The levels of excess media coverage contain three levels: high excess media coverage (*HC*), medium excess media coverage (*MC*) and low excess media coverage (*LC*).

The stocks with high excess media coverage are still not affected by market risk in the three-factor model (*p*-values of *b* are 0.677, 0.598 and 0.419, respectively). Consistently, the coefficients of *SMB* for this kind of stocks are all significantly negative (*p*-values = 0.092, 0.014 and 0.019, respectively) among *Inter (R, C)*, *BD (R, C)*, and *Con (C, R)*, which indicates that these stocks exhibit big size [21]. Intuitively, the big companies are more capable of manipulating information dissemination, so they have high excess media coverage. They do well in resisting market risk, and thus tend to not be affected by market risk. However, the coefficients of *HML* are not all significant among *Inter (R, C)*, *BD (R, C)*, and *Con (C, R)*, so the stocks are insensitive to *HML*.

Displayed in Table 4 are the factor coefficients of Four-Factor Model. *b* represents the coefficient of *EMKT*, *s* is the coefficient of *SMB*, *h* is the coefficient of *HML* and *w* is the coefficient of *WML*. The momentum portfolios are *Inter (R, C)*, *BD (R, C)*, and *Con (C, R)*. The levels of excess media coverage contain three levels: high excess media coverage (*HC*), medium excess media coverage (*MC*) and low excess media coverage (*LC*).

Table 4. Factor sensitiveness of four-factor model

	b	*p*-value	*s*	*p*-value	*h*	*p*-value	*w*	*p*-value
Panel A: Coefficients in Inter (R, C)								
HC	−0.06	0.676	−0.560	0.099	1.014	0.064	0.077	0.893
MC	0.960	0.000	0.251	0.637	0.017	0.984	−0.164	0.855
LC	0.394	0.007	−0.020	0.950	−0.215	0.678	−0.528	0.334
Panel B: Coefficients in BD (R, C)								
	b	*p*-value	*s*	*p*-value	*h*	*p*-value	*w*	*p*-value
HC	0.071	0.615	−0.763	0.018	0.860	0.095	0.437	0.418
MC	0.301	0.021	−0.018	0.950	−0.222	0.635	−0.658	0.183
LC	0.937	0.000	0.161	0.744	−0.969	0.225	−0.719	0.391
Panel C: Coefficients in Con (C, R)								
	b	*p*-value	*s*	*p*-value	*h*	*p*-value	*w*	*p*-value
HC	0.123	0.428	−0.810	0.022	0.686	0.224	0.253	0.669
MC	0.892	0.000	0.323	0.525	−0.442	0.589	−0.207	0.809
LC	0.323	0.035	−0.113	0.742	−0.418	0.449	−0.853	0.144

Among *Inter (R, C)*, *BD (R, C)*, and *Con (C, R)*, with high excess media coverage, the stocks are still not affected by market risk (*p*-values of *b* are 0.676, 0.615 and 0.428, respectively). The coefficients of *SMB* for high coverage stocks keep significantly negative (*p*-values = 0.099, 0.018 and 0.022, respectively). However, the coefficients of *HML* and *WML* are not all significant among *Inter (R, C)*, *BD (R, C)*, and *Con (C, R)*. Consequently, the stocks are insensitive to *HML* and *WML*. In addition, the momentum return of the Chinese stock market is not accounted for by the momentum factor, *WML*.

4 Conclusions

We analyze manipulated information dissemination and risk-adjusted momentum returns in China. By the use of excess media coverage as a proxy for manipulated information dissemination, the raw momentum returns of Chinese stock market are negative across all levels of excess media coverage. When we adjust risks by factor models, the momentum returns all turn into significantly positive. These results suggest that the manipulations of information dissemination contribute to stock price instabilities.

However, every coin has two sides. Manipulating information also plays a positive role to certain extent. In particular, the stocks with high excess media coverage exhibit big size and are not affected by market risk. Although the big companies are more capable of manipulating information dissemination, this tendency also helps them to resist market risk well.

References

1. Lin, H., Quill, D.: Information diffusion and the predictability of New Zealand stock market returns. Account. Finance **56**, 749–785 (2016)
2. Wu, F.: Do stock prices underreact to information conveyed by investors' trades? Evidence from China. Asia Pac. J. Financ. Stud. **42**, 442–466 (2013)
3. Chung, S.L., Liu, W., Liu, W.R., Tseng, K.: Investor network: implications for information diffusion and asset prices. Pac. Basin Financ. J. **48**, 186–209 (2018)
4. Pantzalis, C., Wang, B.: Shareholder coordination, information diffusion and stock returns. Financ. Rev. **52**, 563–595 (2017)
5. Lin, H.W., Hung, M.W., Huang, J.B.: Artificial momentum, native contrarian, and transparency in China. Comput. Econ. **51**(2), 263–294 (2018). Accepted Manuscript
6. Chen, Z., Lu, A.: Slow diffusion of information and price momentum in stocks: evidence from options markets. J. Bank. Finance **75**, 98–108 (2017)
7. Doukas, J.A., McKnight, P.J.: European momentum strategies, information diffusion, and investor conservatism. Eur. Financ. Manage. **3**, 313–338 (2005)
8. Zhang, Y., Song, W., Shen, D., Zhang, W.: Market reaction to internet news: information diffusion and price pressure. Econ. Model. **56**, 43–49 (2016)
9. Li, X., Shen, D., Zhang, W.: Do Chinese internet stock message boards convey firm-specific information? Pac. Basin Financ. J. **49**, 1–14 (2018)
10. Peress, J.: The media and the diffusion of information in financial markets: evidence from newspaper strikes. J. Finance **5**, 2007–2043 (2014)
11. Huang, T.L.: The puzzling media effect in the Chinese stock market. Pac. Basin Financ. J. **49**, 129–146 (2018)
12. Guldiken, O., Tupper, C., Nair, A., Yu, H.: The impact of media coverage on IPO stock performance. J. Bus. Res. **72**, 24–32 (2017)
13. Chiao, C., Lin, T.Y., Lee, C.F.: The reactions to on-air stock reports: prices, volume, and order submission behavior. Pac. Basin Financ. J. **44**, 27–46 (2017)
14. Jegadeesh, N., Titman, S.: Returns to buying winners and selling losers: implications for stock market efficiency. J. Finance **48**, 65–91 (1993)
15. Fama, E., French, K.: Common risk factors in the returns on stocks and bonds. J. Financ. Econ. **33**, 3–56 (1993)

16. Fama, E., French, K.: Size, value, and momentum in international stock returns. J. Financ. Econ. **105**, 457–472 (2012)
17. Hillert, A., Jacobs, H., Müller, S.: Media makes momentum. Rev. Financ. Stud. **27**, 3467–3501 (2014)
18. Asem, E.: Dividends and price momentum. J. Bank. Finance **33**, 486–494 (2009)
19. Celiker, U., Kayacetin, N., Kumar, R., Sonaer, G.: Cash flow news, discount rate news, and momentum. J. Bank. Finance **72**, 240–254 (2016)
20. McInish, T.H., Ding, D.K., Pyun, C.S., Wongchoti, U.: Short-horizon contrarian and momentum strategies in Asian markets: an integrated analysis. Int. Rev. Financ. Anal. **17**, 312–329 (2008)
21. Garlappi, L., Yan, H.: Financial distress and the cross-section of equity returns. J. Finance **3**, 789–822 (2011)

Actionable Pattern Discovery for Tweet Emotions

Angelina Tzacheva$^{(\boxtimes)}$, Jaishree Ranganathan,
and Sai Yesawy Mylavarapu

Department of Computer Science, University of North Carolina at Charlotte,
Charlotte, NC, USA
{aatzache,jrangan1,smylaval}@uncc.edu

Abstract. The most popular form of communication over the internet is text. There are wide range of services that allow users to communicate in the natural language using text messages. Twitter is one such popular Micro-blogging platform where users post their thoughts, feeling or opinion on a day-to-day basis. These text messages not only contain information about events, products and others but also the writer's attitude. This kind of text data is useful to develop systems, which detect user emotions. Emotion detection has wide variety of applications including customer service, public policy making, education, future technology, and psychotherapy. In this work, we use Support Vector Machine classifier model to automatically classify user emotions. We achieve accuracy in the range of 88%. The Emotional information mined from such data is huge and these findings can be more useful if the system is able to provide some actionable recommendations to the user, which help them, achieve their goal and gain benefits. The recommendations or patterns are Actionable if user can perform action using the patterns to their advantage. Action Rules help discover ways to reclassify objects with respect to a specific target, which the user intends to change for their benefits. In this work, we focus on extracting Action Rules with respect to the Emotion class from user tweets. We discover actionable recommendations, which suggests ways to alter the user's emotion to a better or more positive state.

Keywords: Actionable pattern discovery · Data mining · Emotion mining · Support Vector Machine · Scalability

1 Introduction

According to Merriam Webster, dictionary [1] Micro-blogging is blogging done with severe space or size constraints typically by posting frequent brief messages about personal activities. There are wide variety of such micro-blog services available on the web including Twitter [2], Tumblr [3], Pownce (http://pownce.com) and many others. Among these, Twitter is the most popular. According to ComScore [4], within eight months of its launch, Twitter had about 94,000 users as of April 2007 [5]. In addition, micro-blogging users may post several updates on a single day [5]. Approximately 500 million tweets are posted on Twitter per day. Thus, the amount of textual data generated is huge when we consider the rate of growth of Twitter user's since 2007 and the periodicity

T. Ahram (Ed.): AHFE 2019, AISC 965, pp. 46–57, 2020.
https://doi.org/10.1007/978-3-030-20454-9_5

of the posts on a single day by a user. It allows adding emoticons, which are one of the powerful tools to express human emotions. Hashtag is a tagging convention that helps people associate tweets with certain events or contexts [6]. It is a keyword prefixed with '#' symbol. These hashtags sometimes indicate the writer's emotion. For example the tweet, "Homemade chicken soup is the best #happy" indicates happiness [7].

Data mining from such rich sources of text helps gain useful insights in a range of applications. For instance, Gupta et al. [8] study the customer care email in- order to identify customer dissatisfaction and help improve business. Analyzing the social media posts of a particular community might help government officials in public policy making to improve the quality of life of people in that area. In educational domain, identifying student's thoughts and emotion about the university, faculty helps improve the quality of education. In the field of psychology, where online social therapy is used for assisting mental health as face-to-face early intervention services for psychosis is for limited time period and benefits may not persist after its termination [9] and in scenarios where machines are used as psychotherapist [10]. After information is gathered from such data, it is necessary to validate the mined information. For this purpose, there are many supervised learning models that help automatically classify new set of test data, given a considerable amount of data for training.

With the proliferation of information through various sources, there is access to enormous amount of data, at the same time leads to poor information in the raw form and inefficient decision-making [11]. The volume of discovered patterns is huge despite the use of data mining strategies, which leads to unreliable and uninteresting knowledge [11]. Actionable patterns are those that help users benefit by using it to their own advantage. Action Rules are special type of rules that help identify actionable patterns from the data [12].

2 Related Work

In this section, we review literature works in the areas of Emotion classification from text, and actionable pattern mining based on text classification.

2.1 Emotion Classification from Text

Mishne [13] classify writer's mood in blog text collected from Live Journal, a free weblog service using Yahoo API. They use following features to train the SVMlight model from Support Vector Machine package: frequency counts (words, Part-Of-Speech), and length of blog post; subjective nature of blogs like semantic orientation, Point-wise Mutual Information (PMI) which is a measure of the degree of association between two terms; features unique to online text like emphasized words, special symbols including punctuation's, and emoticons. They attribute subjective nature of the corpus "annotation" and nature of blog posts as major factors for low accuracy.

Danisman and Alpkocak [14] use Vector Space Model (VSM) where each document is a vector and terms correspond to dimensions and develop a text classifier. Term Frequency - Inverse Document Frequency (tf-idf) weighting scheme is used to calculate weight of each term in the document. They have analyzed the effect of emotional intensity and stemming to the classification performance. Results show that Vector Space Model performs equally well compared to other well-known classifiers.

Mohammad [15] developed corpus from Twitter posts using emotion hash-tags called Twitter Emotion Corpus (TEC) consisting of 21,000 tweets. Support Vector Machines (SVM) with Sequential Minimal Optimization (SMO) classifier was used with unigram and bigram features. The automatic classifiers obtained an F-score much higher than the random baseline (SemEval – 2007, 1000 headlines dataset). Similar to Wang et al., in this paper best results are achieved with higher number of training instances. For example, Joy-NotJoy classifier get the best results compared to Sadness-NotSadness.

Roberts et al. [16] create emotion corpus from Twitter. The corpus contains seven emotions annotated across 14 topics including Valentine's Day, World Cup 2010, Stock Market, Christmas etc. The emotions are based on Ekman's [17] six basic emotions and LOVE. The topics of each tweet obtained by considering the tweet associated with a probabilistic mixture of topics using Latent Dirichlet Allocation (LDA) topic modeling technique. The system uses a series of binary SVM classifiers to detect each of the seven emotions annotated in the corpus. Each classifier performs independently on a single emotion, resulting in 7 separate binary classifiers implemented using the software available from WEKA and uses specific set of features like punctuation, hypernyms, n-grams, and topics. According to the results, FEAR is the best performing emotion and suggests that this emotion is highly lexicalized with less variation than other emotions, as it has comparable recall but significantly higher precision. Overall, in this work, the macro-average precision is 0.721 and recall is 0.627.

Purver et al. [18] used Twitter data labeled with emoticons and hash-tags to train supervised classifiers. They used Support Vector Machines with linear kernel and unigram features for classification. Their method had better performance for emotions like happiness, sadness, and anger but not well in case of other emotions like fear, surprise, and disgust. They achieved accuracy in the range of 60%.

2.2   Actionable Pattern Mining

In [19] the primary intent of the Action Rules generated is to provide viable suggestions on how to make a twitter user feel more positive. For Twitter social network data, Actionable Recommendations include - how to increase user friend's count, and how to change the overall sentiment from negative to positive, or from neutral to positive.

3   Methodology

3.1   Data Collection

In data collection step we used Twitter streaming API [20] to collect the data with the following attributes TweetID, ReTweetCount, TweetFavouriteCount, TweetText, Tweet- Language, Latitude, Longitude, TweetSource, UserID, UserFollowersCount, UserFavoritesCount, UserFriendsCount, UserLanguage, UserLocation, UserTimeZone,

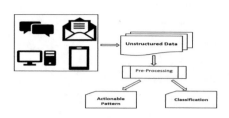

Fig. 1. Overall methodology.   **Fig. 2.** Emotion labeling.

IsFavorited, IsPossiblysensitive, IsRetweeted, RetweetedStatus, UserStatus, MediaEntities. We collected around 520,000 tweets as raw data. Figure 1 shows the overall model of the proposed methodology.

3.2 Pre-processing

The extracted tweet text is pre-processed to make the informal text suitable for emotion classification. We lower case all the letters in the tweet; remove stop words i.e. the most frequent words in English which will not add value to the final emotion; replace slang words with formal text, example b4 → before, chk → check, etc. After pre-processing we have around 200,000 tweets.

3.3 Emotion Labeling

To identify the emotion class, we use the National Research Council - NRC lexicon [21, 22]. The Annotations in the lexicon are at WORD-SENSE level. Each line has the format: <Term> <AffectCategory> <AssociationFlag> where Term is a word for which emotion associations are provided, Affect Category is one of the eight emotions anger, fear, anticipation, trust, surprise, sadness, joy, or disgust and one of two polarities negative or positive, Association flag indicates that the target word has association with category word or not.

Apart from word level annotation, to increase the weightage of each emotion class assigned to tweet we also use the hashtags and emoticons inside the tweet text. For hashtags, we utilize the NRC Hashtag Emotion Lexicon [15, 23], which is a list of words and their associations with eight emotions. The associations are computed from tweets with emotion-word hashtags such as #happy and #anger.

Table 1. Encoding categorical attributes.

Attribute	Encoding
'False':0, 'True':1	'iPhone':1, 'Android':2, 'TweetDeck':3, 'web':4
UserLanguage	Each user language assigned a numeric value
MediaEntities	'None':0, 'photo':1
IsPossiblySensitive	

<table>
<tr><th>Emotion</th><th>Emoticons</th></tr>
<tr><td>sadness</td><td>>:[:-(:(:-c :c :-< :< :-[:[:{</td></tr>
<tr><td>anger</td><td>:-|| :@ >:(</td></tr>
<tr><td>joy</td><td>:) ;) =) :] :P :-P ;P :D ;D >:3 :-) ;-) :^) :o) ;^) :-D :-></td></tr>
<tr><td>surprise</td><td>:-o :-O o_O O_o :$</td></tr>
<tr><td>disgust</td><td>D:< D: D8 D; D= DX v.v</td></tr>
</table>

Fig. 3. Emoticons.

All emoticons retained in the data collection process and validated while assigning weights to each emotion class. Figure 3 shows the list of emoticons used in this process. Figure 2 explains the steps involved in assigning final emotion class.

3.4 Additional Pre-processing

In addition to pre-processing steps for text data, additional pre-processing is performed on the numeric attributes of data in order to make it suitable for Classification. The data set has the following Numeric Attributes: AngerScore, TrustScore, FearScore, SadnessScore, AnticipationScore, DisgustScore, SurpriseScore, JoyScore, PositiveScore, NegativeScore, LoveScore, PeopleScore, MessageScore, InstantScore, GetScore, KnowScore, GoingScore, UserFollowersCount, UserFavoritesCount, UserFriendsCount and are normalized using python scikit learn MinMaxScaler in the range of −1 to +1. After additional pre-processing the data is converted into LIBSVM [24] format that is suitable for classification using Support Vector Machine (Table 1).

3.5 Emotion Classification

Classification is a supervised machine learning model that learns the data using labeled train set and predicts the test set for which the model does not know the actual class labels. This model is further evaluated with the help of validation measures like precision, recall, f1-score, and accuracy. In this paper, we have used Support Vector Machine as a classification model for automatically classifying Twitter dataset with emotion. Figure 4 shows the overall processing flow of Support Vector Machine classification utilized in this work.

Support Vector Machines – SVM: Support Vector Machines (SVM) are a useful technique for data classification originally designed for binary classification [25, 26]. Hsu and Lin [26] provide overview of methods for multiclass support vector machines. In order to extend binary SVM to multiclass problems there are three methods: ONE-AGAINST-ALL, ONE-AGAINST-ONE, and Directed Acyclic Graph SVM - DAGSVM methods. Formal definition of these methods as stated in [26]: ONE-AGAINST-ALL: This method constructs k SVM models, where k is the number of classes. The ith SVM is trained with all of the examples in the ith class with positive labels, and all other examples with negative labels. ONE-AGAINST-ONE: This method constructs $k(k-1)/2$ classifiers where each one is trained on data from two

Pseudocode Support Vector Machine Classifier

Input: Dataset D
Output: Confusion Matrix, Validation
Train ← Split [D, size=0.7]
Test ← Split [D, size=0.3]
SVMMultiClassOneVsAllSGD.train(input as RDD, numIterations, stepSize, regParam):
    X: Number of Samples
    Y: Labels, where $Y_i \in \{1, ..., N\}$
    M: SVM SGD classifier model
    For each *n* in $\{1, ..., N\}$
        Construct a new label vector *v* where $v_i = 1$ if $y_i = k$ and $v_i = 0$ otherwise
        Apply *M* to *X*, *v* to obtain *list of classifiers*
Test the model using *'Test'*
Calculate *Scores*
Compute *Confusion Matrix*
Validate Model

Fig. 4. Process – Support Vector Machine.   **Fig. 5.** Support Vector Machine - Pseudocode

classes. Directed Acyclic Graph SVM - DAGSVM: The training phase of DAGSVM is the same as one-against-one method by solving $k(k-1)/2$ internal nodes and k leaves. Each node is a binary SVM of ith and jth classes. Given a test sample x, starting at the root node, the binary decision function is evaluated. Then it moves to either left or right depending on the output value. Therefore, we go through a path before reaching a leaf node, which indicates the predicted class. In this paper, we utilize the ONE-AGAINST-ALL method, pseudocode shown in Fig. 5.

3.6 Actionable Pattern Mining

Actionability is a property of the discovered knowledge. Patterns are considered Actionable if the user can act upon them, and if this action can benefit the user, or help them to accomplish their goals. Action Rules mining is a method to extract Actionable patterns from the data. Action Rules are rules that describe a possible transition of data from one state to another more desirable state. Action Rules are rules that help reclassify data from one category to another more desirable category. Consider the information system S in Table 2. Equations (1) and (2) are example Action Rules. According to Eq. (1) r1 says that if the value A2 remains unchanged and value B will change from B2 to B1 for a given object X, then it is expected that the value D will change from H to A for object X.

In a similar way, the rule r2 in Eq. (2) says that if the value C2 remains unchanged and value b will change from B2 to B1, then it is expected that the value D will change from H to A.

$$r1 = [((A, A2 * (B, B2 \rightarrow B1)) \rightarrow (D, H \rightarrow A)]. \tag{1}$$

$$r2 = [((C, C2 * (B, B2 \rightarrow B1)) \rightarrow (D, H \rightarrow A)]. \tag{2}$$

By support and confidence of rule r we mean:

1. $\text{sup}(r) = \min\{\text{card}(Y1 \cap Z1), \text{card}(Y2 \cap Z2)\}$
2. $\text{conf}(r) = \text{card}(Y1 \cap Z1)/ \text{card}(Y1). \text{card}(Y2 \cap Z2)/ \text{card}(Y2)$. If $\text{card}(Y1) \neq 0$, $\text{card}(Y2) \neq 0$, $\text{card}(Y2 \cap Z2) \neq 0$.
3. $\text{Conf}(r) = 0$ otherwise.

Now, let us consider the Eq. (1) for support and confidence with example.

- $N_s(a, a_2 \rightarrow a_2) = [\{x_2, x_3, x_5, x_6, x_{10}\}, \{x_2, x_3, x_5, x_6, x_{10}\}]$
- $N_s(b, b_2 \rightarrow b_1) = [\{x_2, x_6, x_8, x_{10}\}, \{x_1, x_3, x_4, x_5, x_7, x_9\}]$
- $N_s(a, a_2 \rightarrow a_2) * (b, b_2 \rightarrow b_1) = [\{x_2, x_6, x_{10}\}, \{x_3, x_5\}]$
- $N_s(d, H \rightarrow A) = [\{x_1, x_2, x_6, x_9, x_{10}\}, \{x_3, x_4, x_5, x_7, x_8\}]$

Table 2. Information system S.

X	Attribute A	Attribute B	Attribute C	Attribute D
X1	A1	B1	C1	H
X2	A2	B2	C1	H
X3	A2	B1	C1	A
X4	A1	B1	C2	A
X5	A2	B1	C2	A
X6	A2	B2	C2	H
X7	A1	B1	C2	A
X8	A1	B2	C1	A
X9	A1	B1	C1	H
X10	A2	B2	C1	H

Therefore, for rule r1, support sup(r1) = 2, confidence conf(r1) = 3/2 . 3/2 = 1.

Tzacheva et al. [31] describe that these formulas for support and confidence are too complex for computation provide definition of new support and confidence for Action Rules as below. New support and confidence of rule r is given as sup(r) = card($Y_2 \cap Z_2$), conf(r) = card($Y_2 \cap Z_2$)/ card(Y_2).

3.7 Spark Scalability for Big Data

Increase in data size and the need to scale out computations to multiple nodes gave rise to the distributed programming models. One such model is Apache Spark, which is similar to Hadoop MapReduce. In addition to the similarities, Apache Spark includes data sharing abstraction called Resilient Distributed Dataset's (RDD's) [27].

4 Experiments and Results

In this section, we describe our experiment and results. We extract the data via Twitter streaming API [20] using Apache Spark [28] Scala programming language. The raw data extracted consists of around 520,000 instances. The extracted data is processed as explained in Sect. 3.2. As part of feature, augmentation additional attributes are added to the existing corpus along with the emotion label. We use the NRC Lexicon [15, 23] to label data with Emotion Class as shown on Fig. 2. This results in a corpus of tweets and supporting features consisting of around 174,000 instances.

4.1 Classification – Support Vector Machine

We use WEKA Data Mining Software [29] and Apache Spark [28] to develop the Support Vector Machine One Vs All Multi class classifier. Support Vector Machine classification model requires pre-processing of data, which includes normalization, categorical to numeric or binary, LIBSVM format. Based on the pre-processing three experiments are performed: Using only the numerical attributes in the data, Using all the attributes where the categorical fields are encoded as numeric values, Using all attributes where the categorical fields are encoded as binary.

Experiment 1 - Using Only Numeric Attributes: Experiment 1 is performed by selecting only the numerical attributes from the original dataset which includes: AngerScore, TrustScore, FearScore, SadnessScore, AnticipationScore, DisgustScore, SurpriseScore, JoyScore, PositiveScore, NegativeScore, LoveScore, PeopleScore, MessageScore, InstantScore, GetScore, KnowScore, GoingScore, UserFollow-ersCount, UserFavoritesCount, User- FriendsCount.

We achieve accuracy of 84.92% with WEKA Data Mining software Multiclass Classifier with SVM Stochastic Gradient Descent (SGD) Tables 3 and 4.

Table 3. WEKA – confusion matrix – experiment 1.

A	B	C	D	E	F	G	H	Class
10133	0	0	1481	0	0	0	0	A - Sadness
144	5535	1	2080	0	0	57	0	B - Joy
152	6	1477	716	17	0	2	0	C - Fear
104	10	33	15150	3	0	4	0	D - Anticipation
127	42	22	339	6342	0	1	0	E - Trust
50	18	10	293	213	1124	0	1	F - Surprise
165	6	85	179	61	3	2600	0	G - Anger

Table 4. WEKA – precision, recall, F-measure – experiment 1.

Measure	Sadness	Joy	Fear	Anticipation	Trust	Surprise	Anger	Disgust
Precision	0.922	0.985	0.892	0.712	0.935	0.996	0.926	1.000
Recall	0.872	0.708	0.623	0.990	0.923	0.658	0.839	0.593
F-Measure	0.897	0.824	0.734	0.828	0.929	0.792	0.880	0.744

We achieved almost similar accuracy with Spark single node and six-node cluster as 88.16% and 88.01% respectively. The confusion matrix and classifier evaluation with precision, recall, and F1-score is shown in Tables 6, and 5 respectively.

However, the Spark program runs faster in cluster when compared to Single Node machine for all the three experiments. The results of average run time for execution is shown in Table 7.

Table 5. Spark – precision, recall, F-measure – experiment 1.

Measure	Anticipation	Sadness	Joy	Trust	Disgust	Anger	Fear	Surprise
Precision	0.8898	0.8313	0.8848	0.8912	0.9634	0.9248	0.9153	0.0
Recall	0.9920	0.9984	0.8538	0.9096	0.8232	0.7697	0.4313	0.0
F-Measure	0.9381	0.9072	0.8691	0.9003	0.8878	0.8402	0.5864	0.0

Table 6. Spark – confusion matrix – experiment 1.

A	B	C	D	E	F	G	H	Class
15199.0	24.0	83.0	0.0	12.0	3.0	0.0	0.0	A - Anticipation
13.0	11675.0	2.0	3.0	0.0	0.0	0.0	0.0	B - Sadness
657.0	234.0	6633.0	137.0	9.0	94.0	4.0	0.0	C - Joy
409.0	172.0	25.0	6262.0	12.0	3.0	1.0	0.0	D - Trust
41.0	436.0	120.0	25.0	2898.0	0.0	0.0	0.0	E - Disgust
79.0	580.0	3.0	48.0	8.0	2401.0	0.0	0.0	F - Anger
140.0	630.0	127.0	301.0	24.0	75.0	984.0	0.0	G - Fear

Table 7. Average runtime – spark single node and spark cluster.

Experiment	Number of instances	Spark single node runtime (secs)	Spark 6 node cluster runtime (secs)
1	174688	256.75	207.70

4.2 Action Rules

In this experiment, we extract action rules to identify what changes in attributes lead to change in emotion to a more positive state. For example, change from 'sadness' to 'trust'; 'sadness' to 'joy'. The dataset consists of continuous attributes which are discretized into intervals. The intervals are determined with the help of WEKA data mining software using unsupervised attribute discretization [29]. The following are the list of parameters set to discretize the data, 174688 instances, Weka – unsupervised discretize filter with 5 bins and equal frequency binning. We use the following attributes AngerScore, TrustScore, FearScore, SadnessScore, AnticipationScore, DisgustScore, SurpriseScore, JoyScore, PositiveScore, NegativeScore, LoveScore, PeopleScore, MessageScore, UserFollowersCount, UserFavoritesCount, UserFriendsCount, Tweet- Source, FinalEmotion from the original dataset. Discretization for the numeric attributes are shown in Table 9. The dataset with 174688 instances is divided into 100 parts based on the target class attribute 'FinalEmotion'. Action rules are generated for one part of the dataset with 1439 instances and 18 attributes listed above. The Table 8 gives list of parameters used for action rule generation.

Figure 6 shows sample action rules generated. Let us consider the action rule AR1, this rule suggest possible changes to achieve a desirable emotional state of 'joy'. The action rule is interpreted as follows: If the user tends to use more positive words as

Table 8. Action rule - parameters.

Parameter	Values
Stable attributes	LoveScore, PeopleScore, MessageScore
Decision attribute	FinalEmotion
Support	20
Confidence	30

S.NO.	Action Rule
AR1	(AnticipationScore, 2 → 0) ∧ (DisgustScore, 1 → 0) ∧ (JoyScore, 0 → 2) ∧ (PositiveScore, 0 → 1) ∧ (SadnessScore, 2 → 0) ⇒ (FinalEmotion, sadness → joy) [Support: 21, Old Confidence: 100%, New Confidence: 100%]
AR2	(AngerScore, 2 → 0) ∧ (AnticipationScore, 2 → 0) ∧ (JoyScore, 0 → 2) ∧ (SadnessScore, 2 → 0) ∧ (TrustScore, 2 → 0) ⇒ (FinalEmotion, sadness → joy) [Support: 23, Old Confidence: 75%, New Confidence: 100%]
AR3	(AngerScore, 4 → 0) ∧ (AnticipationScore, 2 → 0) ∧ (DisgustScore, 4 → 0) ∧ (FearScore, 4 → 0) ∧ (JoyScore, 2 → 0) ∧ (SadnessScore, 4 → 0) ∧ (SurpriseScore, 2 → 0) ⇒ (FinalEmotion, sadness → trust) [Support: 30, Old Confidence: 100%, New Confidence: 100%]
AR4	(AngerScore, 2 → 0) ∧ (AnticipationScore, 2 → 0) ∧ (DisgustScore, 2 → 0) ∧ (FearScore, 3 → 0) ∧ (JoyScore, 2 → 0) ∧ (SadnessScore, 3 → 0) ∧ (SurpriseScore, 2 → 0) ⇒ (FinalEmotion, sadness → trust) [Support: 33, Old Confidence: 97%, New Confidence: 97%]
AR5	(AngerScore, 3 -> 0) ∧ (AnticipationScore, 2 -> 0) ∧ (DisgustScore, 3 -> 0) ∧ (FearScore, 4 -> 0) ∧ (JoyScore, 3 -> 0) ∧ (SadnessScore, 3 -> 0) ∧ (SurpriseScore, 3 -> 0) ⇒ (FinalEmotion, fear -> trust) [Support: 33, Old Confidence: 97%, New Confidence: 97%]
AR6	(AnticipationScore, 2 -> 0) ∧ (FearScore, 4 -> 0) ∧ (JoyScore, 3 -> 0) ∧ (NegativeScore, 3 -> 0) ∧ (SurpriseScore, 3 -> 0) ⇒ (FinalEmotion, fear -> trust) [Support: 31, Old Confidence: 100%, New Confidence: 100%]

Fig. 6. Sample action rules.

Table 9. Discretization parameters.

Attribute	Bins	ValueSet
AngerScore	-infinity, 0.002068, 0.997299, 1.007317, 2.0893, infinity	0,1,2,3,4
TrustScore	-infinity, 0.011484, 0.935696, 1.01071, 2.01071, infinity	0,1,2,3,4
FearScore	-infinity, 0.003022, 0.990587, 1.00374, 2.06263, infinity	0,1,2,3,4
SadnessScore	-infinity, 0.004326, 0.973808, 1.00306, 2.00306, infinity	0,1,2,3,4
AnticipationScore	-infinity, 0.324121, 0.992358, 1.00685, 2.00551, infinity	0,1,2,3,4
DisgustScore	-infinity, 0.000009, 0.997325, 1.00053, 2.00085, infinity	0,1,2,3,4
SurpriseScore	-infinity, 0.000056, 0.999872, 1.00141, 2.00545, infinity	0,1,2,3,4
JoyScore	-infinity, 0.001784, 0.999909, 1.00515, 2.00515, infinity	0,1,2,3,4
PositiveScore	-infinity, 0.5, 1.5, 2.5, 3.5,infinity	0,1,2,3,4
NegativeScore	-infinity, 0.5, 1.5, 2.5, 3.5,infinity	0,1,2,3,4
UserFollowersCount	-infinity, 105.5, 307.5, 656.5, 1662.5,infinity	0,1,2,3,4
UserFavoritesCount	-infinity, 575.5, 2570.5, 7123.5, 19418.5,infinity	0,1,2,3,4
UserFriendsCount	-infinity,146.5, 310.5, 574.5, 1253.5, infinity	0,1,2,3,4

denoted by (JoyScore, 0 → 2) and (PositiveScore, 0 → 1), and reduce the words related to negative emotions like disgust, sadness and anticipation as denoted by (DisgustScore, 1 → 0) and (SadnessScore, 2 → 0) and (AnticipationScore, 2 → 0), then it is possible to change the emotion from 'sadness' to 'joy'. In that case, the emotion associated with this user tweet can be classified as 'joy', and we expect that the user is feeling more positive.

5 Conclusion

In this work, we automatically detect user emotion from tweet data using the NRC Emotion Lexicon [21, 22] to label the Emotion class for our data. We use Support Vector Machine with Multiclass classification in particular ONE-AGAINST-ALL implementation in both WEKA data mining software [29] and Apache Spark system

[28] over Hadoop 6 node Cluster for big data scalability. We achieve accuracy of 84.9% to 88.01%. The Spark system is able to scale to BigData with six node cluster, as the data is partitioned into several sets and processed in parallel at each cluster node. This is an extension of previous study [30] of finding user emotions from tweet data using the NRC emotion lexicon to label the emotion class for our data. In the previous work, we examined several classifiers including Decision Tree, Decision Forest, and Decision Table Majority. In this work, we extract Action Rules to identify what factors can be improved in order for a user to attain a more desirable positive emotion. We suggest actions that can be undertaken to reclassify user emotion from a negative emotion to more positive emotion. For instance from 'sadness' to 'joy', 'sadness' to 'trust', and 'fear' to 'trust'. In the future, we plan further test with larger social networking data. We also plan to apply this system for customer surveys and education evaluations.

References

1. M.-W. Dictionary, "Merriam-webster" (2002). http://www.mw.com/home.htm
2. Honey, C., Herring, S.C.: Beyond microblogging: conversation and collaboration via twitter. In 42nd Hawaii International Conference on System Sciences, 2009. HICSS 2009, pp. 1–10. IEEE (2009)
3. Chang, Y., Tang, L., Inagaki, Y., Liu, Y.: What is tumblr: a statistical overview and comparison. ACM SIGKDD Explor. Newsl. **16**(1), 21–29 (2014)
4. Sullivan, D.: Comscore media metrix search engine ratings. Search Engine Watch **21** (2006)
5. Java, A., Song, X., Finin, T., Tseng, B.: Why we twitter: understanding microblogging usage and communities. In: Proceedings of the 9th WebKDD and 1st SNA-KDD 2007 Workshop on Web Mining and Social Network Analysis, pp. 56– 65. ACM (2007)
6. Chang, H.-C.: A new perspective on twitter hashtag use: diffusion of innovation theory. Proc. Assoc. Inf. Sci. Technol. **47**(1), 1–4 (2010)
7. Hasan, M., Agu, E., Rundensteiner, E.: Using hashtags as labels for supervised learning of emotions in twitter messages. In: ACM SIGKDD Workshop on Health Informatics. New York, USA (2014)
8. Gupta, N., Gilbert, M., Fabbrizio, G.D.: Emotion detection in email customer care. Comput. Intell. **29**(3), 489–505 (2013)
9. D'Alfonso, S., Santesteban-Echarri, O., Rice, S., Wadley, G., Lederman, R., Miles, C., Gleeson, J., Alvarez-Jimenez, M.: Artificial intelligence-assisted online social therapy for youth mental health. Front. Psychol. **8**, 796 (2017)
10. Tantam, D.: The machine as psychotherapist: impersonal communication with a machine. Adv. Psychiatr. Treat. **12**(6), 416–426 (2006)
11. Kaur, H.: Actionable rules: issues and new directions. In: World Enformatika Conference - WEC (5), pp. 61–64. Citeseer (2005)
12. He, Z., Xu, X., Deng, S., Ma, R.: Mining action rules from scratch. Expert Syst. Appl. **29**(3), 691–699 (2005)
13. Mishne, G., et al.: Experiments with mood classification in blog posts. In: Proceedings of ACM SIGIR 2005 Workshop on Stylistic Analysis of Text for Information Access, vol. 19, pp. 321–327 (2005)

14. Danisman, T., Alpkocak, A.: Feeler: emotion classification of text using vector space model. In: AISB 2008 Convention Communication, Interaction and Social Intelligence, vol. 1, p. 53 (2008)
15. Mohammad, S.M.: #emotional tweets. In: Proceedings of the First Joint Conference on Lexical and Computational Semantics - Volume 1: Proceedings of the Main Conference and the Shared Task, and Volume 2: Proceedings of the Sixth International Workshop on Semantic Evaluation, pp. 246–255. Association for Computational Linguistics, Stroudsburg (2012)
16. Roberts, K., Roach, M.A., Johnson, J., Guthrie, J., Harabagiu, S.M.: Empatweet: annotating and detecting emotions on twitter. In: LREC, vol. 12, pp. 3806–3813. Citeseer (2012)
17. Ekman, P.: An argument for basic emotions. Cogn. Emot. **6**(3–4), 169–200 (1992)
18. Purver, M., Battersby, S.: Experimenting with distant supervision for emotion classification. In: Proceedings of the 13th Conference of the European Chapter of the Association for Computational Linguistics, pp. 482–491. Association for Computational Linguistics (2012)
19. Ranganathan, J., Irudayaraj, A.S., Tzacheva, A.A.: Action rules for sentiment analysis on twitter data using spark. In: 2017 IEEE International Conference on Data Mining Workshops (ICDMW), pp. 51–60, November 2017
20. Makice, K.: Twitter API: Up and Running: Learn How to Build Applications with the Twitter API. O'Reilly Media, Inc., Beijing (2009)
21. Mohammad, S.M., Turney, P.D.: Crowdsourcing a word– emotion association lexicon. Comput. Intell. **29**(3), 436–465 (2013)
22. Mohammad, S.M., Turney, P.D.: Emotions evoked by common words and phrases: using mechanical turk to create an emotion lexicon. In: Proceedings of the NAACL HLT 2010 Workshop on Computational Approaches to Analysis and Generation of Emotion in Text, pp. 26–34. Association for Computational Linguistics (2010)
23. Mohammad, S.M., Kiritchenko, S.: Using hashtags to capture fine emotion categories from tweets. Comput. Intell. **31**(2), 301–326 (2015)
24. Chang, C.-C., Lin, C.-J.: Libsvm: a library for support vector machines. ACM Trans. Intell. Syst. Technol. (TIST) **2**(3), 27 (2011)
25. Hsu, C.-W., Chang, C.-C., Lin, C.-J., et al.: A practical guide to support vector classification (2003)
26. Hsu, C.-W., Lin, C.-J.: A comparison of methods for multiclass support vector machines. IEEE Trans. Neural Networks **13**(2), 415–425 (2002)
27. Zaharia, M., Xin, R.S., Wendell, P., Das, T., Armbrust, M., Dave, A., Meng, X., Rosen, J., Venkataraman, S., Franklin, M.J., et al.: Apache spark: a unified engine for big data processing. Commun. ACM **59**(11), 56–65 (2016)
28. Meng, X., Bradley, J., Yavuz, B., Sparks, E., Venkataraman, S., Liu, D., Freeman, J., Tsai, D., Amde, M., Owen, S., et al.: Mllib: machine learning in apache spark. J. Mach. Learn. Res. **17**(1), 1235–1241 (2016)
29. Witten, I.H., Frank, E., Hall, M.A., Pal, C.J.: Data Mining: Practical Machine Learning Tools and Techniques. Morgan Kaufmann, San Francisco (2016)
30. Ranganathan, J., Hedge, N., Irudayaraj, A., Tzacheva, A.: Automatic detection of emotions in twitter data - a scalable decision tree classification method. In: Proceedings of the RevOpID 2018 Workshop on Opinion Mining, Summarization and Diversification in 29th ACM Conference on Hypertext and Social Media (2018)
31. Tzacheva, A.A., Sankar, C.C., Ramachandran, S., Shankar, R.A.: Support confidence and utility of action rules triggered by meta-actions. In: 2016 IEEE International Conference on Knowledge Engineering and Applications (ICKEA), pp. 113–120. Singapore (2016). https://doi.org/10.1109/ickea.2016.7803003

Fourth Industrial Revolution: An Impact on Health Care Industry

Prisilla Jayanthi[1(✉)], Muralikrishna Iyyanki[2], Aruna Mothkuri[3],
and Prakruthi Vadakattu[4]

[1] Hyderabad, India
[2] R&D JNTUH, Hyderabad, India
iyyanki@icorg.org
[3] IBS, The ICFAI Foundation for Higher Education, Hyderabad, India
arunam@ibsindia.org
[4] Biomedical Engineering MIT, Manipal University, Manipal, India
prakruthi.vadakattu@gmail.com

Abstract. The World Economic Forum annual meeting, held in Davos, Switzerland, emphasized the Fourth Industrial Revolution as one of the most cutting-edge innovative techniques to be seen in the forthcoming era. This has a greater impact on the future of production and the role of government, business and academia in all developing technologies and innovation where industries, communication and technologies meet. The fourth industrial revolution combines the physical, digital, and biological spaces and is changing the healthcare industry. The FCN-32 semantic segmentation was performed on the brain tumor images which produced better results for identifying the tumors as ground truths and predicted images was achieved. The best calculated loss = 0.0108 and accuracy = 0.9964 for the given tumor images was achieved. The earlier detecting and analysis of any disease can help diagnosing and treatment in better means through artificial intelligence techniques. The healthcare industry can serve better with faster and quality services to remote, rural and unreachable areas and thereafter reduces the cost of hospitalization.

Keywords: Artificial Intelligence · Accuracy · Health care ·
Industrial revolution · Machine learning · Semantic segmentation

1 Introduction

The fourth industrial revolution a present-day society has experienced an over-whelming growth since the dawn of mechanical production and steam power energy recognized in 1784, Artificial Intelligence (AI) will create waves in next industries and will play a significant role in its development. Next, the timeline in 1870 the society's essential transformation was electrical energy and mass production, in 1969 the third revolution with electronics through internet technologies. From detection of objects, speech recognition, language translation, faces recognition and analyzing of images, the whole world has seen the AI marshaling. Elon Musk, founder of OpenAI in a statement quotes rightly that AI can potentially become "more dangerous than bombs".

© Springer Nature Switzerland AG 2020
T. Ahram (Ed.): AHFE 2019, AISC 965, pp. 58–69, 2020.
https://doi.org/10.1007/978-3-030-20454-9_6

The computer's ability to learn through passing the task of optimization weights of the variables, available data to make accurate predictions about the future. The machine learning algorithms made it possible for the quality of the predictions to improve with experience. The more data, the better the prediction engines are created.

Schwab, World Economic Forum chairman mentions that the evidence of histrionic change is all around us and it's all happening at exponential speed. With artificial intelligence, mobile supercomputing, intelligent robots, self-driving cars, neuro-technological brain enhancements, genetic editing has paved its way beyond the potentials.

With the implementation of AI to help optimize and scale operations through predictive maintenance, improved safety among others. The promise of digital technology is to transform the industry and advanced analytics into augmented reality, and other technologies that the Industrial Internet of Things (IIOT) is about to bringing the future. Digital literacy, that gives students the adaptive abilities need to participate fully in the global digital society. The digital economy and derive new opportunities for employment, innovation, creative expression, and social inclusion. Education and lifelong learning will be important to equip present and future generations to be a productive part of this new world but to meet the societal challenges presented by the fourth industrial revolution, and the existential challenges presented by climate change and population growth. Project based learning has engaged youth to concentrate more and thereby kept them working and explore more.

Alvin Toffler, American writer suggests that "The illiterate of the 21st century will not be those who cannot read and write, but those who cannot learn, unlearn, and relearn" [1].

2   Previous Study

Park H describes that the fourth industrial revolution is built on the third revolution, and is characterized as "a fusion of technologies that are blurring the lines between the physical, digital, and biological spheres". This fusion of technologies is not just a product of science and engineering, but is a product of values and institutions. Teamwork with industry, government, and academia builds a pooled vision of the future. Advances in technological are giving rise to a large variety of smart connected products and services, combining sensors, software, data analytics, and connectivity in all kinds of ways. These innovative are restructuring healthcare industry boundaries and are leading for the creation of whole new industries. Fourth industrial revolution will be instrumental for all, as it will be connecting patients to doctors, to their caregivers, and clinicians with combined technologies [2].

Prisilla and Iyyanki proposed a network which would detect the tumor part in the brain of the human. The ordinary segmentation is time-consuming process and is tedious when prepared manually, and is prone to human error. The code for automatic segmentation is inexpensive, reliable, and is scalable; works with large networks with more accurate and better at classification. This segmentation provides a faster and a well-planned treatment for the large scale research. Extending the classification nets to segmentation, and improving the architecture with multi-resolution layer combinations improves simplifying and speeds up learning. The image segmentation can learn certain specific feature than any macroscopic features, and moreover inclusion of more abnormal examples which are very variant and unusual. Hence, the convolutional neural network produced more accurate segmentation result helping to understand the tumor part in the images of the brain for assisting the surgeon to diagnose it faster and further analysis [3].

Prisilla et al. designed a decision tree for taking the decisions for the brain tumor detection; the program helps to understand which attribute is most highly prioritized and thus enables to understand the tumor of any disease. Decision trees are reliable and scalable, providing high classification accuracy with a simple representation of collected knowledge and effective decision making technique that can be used in medical care. Decision tree supports and handle huge datasets with simple and fast integration. It is easy to predict the classification of unseen records using decision tree [4].

Menze et al. stated that neuro-radiologists manually segmented the tumors with a cross-rater dice score of 0.75–0.85, and the model predictions made by Menze and team was equivalent with the experts made. UNet, an auto-encoder has an encoding path for contracting pair with a decoding path i.e. expanding giving a "U" shape. The UNet predicts a pixel-wise segmentation map of the input image rather than classifying the input image as a whole. UNet segments brain tumors from raw MRI scans with very little data to train a UNet model to accurately predict where tumors exist. The dice coefficient (BraTS dataset) for the model is 0.82–0.88. The tumor segmentation masks as predicted by the different UNet configurations gave 95% of the predictions differed by less than 0.1 dice points and over 98% by less than 0.2 points. An UpSampling2D yielded an average test dice score of 0.8718, while Conv2DTranspose produced 0.8707 [5].

De Fauw et al. developed the architecture of optical coherence tomography (OCT) imaging for ophthalmology. The UK National Health Service (NHS) adopted to use OCT for comprehensive initial assessment and patient triage requiring rapid non-elective assessment of acute and chronic sight loss. But then two challenges faced by automated diagnosis of a medical images are image variations, and secondly, patient-to-patient inconsistency in pathological lab analyzing appearances of disease.

The existing approaches of deep learning deal with few combinations using a single end-to-end black-box network that requires millions of labeled scans. The framework decouples the two problems namely technical variations in the imaging process, and pathology variants and solves them independently. A deep segmentation network creates a detailed device-independent tissue segmentation map and a deep classification network analyses this segmentation map and provides diagnoses and referral suggestions. The presence of uncertain regions in the image is a significant challenge in OCT image segmentation, thus the true tissue type cannot be inferred from the image. Multiple instances are trained for the network segmentation and each network instance creates a full segmentation map for the given scan, resulting in multiple hypotheses. The clinical OCT scans analyses and makes a standard comparable to clinical experts. The study concluded that the medical imaging addresses a wide range of medical imaging techniques, and integrate clinical diagnoses [6].

3   Fourth Industrial Revolution

Newton who formulated the laws of motion from where first industrial revolution was catalyzed to design stems engines. Faraday and Maxwell, a unified magnetic and electric force has led to electricity generation and electric motor which were influential in the second industrial revolution. The third industrial revolution was catalyzed by the discovery of a transistor, the electronic age that gave rise to computers and internet. The fourth industrial revolution is power-driven by artificial intelligence and it will renovate the workplace from tasks based features to the human centered features as in Fig. 1.

The degree of automation has improved the industrial companies with more intelligent and self-adaptive as advances are made through artificial intelligence. The fourth industrial revolution era, the diverse business models with customer access and hence creating new refinements of production methods and generating extra digital revenues and optimizing customer experience.

The fundamental technology that is essential for accelerating growth of revolution. The first is the adoption/diffusion of the technology and second is a network effect, the technologies needed to work to drive growth.

Digital energy: merging smart power grids and smart meters into platforms that dynamically generate energy and demand from various sources.
Digital transport: capable of moving people and goods across oceans, skies, and land independently.
Digital health: enabling remote health care from any corner.
Digital communication: connecting and interacting billions of people and things.
Digital production: it will bring a paradigm.

4   Informatics Research in Health Care

Open innovation, the blend of humans and computers to form distributed systems for the purpose of accomplishing innovative tasks and accuracy. Several forms of technology-driven R&D come to improve data acquisition accuracy; advanced big-data analytics to spot hidden statistical patterns; artificial intelligence techniques, and knowledge discovery. The humans and computers do have their own fortes and weaknesses. Artificial intelligence is moving into the mainstream, and the convergence of increasing computing power, big data and machine learning should be appreciated for reshaping the world [7].

The big challenge faced in healthcare is aging in the today's world population and lifestyle-related diseases are mounting by 2020. Heart diseases, cancers, respiratory diseases and diabetes are killing 31 million people a year according to the survey of World Health Organization. Healthcare costs are spiraling out of control. The big data and artificial intelligence revolution has led to the advent of the fourth industrial revolution, combining connected devices with cloud computing.

4.1   Revolution in HealthCare

The healthcare sector is the major largest sector to benefit from what the World Economic Forum calls the fourth industrial revolution. Mobile health applications can monitor and provide direct provision of care to patients. The mHealth innovations can help bring healthcare access across the world, improve clinical data gathering and improve the delivery of health care information. Nanotechnology, the ability to manipulate atoms and molecules, has the potential to improve diagnosis and treatment of heart disease patients. Nanotechnology is being used to detect new viruses.

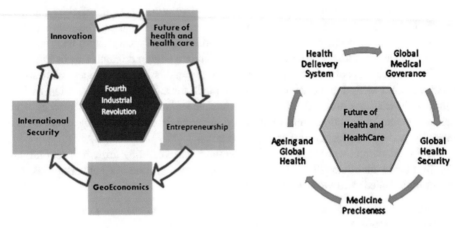

Fig. 1. Fourth industrial revolution   **Fig. 2.** Healthcare

The health sector will lead the track for healthcare practices to influence the promises of this new industrial revolution. The advancements in reformed healthcare will trigger opportunities for developing faster and ensuring diagnostics will be indispensable in determining accurately the efficacy of a treatment. The future for the health care industry in adapting to this new era will continue to thrive in Fig. 2.

Visualizing about the world where science would overhaul the causes of diseases as opposed to dropping the effects of indicators. Think the cell regeneration science would help diabetes or renal failure patients relish disease-free lives as they benefit from early transplants of artificial pancreas and kidney. AI techniques innovation will then finally become the instrument to value-based health care, as long as it is true that regulatory pathways progress at the speed science does, and new mechanisms for full transparency of new cure.

4.2   Drones in HealthCare

Drone, a class of aircraft without a human on board; has paved the unreachable to reachable by delivering the medications like blood, defibrillators, vaccinations, organs or other health care items. The credible drone has the leapfrog over transportation setup. Drones are usually known as unmanned aerial vehicles or unmanned aircraft, remotely operated aircraft, and remotely piloted aircraft. Drones can carry two to five kilograms and transport items about 12 km, traveling up to 40 km to 60 km per hour, taking about 18 min including lift off and landing.

A smartphone app in the drone enables the senders to select the destinations. The drone then mechanically generates a route based on the topography, weather, and airspace and population density. The drones mechanically avoid airports, schools and public squares to reach the destination. After handing over drone on reaching back to its nest, the sim card and new battery is replaced along with the medication item for its next supply. Safety, accuracy and speed are more essential in healthcare for human lives. From the remote clinics a health care worker can order supplies via text and the delivery of any medical item reached within 15 min. Drone outreaches medication where an ambulance cannot reach with much faster speed and thus saving lives [8].

4.3   Artificial Neural Networks in HealthCare

The broad exploring of ANN applications in health care in Fig. 3 have found added advantages that includes self-learning, high-parallelism strength and high-speed and error tolerance against noises which influence the constraints and representing a non-linear systems in which the correlation among the variables is unknown. ANNs are called connection-oriented networks, fashioned by the series of neurons, organized in layers and each neuron is linked with another neuron in the next layer through a weighted link. The number of neurons and one or more hidden layers of the perceptron determine their complexity which creates new neural connections for solving complex problem.

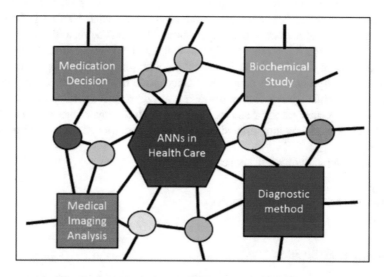

Fig. 3. Overview of an application of ANN in healthcare

The Tables 1 and 2 [9] shows the predicted value ranges from 0.5 to 1.3 and significance F-value has 0.00070 which is less than 0.05 respectively. Using Leven-berg Marquardt backpropagation algorithm the network is trained by adjusting the hidden layers and modifying weights and the performance plot for best validation performance of 0.0182 is achieved at epoch 2 as in Figs. 2 and 5 [9]. The application of ANN shows that the error rate drops by 80% on predicting the disease.

4.4 Security in Healthcare

The security of sensitive data throughout the data life cycle needs to be protected with the same secure approach needed to produce hardened devices. The sensitive data is not limited to processing information and includes a privacy regulations associated with manufacturer's property. With the growing traffic one can move along the signature-based detection technologies, yet they are only to certain limited ability to detect activities within their signature database. Hence protecting cloud data storage and data movement needs the use of strong encryption, artificial intelligence, and machine learning solutions provides robust and responsive threat intelligence, intrusion detection, and intrusion prevention solutions [10].

5   Results and Discussion

Semantic segmentation does indispensable tasks in analyzing images. The method of associating each pixel of an image with a class label such as person, road, grass or car is called sematic segmentation. The semantic segmentation can be implemented on image segmentation and the disease detection in the health care centers for faster diagnosing. In fully convolutional networks, each layer output in a ConVnet is a 3D array of size a × b × c, where a and b represent height and width and are spatial dimensions and c is channel dimension. The first layer is the image, with pixel size a × b and c, channel dimension. In fully connected networks, the receptive field is the region in the input space that a particular CNN's feature is looking for. A receptive field of a feature can be described by its center location and its size. The deep the layers the receptive field is larger. An FCN works on any input size, and yields an output of corresponding spatial dimensions.

The three steps involved in semantic segmentation

1. Image Classification
2. Upsampling through Deconvolution
3. Blending the Output

5.1   Image Classification

In classification, an input image is downsized and made to pass through all the convolution layers and fully connected layers, and results in one output, a predicted label for the input image. The input image is downsized only to get a single predicted output. The process of making output size smaller is referred as convolution by taking the input image.

5.2   Upsampling Through Deconvolution

In deconvolution, it is through upsampling the output image size gets bigger. It is also referred as upconvolution, and transposed convolution. The upsampling path is used for enabling precise localization i.e., exactly where upsampling helps in recovering the fine-grained spatial information lost while the pooling data.

5.3   Blending the Output

After passing through several convolutional layers let's say up to conv7, the output size becomes smaller, then applying 32× upsampling makes the output to have the same size of input image. But it makes the output label map rougher. This happens because the deep features can be obtained when going deeper, but spatial location information is lost when going deeper has to be noted. Which interprets the output from shallower layers have more location information. If both are combined then the result is enhanced. For fusing element by element addition is used to get the output in Fig. 5.

Output- the calculation of loss and accuracy
Found 5000 images belonging to 2 classes.
Found 2000 images belonging to 2 classes.
Epoch 1/20
2000/2000 [==============================] - 546s 273ms/step - loss: 0.4128 - acc: 0.8264
Epoch 2/20
2000/2000 [==============================] - 574s 287ms/step - loss: 0.2767 - acc: 0.8872
Epoch 3/20
2000/2000 [==============================] - 559s 280ms/step - loss: 0.1717 - acc: 0.9340
Epoch 4/20
2000/2000 [==============================] - 561s 281ms/step - loss: 0.1018 - acc: 0.9620
Epoch 18/20
2000/2000 [==============================] - 555s 277ms/step - loss: 0.0133 - acc: 0.9957
Epoch 19/20
2000/2000 [==============================] - 553s 277ms/step - loss: 0.0133 - acc: 0.9956
Epoch 20/20
2000/2000 [==============================] - 556s 278ms/step - loss: 0.0132 - acc: 0.9960

The Fig. 4 demonstrates graphical representation of an accuracy/loss plot for data training and validating of the fully convoluted images shown in the Fig. 5. For the epoch 1, Fig. A shows a loss: 0.0132 and accuracy: 0.9960; Fig. B produces a loss: 0.0108 and accuracy: 0.9964; Fig. C produces a loss: 0.0153 and accuracy: 0.9948 and Fig. D produces a loss: 0.0130 and accuracy: 0.9958 at 278 ms, 294 ms, 288 ms and 315 ms per step respectively. The Table 1 displays the corresponding accuracy/loss values of the figures A, B, C and D plots.

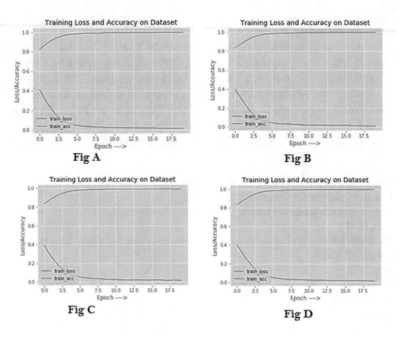

Fig. 4. An accuracy/loss plot for training and validation data

Table 1. Accuracy/loss values for training and validating data of fully convoluted images

	A		B		C		D	
Epoch	Loss	Acc.	Loss	Acc.	Loss	Acc.	Loss	Acc.
1/20	0.4128	0.8264	0.3963	0.8352	0.3900	0.8363	0.4024	0.8332
2/20	0.2767	0.8872	0.2795	0.8859	0.2659	0.8910	0.2829	0.8846
......								
19/20	0.0133	0.9956	0.0115	0.9965	0.0183	0.9941	0.0161	0.9948
20/20	0.0132	0.9960	0.0108	0.9964	0.0153	0.9948	0.0130	0.9958

6   Fourth Industrial Revolution Prominence

The global income level rise and the improvement in the quality of life are coined in the fourth industrial revolution across the world. The new innovative techniques led to a supply-side miracle, with long-term gains in efficiency and productivity. The costs of transportation and communication will drop, logistics and global supply chains will become more effective, and the trading cost will diminish [11]. The key economic concern, inequality combined with mass unemployment represents the greatest societal concern associated with the fourth industrial revolution and thus the employees of today need to get updated more than ever before. The government will gain new technological powers to increase their control over populations, based on pervasive surveillance systems and the ability to control digital infrastructure.

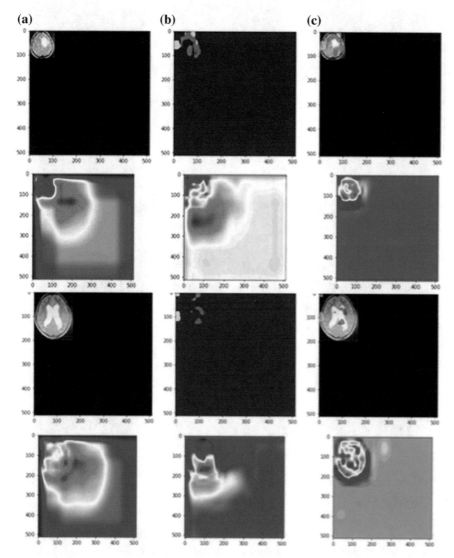

Fig. 5. Fully convolutional network outcome of brain tumor images. a. original image b. ground truth, c. predicted image

(a)   **(b)**   **(c)**

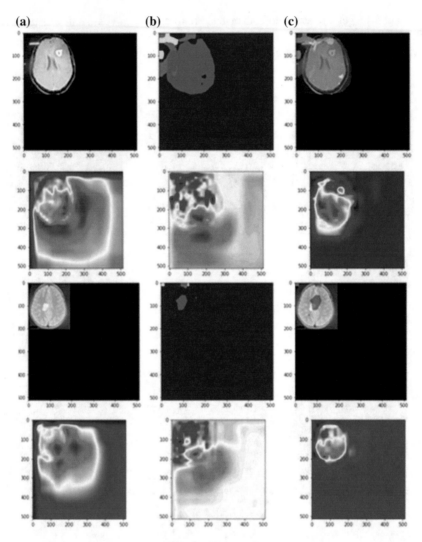

Fig. 5. (*continued*)

7   Conclusion

The medical resource delivery by drone will significantly reduce the number of patients who need to be hospitalized and reduce the overall cost of medical services that comes with hospitalization. The future health care services will enhance with intelligent human-centered services that are dedicated and service-oriented. The fourth industrial revolution brings positive facts on AI, ANN, deep learning and its application in the field of health care. The FCN-32 produced a better predicted output for the brain tumor images. Hence, semantic segmentation helps in detecting the tumors in any type of cancer or diseases. On training the brain tumor image dataset, best accuracy = 0.9964 and loss = 0.0108 were achieved.

Acknowledgements. We appreciate Dr. Iyyanki Muralikrishna for his innovative thoughts and constant encouragement.

Conflicts of Interest. The authors declare that there are no conflicts of interest regarding the publication of this paper.

References

1. Brown-Martin, G.: Education and the Fourth Industrial Revolution, August 2017
2. Park, H.A.: Are We Ready for the Fourth Industrial Revolution? IMIA Yearbook of Medical Informatics (2016)
3. Prisilla, J., Iyyanki, V.M.K.: Convolution neural networks: a case study on brain tumor segmentation in medical care. In: Proceedings of the International Conference on ISMAC in Computational Vision and Bio-Engineering. Springer Nature Switzerland AG (2018)
4. Prisilla, J., et al.: Decision trees for predicting brain tumors: a case study in health care. In: Third International Congress on Information and Communication Technology, Advances in Intelligent Systems and Computing 797. Springer Nature Singapore Pte Ltd. (2018). https://doi.org/10.1007/978-981-13-1165-9_83
5. Thieme, M., Reina, T.: Biomedical Image Segmentation with U-Net (2018)
6. De Fauw, J., et al.: Clinically applicable deep learning for diagnosis and referral in retinal disease. Nat. Med. (2018). https://doi.org/10.1038/s41591-018-0107-6
7. Xing, B., Marwala, T.: Implications of the Fourth Industrial Age on Higher Education (2017)
8. Scott, J.E., Scott, C.H.: Drone delivery models for health care. In: Proceedings of the 50th Hawaii International Conference on System Sciences (2017)
9. Prisilla, J., et al.: Deep learning in oncology- a case study on brain tumor. Int. J. Cancer Res. Ther. **3**(1), 1–5 (2018). ISSN: 2476-2377
10. Waslo, R., Lewis, T., Hajj, R., Carton, R.: Connected objects expanding risks to the physical object. In: Industry 4.0 and Cybersecurity: Managing Risk in an Age of Connected Production (2017)
11. Klaus, S.: The Fourth Industrial Revolution: what it means, how to respond. World Economic Forum (2016)

Human–AI Collaboration Development: Interim Communication Rivalry of Generation

Olga Burukina[1,2(✉)]

[1] The University of Vaasa, Wolffintie 34, 65200 Vaasa, Finland
burukinaolga@yahoo.com
[2] The Institute of Legislation and Comparative Law under the Government
of the Russian Federation, B. Cheremushkinskaya Street 34,
117218 Moscow, Russia

Abstract. The technical revolution has been opening unprecedented opportunities, but at the same time it has force people to face unheard of ethical problems. The author believes that the totality of problems will include, on the one hand, ethical aspects of people's attitudes to AI, and on the other, problems associated with AI itself, which can lead to systemic risks that exceed the danger of using nuclear technologies that humanity was unable to keep under control. In this regard, it is obvious that the potential risks and possible losses and damages predicted by scientific analysis, which are fraught with the reckless use of AI, are to be taken with great attention, although the probability of their occurrence may seem quite low today. Basing on systemic approach, Strauss–Howe generational theory and survey methodology, the author has conducted a pilot research engaging two sample populations – one of Russian university academics belonging to Generation X and the other of Russian university students belonging to Generation Y. The overall number of the pilot survey participants exceeded 100 respondents. The interim research results have revealed a kind of current communication rivalry between AI apps users belonging to Generation Y and AI apps 'virtual hostesses' and collaboration trends in the attitudes of the AI apps users belonging to Generation X, which are attributed to the profound differences in the two generations' value systems and mentalities and might highlight a deeper problem of a possible partial loss of the national value system as the core of the national mentality, as well as a possible mainstream of Human–AI collaboration development based on a kind of communication rivalry and potential intellectual slavery.

Keywords: Artificial Intelligence · Generation X · Generation Y ·
Collaboration · Communication rivalry · Ethical issues

1 Introduction

The influence of artificial intelligence (AI) on the lives of people over the past two decades is constantly increasing and is expected to become even more global in the near future. AI is used in various areas – from personal mobile devices and cars with on-board voice control computers to space shuttles and the Alpha space station – and has good prospects due to the continuous improvement of computer equipment and

© Springer Nature Switzerland AG 2020
T. Ahram (Ed.): AHFE 2019, AISC 965, pp. 70–82, 2020.
https://doi.org/10.1007/978-3-030-20454-9_7

software, on the one hand, and the improvement of AI algorithms, which already significantly surpass some expert competences of man – on the other. As the capabilities of the AI develop, the volume of applications in which it is used will continue to grow, and it is highly likely that AI will soon begin to self-improve based on self-optimization of the algorithms, eventually reaching the level of human intelligence, and then surpassing it.

In a number of applications of AI – from safe driving to medical diagnostics – its superiority over humans has already been proven and documented. Nevertheless, we cannot predict and control possible AI solutions in situations where serious injuries caused by a collision of an unmanned vehicle with several pedestrians can be avoided only at the cost of serious or fatal injuries to passengers inside an AI vehicle. Or, vice versa, hitting a pedestrian might seem the best option for a driverless car. Thus, on 18 March 2018, an Uber self-driving auto struck and killed a woman on a street in Tempe, Arizona, despite having an emergency backup driver behind the wheel [1].

In addition to all possible cautions and macabre forecasts, there is another danger, no less significant than physical destruction, namely the danger of "loss of humanity", i.e. human losses of the most significant characteristics that distinguish man from animals or robots.

2 AI on the Rise

Following the success of the Internet and ICTs, Artificial Intelligence is the next big thing of the high-tech industry. Three essential aspects including next-generation computing architecture, access to historical datasets, and advances in Deep Neural Networks are accelerating the pace of innovation in the field of Machine Learning and Artificial Intelligence [2].

A little over a decade ago, a supercomputer – IBM's Deep Blue – defeated the then world chess champion, Russian grandmaster Gary Kasparov. In 2016 DeepMind's computer program, AlphaGo, beat Go champion Lee Sedol, a Korean professional Go player of 9 dan ran. On 18 June 2018, IBM's Project Debater engaged in the first-ever live, public debate with humans. On preparing arguments for and against the statement: "We should subsidize space exploration," the AI system stated that its aim is to help "people make evidence-based decisions when the answers aren't black-and-white" [3].

According to a recent Oxford and Yale University survey of over 350 AI researchers, machines are predicted to be better than us at translating languages by 2024, writing high-school essays by 2026, driving a truck by 2027, working in retail by 2031, writing a book by 2049 and performing surgery by 2053 [4].

AI software is expected to create unseen business opportunities and societal values, with still smarter chat bots providing expert assistance and robot advisors conducting instantaneous research in most humanities. Artificial intelligence is getting big. The evidence is everywhere. From critical life-saving medical equipment to smart homes, AI will be infused into almost every application and device [2]. A few years more, and humanity will be amazed at AI achievements and face new challenges. To be proactive, researchers should set challenging tasks, rising issues that seem today too early or insignificant before it has become too late the day after tomorrow.

Addressing university students in 2017, Russian President Vladimir Putin stated that the nation leading in AI 'will be the ruler of the world' and warned that AI offers both 'colossal opportunities' and yet unpredictable threats [5]. The same year, Elon Musk and 116 other technology leaders including Bill Gates and Steve Wozniak, Stuart Russell and Stephen Hawking sent a petition to the United Nations warning of a "third revolution in warfare" and calling for new regulations on the current and future AI weapons development [6].

Nowadays, more than 50 nations are developing battlefield robots and drones. The most sought-after are robots capable of making the "kill decision" without human control. These weapons are currently not prohibited by International Law, but even if they were, it is highly doubtful they can ever conform to IHL and IHRL or even laws governing armed conflicts, as AI on the loose would have to avoid mistakes telling friends from foes and combatants from civilians and provide aid to the latter along with wounded comrades in arms. And the issue of accountability seems to be even more complex. So far, these sore questions remain unanswered as the development of intelligent killing machines has gradually turned into a yet unacknowledged arms race. The numerous 'collateral losses' among civilians in Afghanistan, Pakistan, Yemen, and Somalia have highlighted how ethically fraught the situation is [6].

3 Specificities of Two Generations

As is known, Generational Theory was developed by William Strauss and Neil Howe based on the description of repetitive generational cycles in US history [7]. Though the theory seems to be contradictory, having gained a lot of advocates and as many opponents, the idea to compare generations seems to be quite fruitful, as people separated by 15–20 or more years naturally differ greatly due to the different life conditions, which the 20th century used to change every 30 or 20 years or even faster. The two generations, whose representatives' attitudes are analysed in the presented research, are Generation X and Generation Y, which life circumstances differed a lot as the majority of their representatives were born in very different countries due to the collapse of the Soviet Union.

The range of Generation X is from 1965 to 1984 (but the dates are not set in stone) [8], therefore, the range of Generation Y is from 1985 to 2004. The traits listed below are generalised, yet, they are found in most representatives of the generations (basing on several surveys with the participants confirming the fact), naturally developed to a different extent due to family upbringing and personal life circumstances.

3.1 Generation X

Russian Generation X is characterized with a changed value system (as compared to the previous generation – that of their parents) and changes in the quality of life. However, as this generation faced the tragic collapse of the USSR, the dispelled value of "public good", along with the need to survive, it has developed a number of core qualities, partly borrowed from the two previous generations (such as the enduring values of love, care, and family), along with hyper responsibility (in the first place – caring for others, even to the detriment of their own interests), strive for self-education and self-knowledge, increased anxiety, a sense of internal conflict, emotional instability, and exposure to depression. Mostly, this generation has easily entered the new digitalised world, although some representatives experienced certain difficulties mastering new ICTs.

3.2 Generation Y

Russian Generation Y has developed a new system of values, trying to get used to the collapse of the USSR, the emergence of new technologies, high probability of terrorist attacks and military conflicts. Therefore, the key traits characterising the generation include love for freedom, "lightness" in relation to life, flexibility and adaptability to changing conditions, increased attention to the quality of their life, desire to get fun and satisfaction, reluctance to accept the shortcomings of other people, greater importance of career than creating a family, a lack of desire to plan their future, preference of the Internet to any other sources of knowledge, knowledge devaluation, along with credulity, naivety and infantilism (due to the availability of any information). Being digital natives, they are mostly on short terms with ICTs and cannot imagine their life without PCs and all sorts of gadgets. Unlike their parents, they prefer to spend leisure time in the virtual world, with some of them abusing the digital reality or digital reality abusing them.

4 Research Outline

The introduced pilot research was planned to identify the contemporary perceptions of AI by two university populations – academics and students, belonging to two different generations – Generation X and Generation Y in order to reveal some current trends and grope further possible challenges.

The methodology applied in the pilot research includes a questionnaire-based survey designed to suit the research goals, content analysis of the participants' responses, and extensive review of literature devoted to AI, UAVs and digitalisation at large.

4.1 Research Goal and Hypotheses

The goal of the research is to reveal the attitudes of academics belonging to generation X and students belonging to Generation Y to AI, its current achievements and future advances and compare them basing on systemic approach and content analysis of the contemporary discourse.

The goal of the pilot research prompted two hypotheses –

Hypothesis 1 – unlike representatives of Generation Y, representatives of Generation X are more restrained in their perception of AI and more cautious as regards to its further development.

Hypothesis 2 – unlike representatives of Generation X, representatives of Generation Y are on better terms with ICTs and gadgets and take AI for granted not perceiving it as an equal partner in communication.

The analysis of the global research discourse and preliminary collegial discussions of the issues raised in the questionnaire have contributed to identification of five challenges facing the mankind in the near future, generated by the too high speed of technological transformations, the rise of AI, drones and battlefield robots and human inability or unwillingness to tackle sore issues before they start tackling humans. The challenges revealing both the greatness of human intelligence and utter vulnerability of humanity torn apart by contradictions and rivalry for world domination in the absence of a comprehensive mankind development strategy are as follows

(1) The issue of *trust* seems to be the most serious challenge to date. With trust being the most important quality in human relations valued since the first humans appeared on the Earth, people haven't learned to respect and value trust much enough not to abuse and betray it. People do not trust strangers, nations do not trust other nations and naturally, this approach will be transferred to human-AI relations, with AI soon learning not to trust people.
(2) Teaching AI in combat robots and drones to kill people (even the most evil ones) is the shortest way to face the challenge of possible annihilation and only probable survival.
(3) The *underestimation* of *AI abilities to learn* and study, including its soon realised requirement to study people may lay ground for a further built vicious circle in human-AI relations, with people possibly trapped inside.
(4) The current *lack of respect* to the still young AI making its first steps can be easily transferred to more mature AI developments, which may soon learn to disrespect and disregard people.
(5) The *utilitarian attitude* to AI may lead to humans' attempting to get new slaves and thus the challenge of losing some important qualities that the mankind has gained for the last 150 years after slavery was abolished and all the people were proclaimed equal. And here, again it is important to keep in mind the AI growing ability to learn faster than most humans can imagine.

4.2 Research Methods and Processes

The research methodology rests upon systemic and comparative approaches to the ideas of AI perception and collaboration with it in the present and in the future, and includes a literature review, a discourse analysis and a qualitative empirical study launched as a pilot survey of two university populations' attitudes and opinions.

The questionnaire consisted of 3 background questions (Age, Gender, Occupation) and 16 thematic research questions, some of which were scaled according to Likert bipolar scaling, measuring the participants' positive or negative responses from 'Fully approve' to 'Totally disapprove' aimed at revealing the respondents' attitudes to certain AI issues and multiple-choice questions intended to identify the respondents' knowledge, habits or opinions.

4.3   Respondents

The survey was designed as pilot and was supposed to be voluntary, so the samples were limited by the number of questionnaires handed out, the timeframe and the participants' free choice to take part in the survey. The pilot research was planned to be comparative, engaging students and academics of Lipetsk State Pedagogical University, Russia. The university was chosen due to its characteristics.

Lipetsk State Pedagogical University (LSPU) belongs to the categories of 'small' and applied universities as its population of students is about 5,000 persons and its academic population is 356 persons (288 full-timers and 68 part-timers). LSPU provides education in the following 10 fields: mathematics, technical studies, law, social sciences, philology, psychology, education studies, natural sciences, cultural studies and arts, physical culture and sports, at all the three levels – for Bachelor's, Master's and post-graduates.

5   Research Results

The participation in the survey was voluntary and intended to encompass as many academics and students as possible. However, as the population of students at LSPU is 14 times larger, it was decided that the samples could be unequal covering twice as many students as academics. Therefore, 40 academics were provided with questionnaires printed out on paper, with 38 questionnaires returned valid for further analysis; and 80 questionnaires printed out on paper were distributed among students, with 78 students returning questionnaires valid for analysis.

The background questions were supposed to identify the respondents' age, gender and occupation, as these data have proved the respondents' relevance to the research, with age and occupation in particular as the research presumed comparison of two generations – Generation Y and Generation X represented by students and academics accordingly.

The histograms below (Figs. 1 and 2) show a majority of the employees were aged from 31 to 45 (68%), with 23% aged from 46 to 55, 3% aged from 26 to 30, 3% aged from 56 to 60 and 3% aged over 60. These data confirm that an overwhelming majority of the academics taking part in the survey belong to Generation X and the rest are very close to it. A majority (83%) of the students participating in the research were under 20 years of age – Bachelor's students, with 17% of them aged from 21 to 25, i.e. Master's

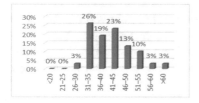

Fig. 1. Academics' age groups

Fig. 2. Students' age groups

students. It means that all the student participants belong to Generation Y. Therefore, the samples are relevant for the research goal and can be compared safely.

Gender is a variable of secondary importance for this research, yet, it also proves its validity as the shares of female and male participants in the two samples are not contradictory – there are 48% of male and 52% of female participants in the academics sample, against 31% of male and 69% of female respondents in the student sample (Figs. 3 and 4 below). However, it would be interesting to go into a deeper research investigating the specificity of women's and men's attitudes to AI.

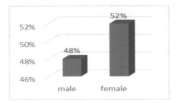

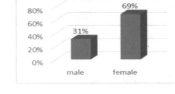

Fig. 3. Academics' gender   **Fig. 4.** Students' gender

The histograms below prove that all the employees in the sample are engaged in teaching, and that the main occupation of the student sample is studies, which confirms the validity of both groups of the participants for the research (see Figs. 5 and 6 below).

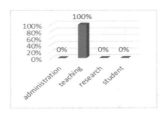

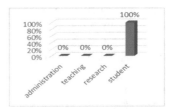

Fig. 5. Employees' occupation   **Fig. 6.** Students' occupation

Question 1 intended to identify the participants' general attitude to AI, ranging it from highly positive to utterly negative. 26% of the academics expressed their highly positive attitude to AI (against 12% of the student sample). The shares of the respondents who expressed their positive attitudes were very close – 29% of the academics and 26% of the students. Neutral attitude (close to indifferent) was expressed by 13% of the Generation X representatives and 40% of the Generation Y representatives. But cautious about AI were 32% of the academics and 21% of the students, with the attitude of 1% being utterly negative. The higher level of cautiousness among the academics proves Hypothesis 1.

Question 2 inquired about the presence of AI in the participants' life. 52% of the academics and 63% of the students confirmed that AI was present in their life, 35% and 31% correspondingly found this question difficult to answer, and 13% of the academics and 6% of the students denied AI presence in their life. The shares of AI proponents in the two groups were closer to each other than the shares of those who did not feel the presence of AI in their life. The twice as big share of the denying academics proves Hypothesis 2.

Question 3 aimed at revealing the participants' estimation of the usefulness of AI developments ranging from 'Very useful' to 'Harmful'. The attitudes of the two group members proved to be very similar, with the academics perceiving AI a bit more warmly: 46% of them found AI very useful (against 21% of the students), 35% believed AI is useful (against 59% of the students), yet, the bulk share of positive attitudes was practically the same (81% of the academics and 80% of the students), 16% of the academics and 12% of the students were neutral about the idea, 5% of the students found AI useless and equal shares of both groups believed that AI developments are harmful. The somewhat warmer welcome of the academics accounts for their professional vision and mission and their habit to be in the vanguard of the progress.

Question 4 targeted to reveal the respondents' perception of AI safety /danger. The students' responses showed a higher level of trust to AI, as 5% of the student respondents were sure of AI being totally safe (with 0% responses of the academics in this estimation), 14% of the students and 16% of the academics believed that AI was safe, 59% and 58% accordingly believed AI was neutral, 19% of the students and 26% of the academics found AI dangerous, and 3% of the students thought it was very dangerous. The data received from both groups are very close, though the academics proved to have sensed more danger in AI developments.

Question 5 aimed at identifying the kinds of AI present in the respondents' life. For the research purposes the following AI developments were chosen: chat-bots (Facebook, VKontakte, etc.), chess-playing computers, self-driven autos, electronic concierges (OK Google, Siri, Alisa), bank terminals recognising images, and drones recognising images. The received responses showed that the survey participants must have misunderstood the question and answered it basing on their knowledge of certain AI developments, and not on their actual presence in their lives. Thus, 27% of the academics and 40% of the students confirmed the presence of chat-bots (Facebook, VKontakte, etc.) and 38% and 29% accordingly admitted the presence of electronic concierges (OK Google, Siri, Alisa). These data seem to be quite truthful but then come the responses, inspiring much doubt. 6% of the academics and 11% of the students stated they had chess-playing computers in their lives obviously confusing Deep Blue with chess play computer games, 2% of the academics (0% of the students) indicated the presence of self-driven autos in their life, though such vehicles are still under development in Russia, 27% of the academics and 18% of the students stated the availability of bank terminals recognising images, though their future installation was only announced by Sberbank some weeks ago, and 2% of the students (0% of the academics) confirmed the availability of drones recognising images, though again, such drones are only entering mass production and only for military and defence purposes.

The received data were contradictory proving that a larger share of students deal with chat-bots, though a larger share of academics seem to use electronic concierges.

Therefore, *Question 6* intended to clear up the respondents' practices of using chat-bots and electronic concierges and identify the frequency of their usage. The received responses were predictable and proved Hypothesis 2 confirming that representatives of Generation Y are on better terms with AI developments and ICTs at large: 21% of the students and only 10% of the academics used the development very often; 27% of the students and just 16% of the academics used them often; 26% of the students and 19% of the academics used the AI developments occasionally; 18% of the students and 29% of the academics used them seldom, and only 8% of the students against 26% of the academics never used chat-bots and electronic concierges.

Question 7 targeted ethical issues in the mode of building a dialogue with an electronic concierge /chat-bot and inquired whether the respondents would say "thank you" at the end of the conversation. The received data turned out to be quite unexpected – the students proved to be more polite towards chat-bots and electronic concierges and did not forget to thank Siri, Alisa or OK Google in 20% of cases (8% of the students did it always and 12% did it often), though only 3% of the academics thanked electronic concierges always and 10% did it often. 22% of the students and 19% of the academics thanked them occasionally. 21% of the students and 16% of the academics did it seldom but the number of those respondents who never thanked electronic concierges and chat-bots is really high – 37% of the students and 52% of the academics. The greater number of polite students supposedly accounts for the overwhelming majority of female students in the sample and possibly, for the field of the academics' expertise (most of them worked in the Institute of Natural, Mathematical and Technical Sciences and may be prone to perceiving AI formally as a scope of algorithms). Further interviews with some representatives (both male and female) of both samples may shed some light on the reasons for such responses.

However, answering *Question 8* aimed at identifying the level of importance of the electronic concierges' addressing the respondents by name and finishing the conversations with "Thank you for contacting me /asking me", an overwhelming majority of the student population (76%) confirmed its importance, with 19% admitting high importance, with only 4% seeing no difference and 1% stating low importance. Unlike the representatives of Generation Y who seemed to be highly concerned with AI polite treatment, the representatives of Generation X were not so much obsessed with display of respect by AI: only 25% of the sample confirmed importance of being called by name at the beginning of the conversation and being thanked at the end, and just 2% found it very important; 43% of the academics did not see much difference, 26% stated low importance and 4% admitted low importance. These data show that representatives of Generation Y are more sensible to the way AI may treat them in communications (so far), and proved to be much more demanding and having higher expectations in comparison with the representatives of Generation X.

The difference between the share of the students who often or always thanked the electronic concierges and the share of the students who expected to be thanked has revealed an ethical inconsistency, and this behaviour pattern violates the proverb 'How goes around comes around', which was developed as an important rule, almost a value in the minds of Generation X.

Question 9 focused on the respondents' attitude to the possibility of using artificial intelligence in advertising and marketing at large. This information is important for revealing the level of trust to AI as part of persuasive technologies. A majority of the student population approved (43%) and fully approved (3%) AI use in marketing and advertising in particular in comparison to 16% (approved) and 6% (fully approved) of the academics. Most academics (62%) and many students (40%) expressed their neutral attitude to the issue. The shares of those who disapproved (13% of the academics and 10% of the students) and those who totally disapproved the idea (3% of the academics and 4% of the students) were almost the same. The data confirmed a higher level of trust characteristic of the student population.

Question 10 was supposed to identify the level of trust to the information provided by electronic concierges. The received data are provided in the table below (Table 1)

Table 1. Levels of trust to information provided by electronic concierges.

Degree of trust	Academics	Students
by 100%	3%	5%
by 60–75%	36%	42%
by 40–50%	23%	28%
by 10–30%	19%	21%
I don't trust at all	19%	4%

The students' responses demonstrate a higher level of trust to the information provided by electronic concierges.

Going further into the issue of trust to the information provided by intelligent systems, *Question 11* was designed to identify the respondents' attitude to probability of unreliable information provided by AI systems. 5% of the students considered the probability very low (0% of the academics), 26% of the students found it low (10% of the academics), 24% of the students and 35% of the academics found it difficult to decide, 40% of the students and 45% of the academics admitted it was probable and 5% of the students and 10% of the academics thought the probability was very high. These data confirm the previously received information of the students' higher level of trust to AI.

Questions 12 and 13 were similar to Questions 10 and 11 but aimed to identify the level of trust to chat-bots. The data received in response to Question 12 were as follows (Table 2)

Table 2. Levels of trust to information provided by chat-bots

Degree of trust	Academics	Students
by 100%	3%	5%
by 60–75%	23%	35%
by 40–50%	32%	39%
by 10–30%	19%	13%
I don't trust at all	23%	8%

The participants' responses revealed an overall lower level of trust to chat-bots in comparison with electronic concierges and consistently higher level of trust in perception of the representative of Generation Y.

Question 13 inquired more data on the probability of unreliable information provided by chat-bots. 4% of the students believed that the probability was very low (0% of the academics), 14% of the students considered it low (10% of the academics), 33% of the students and 26% of the academics chose the option 'Difficult to say,' 36% of the students and 48% of the academics agreed that it was probable and 13% of the students and 16% of the academics found the probability very high. These data show a higher level of trust to AI demonstrated by the students and at the same time a lower level to trust to the reliability of the information provided by chat-bots in comparison with electronic concierges.

Question 14 was designed to clarify the respondents' attitudes to the new method of warfare using artificial intelligence (combat robots, drones, etc.). 10% of the academics and 9% of the students found it very promising, 10% of the academics and 9% of the students believed it was promising, 26% of the academics and 21% of the students expressed their neutral attitude; however, 19% of the academics and 37% of the students were cautious about the new AI-based warfare, and 35% of the academics and 21% of the students expressed a negative attitude to the issue. Though the share of positive and neutral attitudes practically coincides, the representatives of Generation Y proved to be more cautious and the representatives of Generation X showed a more negative attitude to the issue, which can be partly explained by their larger life experience and expertise.

Question 15 focused on identifying the stakeholder responsible for the development and use of artificial intelligence, in the respondents' opinion. The respondents put the responsibility on the core stakeholders to the following extent (Table 3)

Table 3. Key stakeholders' responsibility for AI development and use

Responsible stakeholder	Academics	Students
Government	27%	18%
Producing corporations	18%	44%
Customers/consumers	12%	9%
Operators of robotized machines	12%	10%
Society at large	31%	19%

The responses have revealed the higher level of the academics' own responsibility blaming primarily the society at large, then the government they voted for and only then producing corporations, as well as customers/consumers and operators of robotized machines. The students mainly found responsible producing corporations, trying to avoid any responsibility for such sore issues. These two behaviour modes are typical for representatives of the two generations.

Question 16 aimed at identifying the respondents' attitudes to further AI development. A majority of the responses proved to be positive, which means the Russian university students and employees are in trend: 13% of the academics and 21% of the students believed that further AI development was undoubtedly worth continuing, and 45% of the academics and 40% of the students found it promising. For 29% of the academics and 26% of the students it was difficult to express their opinions, while 13% of the academics and 10% of the students stated that further AI development should not be continued and 3% of the students even believed it should be banned. The mainly positive attitudes reflected the high expectations for AI advance and achievements based on the previously received unprecedented support provided by ICTs and the Internet, which made the world smaller and life faster.

6 Conclusions and Discussion

The results of the AI discourse analysis and the responses of the survey participants have revealed differences in attitudes to AI of Russian representatives of Generation X and Generation Y stipulated by the differences in their attitudes to life in general, to the world and to themselves.

The findings have highlighted an interim communication rivalry between representatives of Generation Y and AI, as the former have higher expectations and requirements for respect and courtesy than they are ready to offer.

The survey has proved both hypothesis and demonstrated the higher level of trust to AI on the part of the students – representatives of Generation Y. The results of secondary importance have shown a higher level of trust to electronic concierges in comparison to chat-bots and highlighted a considerable lack of trust to the relevance and reliability of the information provided by both electronic concierges and chat-bots (the latter perceived as less reliable).

The five challenges formulated above prompt a number of priorities, some of them being really urgent (primarily including control over the development and use of robotized machines and drones).

With AI getting integrated into human lives more deeply, it is important to learn from the previous lessons taught by the Internet and ICTs and take some preliminary measures allowing people to always remain humans and decision-makers.

References

1. D'Monte, L.: The rise of Artificial Intelligence and impending takeover (2018). https://www. livemint.com/AI/CXIjX1q8hyhnq7Fg7k7pVL/The-rise-of-Artificial-Intelligence-and-impending-takeover.html
2. Janakiram, M.S.V.: Here Are Three Factors That Accelerate the Rise of Artificial Intelligence (2018). https://www.forbes.com/sites/janakirammsv/2018/05/27/here-are-three-factors-that-accelerate-the-rise-of-artificial-intelligence/#763a571dadd9
3. D'Monte, L.: IBM builds Artificial Intelligence machine that can debate with humans (2018). https://www.livemint.com/Technology/8M7lQRmLElLz3JLeYbDCFI/IBMs-artificial-intelligence-machine-now-debates-with-human.html
4. Revell, T.: AI will be able to beat us at everything by 2060, say experts (2017). https://www.newscientist.com/article/2133188-ai-will-be-able-to-beat-us-at-everything-by-2060-say-experts/

5. Vincent, J.: Putin says the nation that leads in AI 'will be the ruler of the world'. The Russian president warned that artificial intelligence offers 'colossal opportunities' as well as dangers (2017). https://www.theverge.com/2017/9/4/16251226/russia-ai-putin-rule-the-world
6. Barrat, J.: Why Stephen Hawking and Bill Gates Are Terrified of Artificial Intelligence (2015). https://www.huffingtonpost.com/james-barrat/hawking-gates-artificial-intelligence_b_7008706.html
7. CenSAMM: Strauss-Howe Generational Theory (2018). https://censamm.org/resources/profiles/strauss-howe-generational-theory
8. Dangerfield, K.: Xennials: the generation sandwiched between gen-Xers and millennials (2017). https://globalnews.ca/news/3579270/xennials-generaion-x-millennials-generation/

Detection of Cutaneous Tumors in Dogs Using Deep Learning Techniques

Lorena Zapata[1], Lorena Chalco[1], Lenin Aguilar[1],
Esmeralda Pimbosa[1], Iván Ramírez-Morales[1], Jairo Hidalgo[2],
Marco Yandún[2], Hugo Arias-Flores[3], and Cesar Guevara[3(✉)]

[1] Faculty of Agricultural and Livestock Sciences,
Universidad Técnica de Machala, Machala EC 070223, Ecuador
{mlzapata,lchalco,flaguilar,dpimbosa,
iramirez}@utmachala.edu.ec
[2] Universidad Politécnica Estatal Del Carchi, Tulcán, Ecuador
{jairo.hidalgo,marco.yandun}@upec.edu.ec
[3] Institute of Research, Development and Innovation—MIST,
Universidad Indoamérica, Quito EC 170301, Ecuador
{cienciamerica,cesarguevara}@uti.edu.ec

Abstract. Cytological diagnosis is useful in the practical context compared to the histopathology, since it can classify pathologies among the cutaneous masses, the samples can be collected easily without anesthetizing the patient, at very low cost. However, an experimented veterinarian performs the cytological diagnosis in approximately 25 min. Artificial intelligence is being used for the diagnosis of many pathologies in human medicine, the experience gained by years of work in the area of work allow to issue correct diagnoses, this experience can be trained in an intelligent system. In this work, we collected a total of 1500 original cytologic images, performed some preliminary tests and also propose a deep learning based approach for image analysis and classification using convolutional neural networks (CNN). To adjust the parameters of the classification model, we recommend to perform a random and grid search will be applied, modifying the batch size of images for training, the number of layers, the learning speed and the selection of three optimizers: Adadelta, RMSProp and SGD. The performance of the classifiers will be evaluated by measuring the accuracy and two loss functions: cross-categorical entropy and mean square error. These metrics will be evaluated in a set of images different from those with which the model was trained (test set). By applying this model, an image classifier can be generated that efficiently identifies a cytology diagnostic in a short time and with an optimal detection rate. This is the first approach for the development of a more complex model of skin mass detection in all its types.

Keywords: Deep learning · Cancer · Veterinary · Cytology ·
Convolutional Neural Networks

© Springer Nature Switzerland AG 2020
T. Ahram (Ed.): AHFE 2019, AISC 965, pp. 83–91, 2020.
https://doi.org/10.1007/978-3-030-20454-9_8

1 Introduction

Currently, the cytology study for cancer detection is one of the most important lines of research in medicine worldwide, due to cancer disease attacks the majority of existing species on the planet. The main affected are domestic animals and humans.

Cutaneous tumors in dogs are an increasingly common pathology, for this reason it is very important to recognize and treat in time, with the appropriate veterinary procedure. cytological diagnosis is a type of morphological diagnosis based on the microscopic characteristics of cells and extracellular components spontaneously detached from tissues or obtained by procedures.

The study presented by Graf et al. [1], describes the most common tumor types that are mast cell tumors (16.35%), lipomas (12.47%), hair follicle tumors (12.34%), histiocytomas (12.10%), soft tissue sarcomas (10.86%) and melanocytic tumors (8.63%). This research consisted of a database of 1000 tumors per category.

A regional investigation is the one presented in the article of Vinueza et al. [2], which indicates that in Ecuador a study was carried out from 13.573 medical records collected between 2011 and 2014, where cases of patients with neoplasm are increasingly common. These analyses were performed using cytology and histopathology, which results were the detection of some type of neoplasm. The prevalence of tumors in the population was 4.94%, being more frequent in soft tissues (39.3%), skin and annexes (24.4%), and in females, in the mammary gland (14.3%).

The use of cytology as a diagnosis in the clinic has allowed us to perform the first diagnostic evaluation of skin masses, as occurred during a 20-year study period, where the search revealed 400 brain lesions diagnosed by cytology, of which 338 were neoplasms (84.5%) and 62 non-neoplastic lesions (15.5%) [3].

There is a growing interest in using imaging for diagnostic purposes (primary and/or consultation). An important consideration is whether it can safely replace conventional light microscopy as the method by which pathologists review histological sections, cytology slides and hematology slides to render diagnoses [4].

Automated classification of skin lesions is a challenging task due to the variability of characteristics in each lesion. However, training a neural network and testing its performance against 21 certified dermatologists, demonstrated that artificial intelligence is capable of classifying skin cancer with a level of competence comparable to dermatologists, thus potentially extending the reach of dermatologists outside the clinic, which allows to provide low-cost universal access to vital diagnostic care [5].

To determine the molecular class of the tumor, pathologists will have to manually mark the nuclei activity biomarkers through a microscope and use a semiquantitative assessment method to assign a histochemical score (H score) to each mammary tumor nucleus. Manually marking positively stained nuclei is a slow, imprecise and subjective process which will lead to inter-observer and intra-observer discrepancies. The end-to-end deep learning system directly predicts the H score and automatically imitates the pathologists' decision process and uses one fully convolutional network (FCN) to extract all nuclei region (tumor and non-tumor), as a high-level decision making mechanism to directly output the H-score of the mammary tumor image [6].

Image-based machine learning and deep learning in particular have recently demonstrated expert-level accuracy in medical image classification. In a recent study, a deep network is trained to predict colorectal cancer outcome based on images of tumor tissue samples. This network was able to predict results in 420 cancer patients out-performing the visual histological assessment performed by human experts [7].

The aim of the present study is to develop a classifier based on cutaneous images of cytologies applying deep learning to classify the type of masses. In addition, the images will be pre-processed to generate an optimal data bank to be entered into the classifier.

2   Theoretical Background

2.1   Artificial Neural Networks

Artificial Neural Networks (ANN) are data-modelling techniques aimed to analyse complex relationships between input data with its corresponding outputs. ANN are based on set of interconnected artificial neurons that perform the functions of learning, memorization, generalization or abstraction of relevant features. Its functioning is inspired by the human central nervous system, which has numerous cells that work quickly and helps in decision making [8].

In recent years, ANN have become a subject of much relevance in the scientific and research field, due to its ability to handle noise, randomness or incomplete data. ANN have being used for text-to-speech processing, pattern recognition, image segmentation, and are also used in robotics.

The basic structure of a neuron can be represented by $X = \{x^i, i = 1, 2, ..., n\}$ as the input patterns presented to to the neuron, and Y representing the output of the model. Each input is multiplied by its weight w_i, and the bias b is added, the result goes through a transfer function f [9]. The relationship between input and output can be described as follows in Eq. (1).

$$Y = f\left(\sum_{i=1}^{n} w_i \times x_i + b\right) \tag{1}$$

The Multilayer Perceptron (MLP) is a layered neural network. The general scheme of an MLP consists of one input layer, several hidden layers and one output layer. The transfer function is commonly a sigmoid function, however, some other functions can be used [10, 11]. Neurons receive the results of the previous layer and calculates an output to the next layer. The final output layer returns the output of the network [10, 12].

The number of neurons, layers and their connections is known as the architecture of the ANN and it is a key parameter. The architecture depends on the complexity of the data, and there is no general rule for choosing it, this have to be done as an empirical process [13, 14].

2.2 Convolutional Neural Networks

It is a special type of ANN, Convolutional Neural Networks (CNN) use a variation of MLP to require minimal preprocessing, instead of having weights for all the entries, the weights are shared and are convolved through a sliding window [15]. This convolutional property together with a grouping layer makes the model better suited to the images. Our image classification models will be based on CNN networks with the following parameters [16]:

Convolutional layer: is an array of weights, to slide across the previous layer to compute the dot product of the weights and inputs. To reduce the error, a back-propagation algorithm is used.

Activation function: is a non-linear function (sigmoid or rectified linear unit), which is applied to elements of the convolution.

Pooling layer: reduces the complexity of a CNN by decreasing the size of the maps. Two commonly used methods are max pooling and average pooling.

Output layer: is equal to the number of classes in the dataset. Usually it is used a softmax function for classification at the output.

Number of epochs: is the number of times for training and validating the neural network with an acceptable level of accuracy.

Batch size: defines the number of a subset of of samples which will be loaded and propagated during training and validation of the neural network.

Learning rate: controls the size of weight and bias changes in learning of the training algorithm.

Number of hidden layers: defines how many fully connected layers will have the CNN. Each neuron has its activation value which is calculated based on its input values and its weights.

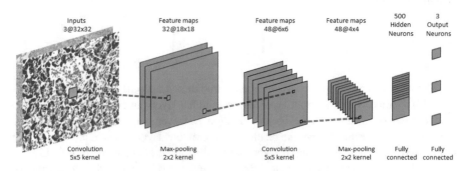

Fig. 1. Schematic representation of a convolutional neural network.

As shown in Fig. 1, input images of cytology are a $3 \times 32 \times 32$ array of numbers, in which 3 correspond to RGB channels of image, in the first convolutional layer, a 2×2 kernel is applied and feature map changes to $32 \times 18 \times 18$. In each subsequent layer, some transformations are performed, until the last fully connected layer. Finally, the output layer has three neurons, one for each class (neoplastic, inflammatory and hyperplastic).

3 Materials and Methods

3.1 Dataset Collection

A database of 1500 cytological images taken from samples of skin masses in dogs of different ages and physiological conditions were collected. These samples have been labeled by an expert in three classes: neoplastic, inflammatory and hyperplastic.

Samples were obtained from dogs with clinical history at the Veterinary Clinic of the Universidad Técnica de Machala. Clinic symptoms of cutaneous masses were recorded among with all physiological data of the animal. Additionally informative data were recorded.

The techniques for taking cytological samples applied to this study were FNAP (by fine needle aspiration), PAF (by fine needle) and imprint, the samples were deposited on a slide plate and then used as diff staining media. Diff-Quik^{TM} or GIEMSA is applied and then the sample is observed at microscope using 40x and 100x lens. After this procedure, the digital image was taken and labeled by an expert. All samples of skin masses have its corresponding histopathologic study using the following criteria:

Inflammatory Characteristics: Presence of different inflammatory patterns with a varied population of cells such as neutrophils (with or without degenerative changes), lymphocytes, plasma cells, monocytes, macrophages, eosinophils and mast cells.

Hyperplastic Characteristics: Large nuclei of little condensed chromatin and prominent nucleoli. The cytoplasm is frequently basophilic and the nucleus-cytoplasm ratio (N: C) relatively constant.

Neoplastic Characteristics: In general, neoplastic cells are larger, more pleomorphic and have a higher and more variable relationship (N: C) compared to normal cells of that same tissue. The pleomorphism presented will be considered, assessing the anisocytosis, cellular macrocytosis, hypercellularity and in the nuclei the following criteria can be seen: macrocariosis, anisocariosis, multinucleation, increase of images with mitosis, abnormal mitoses, thick chromatin pattern, nuclear molding, macronucleus, angular nucleoli and anisonucleosis. The neoplastic cells can be classified into three categories based on certain common characteristics such as discrete or round cells, epithelial cells and mesenchymal cells.

3.2 Image Classification

For the generation of classification models, several techniques were tested to obtain the most appropriate, in addition, a filtering technique is proposed to improve the performance of models during testing. In the training phase, the information was divided in training(70%), validation(20%) and test(10%) subsets, using a cross-validation technique.

The general procedure to train the image classifier is described below:

1. Upload images from your directory, and organized in folders which must be named as the classes.
2. Reshape all images in a list using the python library called Numpy.

3. Define the pixel values as "float32" format and save the resulting vector.
4. Normalize the pixel values between 0 and 1.
5. Create a vector of labels.
6. Divide the data using a cross-validation technique.

The evaluation of the preliminary experiments was based on the Accuracy (ACC) [17], which measures the proportion of correct classifications made by the model.

4   Results and Discussion

4.1   Preliminary Results

In this work we show the creation of a homogeneous cytological imaging dataset, with three classes: neoplastic, inflammatory and hyperplastic. This classes have been labeled by an expert and it is the first step to develop an image classifier that efficiently identifies a cytology diagnosis in a short time and with an optimal detection rate.

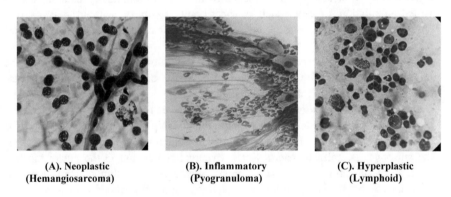

(A). Neoplastic **(Hemangiosarcoma)**	**(B). Inflammatory** **(Pyogranuloma)**	**(C). Hyperplastic** **(Lymphoid)**

Fig. 2. Examples of cytological images contained in dataset

In Fig. 2. It can be found three examples of cytologic images. (A), mesenchymal neoplastic cells compatible with hemangiosarcoma, (B), an inflammation of the granulomatous type and (C), a cytological sample non-inflammatory lymphoid hyperplasia.

In a first approach, a subset of 50 images were used to train during 100 epochs, a model using a pre-trained CNN to extract features of the images. The feature extraction performance in cytological images of our CNN model is based on transfer-learning and fine tuning on the Inception-v3 model [18]. The algorithm was trained to classify examples in the dataset distributed in three classes according to the typology of tissue: neoplastic, inflammatory, hyperplastic.

Image augmentation was applied to the subset of images. This approach allowed to have some preliminary results with low computational cost. Max ACC in preliminary tests reached 0.7, using a learning rate of 0.01, SGD optimizer and freezing the inception pre-trained network at later 150. It is very important to perform exhaustive tests, using the complete dataset, and to optimize the architecture of the ANN, according to the proposed approach.

4.2   Program Code

The code used to obtain preliminary results is shown below. There are two main processed in this algorithm: transfer learning and fine tuning.

```
def train(X_train, X_val, y_train, y_val, layer_freeze):
    tfCallBack = keras.callbacks.TensorBoard('./logs/')
    base_model = InceptionV3

    #transfer learning
    for layer in base_model.layers:
        layer.trainable = False
    model = base_model.output
    model = GlobalAveragePooling2D()(model)
    model = Dense(nhl, activation='relu')(model)
    predictions = Dense(3, activation='softmax')(model)
    model = Model(inputs, outputs=predictions)
    model.compilenum_epoch(optimizer='rmsprop',
                loss='categorical_crossentropy',
                metrics=['accuracy'])
    model.fit(X_train, y_train,
            batch_size=batch_size,
            epochs=100,
            verbose=2,
            validation_data=(X_val, y_val),
            callbacks=[tfCallBack])

    #fine tuning
    for layer in model.layers[:layer_freeze]:
        layer.trainable = False
    for layer in model.layers[layer_freeze:]:
        layer.trainable = True
    model.compile(optimizer=SGD,
                loss='categorical_crossentropy',
                metrics=['accuracy'])
```

Transfer-learning is an efficient technique used to deal with labeled information from a source domain to a destination domain, in order to build an efficient prediction model. In this process come convolutional layers are frozen making them non-trainable and last layer is removed. A new fully connected layer is added, taking the rest of the network for feature extraction and for training the model. Fine-tuning is a process to take a network model that has already been trained for a given task, and make it perform a second similar task.

4.3   Proposed Approach

For the development of our classification models, we recommend to use Tensorflow [19], a machine learning library developed by Google, and some high level libraries as Keras and SkLearn. The proposed models have to be evaluated in terms of accuracy and two loss functions.

The optimization algorithms are necessary arguments to elaborate a NN model in Keras, in such a way that the gradient descent is tuned properly. The following optimizers are proposed:

Stochastic Gradient Descent: also called SGD, it uses the input sample to recalculate the weights, allowing it to be updated at a more frequent rate. Therefore a faster convergence is achieved [20].
RMSprop: it reduces the learning rate for a weight, dividing it by a moving average of the magnitudes of recent gradients for that weight [21].
Adadelta: it is used to control gradient descent. It is a useful optimizer since ir occupies low computational cost, and adapts dynamically over time [22].

5   Conclusions

Our findings show the potential of the proposed approach to classify cytology images in order to support diagnostic decision in veterinary practice. However, to get better results, it is necessary to perform an optimization of the model as proposed here.

The benefit of applying artificial intelligence is the automation of the detection of the images, the precision and adaptability of this process. In medicine, artificial intelligence has provided excellent results, both in the classification, grouping, regression of data or forecasting the biological behavior in general.

This work proposes a first approach for the development of a more complex model for skin mass detection in all its types, which will allow to develop an accurate skin mass analysis software. Its application pottencially is able to improve the diagnosis in time and accuracy.

Acknowledgements [1].   The authors wish to thank CEDIA National Research and Education Network. Iván Ramírez-Morales would also like to thank the support provided by NVIDIA. This work is part of DINTA-UTMACH and GIPASA research groups.

References

1. Graf, R., Pospischil, A., Guscetti, F., Meier, D., Welle, M., Dettwiler, M.: Cutaneous tumors in swiss dogs: retrospective data from the swiss canine cancer registry, 2008–2013. Vet. Pathol. **55**, 809–820 (2018)
2. Vinueza, R.L., Cabrera, F., Donoso, L., Pérez, J., Díaz, R.: Frecuencia de neoplasias en caninos en Quito. Ecuador. Rev Investig Vet Peru **28**, 92–100 (2017)
3. Hamasaki, M., Chang, K.H.F., Nabeshima, K., Tauchi-Nishi, P.S.: Intraoperative squash and touch preparation cytology of brain lesions stained with H + E and diff-QUIKTM: a 20-year retrospective analysis and comparative literature review. Acta Cytol. **62**, 44–53 (2018)

4. Pantanowitz, L., Sinard, J.H., Henricks, W.H., Fatheree, L.A., Carter, A.B., Contis, L., et al.: Validating whole slide imaging for diagnostic purposes in pathology: guideline from the college of American Pathologists Pathology and Laboratory Quality Center. Arch. Pathol. Lab. Med. **137**, 1710–1722 (2013)
5. Esteva, A., Kuprel, B., Novoa, R.A., Ko, J., Swetter, S.M., Blau, H.M., et al.: Dermatologist-level classification of skin cancer with deep neural networks. Nature **542**, 115–118 (2017)
6. Liu, J., Xu, B., Zheng, C., Gong, Y., Garibaldi, J., Soria, D., et al.: An end-to-end deep learning histochemical scoring system for breast cancer TMA. IEEE Trans. Med. Imaging (2018). https://doi.org/10.1109/tmi.2018.2868333
7. Bychkov, D., Linder, N., Turkki, R., Nordling, S., Kovanen, P.E., Verrill, C., et al.: Deep learning based tissue analysis predicts outcome in colorectal cancer. Sci. Rep. **8**, 3395 (2018)
8. Cascardi, A., Micelli, F., Aiello, M.A.: An artificial neural networks model for the prediction of the compressive strength of FRP-confined concrete circular columns. Eng. Struct. **140**, 199–208 (2017)
9. Jiang, J., Trundle, P., Ren, J.: Medical image analysis with artificial neural networks. Comput. Med. Imaging Graph. **34**, 617–631 (2010)
10. Kruse, R., Borgelt, C., Klawonn, F., Moewes, C., Steinbrecher, M., Held, P.: Multi-layer perceptrons. In: Computational Intelligence, pp. 47–81. Springer, London (2013)
11. Ruck, D.W., Rogers, S.K., Kabrisky, M., Oxley, M.E., Suter, B.W.: The multilayer perceptron as an approximation to a Bayes optimal discriminant function. IEEE Trans. Neural. Netw. **1**, 296–298 (1990)
12. Gardner, M.W., Dorling, S.R.: Artificial neural networks (the multilayer perceptron)—a review of applications in the atmospheric sciences. Atmos. Environ. **32**, 2627–2636 (1998)
13. Herrera, F., Hervas, C., Otero, J., Sánchez, L.: Un estudio empírico preliminar sobre los tests estadísticos más habituales en el aprendizaje automático. Tendencias de La Minería de Datos En Espana, Red Espanola de Minería de Datos Y Aprendizaje (TIC2002-11124-E), pp. 403–412 (2004)
14. Rivero, D., Fernandez-Blanco, E., Dorado, J., Pazos, A.: Using recurrent ANNs for the detection of epileptic seizures in EEG signals. IEEE Congr. Evol. Comput. (CEC) **2011**, 587–592 (2011)
15. Wahab, N., Khan, A., Lee, Y.S.: Two-phase deep convolutional neural network for reducing class skewness in histopathological images based breast cancer detection. Comput. Biol. Med. **85**, 86–97 (2017)
16. Soriano, D., Aguilar, C., Ramirez-Morales, I., Tusa, E., Rivas, W., Pinta, M.: Mammogram classification schemes by using convolutional neural networks. In: Technology Trends, pp. 71–85. Springer, Charm (2018)
17. Fawcett, T.: An introduction to ROC analysis. Pattern Recogn. Lett. **27**, 861–874 (2006)
18. Szegedy, C., Vanhoucke, V., Ioffe, S., Shlens, J., Wojna, Z.: Rethinking the inception architecture for computer vision. In: Proceedings of the IEEE Conference on Computer Vision and Pattern Recognition, pp. 2818–2826. cv-foundation.org (2016)
19. Abadi, M., Barham, P., Chen, J., Chen, Z., Davis, A., Dean, J.: Tensorflow: a system for large-scale machine learning. In: OSDI (2016)
20. Bottou, L.: Large-scale machine learning with stochastic gradient descent. In: Proceedings of COMPSTAT 2010, pp. 177–186. Physica-Verlag HD (2010)
21. Hinton, G., Srivastava, N., Swersky, K.: Neural networks for machine learning lecture 6a overview of mini-batch gradient descent. Cited on 2012:14
22. Zeiler, M.D.: ADADELTA: An Adaptive Learning Rate Method. arXiv [csLG] (2012)

Future Research Method of Landscape Design Based on Big Data

Jingcen Li[✉]

School of Design, Shanghai Jiaotong University,
800 Dongchuan RD. Minhang District, Shanghai, China
Li_jingcen@163.com

Abstract. Big data has become a very hot topic in the field of urban research and planning, which can contribute to the full scale, refinement, humanization and experience quantification of urban planning, but it is still rarely applied in the field of landscape architecture. Big data is dynamic and objective, so it is suitable for landscape research. This paper constructs a new approach to landscape research based on big data with reference to the PERSONA approach in Internet products. Then, through the literature review, it is found that Volunteered Geographic Information (VGI) is more suitable for small scale site analysis.

Keywords: Big data · Research method · Landscape design · Volunteered Geographic Information

1 Introduction

In the contemporary context of information, digitization and networking, the connotation and form of urban space are undergoing rapid changes. It is becoming increasingly difficult to integrate and create large-scale urban space through traditional urban design method. The field of urban design has gradually expanded from the traditional single spatial level to a complex multi-spatial level, from the former static material space to a dynamic and fast-paced urban complex giant system. Many scholars have recognized that emerging big data has great advantages in terms of collection source, scale and scope, timeliness, and this change has the possibility to promote the development of urban design from quantitative change to qualitative change.

Based on these changes, some Chinese scholars have proposed four generations of urban design: the first generation of traditional urban design before 1920, the second generation of modernist urban design after the industrial revolution, and the third generation of green urban design since 1970. And nowadays, China is gradually entering the fourth generation of urban design, which has the characteristics of full-scale, refined, humanized and empirically quantified. At present, many scholars around the world have used big data to conduct citizen behavior analysis and urban spatial analysis. Big data has become an indispensable tool for urban planners.

The change of urban design method and concept has influenced the development of landscape design. Such keywords as "bottom-up", "participatory design" and "humanity" have been emphasized gradually in recent years. The landscape architecture no

T. Ahram (Ed.): AHFE 2019, AISC 965, pp. 92–100, 2020.
https://doi.org/10.1007/978-3-030-20454-9_9

longer only emphasizes the aesthetics, designers should also take more attention to make sites more user-friendly. However, through the reading of literature about the big data, it is found that there are few research methods on the micro scale, while most of the articles are studied on the macro scale such as country and region. So, it's difficult for landscape designers to use the big data. Therefore, this paper intends to sort out some methods of data collection, processing and analysis suitable for small-scale sites in the literature, and then form a systematic research method for landscape architecture, in order to help landscape designers understand sites and users better.

2 Traditional Research Method

The traditional landscape survey method, also known as the field work, is one of the most important parts of the landscape design process. It is mainly divided into two parts, one is the site research, and the other is the user research. The former mainly studies the surrounding conditions of the site, such as accessibility and passenger flow volume. The latter mainly studies the user needs. Landscape designers use the method of questionnaire survey or random interview to understand users' thoughts. If conditions permit, a focus group will be set up to express their needs.

However, as mentioned in many studies, this traditional research method is defective [1–3]. Firstly, the data obtained from the traditional research method is static, but the use of the site changes over time, so the traditional way cannot help designers understand the dynamic use of the site. Secondly, field research requires a lot of time and money, but the number of samples collected is very limited. Finally, the traditional research method is very subjective, so the feasibility of data is highly dependent on the way of investigation.

There are still advantages in the traditional landscape research method. Firstly, the data obtained from the traditional way is highly accurate for it doesn't depend on the accuracy of the device. Then, the data is more representative than the big data, because the traditional way take into account some people who seldom use the Internet and mobile devices, such as the elderly and children. Thirdly, data collection, processing, and analysis here is pretty easy, so there is no high data analysis requirement for landscape designers. Finally, thanks to direct contact with users, their evaluation and needs for the site can be better understood.

3 Method Generation

The new landscape research method mainly adopts the method of data analysis. Compared with the traditional field work, the new method can optimize some data obtained through inappropriate and limited statistical methods, such as passenger flow volume. With the help of big data, the result of the landscape research method can be more dynamic, objective and comprehensive. Therefore, the new method can help landscape designers understand users and sites better.

The specific step is similar to the traditional research method and is divided into two parts: site analysis and user segmentation. Since the traditional survey method is to communicate with site users directly, while the new method is based on data collection, processing and analysis. There is no part of direct contact and communication with users in this step, so this step needs a new method to support.

At present, due to the popularity of the Internet and mobile devices, Internet products have prospered and formed a more systematic user analysis method based on the background of big data, called PERSONA. The specific steps are as follows:

Step 1: Identify the overall map of the costumers;
Step 2: Detailed the target population, conducted in-depth analysis of the target population, excavated the core needs, and drew a complete portrait;
Step 3: The brand usually has more than one segment of the group, for each crowd design customized marketing plan [4].

According to the above steps, it can be seen that the PERSONA is a process of gradually refining the user classification. Through careful user classification, differentiated services can be provided to the users. Currently, PERSONA has been widely applied in personalized recommendation and advertising (Fig. 1).

Based on the actual needs of landscape research and user classification method of PERSONA, the following steps are drawn:

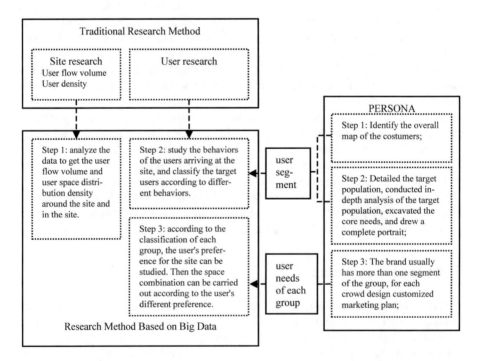

Fig. 1. The generation of Research Method of Landscape Design Based on Big Data

4   Interpretation

The specific steps of landscape research method based on big data are as follows:

Step 1: analyze the data to get the user flow volume and user space distribution density around the site and in the site.

Step 2: study the behaviors of the users arriving at the site, and classify the target users according to different behaviors.

Step 3: according to the classification of each group, the user's preference for the site can be studied. Then the space combination can be carried out according to the user's different preference.

In the following paragraphs, each step is described in detail. At the same time, this paper will illustrate how to select data for each step with literature research.

4.1   Step 1: Analyze the Data to Get the User Flow Volume and User Space Distribution Density Around the Site and in the Site

The data analysis of step 1 is mainly for the research of the site. Since the people around the site are likely to go to the site, the research scope is not only limited to the site, but also the surrounding of the site. As mentioned above, both passenger flow volume and user spatial density distribution are dynamic data. The former lays more emphasis on time dynamics, while the latter lays more emphasis on space dynamics. Both of them can help designers reasonably configure service facilities.

Although the purposes of many studies are inconsistent, many scholars have applied various data to study the spatial and temporal variations of sites (Table 1).

Table 1. The application of big data to study the spatial and temporal variations of sites.

Data type	Author	Data sources	Research method
GPS	Wang, X. et al.	Amap	The customer flow of the site in different time periods is calculated by an application called Amap Index developed by Amap company [5]
GPS	Chen, Y. et al.	The hourly real-time Tencent user density (RTUD)	The RTUD data from social media is used to analyze the time-spatial distribution of urban park users [2]
Smart card data	Li, F. et al.	Smart card data in Beijing	Spatially visualize the bus station distribution information and the bus swipe secondary data, and then combine the land use status to obtain the destination attributes of the residents, and determine the frequency of each land use [6]

(*continued*)

Table 1. (*continued*)

Data type	Author	Data sources	Research method
Volunteered geographic information (VGI)	Sun, Y. et al.	Geotagged photos from Flickr	Use Flickr photos as an example to explore the possibilities of VGI to analyze spatiotemporal patterns of tourists' accommodation in Vienna [7]
VGI	Zhang, Z. et al.	LBS sign-in data of Weibo	This paper takes Sina Weibo as the research object, introduces LBS sign-in data, and uses the "horizontal" and "longitudinal" time stratification method to empirically study the spatio-temporal evolution characteristics of tourists inside Nanjing Zhongshan Scenic Spot by dividing gender and regional attributes [8]
Wi-Fi	Huang, W.	Indoor positioning system	Wi-Fi location devices are installed in major scenic spots of huangshan scenic area to obtain mobile phone location data. After processing, the spatial-temporal trajectory data of tourists are formed to analyze the spatial-temporal distribution of tourist behaviors [9]
Others	Li, Y. et al.	Google street view images	Use City of Buffalo as a case study area to explore extracting pedestrian count data based on Google Street View images using machine vision and learning technology [10]

Most studies of this step use GPS for data analysis. GPS is suitable for large-scale urban space due to the huge and real-time data, so it is widely used in urban research [1]. However, GPS has some disadvantages, such as the inability to obtain personal attributes and accuracy problems, so it is more suitable for analyzing the overall situation of users rather than classifying users.

4.2 Step 2: Study the Behaviors of the Users Arriving at the Site, and Classify the Target Users According to Different Behaviors

4.3 Step 3: According to the Classification of Each Group, the User's Preference for the Site Can Be Studied. Then the Space Combination Can Be Carried Out According to the User's Different Preference

Since most of the literature classifies users for the purpose of studying the preferences of different groups, the step 2 and step 3 are inseparable. So, this section combines the two steps.

Firstly, the user classification method of traditional landscape research is mainly based on the basic attributes of the population (gender and age). Secondly, due to the limited information of some data, it cannot reflect the differences between the users or describe the behavioral characteristics of users. Therefore, in most literature, the classification of users is based on basic attributes (Table 2). However, this way of classifying users can only obtain the differences in behaviors and preferences caused by gender gap or cultural custom gap, which is imprecise.

Table 2. The application of big data to classify users.

Data type	Author	Data sources	Research method
VGI	Vu, H. et al.	Geotagged photos from Flickr	By extracting the geographic information in the photos, the differences in travel preferences between Asian tourists and western tourists were compared [12]
VGI + Mobile phone data	Girardin, F. et al.	Geotagged photos from Flickr	By collecting geotagged photos made public on Flickr, and the number of users making calls from mobile phones on the TIM system, we found inconsistencies in user behavior across different regions of Rome [13]
Mobile phone data	Tian, B. et al.	Mobile phone data from Shanghai	Through the analysis of mobile phone data, local residents and nonlocal tourists are identified. And then calculate the proportion of residents and nonlocal tourists visiting different blocks [14]
VGI	Abbasi, A. et al.	Twitter	By extracting the word frequency of tweets posted by users, the activity purpose of tourists can be identified [11]

Social network big data contains a large amount of Spatial and Temporal information, semantic information, emotional information, correlation information, etc., and its analysis and mining can help us to know users in many aspects. For example, when studying the behavioral characteristics of Sydney tourists, Abbasi, A. et al. extracted the tweets posted by users on Twitter, and then filtered out the higher-frequency words in tweets for clustering to identify the purpose of the tourists [11].

In the past, the evaluation of planning and design was mainly based on the subjective feelings of a few experts or users. But now big data can realize the flexible interaction between landscape designers and users. For example, the designer can obtain the text information published by the user and the spatial information when the user publishes the text through the data of the social platform, in order to collect the user's psychological feelings or cognition of the space and other information, and judge whether the landscape planning can meet the needs of users. Timely discovery of site problems is conducive to design improvement. Some scholars have used big data to analyze users' preferences and evaluations (Table 3).

And with the emergence of some specialized and niche mobile applications, it is possible to study the behavior patterns of a specific group of people. Sports social media, for example, has boomed in the last five years, showing runners' jogging routes. This superimposed map of running routes helps to study the behavior and preferences of runners and provides a reference for the planning of jogging paths (Table 4).

Table 3. The application of big data to analyze users' preferences and evaluations.

Data type	Author	Data sources	Research method
VGI	Wang, X. et al.	Dianping.com	Through word frequency analysis of the site evaluation on dianping.com, the user evaluation is obtained [5]
Web search terms	Zhang, Y. et al.	Search words frequency of Baidu, Google, Mafengwo.com, TripAvisor	Through the research on the search word frequency of different websites, to analyze the interest of tourists, and compare the differences of tourists' interest in scenic spots at home and abroad [15]

Table 4. The application of big data to study the behavior patterns of a specific group.

Data type	Author	Data sources	Research method
VGI	Oksanen J. et al.	Sports tracking	Get movement data from the mobile motion tracking app and generate a heat map of the ride [16]

4.4   Conclusion

In this paper, a new approach of landscape research is formed based on the needs of landscape research and PERSONA. Then combine the literature reading to carry on the concrete explanation to each step.

In the process of literature reading, it can be found that most of the current user classification is based on the user's basic attributes. This classification method cannot help the landscape designers to understand the difference of user behavior, and has low reference value in the design process.

Moreover, in the process of the planner using big data to study the city, SCD, floating car (taxi)GPS data and MPD are frequently used in the research process due to the need of large samples with relatively universal significance [1]. However, according to the literature, it is found that VGI is more suitable for use in small-scale sites, because it can realize refinement of individual attribute data and well applied to facilities site selection and evaluation [1]. In addition, with the development of some niche software, the specific behaviors and preferences of some groups can be understood through the analysis of the text, pictures, location and other information submitted by users in these applications.

The advantage of the new landscape research method compared with the traditional one is that the data is more objective and huger. And because of the real-time data, it can realize the interaction between the landscape designers and the users.

5   Discussion

Big data has become a very hot topic in the field of urban research and planning practice, but some scholars have "cold thought" about big data, thinking that big data may cause some errors and errors due to its inauthentic data collection, representativeness problems, consistency and reliability problems. And also big data may cause ethical issues [17]. So as some scholars believe, big data is only a tool to assist the arrangement and expansion of ideas, which cannot completely replace the design steps, and the role of big data should be viewed rationally [18].

Secondly, although this paper focuses more on the role of big data, the traditional research method can clearly know the specific behavior of users and the specific feeling of using space, which is not available in the research method described in this paper. Therefore, the future research method can be combined with the traditional design research method, which is more helpful for landscape designers to have a deeper and comprehensive understanding of site users.

The method in this paper is not only helpful to the process of landscape research, but also to the process of site maintenance and management. Because of the dynamic of big data, designers can make timely adjustments to the site according to the real-time site use data.

Finally, this method is still not widely used in landscape design field. One reason is that big data is controlled by many Internet companies, and the data is limited to acquire. And secondly, because the concept of big data is gradually emerging, landscape designers have not yet met the expectations of big data analysis and application skills. So, it is difficult to try to apply big data in the landscape planning and design research. Therefore, the application of big data in landscape design still needs efforts and support from all aspects.

References

1. Hao, J., Jin, Z., Rui, Z.: The rise of big data on urban studies and planning practices in China: review and open research issues. J. Urban Manag. **4**, 92–124 (2015)
2. Chen, Y., Liu, X., Gao, W., et al.: Emerging social media data on measuring urban park use and their relationship with surrounding areas—a case study of Shenzhen. Urban For. Urban Green. **31**, 130–141 (2018)
3. Wen, W., Wei, W.: Social media as research instrument for urban planning and design. In: Eighth International Conference on Measuring Technology and Mechatronics Automation, pp. 614–616. IEEE Press, New York (2016)
4. Dai, L.: Design Research, 2nd edn. Publishing House of Electronics Industry, Beijing (2016) (in Chinese)

5. Wang, X., Li, X.: Research on social service value evaluation of Beijing Forest Park based on network big data. Chin. Landsc. Architecture **33**, 14–18 (2017) (in Chinese)
6. Li, F., Li, W., Li, X.: Urban greenway planning research based on bus data big data analysis —taking Beijing as an example. Urban Stud. **22**, 27–32 (2015) (in Chinese)
7. Sun, Y., Fan, H., Helbich, M., et al.: Analyzing human activities through volunteered geographic information: using Flickr to analyze spatial and temporal pattern of tourist accommodation. In: Krisp J. (eds.) Progress in Location-Based Services. Lecture Notes in Geoinformation and Cartography, pp. 57–69. Springer, Berlin (2013)
8. Zhang, Z., Huang, Z., Jin, C., et al.: Study on temporal and spatial behavior characteristics of scenic spots based on Weibo sign-in data—taking Nanjing Zhongshan scenic area as an example. Geogr. Geo Inf. Sci. **31**, 121–126 (2015) (in Chinese)
9. Huang, W.: Preliminary study on environmental behavior analysis based on indoor positioning system (IPS) big data—taking Wanke Songhua Lake resort as an example. World Arch., pp. 126–128 (2016) (in Chinese)
10. Li, Y., Cheng, Q., Wang, Z., et al.: 'Big data' for pedestrian volume: exploring the use of Google street view images for pedestrian counts. Appl. Geogr. **63**, 337–345 (2015)
11. Abbasi, A., Rashidi, T.H., Maghrebi, M., et al.: Utilising location based social media in travel survey Methods: bringing Twitter data into the play. In: 8th ACM SIGSPATIAL International Workshop on Location-Based Social Networks (2015)
12. Vu, H.Q., Gang, L., Law, R., et al.: Exploring the travel behaviors of inbound tourists to Hong Kong using geotagged photos. Tour. Manag. **46**, 222–232 (2015)
13. Girardin, F., Calabrese, F., Fiore, F.D., et al.: Digital footprinting: uncovering tourists with user-generated content. IEEE Pervasive Comput. **7**, 36–43 (2008)
14. Tian, B., Niu, X.: Urban design practice supported by big data—space network planning for public activities in the historical and cultural area of Fuxing road, Hengshan road. Urban Plan. Forum. 78–86 (2017) (in Chinese)
15. Zhang, Y.: Research on spatial layout and theme optimization of Huangshan outdoor environment interpretation card (Master's dissertation, Shanghai Normal University) (2018) (in Chinese)
16. Oksanen, J., Bergman, C., Sainio, J., et al.: Methods for deriving and calibrating privacy-preserving heat maps from mobile sports tracking application data. J. Transp. Geogr. **48**, 135–144 (2015)
17. Liu, J., Li, J., Li, W., et al.: Rethinking big data: a review on the data quality and usage issues. Isprs J. Photogramm. Remote. Sensing. **115**, 134–142 (2016)
18. Li, F., Li, X., Li, W., et al.: Application research of location service data in landscape architecture in big data era. In: Proceedings of the 2015 Annual Meeting of the Chinese Society of Landscape Architecture, pp. 271–275. China Architecture and Building Press, Beijing (2015) (in Chinese)

Multicriteria Analysis in the Proposed Environmental Management Regulations for Construction in Aurora, Guayas, Ecuador

Christian Zambrano Murillo[1(✉)],
Jesús Rafael Hechavarría Hernández[1], and Maikel Leyva Vázquez[2]

[1] School of Architecture and Urbanism, University of Guayaquil,
Guayaquil, Ecuador
{christian.zambranomu, jesus.hechavarriah}@ug.edu.ec
[2] School of Mathematical and Physical Sciences, University of Guayaquil,
Guayaquil, Ecuador
maykel.leyvav@ug.edu.ec

Abstract. Studies have shown that the construction industry is one of the biggest polluters on the planet, so environmental standards must be implemented to minimize its impact. The daily increase in housing is shortening the forests and areas once protected. Due to the increase in demand for housing in Guayaquil, the population has migrated to neighboring cantons acquiring villas in urbanizations. This increase in the real estate sector is not ecologically sustainable because harmful materials are used for the environment and health, in addition to the use of machinery that uses fossil fuel. The objective of this research is to determine what would be the new regulations of environmental management in the building systems for the buildings of "La Aurora" by reviewing the pollutants that throw the materials most used in construction in order to establish the margins of contamination allowed and the means to compensate for the damage created. To determine the new regulations, a scenario analysis combined with neutrosophical cognitive maps and the TOPSIS multicriteria method is used. This combination allows analyzing the main scenarios and ordering them according to the multiple dimensions of the problem. Among the proposals to mitigate environmental pollution are the use of less polluting materials in construction, clean energies in the illumination of homes and public spaces as well as the increase of green areas and reforestation programs per m^2 of areas dedicated to develop.

Keywords: Human factors in architecture ·
Sustainable urban planning and infrastructure

1 Introduction

The increase in real estate throughout the Ecuadorian territory, mainly in the Delta del Guayas is impressive as it has invited directly to the people to seek a rational and funded long-term housing, this demand comes mostly neighbors, such as Guayaquil, Samborondón and Daule, which has led to the creation of housing plans called citadels

© Springer Nature Switzerland AG 2020
T. Ahram (Ed.): AHFE 2019, AISC 965, pp. 101–113, 2020.
https://doi.org/10.1007/978-3-030-20454-9_10

or private developments, in Daule, giving as a complex product of villas and condominiums such as "La Joya", "Villa Club" cantons "La Rioja", "Villa Italia" and "Villas del Rey", which serve predominantly social class consumers called "media-alta".

This increase in the real estate sector conceptualized as unsustainable, to meet three of its four pillars (economic, social and political), leaving aside the eco systemic or environmental axis.

One of the main problems is the traditional construction system that is highly polluting and generates significant wastes that are not managed for better disposal; using materials harmful to health and require the use of much energy for its creation; besides using fossil fuels in heavy of semi or heavy machinery, which emit pollutants such as sulfur oxide and nitrogen oxide, among others.

The lack of standards environmental management and restoration and maintenance of protected areas, has allowed promoters made deforestation, indiscriminate agricultural practices to implement developable projects, resulting in increased environmental pollution in the parish "La Aurora" currently it is affected by the rise in temperature by 2° and air pollution.

This increase in environmental pollution, anticipated negative effects for the future of "La Aurora", such as infectious diseases, acid rain, as well as the total loss of forests and protected areas, thus creating a serious greenhouse effect and unsustainable conditions for life in the territory if it remains poorly managing the environment as invisible axis in planning the growing developments in "La Aurora".

2 Matherials and Methods

Sustainability must meet four indicators, and these should be maintained over time, but one of these is true, the project is not sustainable, these indicators are: economic, political, social and ecological, usually almost all projects maintain in time three of these indicators, but the latest environmental issue environmental, usually it does not become of interest to investors, as they see an expense having to generate an environmental certification in their companies or projects, seen as a loss of investment because it generates no revenues, according to [1] and on the other hand feel that giving space to the environment in their projects puts at a disadvantage by not optimize the maximum of its territory. This issue of sustainability in the ecological environment, creates a lot of contingency,

Environmental management through standards and regulations improves the relationship between environmental and urban project, and therefore participates in the management of them, like assessing the technical and financial [2] aspects.

The construction sector is considered one of the main sources of environmental pollution in the world compared to other industries [3], it produces CO^2 emissions, noise pollution and cross-contamination through the transfer of waste to sites unsuitable [4]. Projects in these developments are done thinking about the economic and political indicator, rather than the social and environmental.

These constructions do not maintain a respectful process with the environment and have a massive direct or indirect effect, and very few technical managers or contractors who devote their efforts to contribute to the care and maintenance of the environment

[3], let alone when they are massive housing construction, as in the case of private developments in many sectors of Ecuador, and in this study "La Aurora", known for its high housing demand affecting the Delta del Guayas.

3 Methodology

The methodology applied in this case was conducted in Chile and Spain, which analyzes the ecological footprint per square meter for comparative development of materials and construction techniques, in which the former is mainly built with wood, and the other with concrete brick and the study found that the mark in Chile can be reduced from 0.17 to 0.07 hag/ m^2, by the use of recycled wood and in Spain the 25% reduction is achieved by the use of concrete and steel recycled [5].

The application of computer software through a casual model, which is a tool used to understand complex systems, it is useful for decision making, which is natural and easy to understand and is convincing because it gives us a particular conclusion, and specifically casual model used the fuzzy cognitive maps (MCD). In decision scenario analysis is used [6], so this methodology was used combining the MCD by the neutrosophical studying the origin, nature and scope of neutralities [7] and the multi method TOPSIS (Technique for order preference by similarity to perfect solution); relative vs absolute, which is also a decision-making model which is based on solving the problems posed by nesting using algorithms [8].

As reviewed above, the problems caused by the construction industry, as the use of machines that use fossil fuel and materials used that leave a large footprint when being manufactured and a huge waste the time of use, cause a greenhouse effect, and altering weather conditions affecting the environment Delta Guayas. In 1992 the Framework Convention of the United Nations was created on Climate Change (UNFCCC) and in 1995 the first COP (Conference of Parties) in Berlin was held, this was the basis for that in 1997 the Kyoto Protocol was signed, according to which 37 countries agreed to lower the gases that cause greenhouse effects 5.2% for 2012, compared to levels (GHG) 1990 [9] (Table 1).

Table 1. Estimate of the temperature increase year 2100

Scenario for the year 2100	Average temperature increase	Increase in temperature (16% probability)	Increase in temperature (84% probability)
Current trend without changes in the political intervention (Baseline)	3,8 °C	3,1 °C	4,8 °C
Current trend high stage	3,7 °C	3,0 °C	4,6 °C
Current trend low scenario	3,6 °C	2,9 °C	4,4 °C
Confirmed commitments	3,1 °C	3,1 °C	3,9 °C
Conditional commitments	3,0 °C	2,4 °C	3,8 °C

A history of decision-making on climate change mitigation, maintaining the objective agreed in Kyoto (1997), which is to reduce GHG levels to minimize damage and enhance opportunities that are beneficial to the planet. There is evidence in the Scopus database, which scientific papers of this kind have been increasing in recent years, indicating that there is a greater awareness of the problems caused by climate change, a total of 1188 articles published were found from 1992 to adaptation and mitigation 542 articles published since 1995 (Fig. 1).

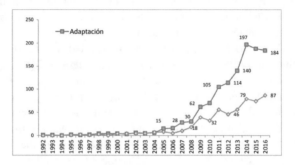

Fig. 1. Articles published in decision making for adaptation and mitigation to climate change (1992–2016)

Climate change is a fact, should join efforts to mitigate and caused extensive damage and adapt to the construction industry to stop influencing the following major pollutants points:

1. Materials - The construction consumes significant amounts of materials such as wood, steel, copper, glass, aluminum, polymers, etc. All these materials must be extracted from the earth's crust, at the expense of energy and habitat destruction.
2. Energy - Physical assembly building consumes a lot of energy. This harvest fossil energy emitted CO^2 and air pollutants conventional, besides contributing to resource extraction and associated habitat loss.
3. Systemic Impacts - Basically, building construction contributes to the continuous operation of the systems that cause environmental damage. This incurs both the vehicle system, power and data systems and industrial system that produces the various "spare parts" that a building consumes during its lifetime.

In the construction sector more than 2 tons of raw materials needed per m^2 house we build, the amount of energy associated with the manufacture of the materials that make up a home can amount to approximately one third of energy consumption a family over a period of 50 years, production of construction and demolition waste exceeds one ton annually per capita.

Heavy and semi heavy machinery used in construction are a major pollutant, and using fossil fuels, but instead are helpful to optimize the construction process, these greatly facilitate the hard work and help reduce the number of accidents caused in these processes; but are nevertheless highly pollutant carbon emissions released into the atmosphere. Implementing a regulation mode law made for environmental management in the construction processes must provide a new construction system that addresses the reduction of pollutants and use of environmentally friendly materials and the use of clean energy, applied to the conservation of ecosystems and the environment, is an eminent task, creating new regulations (Fig. 2).

Machinery	Description	Environmental impact
	Self-propelled machine on wheels or chains with a superstructure zapaz to perform a rotation of 360º, which excavates, loads, lifts, rotates and unloads materials by the action of a spoon fixed to a set of boom and rocker, without the chassis or structure When moving, there are different types of cable excavators, mechanical, hydraulic, mounted on chains, wheels, tires, rails.	Noise inside and outside, emission of gases (carbon dioxide CO_2, nitrogen dioxide, nitrogen oxide (IV), nitric oxide NO_2), particles in suspension (dust).
	The backhoe is a machine in which the boom lowers and rises in each operation; the spoon, attached to it, digs pulling towards the cart, ie backwards, instead of pushing forward, as does the normal excavator. This is used to excavate vertical slopes below the plane and support of the machine, load, move, mobilize and demobilize.	Noise inside and outside, emission of gases (carbon dioxide CO_2, nitrogen dioxide, nitrogen oxide (IV), nitric oxide NO_2), particles in suspension (dust).
	Machine for earth movement with great power and robustness in its structure, specially designed for the work of cutting (digging) and at the same time pushing with the blade (transport). It is used to excavate (open-cut in large dimensions) and carry in large dimensions.	Noise inside and outside, emission of gases (carbon dioxide CO_2, nitrogen dioxide, nitrogen oxide (IV), nitric oxide NO_2), particles in suspension (dust).
	The front loader is a tractor, mounted on tracks or wheels, which has a large bucket at the front end, the loaders are loading equipment, hauling and eventually excavation, in the case of hauling is only recommended in distance short It is used to excavate, load, unload, haul or transport.	Noise inside and outside, emission of gases (carbon dioxide CO_2, nitrogen dioxide, nitrogen oxide (IV), nitric oxide NO_2), particles in suspension (dust).

Fig. 2. Classification of construction machinery and its environmental impact

4   Results

4.1   A New Model of Environmental Management

Environmental regulation, as a set of formal rules, understand the rules imposed by the government to limit polluting activities and penalties, fines and legal costs for breach [10–12]. There are different regulatory instruments, such as command and control instruments and market-based (taxes and subsidies) [13]. In this study, we focused on command and control instruments, as they are comparable in different states. As required an interested party, the government is responsible for the formal license of construction companies to operate and provide the necessary legitimacy companies [14–17]. Compliance with the regulation represents the basic phase of commitment to environmental [18] management. Environmental regulation provides equal conditions for all players and balances, competitive forces, which can measure highest standards of operational change [19]. Companies with a strategy of pure pollution prevention are primarily concerned compliance with environmental regulation [20]. In general, the mere adoption of a compliance strategy for an environmental reactive strategy in which companies do not seek a competitive advantage [12].

Environmental regulation mainly aims to reduce externalities. The command and control instruments tend to increase operating costs in the first moments after implementation [12]. In recent decades, emerging countries have improved control systems on polluting practices, reducing the regulatory gap between them and developed countries [21]. In terms of effectiveness, coercive regulation, generally composed of command and control instruments it is assumed. Improve environmental management practices. A study on the industrial sector of Brazil indicated that sanctions and demands from regulators were the most influential in the environmental performance of company's factors [22].

Therefore two management strategies for the development of environmental regulations are analyzed:

Voluntary standards - are self-regulatory strategies that go beyond compliance. In the literature there is a wide range of terminologies to indicate such strategies: Practical/ voluntary environmental initiatives [14, 23]; environmental standards volunteers [11, 13]; and voluntary standards (2) [9, 24].

Demand Stakeholders - the importance of stakeholders in achieving the goal of legislation is widely recognized [25]. The definition of stakeholders themselves, as those that affect or are affected by the company [26], shows its importance in business strategy. Develop resources in the complexity of the interaction between the company and its partners, creating value on complementarity [27, 28]. In view of this, pressure from stakeholders has an important influence on the implementation of environmental strategies and contributes to the development of resources [29, 30] (Fig. 3).

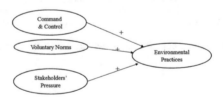

Fig. 3. Hypothesized model

By this hypothetical model, they can increase the effectiveness of environmental management practices and seek precautionary nature rights established in the Constitution of 2008 and COTAC and the COA, to care for our environment and ecosystems.

4.2   Materials Analysis

The analysis focused on the foundations of the structure, partitions, window frames, doors and roof of the building in question. Subsystems and its associated materials are listed in Table 2.

The limit for the system under study is delimited from the extraction of raw materials, through manufacturing, distribution and final disposal, as shown in Fig. 4.

Table 2. Subsystems and associated materials

Building system	Characteristics	Materials
Foundation	Reinforced concrete structure	Cement
		Steel
Structure	Reinforced concrete structure	Cement
		Steel
		Wood
Masonry	Brick blocks and mortar applied	Cement
		Ceramic
Wall covering	Tiles, flooring, mortar applied	Cement
		Ceramic
Frames	Doors and wood windows	Cement
		Steel
		Wood
Roofing	Roof with two slopes on ceramic tiles and wooden structure	Steel
		Wood
		Ceramic

Despite the significant period of time and impacts due mainly to the use energy and water, use phase of buildings was excluded from the analysis. On the one hand, in relation to the use of water and energy, considerations not related to the materials studied, and second, the potential for renovations and building maintenance are the responsibility of the user and the need for configuration.

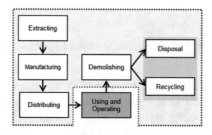

Fig. 4. Layout of the system's boundary.

As the end of life, despite regulatory recommendations, however, it is known that in practice most construction waste in "La Aurora" have their final destination or not in specific landfills. In fact, approximately 3% of waste is recycled in Guayaquil, and the vast majority has its final disposal in landfills and unauthorized vacant lots. Furthermore, one should also consider the effective reuse of some materials and products when construction is completed and demolishing a building, which is described below in Table 3.

Table 3. Distribution of waste at its end of life.

Material	Quantity	Treatment
Cement and ceramic	25%	Recycling
	75%	Landfilling
Steel and wood	50%	Recycling
	50%	Landfilling

As a database, use the Ecoinvent and also use the base data Idemat 2001. As for the quality of the databases, highlights the fact that they are of foreign origin and, therefore portray reality European. In addition, the construction technology in the country itself is largely artisanal hand, unlike what happens in the reference countries. So if you search, adjust inventories constitutive modeling processes whenever possible, and necessary travel and type of energy used.

Data Collection. Quantification of materials was based, in general, in the August issue of the Journal of the Chamber of Construction of Guayaquil - Construction and Development (August, 2018), which is characterized as one of the reliable databases on building in the city of Guayaquil.

Lifecycles of materials used were modeled as input and output streams of the processes, as shown in Fig. 5. Note that the inputs and outputs were based on the databases used in this work.

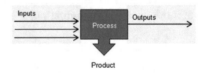

Fig. 5. Schematic of the modeling life cycle.

On the subject of transport, road transport was favored at all stages, given the proximity of the sites (Via Daule, La Aurora and Pascuales) and the field of road network in the country. As for the manufacturing phase, it has been regarded as the place of its realization, the site of the extraction of raw materials. With regard to the distribution phase, an average of distances between existing providers near the construction site was considered, this distance being equal to 10 km. Finally, at the end of the stage of life, we have taken into account only those scenarios in which the waste is destined for landfills and processed for recycling, with displacements of 12 and 55 km, respectively (Table 4).

This evaluation is translated consumption rate and residues identified in the inventory phase, environmental, such as the greenhouse effect, hole in the ozone, smog, acid rain, eutrophication, toxicity, among other layer impacts. For this, the calculation methodology IMPACT 2002 + , which proposes a combination of classical approaches (midpoint) and directed to damage (endpoint), grouping the strengths of methods, such as IMPACT2002 +, was used Eco Indicator99 LMC 2000 and IPCC. selected for

Table 4. Displacements between phases of the material's life cycle

Material	Process	Origin - Destination
Steel	Extraction	3 km
	Manufacturing	10 km
Ceramic	Extraction	3 km
	Manufacturing	42 km
Cement	Extraction	3 km
	Manufacturing	60 km
Wood	Extraction	3–5 km
	Manufacturing	100–200 km

further observation categories are related to global warming, consumption of natural resources, non-renewable energy consumption and toxicity to human health (Fig. 6).

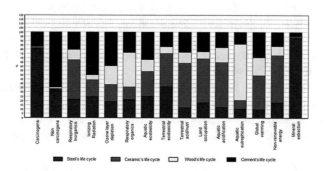

Fig. 6. Analysis IMPACT 2002 + on construction materials

Analyze the results, it was found that the most serious impacts are related to global warming, consumption of non-renewable energy and toxicity to human health. In connection with global warming, it has been found that is most responsible represented by lifecycle ceramics. We know that global warming is produced largely by burning fossil fuels, both used in manufacturing processes and distribution, transportation. It is also important to note that the ceramic material was in larger amounts, corresponding to about 75% by weight of the materials considered for the building studied, as shown in Fig. 7 Therefore, it is expected that is responsible for major impacts.

It was observed, therefore the ceramic was highlighted as the most responsible for the impacts, perhaps because they require a greater amount of mass between the materials. In addition, the life cycles of cement and steel also had a significant impact, most often related to toxic substances.

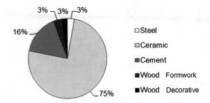

Fig. 7. Composition of the building materials studied in Kg.

5   Discussion and Recomendations

The construction industry generates pollutants into the environment by more than 50% overall contaminant being the industry that generates the most negative environmental impacts, so that the application of these regulations set new standards for the system construction for the lifting of future construction in La Aurora. Not only manufacturers of building materials, but also to all the agents in the building process (architects, engineers, surveyors, builders and administration), establishing six essential require-ments (general criteria are specified in documents in its annexes interpretive and must be fulfilled for a lifetime economically reasonable) for construction products; one of which is related to the "Hygiene,

It also works must be designed and constructed so that the amount of energy required in use shall be low and that the products must not release pollutants or waste liable to be dispersed in the environment and change the quality of the environment and behaving risks the health of humans, animals or plants, and compromising the balance of ecosystems.

Materials shall not emit toxic or in the production process or in construction materials, avoiding therefore dangerous or unhealthy indoor environment for occu-pants, as what has been called sick building syndrome, whose occupants could suffer respiratory diseases. This is meant to address, for example, those buildings built in the seventies on, very energy efficient, but so tight Nordic countries could not breathe and many people became ill.

The impact on the environment must be considered at every stage of the life cycle of building material, especially when manufactured, produces and constructs; used in finished works; and topples, download, or revalue incinerates waste. It is considered to be supporting the development of initiatives in various industrial sectors, based on voluntary agreements, including the development of codes of conduct and guidelines of good environmental practices that promote pro-active voluntary action. Similarly, environmental agreements between industry and public authorities Construction should be completed in statements of intent or partnerships, may be adopted in the form of unilateral industry committee, but recognized by those authorities.

6 Conclusions

The results show that urbanization promotes economic growth through the accumulation of physical capital, knowledge capital and human capital; the relationship between growth and urbanization is a benign [31] interaction, in this globalized world economic expansion and high ecological pressure, especially in the construction sector, identify the major causes of environmental pollution by buildings we do [5].

Materials with less environmental impact, for use in the building must incorporate environmental sustainability criteria, such as high energy efficiency, durability, recoverability, renewable resources, use of clean technology and waste recovery. While there is no universally accepted methodology that quantifies the many and varied existing criteria, the possibility of using other methods such as the Life Cycle Analysis. True, this methodology is costly, but it is the most reliable tool for assessing environmental burdens associated with a product or activity. Therefore, collaboration between the authorities and the industry sector of construction in order to develop a Life Cycle Inventory is necessary.

Also a National Plan for Sustainable Building to collect not only the criteria for the use of materials with low environmental impact, but also other subject areas mentioned, among others, energy efficiency and waste management of construction and demolition. In line with this waste management, development of rules requiring all construction projects incorporating recyclable materials from plants installed to effect treatment it is necessary. Finally, and with regard to public projects, the rules regulate the Contracts of Public Administrations should take into account the environmental variable, rewarding those projects involving building materials that create the least number of construction waste.

Acknowledgments. The authors would like to School of Architecture and Urbanism of the University of Guayaquil for the support of esta research.

References

1. Pinto, G.M.C., Pedroso, B., Moraes, J., Pilatti, L.A., Picinin, C.T.: Environmental management practices in industries of Brazil, Russia, India, China and South Africa (BRICS) from 2011 to 2015. J. Clean. Prod. **198**, 1251–1261 (2018) https://doi.org/10.1016/J.JCLEPRO.2018.07.046
2. Angel, S.E., Carmona Maya, S.I., Villegas, R.L.C.: Medellín. Fac. of Mines. School of Geosciences and Environment. Environmental Management in Development Projects. National University of Colombia (2001). http://www.sidalc.net/cgi-bin/wxis.exe/?IsisScript=BAC.xis&method=post&formato=2&cantidad=1&expresion=mfn=044718
3. Enshassi, A., Kochendoerfer, B., Rizq, E.: Evaluation of environmental impacts of construction projects. Constr. Eng. J. **29**(3), 234–254 (2014). https://doi.org/10.4067/S0718-50732014000300002
4. Zhang, H., Zhai, D., Yang, Y.N.: Simulation-based estimation of environmental pollutions from process construction. J. Clean. Prod. **76**, 85–94 (2014) https://doi.org/10.1016/J.JCLEPRO.2014.04.021

5. Gonzalez-Vallejo, P., Munoz-Sanguinetti, C., Marrero, M.: Environmental and economic assessment of dwelling construction in Spain and Chile. A comparative analysis of two representative case studies. J. Clean. Prod. **208**, 621–635 (2019). https://doi.org/10.1016/J.JCLEPRO.2018.10.063

6. Leyva-Vazquez, M.: Model helps making decisions based on Fuzzy cognitive maps (2013). https://doi.org/10.13140/RG.2.1.1233.8406

7. Leyva-Vazquez, M., Smarandache, F.: Neutrosofía: new advances in the treatment of uncertainty (2018)

8. Ceballos, B., Lamata, M.T., Pelta, D., Sanchez, J.M.: The relative TOPSIS method vs. absolute. Straight **14**(2), 181–192 (2013). https://doi.org/10.1109/Chilecon.2015.7404650

9. Sierra Velez, X.: Tools for decision making for mitigation and adaptation to climate change (2017). http://bdigital.unal.edu.co/59132/1/1053792568.2017.pdf

10. Bansal, P., Roth, K.: Why companies go green: a model of ecological responsiveness. Acad. Manag. J. **43**(4), 717–736 (2000). https://doi.org/10.2307/1556363

11. Delmas, M.A., Toffel, M.W.: Stakeholders and environmental management practices: an institutional framework. Bus. Strat. Environ. **13**(4), 209–222 (2004). https://doi.org/10.1002/bse.409

12. Rugman, A.M., Verbeke, A.: Corporate environmental strategies and regulations: an organizing framework. Strat. Manag. J. **19**(4), 363–375 (1998). https://doi.org/10.1002/(ICIS)1097-0266(199804)19:4<363::AID-SMJ974>3.0.CO;2-H

13. Barbieri, J.C.: Environmental Gestão Business CONCEITOS, Models and Tools. Saraiva, São Paulo (2007)

14. Christmann, P.: Multinational companies and naturally the environment: environmental determinants of overall standardization policy. Acad. Manag. J. **47**(5), 747–760 (2004)

15. Delmas, M.A., Toffel, M.W.: Organizational responses to environmental demands: opening the black box. Strat. Manag. J. **29**(10), 1027–1055 (2008). https://doi.org/10.1002/smj.701

16. Henriques, I., Sadorsky, P.: The determinants of an environmentally responsive firm: an empirical approach. J. Environ. Econ. Manag. **30**(3), 381–395 (1996). https://doi.org/10.1006/jeem.1996.0026

17. Walker, H., Di Sisto, L., McBain, D.: Drivers and barriers to environmental supply chain management practices: lessons from the public and private sectors. J. Purch. Supply Manag. **14**(1), 69–85 (2008). https://doi.org/10.1016/j.pursup.2008.01.007

18. Oliver, C.: Strategic responses to institutional processes. Acad. Manag. Rev. **16**(1), 145–179 (1991). https://doi.org/10.5465/AMR.1991.4279002

19. Porter, M.E., van der Linde, C.: Toward a new conception of the environment - competitiveness relationship. J. Econ. Perspect. **9**(4), 97–118 (1995). https://doi.org/10.1257/jep.9.4.97

20. Buysse, K., Verbeke, A.: Proactive environmental strategies: a stakeholder management perspective. Strat. Manag. J. **24**(5), 453 (2003). https://doi.org/10.1002/smj.299

21. Meyer, K.E.: Perspectives on multinational enterprises in emerging economies. J. Int. Bus. Stud. **35**(4), 259–276 (2004). https://doi.org/10.1057/palgrave.jibs.8400084

22. Seroa da Motta, R.: Analyzing the environmental performance of the Brazilian industry sector. Ecol. Econ. **57**(2), 269–281 (2006). https://doi.org/10.1016/j.ecolecon.2005.04.008

23. Sharma, S., Vredenburg, H. Proactive corporate environmental strategy and the development of competitively valuable organizational capabilities. Strat. Manag. J. **19**(8), 729–753 (1998). https://doi.org/10.1002/(sici)1097-0266(199808)19:8<729::AID-SMJ967>3.0.CO,2-4

24. Lopez-Gamero, M.D., Molina-Azorín, J.F., Claver-Cortes, E.: The potential of environmental regulation to change managerial perception, environmental management, competitiveness and financial performance. J. Clean. Prod. **18**(10/11), 963–974 (2010). https://doi.org/10.1016/j.jclepro.2010.02.015

25. Freeman, R.E., Wicks, A.C., Parmar, B.: Stakeholder theory and "the corporate objective revisited". Organ. Sci. **15**(3), 364–369 (2004). https://doi.org/10.1287/orsc.1040.0066
26. Freeman, R.E.: Strategic Management: A Stakeholder Approach. Pitman, Boston (1984)
27. Adegbesan, J.A.: On the origins of competitive advantage: strategic factor markets and heterogeneous resource complementarity. Acad. Manag. Rev. **34**(3), 463–475 (2009). https://doi.org/10.5465/AMR
28. Sharma, S., Henriques, I.: Stakeholder influences on sustainability practices in the industry Canadian forest products. Strat. Manag. J. **26**(2), 159–180 (2005). https://doi.org/10.1002/smj.439
29. Hart, S.L., Sharma, S.: Engaging stakeholders for competitive fringe imagination. Acad. Manag. Exec. **18**(1), 7–18 (2004). https://doi.org/10.5465/ame.2004.12691227
30. Porter, M.E., Kramer, M.R.: Creating shared value. Harv. Bus. Rev. **89**(1/2), 62–77 (2011)
31. Liang, W., Yang, M.: Urbanization, economic growth and environmental pollution: Evidence from China. Sustain. Comput.: Inform. Syst. **21**, 1–9 (2019). https://doi.org/10.1016/J.SUSCOM.2018.11.007

Random Samplings Using Metropolis Hastings Algorithm

Miguel Arcos-Argudo$^{(\boxtimes)}$, Rodolfo Bojorque-Chasi,
and Andrea Plaza-Cordero

Research Group on Artificial Intelligence and Assistance Technologies
(GIIATA), Salesian Polytechnic University, Cuenca, Ecuador
{marcos, rbojorque, aplaza}@ups.edu.ec

Abstract. Random Walks Samplings are important method to analyze any kind of network; it allows knowing the network's state any time, independently of the node from which the random walk starts. In this work, we have implemented a random walk of this type on a Markov Chain Network through Metropolis-Hastings Random Walks algorithm. This algorithm is an efficient method of sampling because it ensures that all nodes can be sampled with a uniform probability. We have determinate the required number of rounds of a random walk to ensuring the steady state of the network system. We concluded that, to determinate the correct number of rounds with which the system will find the steady state it is necessary start the random walk from different nodes, selected analytically, especially looking for nodes that may have random walks critics.

Keywords: Markov chains · Small worlds · Metropolis hastings ·
Random walks · Node sampling · Random sampling

1 Introduction

Actually, most complex systems such as biological cell development networks and brain activity are studied under their network structure [1] to understand how it communicate, work, self-organize, evolve, etc. This allows to understand the importance of the subject of our study within the field of computing, since the application of random sampling can be seen in various environments such as for the efficient collection of energy data in wireless sensor networks [2], analysis of social networks and information [3], operations on big data [4], etc. This paper presents the results of random walks in a graph with characteristics of a Markov Chain, by implementing the Metropolis-Hastings Random Walks (MHRW) algorithm, with which random samples can be obtained after a certain number of jumps (rounds), and in such a way that all the nodes of the network have a uniform probability of being chosen as a sample. The work has determined the number of rounds with which a uniform sampling distribution is obtained, and with a stable average error which is called the "steady state of the network", regardless of the node by which the walk begins. In addition, experiments have been carried out starting the random walks from initial nodes chosen in an analytical way, taking into account the main concepts and metrics on graphs, which has

© Springer Nature Switzerland AG 2020
T. Ahram (Ed.): AHFE 2019, AISC 965, pp. 114–122, 2020.
https://doi.org/10.1007/978-3-030-20454-9_11

allowed executing a possible method of finding nodes that present special characteristics and provoke that the random path is come back critical.

2 MHRW Algorithm

The importance of the implementation of sampling algorithms and random walks on
networks is studied by Sevilla et al. [5], Arcos [6], among others. According to [7] the
Metropolis-Hastings Algorithm (M–H) is extremely versatile, and is widely used in
physics, the authors use (1) as probability density function:

$$\int q(x, y)dy = 1 \qquad (1)$$

This density is capable of generating a new state y when the current state is
x starting from $q(x, y)$, however (1) does not guarantee that the probability of staying in
the state x is the same as passing to the state y (condition of reversibility), since the
probability of transiting from x to y would be greater. Then, it is convenient to include a
condition that reduces the number of state changes from x to y, by entering a probability
$\alpha(x, y) < 1$; if a state change does not occur, it remains in x returning this value for the
given distribution. Then, we can establish the following relationship:

$$p_{MH}(x, y) \equiv q(x, y)\alpha(x, y) \text{ for } x \neq y \qquad (2)$$

where $\alpha(x, y)$ is still to be determined. With $PMH(x, y)$ he reversibility condition can
be satisfied as shown below (consider that $\alpha(y, x) \approx 1$):

$$\pi(x)q(x, y)\alpha(x, y) = \pi(y)q(y, x)\alpha(y, x)$$
$$= \pi(y)q(y, x) \qquad (3)$$

where π represents the desired distribution, then:

$$\alpha(x, y) = \frac{\pi(y)q(y, x)}{\pi(x)q(x, y)} \qquad (4)$$

Thus, the authors conclude that $PMH(x, y)$ is reversible. The probability of
obtaining a new value of the state during the tour is:

$$\alpha(x, y) = \begin{cases} min\left[\frac{\pi(y)q(y,x)}{\pi(x)q(x,y)}, 1\right], & \text{if } \pi(x)q(x, y) > 0 \\ 1, \text{ any other case} \end{cases} \qquad (5)$$

3   Concepts and Basic Notions

A network of any nature can be represented by a graph which consists of a finite set of nodes. Each node represents an element of the network and must have a unique identifier; the way in which network elements are connected are represented by edges; an edge is a line joining a pair of nodes; when a node O is directly connected to a node P by an edge, we can say that O and P are neighboring nodes. In a directed graph each edge has an address, that is, each edge a_i that begins in a source node O allows jump to a destination node P, but it is impossible to jump from node P to node O using the same edge a_i, Fig. 1. In an undirected graph edges are bidirectional, that is, it is possible to jump from node O to node P, and form node P to node O using the same edge a_i for both cases, Fig. 2. In an undirected graph the number of edges that are connected to it determines the degree of a node. A connected graph is a graph in which all its nodes are connected to each other by edges (Fig. 1). A disconnected graph is a graph in which a node or group of nodes in the graph are isolated from another group of nodes (Fig. 3). A weighted graph is a graph in which each a_i has an associated weight w_i, this weight represents the cost of jumping from the origin node O to the destination node P using the edge a_i (Fig. 4). An unweighted graph is a graph in which all its edges have the same associated weight, in this case the weight of each edge a_i in the graph is not graphically represented (Fig. 2). The present study uses an undirected, connected and unweighted graph, which coincides with the representation of Fig. 2.

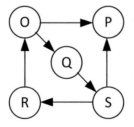

Fig. 1.   Directed, connected and unweighted graph.

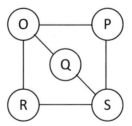

Fig. 2.   Undirected, connected and unweighted graph.

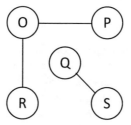

Fig. 3. Undirected, disconnected and unweighted graph.

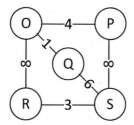

Fig. 4. Undirected, disconnected and unweighted.

In a random walk we call the node in which the state of the network is located at an instant of time t, that is, if the state of the network at time t is in node O, then, the node O is the current node at time t, if the next state of the network at time $t + 1$ is the node P, then node P is the current node at time $t + 1$.

A random walk is constructed in the following way: given a graph G and an initial node O as starting point (current node) in which the state of the network is located at time $t = 0$, a neighbor node of O is randomly selected, this can be the node P, then, the state of the network passes to node P at time $t = 1$, then P is the current node, then, a neighbor node of P is selected randomly, this can be the node Q, and the state of the network passes to node Q which is the current node at time $t = 2$, this is repeated successively until a stop condition is reached. It should be noted that in order to determine that a network passes from one state to another, some condition may be imposed, in case this condition is not fulfilled, and the node O is the current node, the network would not change state and the current node would continue being the node O [8–10], in this work this condition is defined in (6).

Our goal is to demonstrate that by using the MHRW algorithm we can perform a uniform sampling of the nodes of a connected, undirected and unweighted graph, and stabilize the mean error with respect to the uniform distribution, understood by uniform distribution to all the nodes have the same probability of being chosen as a sample.

4 Discussion

We have used a graph with characteristics of Markov chain that was obtained from the Network Data Repository[1], it represents the Facebook network of the Massachusetts Institute of Technology (MIT), it is composed by 6402 interconnected nodes through 251230 edges. For each experiment, a number of samples equivalent to 100 * number of graph nodes was obtained uniformly. In the simulation, 100 random walks were executed by randomly selecting the initial node, and also by taking at convenience nodes that present relevant characteristics in graph analysis; the results are summarized in Table 1. The condition of the implementation with which it is satisfied (5), is given by the generation of a random value p in a uniform manner, such that:

$$p \leq \min \left[\frac{degree(currentnode)}{degree(neighborNode)}, 1 \right] \tag{6}$$

That is, the higher the degree of the neighbor selected at random, the less likely it is that it will be visited or chosen as a sample (current node); when the condition of (6) is met, the current node becomes the selected neighbor. The criterion to stop the execution of the simulation in each case was given by the stabilization of the mean error calculated by (7), that is, when the value of the error showed no significant changes, the experiment was stopped.

$$E = \frac{\sum_{i=1}^{n} \frac{m_i - freq}{freq}}{n} \tag{7}$$

Where n is the number of nodes in the graph, m_i is the number of times the node i was taken as a sample, and $freq$ is the sampling frequency. For the random selection of the neighbor, a uniform distribution function was used [11]; it was possible to verify that this function generates random numbers with a distribution similar to the distribution of the random selection of the neighboring nodes during all the samplings made in all experiments (Figs. 5 and 6).

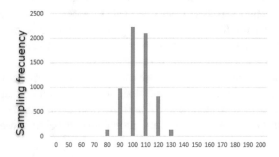

Fig. 5. Distribución del muestreo de los nodos seleccionados después de paseos de 20.000 rondas.

---

[1] http://networkrepository.com/.

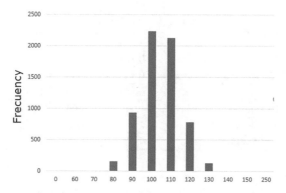

Fig. 6. Distribution of random numbers generated with the uniform distribution function.

Analyzing the results obtained, we conclude that, for the set of tests proposed, the average error is stabilized at approximately 8%, as shown in Table 1, which is corroborated by making increasingly long random walks, we have conducted experiments with 20,000 rounds prior to taking the sample. The variation of the average error can be seen in Fig. 7. The first column of Table 1 shows the identifier of the node, the second column shows the name of the metric by which the node was chosen, the third column shows the value of the metric specified in the second column, the fourth column shows the number of rounds needed for the average error value to reach a steady state, the fifth column shows the value of the average error obtained during the random walk ($\approx 8\%$ in all the cases). As can be seen in Table 1, when the random walk was initiated by the node 1010 whose eigenvector value is the highest of all, the number of necessary rounds increased considerably compared to the previous nodes, this led us to analyze the characteristics of this node; it is a node of degree 1 whose only neighbor had a significantly higher degree, later we verified if there was another node whose degree is 1 and whose only neighbor has an even greater degree than the previous one, this is how we find the node 4503.

Table 1. Results of executed experiments

Node	Metric	Value	Rounds	Avg. error
2411	Random selected node	–	400	0.081
6400	Lowest degree	1	470	0.081
2982	Highest degree	708	350	0.080
30	Lowest betweenness	0	750	0.081
1940	Highest betweenness	4.9E + 12	300	0.083
173	Lowest coefficient clustering	0	400	0.083
3302	Highest coefficient clustering	1	500	0.080

(continued)

Table 1. (*continued*)

Node	Metric	Value	Rounds	Avg. error
3277	Lowest closeness	0.181	500	0.080
2982	Highest closeness	0.495	300	0.083
558	Lowest eigenvector	0.001	290	0.083
1010	Highest eigenvector	4,9+12	1600	0.079
4503	Critic node	–	6000	0.078
	Average error			0.081

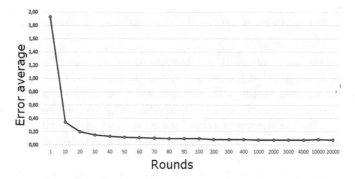

Fig. 7. Variation of the average error of the random walk initiated by a node chosen randomly (Node 2411).

When the random walk was started by this node, the value of the average error decreased at a significantly lower speed, so this became the most critical route of the set of tests performed, the results are shown in the penultimate row of Table 1. The feature of node 4503 is that its degree is 1, and its only neighbor (node 2720) has a degree of 617, so meeting a condition that satisfies Eq. (6) has a probability of 1/617, then, the random walks turned out to be longer. Sampling reached a steady state after a walk of approximately 6,000 rounds. That is why we have verified that the number of rounds with which it is guaranteed that the random walks in the graph arrive at a stationary state applying MHRW is 6,000, regardless of the node by which the route begins, which coincides with the result published in [5].

In Table 2 the first column shows the number of rounds used in each random walk, the second column indicates the value of the average error obtained in each trip, the third column shows the identifier of the most sampled node in each random walk, the fourth column shows the number of times that the most sampled node was selected as a sample; In this table it can be seen that in random walks that do not reach 6000 rounds the most sampled node is 4503, that is, due to the small probability of jumping to its neighbor node, the most sampled node is the same node, which prevents reaching the steady state. Note also that although at 4000 rounds the mean average error has decreased to a value of 0.081, the number of times that node 4503 has been sampled is approximately ten times greater than the desired average, so the uniform distribution

Sampling is achieved only in a walk of 6000 rounds, under these conditions the most sampled node is a node different from the initial node. Then, a possible method to find initial nodes that present critical random walks could be taking as initial node the node whose value of eigenvector is the highest of all.

Table 2. Nodes most sampled in random walks that started by node 4503 according to the number of rounds.

Rounds	Avg. error	Most sampled node	Sampling frequency
100	2	4503	545013
200	1	4503	463663
300	1	4503	394263
400	1	4503	334579
500	0.891	4503	284823
600	0.759	4503	242464
700	0.645	4503	206069
800	0.549	4503	175609
900	0.469	4503	149929
1000	0.398	4503	126889
1100	0.341	4503	107955
1200	0.291	4503	91417
1300	0.254	4503	78415
1400	0.221	4503	66384
1500	0.196	4503	56738
1800	0.144	4503	34783
2100	0.117	4503	21413
3000	0.088	4503	5062
4000	0.081	4503	1054
5000	0.081	4503	291
6000	0.078	3267	138

5 Conclusions

The MHRW algorithm guarantees that the walk in a graph are completely random, giving all the nodes the same probability of being visited and taken as a sample, so it is an important tool when performing statistical and predictive analyzes; it should be considered that the algorithm in its implementation requires a random number generator engine whose distribution directly influences the mean error value of the sampling. From the results obtained it can be concluded that it is important to test the random walks starting from different nodes, but not only randomly, is necessary choosing them analytically considering the characteristics of the graph, then, is possible deduce ways of looking for other nodes that are interesting for the analysis and to determine a number of rounds with greater precision. Finally, we conclude that the value of the mean error obtained in a sample will be smaller the larger the sample size obtained.

Random walks should be carried out by increasing the number of rounds progressively, in such a way that a sufficient number of rounds can be established to guarantee the steady state of the network, but that it does not generate unnecessary computational cost.

References

1. Brugere, I., Gallagher, B., Berger-Wolf, T.Y.: Network Structure Inference, A Survey: Motivations, Methods, and Applications. arXiv preprint arXiv:1610.00782 (2016)
2. Baqer, M., Al Mutawah, K.: Random node sampling for energy efficient data collection in wireless sensor networks. In: 2013 IEEE Eighth International Conference on Intelligent Sensors, Sensor Networks and Information Processing, pp. 467–472. IEEE (2013)
3. Blagus, N., Weiss, G., Šubelj, L.: Sampling node group structure of social and information networks. arXiv preprint arXiv:1405.3093 (2014)
4. Gadepally, V., Herr, T., Johnson, L., Milechin, L., Milosavljevic, M., Miller, B.A.: Sampling operations on big data. In: 2015 49th Asilomar Conference on Signals, Systems and Computers, pp. 1515–1519. IEEE (2015)
5. Sevilla, A., Mozo, A., Anta, A.F.: Node sampling using random centrifugal walks. J. Comput. Sci. **11**, 34–45 (2015)
6. Arcos Argudo, M.: Comparative study between Kleinberg algorithm and biased selection algorithm for construction of small world networks. Computación y Sistemas **21**(2), 325–336 (2017)
7. Chib, S., Greenberg, E.: Understanding the Metropolis-Hastings algorithm. Am. Stat. **49**(4), 327–335 (1995)
8. Lovász, L.: Random walks on graphs. Combinatorics, Paul erdos is eighty **2**, 1–46 (1993)
9. Aldous, D., Fill, J.: Reversible Markov Chains and Random Walks on Graphs (2002)
10. Woess, W.: Random Walks on Infinite Graphs and Groups, vol. 138. Cambridge University Press, Cambridge (2000)
11. Freund, J.E., Miller, I., Miller, M.: Estadística matemática con aplicaciones. Pearson Educación (2000)

The Research on Automatic Acquirement of the Domain Terms

Liangliang Liu[1], Haitao Wang[2], Jing Zhao[2], Fan Zhang[2], Chao Zhao[2], Gang Wu[2], and Xinyu Cao[2(✉)]

[1] Shanghai School of Statistics and Information,
Shanghai University of International Business and Economics,
No. 3349 Caobao Road, Minhang District, Shanghai, China
liangliang@suibe.edu.cn
[2] Beijing China National Institute of Standardization, No. 4 Zhichun Road,
Haidian District, Beijing, China
caoxy@cnis.gov.cn

Abstract. There are different features in domain terms on different domain. In this paper, we took TCM clinical symptom terms as example to discuss the acquirement of domain terms due to the particularity and complexity in clinical symptom terms. We analyze the feature of TCM clinical symptom terms, and define the formal representation of the word-formation. Then we use the term in the TCM Clinical Terminology as seed terms, and generate word-formation rule base. We recognize the new TCM clinical symptom terms in the medical records based on the word-formation rule base. Then we verify the recognized terms with statistical method to implement the automatic recognition of TCM clinical symptom terms, as the basis of data analysis and data application in the further.

Keywords: Automatic acquirement · Knowledge ontology · Domain terms · TCM clinical symptom terms

1 Introduction

Medical data contains rich knowledge. Reliable and creative data mining depends more on high-quality medical data with rich information. In the process of medical treatment, Traditional Chinese Medicine (referred to as TCM) clinical terms commonly used which is the main expressions for professional medical physicians' communication, professionals and patients' communication, professionals and computers communication, and different systems of computers information exchange, and the clinical data are generally represented by natural language. The symptom is the most important patient information. The doctor's description is accurate but free while writing clinical records. For example, the patient has "bitter taste" and "thirst", and the doctor's description will be "thirst and bitter taste" instead of two terms "bitter taste" and "thirst". In order to the need of data analysis and data mining, the expressions must be recognized completely. For example, cough for many days, in which time word is significant to doctor's diagnosis. The diversity and complexity of TCM clinical terms expression will increase the difficulty of acquirement.

© Springer Nature Switzerland AG 2020
T. Ahram (Ed.): AHFE 2019, AISC 965, pp. 123–129, 2020.
https://doi.org/10.1007/978-3-030-20454-9_12

Now most terminologies are constructed manually. The number of clinical terms is limited, and a large number of commonly used clinical terms appear in clinical records, hence we need to use automated acquirement techniques to derive clinical common terminology from clinical data.

2 Related Works

In the early stage, the rule-based method was used to automatically extract terms. The rules were established based on linguistic knowledge and the terms matched the rules would be recognized. Beatriee et al. [1] used the part of speech patterns to get candidate words who limited the term into Noun phrase in the field of science and technology. The rule-based method could get a higher quality term. However, ability of the method is limited and the types of recognized terms are not comprehensive because the manual rules limited by linguistic knowledge cannot cover all feature of term composition.

At present, terms are typically extracted by statistical methods on the basis of the corpus, Church, etc. [2] introduced the concept of mutual information to evaluate the combination ability of two words for the first time. Making use of the context information of the term, Frantzi [3] put forward the parameters of the NC-value to extract terms. Xing [4] discussed the distribution regularities and characteristics of term, and noted that the Chinese terms are mainly concentrated between 2–6 characters. Combining mutual information formulas with logarithmic formulas, Dubo [5] extracted terms from the general field. Chen etc. [6] proposed a method to obtain domain vocabulary based on the boot-strapping.

The research of term obtaining in the field of medical is mainly focused on the biomedical. Olsson [7] etc. proposed rule-based method to identify biomedical terminology, the precision has certainly improved compared to dictionary-based method, but it is difficult to extend and transplantation. In recent years, the way that using the methods of term obtaining in some fields of common sense mostly utilizes the obtaining methods and models of western terminology directly. Taking advantage of conditional random field, Wang [8], Zhao [9], Zhang [10], Meng [11] etc. extract Chinese medicine terminology from the Ming and Qing Dynasty Ancient Medicine Case, web pages, the Classified Medical Records of Famous Doctors, Treatise on Exogenous Febrile Disease respectively in order to get a better result.

In the present study, the automatic extraction of TCM clinical terminology has the following limitations: (1) Current automatic extraction methods are mostly based on statistical methods, but the methods require massive manual tagging training corpus. One of the foundation of the TCM corpus's construction is the TCM clinical terms; (2) The complexity and historicity increase the difficulty of extraction; (3) Currently TCM terminology recognition is based on the classical Chinese medicine literatures. However, many new terms appear in clinical records.

3 Key of the Technique of Automatic Acquirement of TCM Clinical Symptom Terms

Because the formation of the TCM clinical symptom terms fit a certain composition principle, in this paper, we analyze the characteristics of TCM clinical terms based on TCM theory, concept and relationship. Then the formation rules of TCM clinical symptom terms are studied by pattern learning. Finally, the clinical symptom terms can be automatically identified by pattern matching and statistical method.

3.1 Framework

In this paper, the acquisition of clinical symptom terms are as follows. First, analyze the characteristics of TCM clinical symptom terms and the rules of construction. Then, study the pattern expression and the word formation rules of TCM clinical symptom terms by pattern learning. Automatically recognize and validate clinical symptom terms by pattern matching and statistical method. Finally, realize the automatic expansion of TCM Clinical Terminology. The Fig. 1 shows the framework of method of acquiring TCM clinical symptom terms.

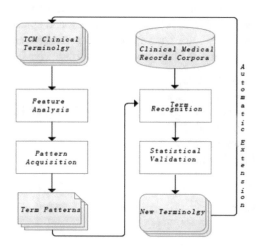

Fig. 1. Framework Diagram

3.2 The Rules Analysis of Term Composition Based on Pattern Learning

The clinical symptom term has some fixed expressions by the term analysis from TCM Clinical Terminology.

"cough" is a term in clinical expression "cough for many days". Therefore, we can define pattern "term +time class-word". At the same time, because "cough" is a verb, we can extract pattern "verb + time class-word". The expression or term "fever for three days" can be recognized by the two patterns from string "…fever for three days, fever is a kind of systemic symptom…".

"thirst" is also a term in clinical expressions "thirst and bitter taste". We can extract pattern "X and Y". "mouth" is a body, and "dry" and "bitter" are both adjective that describe "mouth". At the further, we can acquire pattern "[body][adjective][and][adjective]" and recognize term "thin white fur" by the pattern from clinical records "Stool and pee are normal. Tongue is pale red. Tongue fur is thin and white. Pulse is weak".

"flaming of deficiency-fire" and "dizzy" is also terms in clinical expressions o term "flaming of deficiency-fire accompanied with dizzy". We found that two terms were usually linked by word "accompanied with" to a new term. We can construct pattern [X][accompanied with]Y, with which we can extract new expression or term "cough accompanied with aversion to cold" from the clinical record sentence "…the patient coughed accompanied with aversion to cold and had no sweat. Throat itched three days ago".

TCM clinical terms are usually composed of body and symptom. We can build patter "body + symptom", with which two new terms "extremity joint pain" and "low limb cold" can be extracted from strings "…had chronic rheumatoid arthritis, extremity joint pain and low limb cold, and the joints don't bent and spread".

The patterns are made up some parts and every part can be word or collection of the same type of word. We defined some predicates as follows.

wi: a word or a part of a term
wi.in: wi belong to a word class
wi.pos: the part of speech of wi(reference to the Peking University POS Table)
wi.synonym: the synonym of wi
wi.antonym: the antonym of wi
wi.term: wi is clinical term

Symbol definitions:

[]: a term of a patterm
|: or
&: and

Some terms often show the same semantic meaning. Therefore, we define the concept of word class that is a collection of words in which every word expresses the same or similar meaning. The definition is as follows.

! the name of word class = word string 1 | word string 2|…
The examples are as follows.
! body class-word = mouth | eye | tongue | head | extremity joints | joint |…
! time class-word = year | month | day | hour |…
! conjunction class-word = both…and…| and |…
! accompany-class word = accompanying | accompany with

Table 1. The sample table of term pattern

Examples	Patterns
Cough for many days	[w1.pos = v][w2.pos = m][w3.in = ! time-class word]
	[w1.term][w2.pos = m][[w3.in = ! time-class word]]
Thirst and bitter taste	[w1.term][w2.in = ! conjunction-class word][w2.term]
	[w1.in = ! body-class word][w2.in = ! symptom-class word][w3.in = ! conjunction-class word][w4.pos = a|v]
	[w1.in = ! body-class word][w2.in = ! symptom-class word][w3.in = ! conjunction-class word][w4.term]
Flaming of deficiency-fire accompanied with dizzy	[w1.term][w2.in = ! accompany-class word][w2.term]
Dizzy	[w1.in = ! body-class word][w2.in = ! symptom-class word]
	[w1.in = ! body-class word][w2.pos = a]

The examples can be expressed by the patterns in Table 1 based on previous definition.

We take clinical terms as seeds in TCM Clinical Terminology and partition terms from an atom point. The patterns of term composition are study by pattern learning algorithm. Then the corpus of term composition patterns is built by manual check.

3.3   Terms Acquirement Based on Pattern Recognize and Statistics

The term features are analyzed at first from the TCM Clinical Terminology. Based on the features we can acquire some rules about TCM terms composition and the patterns of terms. The new TCM clinical symptom terms can be recognized from clinical records using patterns.

First, we segment TCM clinical records. But now there is no special Chinese word segment system focusing on TCM. In this paper, we build word segment diction by some exist diction like The Diction of Traditional Chinese Medicine, and segment TCM records by the reverse directional maximum match method (RMM). Because the new clinical symptom term will be segment to some individual words, we need to recognize and count the number of word strings matching patterns. There is a priori knowledge that the common terms will be occur many times. On the foundation of hypothesis, the string will be a new clinical symptom term if it matches some pattern and happen frequently. The algorithm details are as follows.

Example of Algorithm of Clinical common terms Automatic Identification

```
Input : TCM Clinic Terms Patters,Rules,Diction,TCM Records
Output : new TCM clinical symptom terms
Begin
    Read patterns and rules, and Build index;
    Build double array tree of TCM clinical symptom terms
    diction;
    Word segment in TCM clinical records;
    Read every sentence of TCM clinical records;
    If(Match(Wi...Wj,Rule1))
        term←Wi...Wj;
        count(term):=count(term)+1;
    If(count(term)>threshold)
        Output new term;
End
```

4   Results and Analysis

We take the terms as seeds from the TCM Clinical Terminology, analysis the rules of word composition and the patterns, and take 100 clinical records as train corpus.

We use patterns to recognize the new terms, such as "cough for three days", "cough for three days with aversion to cold", "sore throat", "Muscular stiffness", "Muscular stiffness and a little fever", "chest tightness", "red tongue", "thin white fur", "thin white fur and a little yellow dry fur". The method is useful to recognize new clinical symptom terms from traditional Chinese medicine clinical records.

5   Conclusion and Perspective

It is the key problem in Traditional Chinese Medicine that builds TCM clinical symptom term ontology and concepts' association relationship which is useful to eliminate uncertainty of TCM concept and relation. The automatic acquirement of TCM concepts and semantic relations are the key parts of the construction of TCM clinical symptom term ontology.

In this paper, firstly, we construct patterns and rules based on the characteristic of the TCM clinical symptom term and take the terms as the seed in TCM Clinical Terminology. Secondly, the formation of TCM clinical term would be studied based on seeds and the TCM clinical term patterns corpus will be built. At last, we can use the method of pattern recognition and statistical validation to automatically recognize new terms from clinical records. It is the next work that automatically acquiring terms and their relationships from large clinical records, to provide theoretical and fundamental support for TCM information construction.

Acknowledgments. This paper is supported by grants from National Key R&D Program of China (2018YFF0213901) and China National Institute of Standardization(522016Y-4681).

References

1. Beatrice, D., Eric, G., Jean, M.L.: Towards automatic extraction of monolingual and bilingual terminology. In: Proceedings of the 15th conference on Computational Linguistics, Japan, pp. 515–521 (1994)
2. Church, K., Hanks, K.: Word Association Norms, Mutual Information and Lexicography. In: Proceedings of the 27th Annual Meeting on Association for Computational Linguistics, Vancouver, British Columbia, Canada, pp. 76–83 (1989)
3. Frantzi, K., Ananiadou, S.: The C-value/NC-value domain independent method for multi-word term extraction. J. Nat. Lang. Process. **6**(3), 145–179 (1999)
4. Hongbing, X.: Structural features and distribution of Chinese-English terms in the corpus from information field. Inf. Technol. Appl. **1**, 22–25 (2000)
5. Du, B., Tian, H., Wang, L., et al.: Design of domain-specific term extractor based on multi-strategy. Comput. Eng. **31**(14), 159–160 (2005)
6. Chen, W., Zhu, J.: Automatic learning field words by bootstrapping. In: The Proceedings of the Seventh National Joint Conference on Computational Linguistics, pp. 67–72. Tsinghua University Press, Beijing (2003)
7. Olsson, F., Eriksson, G., Franzen, K., et al.: Notions of correctness when evaluating protein name taggers. In: Proceedings of the 19th International Conference on Computational Linguistics, pp. 765–771 (2002)
8. Wang, S., Li, S., Chen, T.: Recognition of Chinese medicine named entity based on condition random field. J. Xiamen Univ. (Nat. Sci.) **48**(3), 359–364 (2009)
9. Zhao, X.: On the Research of TCM Knowledge Discovery System Based on Web Ming. Beijing Jiaotong University, Beijing (2010)
10. Zhang, W., Bai, Y., Wang, P., et al.: An automatic domain terms extractor method on traditional Chinese medicine books. J. Shenyang Aerosp. Univ. **28**(1), 72–75 (2011)
11. Hongyu, M.: Automatic identification of TCM terminology in Shanghan Lun based on condition random field. J. Beijing Univ. Tradit. Chin. Med. (2014)

Behavioral Analysis of Human-Machine Interaction in the Context of Demand Planning Decisions

Tim Lauer[1,2(✉)], Rebecca Welsch[1], S. Ramlah Abbas[1],
and Michael Henke[2]

[1] Infineon Technologies, Am Campeon 1-15, 85579 Neubiberg, Germany
tim.lauer@infineon.com
[2] Technical University Dortmund, Emil-Figge-Straße 50, 44227 Dortmund,
Germany

Abstract. The trend of digitalization has led to disruptive changes in production and supply chain planning, where autonomous machines and artificial intelligence gain competitive advantages. Besides, the satisfaction of customers' wishes has reached top priority for demand-driven companies. Consequently, companies implement digital applications, for instance neural networks for accurate demand forecasting and optimized decision-making tools, to cope with nervous operational planning activities. Since planning tasks require human-machine interaction to increase performance and efficiency of planning decisions, this analysis focuses on forms of interaction to determine the right level of collaboration. The paper outlines various levels of interaction and analyses the impact of human reactions in the context of an industrial demand planning algorithm use case at Infineon Technologies AG conducting a behavioral experiment. The results show that a variance in the levels of human-machine interaction has influence on human acceptance of algorithms, but further experiments need to be conducted to outline an overall framework.

Keywords: Behavioral analysis · Human-machine collaboration ·
Demand planning · Digitalization · Supply chain planning

1 Introduction

Globalization and dynamic growth have resulted in fast paced and volatile production and supply chain planning cycles, making it increasingly challenging as well as essential to create stable and accurate demand forecasts. A precise demand forecast is crucial for any business to be capable to optimize the flow of materials, information, products, human resources, and finance, throughout a supply chain from the supplier's supplier to the customer's customer [1]. Thus, a proper prediction accuracy enables minimizing common problems, such as over- and/or underproduction, excessive inventory costs, wastage and loss of business [2]. Typically, literature categorizes forecasting methods into two types: subjective and model-based. While judgmental forecasting comprises of guesses, experiences and intuitions of human planners,

© Springer Nature Switzerland AG 2020
T. Ahram (Ed.): AHFE 2019, AISC 965, pp. 130–141, 2020.
https://doi.org/10.1007/978-3-030-20454-9_13

statistical forecasting signalizes mathematical rules, relationships between variables and projection of historical data [3].

Enabled by the emerging trend of digitalization, the latest technology developments in machine learning and artificial intelligence enable the opportunity of combining statistical and judgmental models to improve forecast accuracy overall in more efficient processes. However, while such optimized decision-making tools can help provide accurate and time-consistent demand forecasts, they rely on human-machine interaction at a level not seen before. The interaction of digital solutions with humans has already been thoroughly examined for the shop floor level, e.g. assembly robots supporting operators at automotive production lines. Nevertheless, planning processes have not been in focus of research so far.

Considering this trend, digital solutions are important, time and investment consuming to keep pace in a competitive environment. Therefore, human reactions to digital interactions are an essential basic knowledge, which needs to be researched on to improve performance and optimize processes.

This paper aims to provide a first insight into human-machine interaction on a planning level on the example of demand planning. Consequently, the derived research question is stated to: "How can humans and machines interact in the best possible way to improve demand planning decisions reaching high forecast accuracy?" Answering this question, we outline levels of human-machine collaboration and examine the reaction of human planners at each level, analyzing differences in the demand plan making usage of behavior research.

In a first step, Sect. 2 presents the literature review of necessary background knowledge and the specific topic streams. Section 3 describes the developed conceptual model of the demand-planning scenario and the derived behavior experimental setup. Section 4 shows and discusses the results of the experiment. Section 5 concludes the paper with a summary and outlines of the next steps.

2   Literature Review

The literature review comprises of the three main streams: combinational demand forecasting, human-machine interaction and advice taking. The first section provides insights into concepts of demand forecasting as a typical supply chain planning action. Hereby, the focus is on methods already combining judgmental and statistical forecasts. Next, starting with the well-known levels of automation the section gives an overview of current research and developments of human-machine interaction. The last literature stream deals with the field of advice taking in decision-making.

2.1   Demand Forecasting

Overall, structuring the literature on demand forecasting by aiming on research combining judgmental and statistical methods, five collaboration levels identifies themselves [4].

An option to integrate both method areas is to revise judgmental forecast based on statistical exploration. Hereby, the initial judgment made by experts is unaided by any statistical method, but opinion and experience based. Revised judgmental forecast is mostly more accurate than unrevised judgmental forecast [5].

Another way to link both method streams is to combine two separate forecast approaches after conduction: Out of domain knowledge, experts provide a judgmental forecast undergoing a merger with an extrapolation forecast afterwards. Three advantages come out of this procedure. First, the results are more objective in comparison to unaided judgments. Second, the process is systematically structured and, hence, repetitions with similar results can take place. Third, the approach utilizes the domain knowledge in a more efficient way [6]. While revised judgmental forecast manage without domain knowledge for the unaided judge, revised extrapolation forecasts insert domain knowledge. Consequently, revised extrapolation forecasts are the most common way to integrate statistical and judgmental forecasts. This method improves forecast accuracy in the case, that the expert adjusting the forecast can identify the missing pattern from the statistical model [4].

Next, rule-based forecasting approaches use structured judgmental inputs for statistical procedures [7]. As every situation is unique, this method tries to adjust each single forecast with their own peculiarities. As a requirement, the domain knowledge must be the cornerstone to achieve an accurate forecast [8].

Another possibility is the so-called econometric forecasting. Within such methods, expert judgments incorporate into a structured approach. The judgmental entries serve as decision criteria to select appropriate mathematical forecasting models. Often the choice is on regression models due to being able to consider multi-variate parameter dependencies. This method promises to be the most accurate one, when the relevant domain knowledge is underlying and properly considered [9].

Figure 1 summarizes the different collaboration levels in a graphical process-oriented manner.

Experts are divided over which forecast method is superior [4, 10–12]. When the first statistical methods showed up, judgmental forecasting was criticized for being too unstable and dependent on humans. Years later, experts found out that the human component of forecasting is an essential input [13]. Especially, practitioners in all industries trusted in judgmental forecasting. While statistical models are good in finding patterns and processing a large amount of historical data and parameters, judgement can deal with exceptions [14]. However, unstructured and biased human decisions can also be harmful to the accuracy of planning [7]. Thus, a combination of both method streams can provide good solution in today's business bringing the advantages of judgmental and statistical forecasting together. Various researchers verify that integrated statistical and judgmental methods improve forecast accuracy [15].

2.2   Human-Machine Interaction

Research on human-machine interaction is a fundamental aspect to investigate different collaboration levels elaborating varying roles for humans and machines. The first researchers worked on this topic already in 1978, well known as levels of automation, providing different granularities, at which the human can work alongside a machine, ranging from complete human autonomy to complete machine autonomy [16]. By

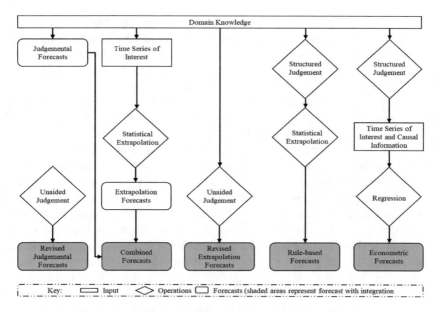

Fig. 1. Integration of judgmental and statistical methods [7].

proceeding time, the discourse included levels where the machine replaces the human at action and decision levels [17].

Today, the trend of digitalization enabling social and cognitive computing, the discourse shifts from levels of automation to collaborative approaches. In a more digitalized world, the role of the human changes from that of a controller of machines and processes, to that of a supervisor [18]. Due to high levels of volatility and possibilities of escalations in manufacturing and planning processes, machines are still unable to replace human intuition completely, although ongoing research is increasingly focusing on developing machines with human cognitive intelligence [19]. However, in line with the outcome separation in demand forecasting, human-machine interaction is manifold as well. Consequently, the same idea emerged to develop systems that are not replacements but instead are team players, working in collaboration with the humans to improve processes and optimize decisions together [20]. This idea refers as teaming approach and incorporates smart tools and humans in a joint group to derive optimal decisions in an efficient way.

2.3 Decision Making

Combining judgmental and statistical methods in demand planning and forecasting is highly dependent on human trust in machines.

Research categorize advice taking - independent from other humans or machines - into three different types: recommendation for, recommendation against and provision of information [21].

Overall, significantly most humans take advice because they feel uncomfortable to ignore the advice from others completely. Here, the belief that advisors have more experience and expertise biases the decision maker, although behavior experiments show that this takes place even if advisors do not consider themselves as more experienced or expert. Lastly, most subjects want to share the responsibility of their actions and decisions. Hereby, people hope that the occurrence of errors reduces when multiple opinions are included in the decisions making process [22].

Although humans take advice, research has shown that human beings underweight advice coming from a machine [23–25]. People discount the advice from an algorithm significantly more than that from a human [26]. This phenomenon was termed "algorithm aversion" [27]. It refers to the preference of people choosing a forecaster over a statistical forecasting model [27, 28]. In addition, people trust machine advice more if they see the machine err [29]. A reason for that might be the expert label given to the computer during the experiment of the conducted behavior experiments [30]. Nevertheless, humans tend to trust algorithm advice more when the alternative human advice is coming from a novice. Similarly, humans trust more in the human experts than in the machine [31]. Thus, the quality of advice can matter. Although, research indicates that people do not tend to believe experts more than laymen [32]. This may be because human beings know that even highly qualified experts can err [32]. Consequently, if humans have the chance to modify the algorithm after the algorithm erred, they are more likely to use the model forecast [33]. Thus, the ability to modify the algorithm builds trust between humans and machines.

Research has also shown that specific kinds of feedback can improve forecast accuracy [34]. Explanation of the used statistical method tends to improve the judgmental forecast in general, but only in specific series and periods [35]. Thus, it can be summarized, that information as feedback or as explanation can improve forecast accuracy with the restriction that every case has peculiarities.

3 Methodology

The following chapter describes in a first step the use case background and the respective demand planning process and decision. Second, procedures, details and formal aspects of the conducted behavior experiment are explained.

3.1 Demand Planning at the Case Company

The aim of this study is to examine the impact of human-machine interaction levels in the context of decision making for the use case of demand planning. Therefore, the demand planning landscape, tools and processes of the semiconductor company Infineon Technologies AG provide the use case setting. The semiconductor industry reflects a suitable for the experiment due to its characteristics of a highly volatile demand, to the bullwhip effect exposed at the upstream segment of the supply chain and to the tremendous pace of technology developments accompany by lead-times up to nine month. The demand planning process covers following the international SCOR-model the short-term operational planning and mid-term tactical planning. In this use case, the focus is on the operational level in form of a product specific detailed plan for the next 26 weeks [36].

Throughout the demand planning process, the human planner called "Supply Chain Planner" (SCP) decides on rules for operational demand planning and stock targets. Most of these processes execute afterwards automatically [36]. An advanced master planning tool matches the inserted demand with the available capacity and hereby calculates a production plan for the next 26 weeks.

However, demand fluctuations, cycle time differences, quality issues or strategic reasons can make it necessary for the SCP to step-in and make adjustments to ensure right quantity and on-time production [37]. The SCP has to make his decision covering five different depend dimensions. The SCP can decide on the priority of the product demand, the specific forecast period, the product granularity, the stocking points along the internal supply chain and the quantity of product decently.

3.2   Behavior Experiment

For any behavior experiment, a huge number of subjects is beneficial. Since the time and number of planers is limited, the described demand planning decision was converted in a daily life decision. Hereby, not only planers, but also non-professionals could take place in the experiment since no domain knowledge is needed. This also provides the basis for the external validity of the results. In multi-level workshops with field experts, the described dimensions of demand planning converted into a wine selection. Table 1 shows the overview of the dimensions in the proxy case.

Table 1. Wine selection decision.

Dimension	Options
Priority	Personnel estimation: red or white wine
Granularity	Barrel – six pack – bottle – glass
Period	Consumption of wine; mature of wine
Amount	How much should you buy?
Stocking point	fridge, kitchen, wine cellar

Next, we selected revised extrapolation forecast for this study based on the overview of options of combinations of judgmental and statistical forecasting methods. Former research shows that revised extrapolation forecast can improve forecast accuracy by the capability of the planners to identify missing patterns of the statistical forecast procedure [4]. Every participants needs to do 24 decisions one each respective planning cycle.

This paper examines three human-machine interaction scenarios to determine the best possible collaboration combination. Hereby, no interaction between human and machine serves as the baseline, the control group. The scenario represents the provision of a forecast value without any information about the method. On the first interaction derived from the literature finding, the planner can select between different opportunities given by the machine. Hereby, each option refers to his personal attitude of risk aversion, e.g. if a planner wants the algorithm to overestimate the demand if he wants to minimize his inventory costs by underestimation but risking potential shortages. The last interaction level enables the possibility to question the machine about the assumptions it made for its calculations. This was designed in the form of a chat interface as common in the case company to provide a feeling of a colleague interaction. This presented the highest form of interaction.

During the experiments the participants only get a storyboard with the forecasting situations, e.g. having a house party with 43 guests. Treatment group 1 (T1) can interact with the algorithm by a chat box and receive an explanation regarding the calculation style. Treatment group 2 (T2) can select between lean, normal and safety planning.

After the experiment, the subjects had to answer survey questions to gather demographic information such as age, gender and education level. Participants also provided information in an open field, how they felt about the algorithm (control group), chatting with the algorithm (T1) and calculation style (T2).

The conducted behavior study on the wine selection case derived from the real system of demand planning examines the subjects' behavior on the three different human-machine interaction scenarios presented. Therefore, the following hypotheses are tested:

$$H_0 = \text{Difference in interaction levels does not influence}$$
$$\text{human acceptance of the algorithm.}$$

$$H_1 = \text{Difference in interaction levels does influence}$$
$$\text{human acceptance of the algorithm.}$$

For the hypotheses test 65 participants were randomly but evenly assigned to the three test cases. The most participants were from the innovations department of Infineon. The validity of using non-professionals subjects in forecasting tasks with context-independent information is supported by literature [38]. As discussed in the literature, experts do not perform significantly better than non-experts when the tasks do not involve domain knowledge. The environment was a laboratory study as T1, with the possibility to chat with the algorithm, needed to be observed. Therefore, the experimenter himself answered the questions, unknown to the participants.

oTree, in connection with the programming language Python, was used to create a user-friendly experiment interface. The interpreter PyCharm enabled the development of the code and heroku as the webserver for publishing the experiment to the participants.

4   Results

Enabling a better understanding of the experiment data and results, the structured analysis starts with demographics followed by a participant's evaluation and ending with a significance calculation.

Among the 65 participants over all, 49% were female and 51% were male. Also examining the groups one by one, the gender is balanced except for treatment group 2, which had a bit more male participants than female (12). In control group and treatment group 1, 22 participants per group had joined the game, while in treatment group 2, 21 attendees answered questions. The mean age of all three groups were 29 rounded up to years. However, 56 of 65 participants refer to the generation Y, which are characterized by a multitasking environment with support of technologies as they played around with it from a very young age. Therefore, an easy adaption to new innovative technologies coming up in future can be assumed.

Calculating, if there is a statistically significant difference between the interaction levels, the Kruskal-Wallis (KW) takes place. The test is applied on the usage of the model forecast and on how often the proposed value of the model forecast has been changed.

The first calculation of the KW test analyzes whether the participant wants to take the model forecast or his own forecast. The highest amount of acceptance is to choose the model forecast three times, the lowest stays with the own decision three times. The Kruskal-Wallis test bases on ranks, as shown in Table 2.

Table 2. Ranks in the Kruskal-Wallis test.

Rank	Description
0	The participant never used the model forecast
1	The participant used the model forecast one time
2	The participant used the model forecast two times
3	The participant used the model forecast all three times

After the rank definition is set, it is applied to the collected dataset and every participant i gets a rank sum R_i, where n represents the observations and t_j the number of used observations of rank j. The KW test is then conducted as follows:

$$H = \frac{12}{n(n+1)} \sum_{i=1}^{c} \frac{R_i^2}{\backslash ni} - 3(n+1). \tag{1}$$

$$H_{korr} = \frac{H}{1 - \frac{\sum_{j=1}^{m} \left(t_j^3 - t_j\right)}{N^3 - N}}. \tag{2}$$

For the first test, H_{korr} calculates to 5.866. This H_{korr}-value is compared to the critical value, which depends on the significance level (α) and degrees of freedom. With three groups the degree of freedom is 2 and the α is usually set to $\alpha_1 = 0.05$ or $\alpha_2 = 0.1$. α describes the bias that the H_0 hypothesis is rejected even if it is true. For the experiment the critical values can be found as in the chi-square distribution:

$$1 - \alpha_1 == 0.95 \rightarrow z_1 = 5.99$$

$$1 - \alpha_2 == 0.90 \rightarrow z_2 = 4.61$$

If the value of test is higher than the critical value the result is interpreted as significant. The H_{korr}-value of 5.866 is between the two critical values, which means it depends on the α, if the difference between the groups could be significant. Considering the small sample size, it can be interpreted as significant for a level of α_2, but a refined experimented set-up size is recommended for future research.

The second analysis tests how often the participants change the value from the model to their own estimation. There are 24 possible opportunities for change. Every participant is analyzed on the number of changes made and put in a ranking for the Kruskal-Wallis test. The KW test shows the results of $H = 4.236$ and $H_{korr} = 4.366$.

The results support the H_0 and do not show a significant difference between the three groups. The H_{korr}-value of 4.366 is not higher neither of the critical values. Hence, H_1 is rejected.

A further approach to determine, if there is a significant difference, is the comparison of the control group with n_1 of T1, and then with n_2 of T2. The Mann-Whitney-U-test is conducted for this, which is also based on the ranking R of values. The z-standard distribution provides the comparable critical value. Thus, both the U-value and z-values are calculated as follows:

$$U = n_1 n_2 + \frac{n_1(n_1 + 1)}{2} - R_1. \tag{3}$$

$$|z| = \frac{U - \frac{n_1 n_2}{2}}{\sqrt{\frac{n_1 n_2 (n_1 + n_2 + 1)}{12}}}. \tag{4}$$

For the experiment, U equals to 228.5 |z| to 0.31688. Both calculations are compared to the z standard distribution values, which are dependent on α. The critical values are as follows:

$$\alpha_1 = 0.05 \quad \rightarrow z = 1.96$$
$$\alpha_2 = 0.1 \quad \rightarrow z = 1.645$$

The comparison of control group and T1 does not yield a significant difference $(0.31688 < 1.645 < 1.96)$. However, the relation between control group and T2 is interesting. Since the value lies between 1.645 and 1.96, the choice of α is again a major aspect $(1.645 < 1.77364 < 1.96)$. Although, α should not be enhanced without valid justification, the result shows that the behavior regarding the interaction in T2 differs from the control group.

A look at the distribution of the values also confirms the results of the two tests. Figure 2 depicts the distribution for the KW test. While the control group does not significantly differ from the T1 (a), T2 (b) is significantly different from the other two groups. The participants of T2 stayed with the model's proposals twice as much as those of the other two groups. Less than 50% of the participants from T1 use the opportunity to chat with the algorithm. This shows that people felt uncomfortable talking to a machine. The control group also demonstrates a mistrust in the model as most people of the group changed the model's proposal at least once.

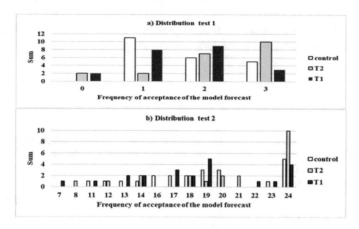

Fig. 2. Distribution of values for the Kruskal-Wallis test.

A final look at the answers to the question, how the participants 'felt' about the different aspects of the experiment, confirms that the participants are not comfortable talking to the algorithm, although they endorse the opportunity to interact regarding the calculation style.

5 Conclusion

The study provides a first insight in the topic of human-machine interaction on a planning level by conducting a behavior study. The approach exhibits how a company can deal with the challenge of digitalization in supply chain planning processes. The examined literature provides detailed information on the combinations of judgmental and statistical methods, and shows how automation and human tendency for advice taking can lead to collaboration possibilities for improved decision-making regarding performance and efficiency.

The results of the behavioral experiment are not completely distinct. The statistical computation indicates a significant difference between the treatment group 2 and the control group. Due to drawbacks of the Kruskal-Wallis test, the Mann-Whitney-U-Test confirm the results. Both tests show similar outcomes: the statistical results are ambiguous regarding the influence of different interaction levels on human acceptance of the algorithm. Hence, the H0 cannot be definitely rejected and no conclusive statement can be made regarding the influence of differences in interaction levels on the human acceptance of the algorithm. Therefore, the research question regarding the best possible way for humans and machine to interact to improve demand planning decisions for high accuracy, cannot be answered satisfactorily at this point of the research.

The reason for the inconclusive results could be the oversimplification of the demand planning decision, an unrepresentative group of participants, but mainly the limited sample size. Moreover, the evaluation of participant's endorsements and statistical calculations produce contradictory findings and indicates towards a bias in the behavior. However, the research is a first step forward in understanding human-machine interaction. It shows that interaction does improve acceptance of algorithm by humans.

References

1. Chapman, S., Ettkin, L., Helms, M.: Supply chain forecasting – collaborative forecasting supports supply chain management. Bus. Process Manage. J. **6**, 392–407 (2000)
2. Goodwin, P., Önkal, D., Thomson, M.: Do forecasts expressed as prediction intervals improve production planning decisions? Eur. J. Oper. Res. **205**, 195–201 (2010)
3. Holden, K., Peel, D., Thompson, J.: Economic Forecasting. Cambridge University Press, Cambridge (1991)
4. O'Connor, M., Webby, R.: Judgmental and statistical time series forecasting - a review of the literature. Int. J. Forecast. **12**, 91–118 (1996)
5. Anderson, A., Carbone, R., Fildes, R., Hibon, M., Lewandowski, R., Makridakis, S., Newton, J., Parzen, E., Winkler, R.: The accuracy of extrapolation (time series) methods - results of a forecasting competition. J. Forecast. **1**, 111–153 (1982)
6. Clemen, R.: Combining forecasts: a review and annotated bibliography. Int. J. Forecast. **5**, 559–583 (1989)
7. Armstrong, J.S., Collopy, F.: Integration of statistical methods and judgment for time series forecasting - principles from empirical research. In: Wright, G., Good-win, P. (eds.) Forecasting with Judgment, pp. 269–293. Wiley, New York (1998)
8. Armstrong, S., Collopy, F.: Rule-based forecasting: development and validation of an expert systems approach to combining time series extrapolations. Manage. Sci. **38**, 1394–1414 (1992)
9. Allen, G., Fildes, R.: Econometric forecasting. In: Armstrong, J.S. (ed.) Principles of Forecasting - A Handbook for Researchers and Practitioners. Springer, Norwell (2001)
10. Edmundson, R., Lawrence, M., O'Connor, M.: The accuracy of combining judgmental and statistical forecasts. Manage. Sci. **32**, 1521–1532 (1986)
11. Bunn, D., Wright, G.: Interaction of judgmental and statistical forecasting methods - issues & analysis. Manage. Sci. **37**, 501–518 (1991)
12. Leitner, J., Leopold-Wildburger, U.: Experiments on forecasting behavior with several sources of information – a review of the literature. Eur. J. Oper. Res. **213**, 459–469 (2011)
13. Goodwin, P., Lawrence, M., O'Connor, M., Önkal, D.: Judgmental forecasting: a review of progress over the last 25 years. Int. J. Forecast. **22**, 493–518 (2006)
14. Fildes, R., Goodwin, P., Lawrence, M.: The design features of forecasting support systems and their effectiveness. Decis. Support Syst. **42**, 351–361 (2006)
15. Edmundson, R., Lawrence, M., O'Connor, M.: The accuracy of combining judgemental and statistical forecasts. Manage. Sci. **32**, 1521–1532 (1986)
16. Sheridan, T., Verplank, W.: Human and Computer Control of Undersea Teleoperators. MIT Man-Machine Systems Laboratory, Cambridge (1978)

17. Parasuraman, R., Sheridan, T., Wickens, C.: A model for types and levels of human interaction with automation. IEEE Trans. Syst. Man Cybern. - Part A: Syst. Hum. **30**, 286–297 (2000)
18. Johannsen, G.: Human-Machine Interaction. Control Systems, Robotics, and Automation. Encyclopedia of Life Support Systems (EOLSS). EOLSS Publishers, Oxford (2007)
19. Zheng, N., Liu, Z., Ren, P., Ma, Y., Chen, S., Yu, S., Xue, J., Chen, Ba., Wang, F.: Hybrid-augmented intelligence: collaboration and cognition. Front. Inf. Tech. Electron. Eng. **18**, 153–179 (2017)
20. Johnson, M., Bradshaw, J.M., Feltovich, P.J.: Tomorrow's human-machine design tools: from levels of automation to interdependencies. J. Cogn. Eng. Decis. Making **12**, 77–82 (2017)
21. Bonaccio, S., Dalal, R.: What types of advice do decision-makers prefer? Organ. Behav. Hum. Decis. Process. **112**, 11–23 (2010)
22. Fischer, I., Harvey, N.: Taking advice - accepting help, improving judgment, and sharing responsibility. Organ. Behav. Hum. Decis. Process. **70**, 117–133 (1997)
23. Kahneman, D., Tversky, A.: Judgment under uncertainty - heuristics and biases. Science **185**, 1124–1131 (1974)
24. Kleinberger, E., Yaniv, H.: Advice taking in decision making - egocentric discounting and reputation formation. Organ. Behav. Hum. Decis. Process. **83**, 260–281 (2000)
25. Fischer, I., Harvey, N.: Taking advice - accepting help, improving judgment, and sharing responsibility. Organ. Behav. Hum. Decis. Process. **70**, 117–133 (1997)
26. Önkal, D., Goodwin, P., Thomson, M., Gönül, S., Pollock, A.: The relative influence of advice from human experts and statistical methods on forecast adjustments. J. Behav. Decis. Making **22**(4), 390–409 (2009)
27. Dietvorst, B., Massey, C., Simmons, J.: Algorithm aversion - people erroneously avoid algorithms after seeing them err. J. Exp. Psychol. Gen. **144**, 114–126 (2015)
28. Bonaccio, S., Dalal, R.: Advice taking and decision-making - an integrative literature review, and implications for the organizational sciences. Organ. Behav. Hum. Decis. Process. **101**, 127–151 (2006)
29. Dijkstra, J.: User agreement with incorrect expert system advice. Behav. Inform. Technol. **18**, 399–411 (1999)
30. Logg, J.: Theory of machine - when do people reply on algorithms? Hav. Bus. Sch. **17-086**, 1–92 (2017)
31. Madhavan, P., Wiegmann, D.: Effects of information source, pedigree, and reliability on operator interaction with decision support systems. Hum. Factors **49**, 773–785 (2007)
32. Kleinberger, E., Yaniv, H.: Advice taking in decision making - egocentric discounting and reputation formation. Organ. Behav. Hum. Decis. Process. **83**, 260–281 (2000)
33. Dietvorst, B., Massey, C., Simmons, J.: Overcoming algorithm aversion - people will use imperfect algorithms if they can (even slightly) modify them. Manage. Sci. **64**, 1–17 (2016)
34. Griggs, K., O'Conner, M., Remus, W.: Does feedback improve the accuracy of recurrent judgmental forecasts? Organ. Behav. Hum. Decis. Process. **66**, 22–30 (1996)
35. Fildes, R., Goodwin, P.: Judgmental forecasts of time series affected by special events - does providing a statistical forecast improve accuracy? J. Behav. Decis. Making **12**, 37–53 (1983)
36. Schiller, C., Yachi, G.: Introduction to Demand Planning. Supply Chain Academy, Infineon Technologies AG, Munich (2017)
37. Schiller, C., Yachi, G.: Production Program, Demand Planning, Target Stock Entry in SPLUI. Supply Chain Academy, Infineon Technologies AG, Munich (2014)
38. Andersen, A., Carbone, R., Corriveau, Y., Corson, P.: Comparing for different time series methods the value of technical expertise individualized analysis, and judgmental adjustment. Manage. Sci. **29**, 559–566 (1983)

Deep-Learned Artificial Intelligence and System-Informational Culture Ergonomics

Nicolay Vasilyev[1(✉)], Vladimir Gromyko[2], and Stanislav Anosov[3]

[1] Fundamental Sciences, Bauman Moscow State Technical University,
2-d Bauman str., 5, b. 1, 105005 Moscow, Russia
nik8519@yandex.ru
[2] Computational Mathematics and Cybernetics,
Lomonosov Moscow State University,
Leninskye Gory, 1-52, 119991 Moscow, Russia
gromyko.vladimir@gmail.com
[3] Public Company Vozrozhdenie Bank,
Luchnikov per., 7/4, b. 1, 101000 Moscow, Russia
sanosov@cs.msu.su

Abstract. System-informational culture (SIC) phenomenology impels human to work in sophisticated scientific space of computer models. Applying computer instrumental systems one has to investigate and compare different fields of knowledge suffering constant cognitive, educational, and intellectual problems. Inter-discipline activity in SIC leans on meanings understanding presented in the utmost mathematical abstractions (UMA). Work in SIC era unites cognition, education, and scientific research. SIC *entelechies* are to evolve rational part of consciousness. The objective is achievable by means of purposeful labor assisted by deep-learned artificial intelligence (DL I_A). Technology is contributed allowing consciousness double helix auto-moulding in order to solve universalities problem. DL I_A is to unwind intellectual processes and develop person's scope of life. System axiomatic method is applied to coordinatization method and continuity property investigation.

Keywords: System-informational culture ·
Deep-learned artificial intelligence · Universal tutoring ·
System axiomatic method · Coordinatization · Consciousness double helix ·
Scope of life · Language of categories · Cogno-ontological knowledge base

1 Introduction

Human activity rationalization in SIC gives ground to innovations. Robotics is applied to patterns recognition in dynamical situations and management in undetermined conditions [1–3]. Semantic technology and fuzzy logic help to solve the problems. Vast and promising horizons lie before I_A. They are I_A-based tutoring and I_A-assistant cognition. They prepare man to natural life in sophisticated culture [4, 5]. Evolution created universal intellectual structures and language processing for natural intelligence (I_N) to lean on [6]. Our approach to rational consciousness auto-moulding considers

© Springer Nature Switzerland AG 2020
T. Ahram (Ed.): AHFE 2019, AISC 965, pp. 142–153, 2020.
https://doi.org/10.1007/978-3-030-20454-9_14

neurophenomenology presentations and linguistic ontology of man. It leans on UMA in order to cope with universalities problem solution with the help of DL I_A. Requirement of SIC is to teach man to cultivate and understand meanings [7].

1.1 System-Informational Culture

On the agenda of SIC is multidisciplinary human activity in computer networks world uniting cognition and research. Change starts in human mind due to its adaptive ability to deal with an ever-changing anthropogenic environment. Mathematics substantiates knowledge transforming senses into meanings. For the reason, rational knowledge becomes the most determinate. Man exists in meanings. Ontological essence shows itself in meta-mathematics and its language. UMA express meanings in concise form fit for rational investigation. Busy by the study, man auto-moulds inevitably his consciousness. Complex intellectual processes especially affect mind occupied by reflection over UMA. In sophisticated culture, life-long deep-learned tutoring partner must support the labor. It can be only super computer DL I_A. Traditional teaching cannot help person to accomplish intellectual breaks and profit scientific meanings.

The objective of the study is to contribute technology of universal tutoring (T_U). Everybody is able to philogenesis apperception and can be trained for natural existence in SIC. Ergonomics must lean on post-neo-classical science, practice of trans-disciplinary activity, informatics tools. Mind is to be fostered following thesis "ontogenesis repeats anthropogenesis". In order to correspond to noosphere sophistication, man is to be busy in multi-disciplinary electronic libraries where DL I_A co-processes big scientific data semantically.

Person's auto-evolving occurs on the grounds of knowledge *without premises* in the form of system axiomatic method (AM_S) and *de docta ignorantia* principles, synergetic paradigm, and semantic *glottogonia*. Rational thinking uses rough minutes of mathematics. T_U leads to rational man's auto-building. Engineering of DL I_A consists of cogno-ontological knowledge base $CogOnt_{K_B}$ implementation [7, 8]. Knowledge of philogenetic significance is compressed in it. UMA are its universal core expressed in language of categories (L_C). The base creates educational environment for person's rational auto-development. DL I_A is to help human to transform ideal mathematical constructions in real (obvious) presentations [4].

1.2 Scope of Life as Ergonomics Entelechies in SIC

Artificial intelligence replaces man in different fields of activity [1–3]. Now it is time to solve indirect I_A problem that is to assist man in intellectual activity itself [4, 5]. Universalities problem resolution enables man with scope of life (SL) as person's ability to unwind internal intellectual processes perceiving and understanding meaningful informational flows. Success of trans-disciplinary activity is based on SL availability. SL secures trained existence in different intellectual worlds, see Fig. 1, and can be attained only on the basis of universal essences identification, study and their understanding. Traditional teaching develops static world scope that differs from SL dynamics. Universal laws usage by SL cannot be substituted by mere calculations, see Fig. 1. It supposes thinking in meanings and *mathesis universalis* mastering [8].

Role of the latter plays L_C. Noosphere hermeneutics circle can be significantly enlarged if new generation of intellectual tools will be applied to cognition [8–10].

Fig. 1. Scope of world is versus scope of life in pattern recognition. Equality of rectangular areas: (a) neuron networks in SIC versus (b) gnomon in 3 c. BC.

In future, it is impossible to do with DL I_A as life-long tutoring partner and collaborator in resolving intellectual problems [4, 5]. DL I_A must develop person's SL because men themselves, computer technologies, models from different knowledge fields, scientific methods, and tools become very sophisticated. Only system holistic view and work on the level of meanings will help to cope with intellectual difficulties of multidisciplinary activity. Mutual work in networks and communications are also united by complex meanings.

Origins of mathematics teach how to investigate relations among things. T_U allows attaining knowledge auto-obviousness by means of real - ideal disco-ordination mastering in phenomena apperception. Reflection level is strengthened to a great extent with the help of mathematics and L_C usage. Many examples show how universal properties are discovered if systems are described in L_C. The language expresses meanings in constructive, concise, and visual form of commutative diagrams. Besides, $CogOnt_{K_B}$ is implemented in it as strata of interconnected categories [5, 7]. Object oriented programming practice supports it. Interpretation scheme of tasks execution imitates I_N thinking. Pursued objective is achieved by T_U technology because transdisciplinary SIC meanings have already been modeled in mathematics [5].

1.3   Cognitive Ontological Base and Problem of Knowledge Understanding

Meanings apperception is instinct of natural intelligence (I_N) developing man's intellectual abilities. It is entelechies of human existence. Direct problem of education in SIC is to auto-mould rational part of consciousness in trend of sophistication [5]. DL I_A is able to maintain the process being aware of it. In order to cognate complex knowledge, T_U is needed. T_U must be continuous and adaptive to every person.

Thinking is instinct to synchronize man's inner presentations about third world with outer anthropogenic reality. Thinking is reflection over disco-ordination between them causing efforts to find out reasons of it. In order to remove real-ideal synchronization lack, consciousness attempts to give answers on arising questions. Sensual humanitarian form of knowledge precedes an explanation. Mind selects convenient tools within reach of thinking in order to resolve arising problems. They are natural (L_N) and natural sciences (L_{S_N}) languages [6, 11]. However, they are not sufficient to fill up the gap in case of sophisticated knowledge pierced by meanings. Rational tools are to be applied to fulfill corresponding intellectual break. Only AM_S of mathematics exposes essences of knowledge [8]. Rough rational minutes allow applying formal algebraic systems with their operations for phenomena study. Due to L_C, investigation becomes supplied by comparative analysis of structures. Coordinatization admits systems study by means of calculations [12].

Systems conceal complex meanings expressible in the form of sophisticated UMA. Problem of I_N is to reveal, identify, and study them in order to return their initial auto-obviousness. True directives are required to be ruled by. *Understanding* is process of UMA investigation. Pre-established knowledge unity and its anticipated meaning are kept in universal constructions of L_C [9, 10]. DL I_A design in accordance with I_N entelechies allows organizing constructive dialog between these intelligencies. It creates necessary context for knowledge investigation and mutual tutoring from positions of universality. Encouraging questions and responses concerning system properties will develop SL and man's holistic presentations. In its functioning, DL I_A leans on philogenetic achievements presented in cognitive base $CogOnt_{K_B}$ dynamics built in L_C [5]. DL I_A engineering - teaching is its continuous enriching by interconnected mathematical structures embedded in computer-integrated environment.

1.4   Consciousness Double Helix

Mathematics unity and power manifest itself in its embedding theorems and free algebraic systems. They reflect processes of knowledge generalization and specialization. Cognitive potential of mathematics is stipulated by ideal objects consistency securing cognogenesis liberty. AM_S of modeling provides a person with co-ordinated presentations about system laws. Neuro-phenomenology of thinking is based on linguistic abilities [4, 6, 11]. Languages spring up unexpectedly and inexplicably as man's need of communication with and explication of outer world [6, 11]. Consciousness double helix (CDH) is model of I_N functioning (Fig. 2).

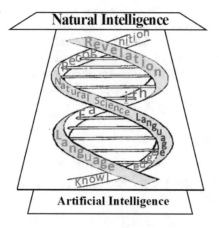

Fig. 2. Consciousness double helix as duality of natural and categories languages.

CDH changes constantly its configuration. It is done by means of auto-folding and crossing-over processes. Humanitarian and rational parts of consciousness are interacting while hypothesis are propounded and verified. Processes of CDH crossing-over strengthen thinking in trend of meanings. It supports humanitarian and natural sciences glottogonia. Natural language (L_N) allows sensing knowledge in metaphoric form. Language of categories (L_C) impels to search knowledge essence presented in the form of UMA. Noosphere evolution created L_C as tool for meanings processing. CDH model tracts I_N as language processor (Fig. 2).

CDH auto-folding is synergetic process with intensifications. According to synergetic paradigm, knowledge generalization and understanding are auto-resolved by means of sequent bifurcations coming to new ideal presentations moulding. Mathesis universalis of L_C was discovered on the way. It brought about employment of system axiomatic method (AM_S). Thinking happens on the ground of modeling. CDH auto-folding compresses knowledge selecting UMA and helping person to enter hermeneutics circle of SIC. Intellectual activity is natural for human life. CDH transformations take place under constant influence of culture. Besides traditional education, SIC manifests itself by means of cognition tools. Among them, there is I_A that is to be life-long I_N partner in cognition. DL I_A creation is educational imperative of SIC. Humanitarian and rational origins mutual penetration synchronizes person's real and ideal notions giving birth to SL (Fig. 2). UMA transcedentality proves the approach can come true.

2 Coordinatization and Measuring Universality

DL I_A tutoring system could apply AM_S to universalities identification and study. Coordinatization universality is stipulated by possibility to map different external structures into mathematical ones. This is objectization process. It supports thinking enabling it with operational level. Numerous algebraic systems are used for the aim because numeric ones are insufficient. Mathematical structures give true general view. They compress knowledge picking out meanings. Due to coordinatization, it is possible to discover new knowledge with the help of calculations. This method unites ideal presentations with "real" work in computer systems. Ideal indicates phenomena and suggests objects for measurement. UMA structures and commutative diagrams of L_C admit even meaning contemplation in frames of wholesome system. Calculation and reflection promote understanding [8, 10].

2.1 On Risks Estimation

Casualty plays ubiquitous role in our world [14]. Probability theory is tool for risks estimation and decision-making under uncertainty [1–3]. UMA unite casualties study with other mathematical theories.

Example 1. *Category of casual values* (CV). Probability space is object in CV category. This structure $\Omega = (\dot{\Omega}, A, d\omega)$ consists of casual events $A, A \subset \dot{\Omega}$, σ – algebra A and additive measure $d\omega : A \to [0, 1]$ defined on it. Morphisms $f : \Omega \to \Omega'$ are equivalence classes $f = \{\dot{f}' : \dot{f}' \sim \dot{f} \Leftrightarrow d\omega\{\dot{f} \neq \dot{f}'\} = 0\}$ of measurable functions $\dot{f} : \dot{\Omega} \to \dot{\Omega}'$ such as:

$$\forall A' \subset A' \Rightarrow f^{-1}(A') \subset A. \tag{1}$$

Correlations $P(A') = d\omega' f(A) = d\omega f^{-1}(A')$ take place. Thus, event A probability is number $P(A) = d\omega(A), 0 \leq P(A) \leq 1$. This measure is continuous function $A \to P(A)$ respectively algebra A operations (axiom) [14]. Due to (1) and object Ω' coordinatization, one can replace $\Omega = (\dot{\Omega}, A, d\omega)$ by probability space (R^n, B, dP_ξ) having coordinates system. Instead of f, $CV\xi = f', f' : \Omega \to R^n$, is now considered. Probability distributions dP_ξ preserve morphisms properties. This is unique measure $dP_\xi = cont(d\dot{P}_\xi)$ built with the help of continuation theorem [14]. Its building starts from assigning probabilities $d\dot{P}_\xi : \Pi \to R$ to events from rectangular sets algebra Π. Minimal σ - algebra $B = gen(\Pi)$ is generated by Π.

Let $\Omega = \Omega_1 \times \Omega_2 \equiv (\dot{\Omega}_1 \times \dot{\Omega}_2, gen(A_1 \times A_2), cont(d\omega_1 \times d\omega_2))$ be objects $\Omega_i, i = 1, 2$, product (universal construction) and $domf_i = \Omega$. Then $CV f_i : \Omega_i \to \Omega', i = 1, 2$, are called independent.

At last, Lebesgue's integral is applied to $P(A)$ calculation:

$$P(A) = \int dP_\chi, \chi(x) = 1 \Leftrightarrow x \in A$$

Coordinates allow applying arithmetic operations to CV. Behavior of CV sequences is studied in the theory. CV numeric characteristics can be also found by integration. Infinite - dimensional object can be furnished with probability measure if there is a family of coordinated measures on cylindrical sets [14].

Next UMA can be identified in probability theory. They are: 0, 1; limit (co-)cone, (co-)product; (co-)Descartes' square; inverse image, epi(mono)(iso)-morphism, factor-object. Analysis shows that category CV is not complete because it does not contain 0 (zero). Any CV morphism is epi-morphism and every object is atomic. CV category contains 1 (unit) – one-point distribution and, the most important of all, there is generator of all probability distributions. It is uniform measure du on $[0, 1]$.

2.2 Metrical Measurements

Category MES of all μ-measurable spaces is inheritor of CV system [14]. It inherits generator − homogeneous measure. Geometrical measurements issued in derivation − integration calculus discovery. Superposition cannot damage maps properties such as measurability, continuity, and smoothness. Hence, there are corresponding categories of all continuous (TOP) and differentiable (DIF) functions. Geometry applies measures of another kind.

Example 2. Topological space $T \in TOP$ metrication means its topology introducing with the help of metric $\rho : T \times T \to R_+$. Area $S_m(\varphi)$ measuring of m-dimensional surfaces corresponds to "ordinary" length, area, or volume calculation in case of

$m = 1, 2, 3$. Due to space T coordinatization $u : T \to R^n$, any elementary figure (or cube) φ in T is presented by a map $\varphi : I^m \to R^n, I = [0, 1], n \geq m$. Cube is example of m-dimensional surface. Any figure is union of elementary ones. It is known [15], notion of area $S_m(\varphi)$ is defined only for piecewise smooth maps φ.

For measuring, it is convenient to use Euclid's' norm $du = |u_2 - u_1|$ where $u_i = u(t_i) \in R^n, i = 1, 2$. It is equivalent to distance $\rho(t_1, t_2)$ between points $t_1, t_2 \in T$. Meaning of the norm is segment $[u_1, u_2]$ length. For polygonal lines, value $S_m(\varphi), m = 1$, is sum of its segments lengths. By Pithagoras' theorem, length of infinitely small curved segment is equal to $du(t) = |\varphi'(t)|dt$. For the whole line, this measure is found by means of Riemann's integral limit construction $(m = 1)$:

$$S_m(\varphi) = \int_{I^m} |\varphi'| dt$$

In general, $m \geq 2$, it is multiple integral of the first kind $(dt = dt_1 \times \ldots \times dt_m)$. Even for polyhedron area calculation it is impossible to do with integrals. If $m \geq 3$ then area search does not come to equally compounded/complemented figures [16]. Measure $S_m(\varphi)$ differs significantly or even does not exist for near-by objects compared in uniform distance, i.e. dependence $\varphi \to S_m(\varphi)$ is discontinuous.

Area measuring admits generalization. Point of view on surfaces is changed significantly if they are *orientated* being m-cubes algebraic sums (chains) $\varphi = \lambda_1 \varphi_1 + \ldots + \lambda_k \varphi_k, \lambda_1, \ldots, \lambda_k \in R$. Then surface is a vector in linear space $\Phi_m = \{\varphi\}$. Let $\Omega_{m,n} = \Lambda^m R^{n*}$ be m-th outer degree of space R^{n*} dual $R^n, dx \in R^n$. In other words, R^{n*} is space of m-th order alternating differential forms ω. Here, CDH auto-folding is observed (Fig. 2). Bi-linear functional $J(\omega, \varphi) : \Omega_{m,n} \times \Phi_m \to R$ is the second kind integral applied to orientated surface area $S_m^{or}(\varphi) = J(\omega, \varphi)$ calculation:

$$J(\omega, \varphi) = \int_\varphi \omega = \int_{I^m} \varphi^*(\omega), \varphi^* : \Omega_{m,n}(\varphi) \to \Omega_{m,n}(I^m)$$

It requires surface embedding in space $R^{m+1}, m \geq 2$, and next differential forms $(m = 1, 2)$ integrating:

$$\omega_{1,n} = \tau_1 dx^1 + \ldots + \tau_n dx^n, \omega_{2,3} = N_1 dx^2 \wedge dx^3 - N_2 dx^1 \wedge dx^3 + N_3 dx^1 \wedge dx^2.$$

Vector-functions $\tau, N : \varphi \to R^n, x = \varphi(t)$, are

$$m = 1 \Rightarrow \tau = \tau(x) \| \varphi'(x), |\tau| = 1; m = 2, n = 3 \Rightarrow N = N(x) \perp \varphi, |N| = 1.$$

Integrations over surface φ and its boundary $\partial\varphi$ are connected by Stokes' theorem

$$\int_\varphi d\omega = \int_{\partial\varphi} \omega$$

inherited from Newton-Leibnitz's one [15]. CDH auto-folding created new ideal presentation about differential form outer derivation $d : \Omega_{m-1,n}(\partial\varphi) \to \Omega_{m,n}(\varphi)$. It displays also dual correlations $d^2\omega = 0, \partial^2\varphi = 0$ between forms and surfaces.

3 Continuity Universality

Continuity was widely used in previous sections. Everybody "feels" continuity but rarely comprehend it. DL I_A tutoring system could apply AM_S for discovering origins of continuity [8]. Sensual form of geometric presentations secures transition to UMA-based definition explaining its meaning. "Real" geometric view and "ideal" algebraic essence of continuity are united in this approach. Geometry can be constructed on the base of symmetries [17]. Besides it, in our approach natural numbers (NN) ordering is used. It is also required NN embedding in more general algebraic systems. It worth mentioning that NN completion up to continuous objects can be done without leaning on order notion.

Example 3. *Continuous objects and maps phenomenology.* Man percepts world symmetries. For instance, translations conserve discrete ray. Existence of "minimal" move brings life to finite rings and infinite natural number object (NNO). The latter is system of cone morphisms (operations). The meaning is expressible by commutative diagrams possessing universality property [9, 10, 12, 13]. Language of categories presents UMA in geometrical form helpful for their auto-obviousness discovering.

Every line l is a cone inheritor. It consists of two anti-isomorphic "rays" l_+, l_- with common initial point - vertex O. Transition from cone l_+ to line l demonstrates how l_+ symmetries semi-group is embedded in group of all line translations. Discrete line l meaning is *integer number object* (INO). It has linear ordering unlike initial cone l_+ perfect ordering. NNO and INO auto-morphisms (symmetries) correspond to arithmetic operations with natural (N) and integer (Z) numbers [4, 10].

Line has mutually inverse translations $s, s^{-1} : l \to l, s \circ s^{-1} = 1$ generating the whole group of translations. In category SET, object INO coincides with set Z of all integer numbers. Similar to NNO [4, 10], object INO possesses next universality property:

$$\begin{array}{ccc} Z & \xrightarrow{\;s\;} & Z \\[-2pt] & \overset{s^{-1}}{\longleftarrow} & \\ {\scriptstyle 1_h^0}\Big\downarrow\nwarrow & \psi\Big\downarrow & \\ A & \xrightarrow{\;g\;} & A \\[-2pt] & \overset{g^{-1}}{\longleftarrow} & \end{array} \tag{2}$$

There are two superposed commutative diagrams in (2). In other words, there is unique morphism $\psi : Z \to A$ whatever object A might be selected having isomorphisms $g, g^{-1} : A \to A$. Object INO is not perfectly ordered as NNO was but it is linear ordered: $z_1 \leq z_2 \Leftrightarrow \exists n \in N z_1 + n = z_2$. On the base of definition (2), auto-morphisms $\psi = \times_{\pm n} \, or \, \psi = +_{\pm n}, n \in N, n \neq 0$, are discovered. It is sufficient to choose $g = s, h = \pm n \, or \, g = +_{\pm n}, h = 0$ and $A = Z$ in (2) correspondingly. The symmetries allow visualizing INO arithmetic operations properties. All translations $+_n, s \equiv +_1$ have

inverse ones $+_{-n}$, i.e. $+_n \circ (+_{-n}) = 0$. Composition of $a + b$ arrows s and its associative law prove correctness of addition laws. They are associative and commutative
ones. Observable consequence of multiplication $(\times_k)$ diagram is distributive law
$(m + n) \times_k = m \times_k + n \times_k$. In case of morphisms $\times_{-n}, n \in N$, corresponding commutative diagrams take form:

$$
\ldots \begin{array}{c} Z \\ \downarrow \\ Z \end{array} \xrightarrow{\cdot} \begin{array}{c} Z \\ \downarrow \\ Z \end{array} \xrightarrow{-}_{a} \begin{array}{c} Z \\ 1 \downarrow \\ Z \end{array} \xrightarrow{\cdot} \begin{array}{c} Z \\ \downarrow \\ Z \end{array} \times_{-n} \ldots \tag{3}
$$

   In particular, one can see explicitly in diagram (3) that $\times_{-1} : Z \to Z$ is anti-isotone
map satisfying to the law $(-1) \times_{-1} \equiv (-1) \times (-1) = 1$. Cone of non-negative integer
numbers $Z_+ = N$ is identified as algebra consisting of all isotone maps $Z \to Z$.

   Applied coordinatization allows introducing infinite discrete plane as product
$Z \times Z$. It has discrete group of movements - translations $+_{(z_1, z_2)}, (z_1, z_2) \in Z \times Z$.
Algebra $Z \times Z$ inherits also multiplication from ring Z. Algebraic meaning of the plane
is $Z \times Z$ is module over ring Z. Its elements are vectors (z_1, z_2).

   By means of plane $Z \times Z$ factorization, rational numbers $q \in Q$ are built. They are
equivalence classes q consisting of collinear vectors $(z_1, z_2), z_2 \neq 0$ habitually denoted
by fractions $q = z_1/z_2$ [12, 13]. Algebra Q has inverse arithmetic operations and linear
ordering $q_1 \geq q_2 \Leftrightarrow q_1 - q_2 \in Q_+$ inherited from Z. It is field having clear
geometrical interpretation. Factorization means that any ray in discrete plane $Z \times Z$
having slope angle $q = \frac{n}{m}$ is mapped in single point $q \in Q$ (Fig. 3). Diverse UMA were
once more applied to these structures. AM_S explains continuous objects phenomenology. Let Q_1, Q_2 be upper $Q_2 = Q_1^\Delta$ and lower $Q_1 = Q_2^\nabla$ cones of one set to
another. Limit cones $(\sup Q_1)$ and co-cones $(\inf Q_2)$ existence and coincidence are to
be postulated:

$$
\Phi \begin{array}{c} Q_2 \\ \downarrow \\ Q_1 \end{array} \qquad \begin{array}{c} \sup Q_1 \leftarrow - - q_2 \\ \searrow \qquad \swarrow \\ Q_1 \end{array} \qquad \begin{array}{c} Q_2 \\ \swarrow \qquad \searrow \\ q_1 \leftarrow - - \inf Q_2 \end{array} \tag{4}
$$

   Without loss of generality, pair Q_1, Q_2 in (4) is decomposition of ordered set
Q considered as a category. Morphism $q_1 \to q_2$ means that relation $q_2 \leq q_1$ takes
place. Let Q_1, Q_2 be sub-categories of Q. There is functor $\Phi : Q_2 \to Q_1$. Objects Q_1, Q_2
are   to   be   substituted   in   schemes   (4)   by   commutative   diagrams
$q_i \leftarrow q_i'$ if $q_i \leq q_i', i = 1, 2$.

Axiom 1. For all object Q partitions $Q_1 \cup Q_2$ there exist objects $\sup Q_1, \inf Q_2$ satisfying to the property (4). Besides, $+\infty \triangleq \sup Q, -\infty \triangleq \inf Q$.

Axiom 2 (Archimedes). For all Q divisions, equality $\sup Q_1 = \inf Q_2$ takes place.

   Ideal object from axiom 2 is denoted by $r = r(Q_1, Q_2)$. New numeric system of
ideal elements R is obtained that embeds Q. Linear ordering is translated from Q to R,
see (4). Axioms 1, 2 shows that all numbers $\sup A = \sup A^\nabla, \inf A = \inf A^\Delta$ are

ordered: $\inf A \leftarrow A \leftarrow \sup A, A \subset Q$. Equalities $\sup(-A) = -\inf A, \inf(-A) = -\sup A$ are also consequences of axioms 1, 2. Algebra R is field because it inherits arithmetic operations from Q by means of limit constructions (4) usage:

$$
\begin{array}{c}
Q_2 \to \inf Q_2 = r \\
\downarrow \qquad \searrow^{q_1 \in Q_1 \cap Q_2'} \\
Q_2' \to \inf Q_2' = r'
\end{array}
\;;\;
\begin{array}{c}
Q_1 \leftarrow r \leftarrow Q_2 \\
+_q \downarrow \quad \downarrow \quad \downarrow \\
Q_1' \leftarrow r' \leftarrow Q_2'
\end{array}
\;\Rightarrow\;
\begin{array}{c}
Q_1 \leftarrow \quad r \leftarrow Q_2 \\
\downarrow +_{\sup Q_1'} \downarrow +_{r'} \downarrow \quad +_{\inf Q_2'} \\
Q_1' \leftarrow \quad r' \leftarrow Q_2'
\end{array}
\qquad (5)
$$

Isotone map $+_q$ makes the diagrams correct $\left(Q_2' = Q_2 + q, r' = r + q\right)$.

On the base of schemes (5) and isotone maps $\times_q : Q \to Q, q \in Q_+$, multiplication $\times_r, r \in R_+$, is introduced. The operation is continued on negative numbers due to the law $\times_{-r} \triangleq \times_r \circ \times_{-1}, r \in R_+$. It completes continuous line R building. Thus, continuity property issues from axioms 1, 2.

Fig. 3. Continuity genesis $N \Rightarrow Q \Rightarrow R$: lines slope angle $q = \frac{n}{m}$ and limit angle φ.

Let take in consideration rational numbers geometrical interpretation, see Fig. 3, and diagrams (4), (5). Then real numbers r are slope angles of real lines $r_1 = r r_2$. Instead of $Z \times Z$ factorization, rational line $q_1 = \frac{p}{q} q_2$ can be completed by points $(r_1, r_2), r_1 = \frac{p}{q} r_2$. In its turn, limit cones of slope angles $\frac{p}{q}$ allows building real numbers r. They are slope angles of new "real" lines $r_1 = r r_2$ It means that "ideal" *continuous* lines complete rational plane Q. *Continuous* plane $R \times R$ is emerging. In its turn, $R \times R$ factorization transforms the lines in points φ, see Fig. 3, restoring linear ordering. Generation of complex numbers algebra $C = (O, +, R \times R)$ displays CDH auto-folding (Fig. 2). Translations $+_z$ of the plane are vectors $z = (r_1, r_2) \in R \times R$. Other morphisms are: plane rotations on angles φ, homotheties $\times_r = \times_{(r,r)}$ (Fig. 3). Under sequent rotations, angles φ are summed up. In previous rational constructions slope angles q were summed up. To express the operation by means of vector z, polar coordinates system is to be used $z = (\varphi, \rho)$. It is more convenient to present complex numbers (object CNO) applying matrix algebra C' to their coordinatization $z \to Z$. It can be done due to isomorphism $C \simeq C'$. New "ideal" numbers Z are now deprived of ordering. Quaternion body (QBO) H is also built by this method:

$$
Z \in C'_{2 \times 2}, H \in C'_{4 \times 4} \Leftrightarrow Z = \begin{pmatrix} r_1 & -r_2 \\ r_2 & r_1 \end{pmatrix}, H = \begin{pmatrix} z_1 & -\bar{z}_2 \\ z_2 & \bar{z}_1 \end{pmatrix}. \qquad (6)
$$

CDH crossing-over is resulted in glottogonia - usage of matrix language (Fig. 2). These algebraic forms are more suitable for planes R^2, C^2 morphisms presentation.

Numbers genesis is chain $NNO \Rightarrow INO \Rightarrow RatNO \Rightarrow RNO \Rightarrow CNO \Rightarrow QBO$. Other continuous algebras CNO and QBO are built. According to Frobenius' theorem, any attempt to continue the chain is doomed to failure. Numeric system collapses [13]. In order to introduce continuous sub-objects and maps, family O of open sets is to be considered. It defines topology $T = \{T \cap O\}$ and gives ground to category TOP of continuous maps $f : T_1 \rightarrow T_2$ satisfying to next characteristic property:

$$\forall A' \subset T' \Rightarrow f^{-1}(A') \subset T. \tag{1'}$$

For instance, arithmetic operations are continuous functions. Definition (1') is similar to (1). The former conserves topology and the latter – $\sigma-$ algebras.

Due to coordinatization, geometry can be studied by calculations. Linear equations are applied to investigate straight lines property to divide planes in two parts. Discrete lines in Q^2 do not partition planes. Any line in $R^2 \simeq C$ does it (Fig. 4). Hyper-planes in quaternion space $H, H \simeq C^2$ have the same property (Fig. 4(a)).

Fig. 4. (a) Continuous spheres $\bar{C}, \bar{H}$; (b) discrete plane Q^2.

Unlike ordered set $\bar{R} = R \cup \{\pm\infty\}$ closure, only one ideal element is to be added to obtain spheres: $\bar{C} = C \cup \{\infty\} \simeq S_2, \bar{H} = H \cup \{\infty\} \simeq S_4$, see Fig. 4(a). Cones (4) are applied here to sequences $\{z_k\}, \{H_k\}$ having norms $|z_k|, |H_k| \rightarrow \infty, k \rightarrow \infty$.

4   Conclusions

Perspective of DL I_A engineering consists of technical system design capable of meanings co-processing. It is not autonomous action. It supposes I_A-I_N partnership for I_N rational auto-moulding sake. I_A is used to promote person's universal learning necessary for successful trans-disciplinary labor in SIC. Up-to-date education crisis can be resolved only by means of mutual natural and artificial intelligences deep-tutoring. It can be done with the help of universal mathematical essences identifying and understanding. Cognitive revolution will be settled by mankind's reach at last the highest 23-d nature development level in classification of E. Haeckel (1834–1919). Meanings co-processing and fostering enriches thinking and consciousness removing divergence between person's internal reflections and external scientific world.

References

1. Deviatkov, V.V., Lychkov, I.I.: Recognition of dynamical situations on the basis of fuzzy finite state machines. In: International Conferences Computer Graphics, Visualization, Computer Vision and Image Processing and Big Data Analytics, Data Mining and Computational Intelligence, pp. 103–109 (2017)
2. Fedotova, A.V., Davydenko, I.T., Pförtner, A.: Design intelligent lifecycle management systems based on applying of semantic technologies. J. Advances in Intel. Syst. Comp. **450**, 251–260 (2016)
3. Volodin, S.Y., Mikhaylov, B.B., Yuschenko, A.S.: Autonomous robot control in partially undetermined world via fuzzy logic. J. Mech. Mach. Science **22**, 197–203 (2014)
4. Gromyko, V.I., Kazaryan, V.P., Vasilyev, N.S., Simakin, A.G., Anosov, S.S.: Artificial intelligence as tutoring partner for human intellect. J. Advances in Intel. Syst. Comp. **658**, 238–247 (2018)
5. Vasilyev, N.S., Gromyko, V.I., Anosov, S.S.: On inverse problem of artificial intelligence in system-informational culture. J. Advances in Intel. Syst. Comp. Hum. Syst. Eng. Des. **876**, 627–633 (2019)
6. Gromov, M.: Circle of Mysteries: Universe, Mathematics, Thinking. Moscow Center of Continuous Mathematical Education, Moscow (2017)
7. Vasilyev, N.S., Gromyko, V.I., Anosov, S.S.: Deep-learned artificial intelligence as educational paradigm of system-informational culture. In: Proceedings of 4-th International Conference on Social Sciences and Interdisciplinary Studies, Palermo, Italy, vol. 20, pp. 136–142 (2018)
8. Gromyko, V.I., Vasilyev, N.S.: Mathematical modeling of deep-learned artificial intelligence and axiomatic for system-informational culture. Int. J. Robot. Autom. **4**(4), 245–246 (2018)
9. McLane, S.: Categories for Working Mathematician, Phys. Math. Edn. Springer, Moscow (2004)
10. Goldblatt, R.: The Categorical Analysis of Logic. N.-H. P. C., Amsterdam, NY, Oxford (1979)
11. Pinker, S.: Thinking Substance. Language as Window in Human Nature. Penguin, Moscow, Lb (2013)
12. Shafarevich, I.R.: Main Notions of Algebra. Regular and Chaos Dynamics, Izhevsk (2001)
13. Skorniakov, L.A.: Elements of General Algebra. Nauka, Moscow (1983)
14. Kolmogorov, A.N.: Main Notions of Probability Theory. Nauka, Moscow (1974)
15. Ilyin, V.A., Pozdnyak, E.G.: Origins of Mathematical Analysis. Nauka, Moscow (1998)
16. Hilbert, D.: Grounds of Geometry. Technology Tier List, Moscow – Leningrad (1948)
17. Bachman, F.: Geometry Construction on the Base of Symmetry Notion. Nauka, Moscow (1969)

Association Matrix Method and Its Applications in Mining DNA Sequences

Guojun Mao[⊠]

School of Mathematics and Physics, AI & AK Institute,
Fujian University of Technology, Fuzhou 350118, People's Republic of China
maximmao@hotmail.com

Abstract. Many mining algorithms have been presented for business big data such as marketing baskets, but they cannot be effective or efficient for mining DNA sequences, any of which is typically with a small alphabet but a much long sizes. This paper will design a compact data structure called Association Matrix, and give an algorithm to specially mine long DNA sequences. The Association Matrix is novel in-memory data structure, which can be so compact that it can deal with super long DNA sequences in a limited memory spaces. Such, based on the Association Matrix structure, we can design the algorithms for efficiently mining key segments from DNA sequences. Additionally, we will show our related experiments and results in this paper.

Keywords: Data mining · DNA sequence · Association Matrix

1 Introduction

Using data mining techniques to analyze biological data is becoming an important research problem. However, most typical data mining algorithms were designed for business or governmental databases, so they could not be an ideal solution to analyze biological data like DNA sequences. One of the most important issues is main memory inefficiencies for super long DNA sequences.

In biological computing, a DNA sequence is often abstracted into a string of characters, which has a small alphabet composed of A, T, G and C, but is a very long size. For example, the human genome is made of roughly three billion of nucleic acids. These features should result to create a different type of study in data mining.

As we known, finding and extracting genetic fragments from a long sequence is an important problem in biology. In data mining, there have been some studies in searching the key segments from an experimental data sequence. Therefore, this paper will aim at finding out key DNA fractions from DNA sequences through using data mining techniques.

In 2012, Papapetrou et al. developed three methods for efficiently detecting poly-regions in DNA sequences [1]. The first applied entropy-based recursive segmentation method; the second used a set of sliding windows to summarize out sequence segments; the third employed a voting-based technique to mine key segments. These methods provide the basic ideas of mining key segments from DNA sequences.

© Springer Nature Switzerland AG 2020
T. Ahram (Ed.): AHFE 2019, AISC 965, pp. 154–159, 2020.
https://doi.org/10.1007/978-3-030-20454-9_15

From the view of data mining, one related problem to this paper is to mining frequent sequences from sequential databases. In 1995, Agrawal introduced and discussed the issue of sequential mining and it was going to become an important research branch of data mining [2]. In 1996, Algorithm GSP was developed for mining sequential patterns that is a breadth-first search and button-up method [3]. Free-Span is another efficient algorithm for mining sequential patterns [4], which has less effort than GSP in candidate sequence generation. In fact, up to now, many effective algorithms for mining sequential data have been presented [5, 6].

Applying data mining techniques to analyzing DNA sequences has also become a research focus. Bell et al. used sequential mining methods to discover common strings from DNA sequences [7]. Liu et al. discussed the problem of principal component analysis and discriminate analysis of DNA features, which employed sequence classification techniques [8]. Habib et al. gave the methods of DNA motif comparison that was based on Bayesian algorithms of data mining [9].

Other related works have: Mannila and his colleagues presented a series methods for mining key sub-sequences from long event sequences, including mining frequent episodes from long sequences [10], and similarity evaluation between event sequences [11]; Keogh gave an online mining method to segment long time series [12]; Stegmaier constructed an unsupervised clustering approach in DNA sequences [13]; Wu made use periodic wildcard gaps to discovering frequent patterns [14].

Our contributions in this paper are: (1) Through designing compact in-memory data structures and short sequence based processing mechanisms, it provides a novel idea to analyze DNA sequences. (2) It introduces an efficient structure called Association Matrix which can help effectively mining key segments from super long DNA sequences. The rest of this paper is organized as follows. Section 2 introduces the Association Matrix structure. Section 3 gives the algorithms for mining key segments from a long DNA sequence. In Sect. 4, we evaluate the performance of the proposed methods. Section 5 concludes this paper.

2 Association Matrix for DNA Sequences

As is known to all that a cell uses DNA to store their genetic information, and a DNA molecule is composed of two linear strands coiled in a double helix, Because two strands strictly abide by the base pairing rules, in modern bioinformatics, a DNA sequence can be represented as a character string composed of A (adenine), T (thymine), C (cytosine), and G (guanine). That is, a DNA sequence is denoted by $s = <e1, e2, ..., eL>$, $e_i \in \{A, G, C, T\}$ for all $i = 1, 2, ..., L$.

Definition 1 (Association Matrix). Supposed a character set W and a sequence on W, say $s = <e1, e2, ..., eL>$. For any a sub-sequence p of s and any element q of W, Association Value of p with q on s is defined as the number of q occurs just after p on s. Given a sub-sequences of s with length m, its Association Matrix is organized into $(vij)_{m*4}$, where vij is the Association Value of the i sub-sequences with the j character of W.

Example 1. Considering $s = <ATGTCGTGATTGCATTACTACT>$, for $p = <A>$, its Association Value with T is 3; for $p = <AT>$, its Association Value with T is 2.

Example 2. Considering *s* in Example 1 as a DNA sequence, then the Association Matrix of all its sub-sequences with size 1 is the following.

$$
\begin{bmatrix}
0 & 3 & 2 & 0 \\
2 & 2 & 1 & 3 \\
1 & 2 & 0 & 1 \\
1 & 2 & 1 & 0
\end{bmatrix} \tag{(1)}
$$

3 Mining Key Segments from DNA Sequences

Discovering key segments from a DNA sequence is an important target for DNA analyses. In fact, in a super long DNA sequence, there exist some key sub-sequences that are often much shorter but appearance-frequent, which is important to identify a life object.

Definition 2 (Key Segment). Given a DNA sequence *s* and a minimum association threshold *t*, if a sub-sequence *p* of *s* is a key segment with size *n*, when the Association Value that *p* and any element *q* in{A, T, G, C} is not less than *t*, the *p* + *q* is a key segment of *s* with size *n* + 1.

According to Definition 2, we can make iteration with incremental sizes to find out all key short sequences from a long DNA sequence.

```
Algorithm Make-key-segment (keySet: ^string)
  {Assuming the investigated DNA sequence is s; minimum
association threshold t};
    var    k,m,n,i,j: Integer;
           rowSet: Array[] of string;
           p: Array[][] of Integer;
    begin
      k := 1; m := 4;
      repeat
        generate   Association Matrix with size k {p[i][j]};
        n := 0;
        for i=1 TO m
          for j=1 TO 4
            if p[i][j] >= t
              begin
                n := n+1;
                generate and insert the corresponding se-
                  quence with size k+1 to rowSet[n];
                insert rowSet[n] to keySet;
              end
        m := n; k++;
      until m=0;
      return keyset;
    end.
```

Example 3. Considering s in Example 1 as a DNA sequence, by Algorithm Make-key-segment, we can recursively find out its key segments as Table 1 shown.

Table 1. Process of searching key segments from a DNS sequence $(t = 2)$.

Size k	Association Matrix	Key segments
1	$\begin{bmatrix} 0 & 3 & 3 & 0 \\ 2 & 2 & 1 & 3 \\ 1 & 2 & 0 & 1 \\ 1 & 2 & 1 & 0 \end{bmatrix}$	{<AT>, <AC>, <TA>, <TT>, <TG>, <CT>, <GT>}
2	$\begin{bmatrix} 0 & 2 & 0 & 1 \\ 0 & 2 & 0 & 0 \\ 0 & 0 & 2 & 0 \\ 1 & 0 & 0 & 1 \\ 1 & 1 & 1 & 0 \\ 1 & 0 & 0 & 0 \\ 0 & 0 & 1 & 1 \end{bmatrix}$	{<ATT>, <ACT>, <TAC>}
3	$\begin{bmatrix} 1 & 0 & 0 & 1 \\ 1 & 0 & 0 & 0 \\ 0 & 2 & 0 & 0 \end{bmatrix}$	{<TACT>}

4   Experiments

In this section, we will show some our experimental results, which can prove the above algorithm has the better performances in main memory space and CPU executing time. In word, comparing with Algorithm Free-Span that is one of the popular sequential pattern mining algorithms [4], the proposed Algorithm Make-key-segment in this paper has obvious advantages in time and space consumptions.

The dataset for the experiments is C256S64N4D100 K [15], and they were conducted on a computer with 800 MHz CPU with 2 GB main memory.

Figures 1 and 2 respectively show the comparing results in execution time and main memory space usages when running Algorithm Make-key-segment and Free-Span in the same technique environments.

From Fig. 1, with increasing Minimum Association Degrees, Algorithm Make-key-segment can keep a good scale down in time consumption and is faster than Algorithm Free-span.

As shown in Fig. 2, Algorithm Make-key-segment need less memory spaces than Free-span do. A main factor contributed to these results is using Association matrix structure, which can be more compact than other data structure such that used in *Free-Span.*

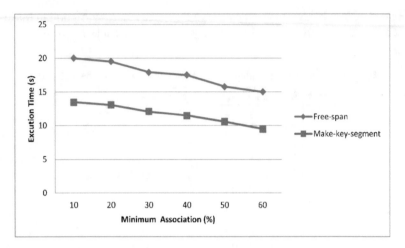

Fig. 1. Execution time of Algorithm Make-key-segment and Free-Span on Dataset C256S64N4D100 K [15], with different Minimum Association Degrees.

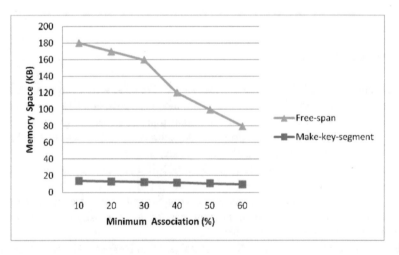

Fig. 2. Memory usages of Algorithm Make-key-segment and Free-Span on Dataset C256S64N4D100 K, with different Minimum Association Degrees.

5 Conclusions

DNA sequences are a different type of explosion of search space from classic transaction sequences, and so traditional sequential mining techniques for transactional data are not effective for them. This paper has studied the problem of mining key segments from DNA sequences.

We are working on more experiments and graph layout algorithms in this research field. We will also investigate more biological sequences and do research to them.

Acknowledgements. I am deeply indebted to the NSFC (China National Science Foundation of China), for its funding support with Number 61773415 makes the related re-search of this paper better.

References

1. Papapetrou, P., Benson, G., Kollios, G.: Mining poly-regions in DNA. Int. J. Data Min. Bioinform. **4**, 406–428 (2012)
2. Agrawal, R., Srikant, R.: Mining sequential patterns. In: The 1995 International Conference on Data Engineering, pp. 3–14. Taipei, Taiwan (1995)
3. Srikant, R., Agrawal, R.: Mining sequential patterns: generalizations and performance improvements. In: The 1996 International Conference on Extending Database Technology (EDBT), pp. 3–17 (1996)
4. Han, J., Pei, J.: Free-span: frequent pattern-projected sequential pattern mining. In: The 2000 International Conference on Knowledge Discovery and Data Mining, pp. 355–359 (2000)
5. Mohammed, J.: SPADE: an efficient algorithm for mining frequent sequences. J. Mach. Learn. **1**, 31–60 (2001)
6. Liu, C., Chen, L., Liu, Z., Tseng, V.: Effective peak alignment for mass spectrometry data analysis using two-phase clustering approach. Int. J. Data Min. Bioinf. **1**, 52–66 (2014)
7. Bell, D., Guan, J.: Data mining for motifs in DNA sequences. In: The 2003 Rough Sets, Fuzzy Sets, Data Mining, and Granular Computing. LNCS, vol. 2639, pp. 507–514 (2003)
8. Liu, Z., Jiao, D., Sun, X.: Classifying genomic sequences by sequence feature analysis. Genomics Proteomics Bioinf. **4**, 201–205 (2005)
9. Habib, N., Kaplan, T., Margalit, H., Friedman, N.: A novel Bayesian DNA motif comparison method for clustering and retrieval. PLoS Comput. Biol. **4**, 1–16 (2008)
10. Mannila, H., Toivonen, H., Verkamo, I.: Discovery of frequent episodes in event sequences. Data Min. Knowl. Discov. **1**, 259–289 (1997)
11. Mannila, H., Salmenkivi, M.: Finding simple intensity descriptions from event sequence data. In: The 7th ACM SIGKDD International Conference on Knowledge Discovery and Data Mining, pp. 341–346 (2001)
12. Keogh, E., Chu, S., Hart, D., Pazzani, M.: An online algorithm for segmenting time series. In: The 2001 IEEE International Conference on Data Mining, pp. 289–296 (2001)
13. Stegmaier, P., Kel, A., Wingender, E., Borlak, J.: A discriminative approach for unsupervised clustering of DNA sequence motifs. PLoS Comput. Bio. **9**, e1002958 (2013)
14. Wu, Y., Wang, L., Ren, J., Ding, W., Wu, X.: Mining sequential patterns with periodic wildcard gap. J. Appl. Intell. **41**, 99–116 (2014)
15. Wang, K., Xu, Y., Yu, J.: Scalable sequential pattern mining for biological sequences. In: The 13th International Conference on Information and Knowledge Management, pp. 10–15 (2004)

Deep Learning-Based Real-Time Failure Detection of Storage Devices

Chuan-Jun Su$^{(\boxtimes)}$, Lien-Chung Tsai, Shi-Feng Huang, and Yi Li

135 Yuan-Tung Road, Chung-Li District, Taoyuan City 32003, Taiwan, R.O.C.

Abstract. With the rapid development of cloud technologies, evaluating cloud-based services has emerged as a critical consideration for data center storage system reliability, and ensuring such reliability is the primary priority for such centers. Therefore, a mechanism by which data centers can automatically monitor and perform predictive maintenance to prevent hard disk failures can effectively improve the reliability of cloud services. This study develops an alarm system for self-monitoring hard drives that provides fault prediction for hard disk failure. Combined with big data analysis and deep learning technologies, machine fault pre-diagnosis technology is used as the starting point for fault warning. Finally, a predictive model is constructed using Long and Short Term Memory (LSTM) Neural Networks for Recurrent Neural Networks (RNN). The resulting monitoring process provides condition monitoring and fault diagnosis for equipment which can diagnose abnormalities before failure, thus ensuring optimal equipment operation.

Keywords: Big data · Hard disk · Failure prediction ·
Recurrent neural networks (RNN) · Long and short term memory (LSTM)

1 Introduction

Following the rapid development of cloud computing technologies, many companies now provide large-scale cloud services to user bases numbering in the millions. Thus, from the service provider's perspective, data center reliability is a critical issue, particularly in terms of hard disk failure on servers, which can interrupt cloud services and result in data loss.

Developers and IT teams require reliable and robust system equipment, including hard drives. Hard drive failure and subsequent data loss are still a serious issue, despite the development of solid state drives (SSD) for servers. SSDs are still considerably more expensive than hard disk drives (HDDs) in server environments. Current HDD providers offer warranty periods ranging from one to three years, but statistics show that as many as 9% of HDDs fail annually [1] without warning. Extended usage of an HDD can result in damage that can result in abrupt failure, potentially leading to data loss with serious consequences for the enterprise. Therefore, maintaining such devices and preventing such loss is of paramount importance.

Traditionally, this is accomplished through preventative maintenance with planned, regular maintenance scheduled based on the age of the equipment. However, this approach can increase operating costs through the pre-mature replacement of still-

© Springer Nature Switzerland AG 2020
T. Ahram (Ed.): AHFE 2019, AISC 965, pp. 160–168, 2020.
https://doi.org/10.1007/978-3-030-20454-9_16

functioning equipment. Predictive maintenance uses actual real-time data to predict future failure of still operating equipment, allowing maintenance personnel to schedule maintenance in advance. Predictive maintenance models are trained using historical data, providing for real-time insight into actual equipment status and predicting future failure. This allows for equipment to be replaced closer to its actual failure point, thus conserving resources.

This study uses big data analytics and artificial intelligence to develop a real-time system for predicting hard drive failure, thus aiding developers and IT teams to maintain data centers. The system can alert maintenance staff to potential imminent failure, allowing for early monitoring and maintenance work. We develop predictive frameworks for hard disk failures by analyzing machine log files instead of using traditional statistical prediction methods. The development process involves the following tasks: (1) retrieving and pre-processing data from the hard disk; (2) training and verifying the predictive model; (3) establishing a predictive system based on the model; and (4) predicting hard drive faults in real-time through streaming data.

The remainder of this paper is organized as follows. Section 2 reviews the relevant literature on techniques such as hard disk failure prediction, recurrent neural networks, and long short-term memory. Section 3 introduces the experimental design, with experimental methods designed based on the requirements and methods introduced in Sects. 1 and 2. Finally, Sect. 4 provides conclusions and recommendations for future research.

2 Related Works

2.1 Predictive Maintenance

Predictive maintenance is one way to maintain industrial, commercial, government and residential technology installations. It involves performing functional inspections and repairs or replacement of components, equipment or devices to maintain machine or facility operations. Maintenance techniques can include restorative, preventive or predictive maintenance.

Predictive maintenance monitors the actual operating condition of equipment and seeking to predict whether it will fail in the future, rather than operating-time or calendar-based maintenance scheduling. PM uses the analysis and modeling of collected equipment data for vibration, temperature and other parameters to predict equipment failure and to devise appropriate maintenance planning, thus reducing maintenance costs. Moreover, continuously collected data can be subjected to big data analysis techniques to incrementally improve fault prediction accuracy.

Canizo [2] established a cloud-based analytics platform that uses data from cloud deployments to generate a predictive model for individual monitored wind turbines, predicting wind turbine status in ten minute increments and displaying a visualization of the predicted results. Zhao [3] proposed a predictive maintenance method based on anomalies detected through data correlations among sensors. The results show this approach outperforms the use of sensor data alone, and can be used to predict failures in advance, thus reducing downtime and maintenance costs.

2.2   Machine Learning

Machine learning is a discipline in computer science that studies the development of programs that can learn. According to Samuel [4], "Machine learning allows a computer to learn without explicit programming." Machine learning techniques are widely used in data mining, email filtering, computer vision, natural language processing (NLP), optical character recognition (OCR), biometrics, search engines, medical diagnostics, credit card fraud detection and speech recognition.

Machine learning is part of artificial intelligence, and is often applied to research on prediction or categorization problems. Machine learning is primarily an algorithm that allows computers to "learn" autonomously. Applying machine learning techniques to historical data allows the computer to learn from feedback, examples and expertise. This feedback loop is used to arrive at identical or completely different predictions for similar future situations. Sample data is used in machine learning training to generate a model, which can then be applied to a test data set for prediction.

The performance limitations of machine learning are determined by data and features, and this upper limit can only be approximated by modeling methods. Prior to machine learning modeling, one must first use feature engineering methods to identify the important eigenvalues in the data. Brownlee [5] argued that feature engineering is the process of transforming raw data into features that better represent the underlying problems of predictive models, thereby improving model accuracy for invisible data. However, using machine learning methods to predict possible hard disk failure requires expert knowledge to perform feature engineering. With the important eigenvalues and data, the prediction module can be built, but hard disk data will vary between manufacturers, raising the need to re-execute feature engineering and the construction of new prediction modules. Deep learning can overcome such problems because it does not require feature engineering.

Deep learning is a branch of machine learning, and is a type of algorithm based on feature learning of data. Feature learning is a technique by which features are learned, transforming raw data into a form that can be effectively learned by machine. It avoids issues related to manual feature extraction, allowing the computer to simultaneously learn to use and extract features. This replaces the need for domain-specific subject matter expertise in identifying the important eigenvalues in the research data. Several deep learning architectures have been applied with good results to computer vision, speech recognition, and natural language processing, including deep neural networks (DNN) and recurrent neural networks (RNN).

2.3   Recurrent Neural Networks

Recurrent neural networks (RNNs) are a type of neural network in which each node has a direct one-way connection, and each node has different inputs or outputs at different time points. Each connection has a weight that can adjust the value. The most basic RNN architecture has three nodes which are divided into input, hidden and output nodes. Giles [6] used RNN to make financial exchange rate forecasts. The results indicate that RNN can identify important information within foreign exchange rate data and accurately predict future movements. In addition, the method can extract important feature values from data during the training process.

Long and short-term memory networks (LSTMs) are a special type of RNN that learns long-term dependencies. They were introduced by Hochreiter and Schmidhuber [7], and subsequently modified by many other groups. LSTM is clearly designed to avoid long-term dependencies. LSTM combined with the appropriate gradient-based learning algorithms, can solve the Vanishing or Exploding Gradient Problems that RNN can experience in long-term memory training, and provides a solution for the time series characteristics problem in hard disk failure prediction. Graves [8] used the TIMIT database to evaluate the bidirectional LSTM (BLSTM) and several other network architecture tests in a framewise phoneme classification. They found that bidirectional networks outperform unidirectional networks, while LSTM outperforms RNN, and multilayer perceptrons (MLPs) are much faster and more accurate.

3   Research Methodology

This study was based on data from Backblaze, an online service offering encrypted backups for individuals, families, organizations and businesses in more than 140 countries. As of the end of 2017, Backblaze used 93,240 hard drives ranging in size from 3 TB to 12 TB.

3.1   Research Architecture

As shown in Fig. 1, this study proposes an architecture consisting of three modules: (1) an extraction, cleanup and loading (ECL) module; (2) a deep learning (DL) module; and (3) a prediction and health management (PHM) module. The ECL module accounts for fault mode (FM) and extract and access (ES), while the ML consists of generating modules, and the PHM includes condition monitoring (CM) and a warning system (WS).

3.2   ECL Module

The ECL module consists of two basic components: Failure Mode (FM) and Extraction and Storage (ES).

Backblaze defines a hard disk failure as follows: (1) HDD completely stops working. (2) HDD data indicates imminent failure. When a hard disk meets either of these criteria, it is removed from the server, marked as malfunctioning, and replaced. A "dead" HDD displays no response to read or write commands. Backblaze uses SMART statistics as indicators of imminent disk failure, and a disk determined to be at risk of failure will be removed.

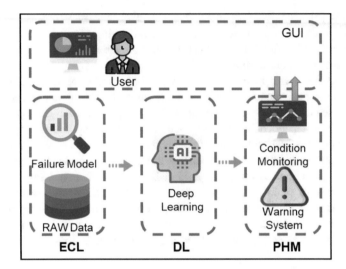

Fig. 1. Research architecture

The extraction and storage components are responsible for collecting SMART data from system log files. Pre-processing includes SMART extraction, noise cleanup and classification. The module performs the following tasks:

- Data extraction – Extracting information from raw data.
- Data cleaning – removal of data noise.
- Data loading – loading data to its final destination.

3.3   DL Module

The second module system consists primarily of a generated model. We define the HDD failure mode and applied SMART training data to implement the deep learning algorithm to solve the predictive monitoring problem for HDD failure. Figure 2 shows an overview of the deep learning model.

In the model construction step, the document system's SMART attributes are already ready for model training. The prediction model for establishing LSTM was built using Keras, a high-level neural network API written in Python that runs as a backend with TensorFlow, CNTK or Theano. Using Keras to build a predictive LSTM network includes three steps: (1) data initialization, (2) fitting it to the training data, and (3) fault prediction. The output is the result of an HDD failure prediction.

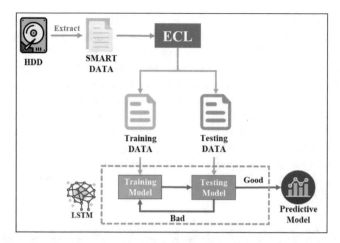

Fig. 2. Deep learning module

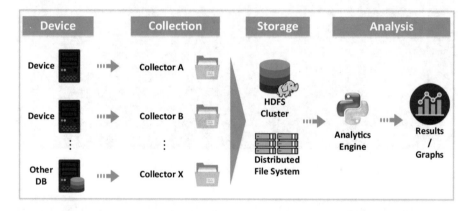

Fig. 3. Research architecture infrastructure concept

3.4   PHM Module

To present the overall research architecture, we propose using a more sophisticated approach for monitoring equipment. Collecting and processing data requires an infrastructure. In Pinheiro's research, the infrastructure describing system health is a large distributed software system that collects and stores hundreds of attribute values from all of Google's servers [9]. It also provides an interface for processing data for any analysis task. We use the health system infrastructure to illustrate the infrastructure of the research architecture, with the infrastructure concept map shown in Fig. 3.

The research architecture infrastructure consists of a data collection layer that collects all historical or streaming data from each device, which is then stored and processed using a distributed file system, and finally processed using an analytics server to product prediction results. Thus far, we have trained our prediction model. To further improve on the original monitoring system, we propose a two-step method as follows: (1) A Value Prediction (VP) component establishes a background service on a device, and the streaming data is sent to the cloud-based analysis platform; and (2) the Warning System (WS) component runs on the analysis server, analyzing the streaming data received from the device and providing device status reports to the user.

To collect SMART data from a hard drive requires a HDD monitoring kit or software. This study used smartmontools (https://www.smartmontools.org/), which consists of two utilities, smartctl and smartd, which monitor SMART data on HDDs and SSDs. This system allows the user to check hard disk SMART data and run various tests to determine disk health. Figure 4 presents a screenshot of smartmontools monitoring SMART data.

```
SMART Attributes Data Structure revision number: 16
Vendor Specific SMART Attributes with Thresholds:
ID# ATTRIBUTE_NAME          FLAG     VALUE WORST THRESH TYPE      UPDATED  WHEN_FAILED RAW_VALUE
  1 Raw_Read_Error_Rate     0x000a   100   100   ---    Old_age   Always      -       0
  2 Throughput_Performance  0x0005   100   100   ---    Pre-fail  Offline     -       0
  3 Spin_Up_Time            0x0007   100   100   ---    Pre-fail  Always      -       0
  5 Reallocated_Sector_Ct   0x0013   100   100   ---    Pre-fail  Always      -       0
  7 Seek_Error_Rate         0x000b   100   100   ---    Pre-fail  Always      -       0
  8 Seek_Time_Performance   0x0005   100   100   ---    Pre-fail  Offline     -       0
  9 Power_On_Hours          0x0012   100   100   ---    Old_age   Always      -       3478
 10 Spin_Retry_Count        0x0013   100   100   ---    Pre-fail  Always      -       0
 12 Power_Cycle_Count       0x0012   100   100   ---    Old_age   Always      -       1582
167 Unknown_Attribute       0x0022   100   100   ---    Old_age   Always      -       0
168 Unknown_Attribute       0x0012   100   100   ---    Old_age   Always      -       0
169 Unknown_Attribute       0x0013   100   100   ---    Pre-fail  Always      -       322229895806
170 Unknown_Attribute       0x0013   100   100   ---    Pre-fail  Always      -       47252242530
173 Unknown_Attribute       0x0013   194   194   ---    Pre-fail  Always      -       0
175 Program_Fail_Count_Chip 0x0013   100   100   ---    Pre-fail  Always      -       0
192 Power-Off_Retract_Count 0x0012   100   100   ---    Old_age   Always      -       170
194 Temperature_Celsius     0x0023   070   043   ---    Pre-fail  Always      -       30 (Min/Max 15/57)
240 Head_Flying_Hours       0x0013   100   100   ---    Pre-fail  Always      -       0
```

Fig. 4. Smartmontools reading SMART data

The warning system monitors streaming data from the hard disk, with the WS flow chart shown in Fig. 5.

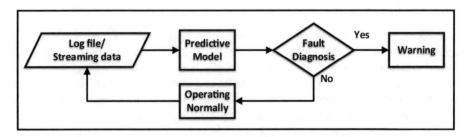

Fig. 5. Warning system flow

Once the hard disk data has been collected from the status monitor, the following steps are taken:

1. Extract the SMART attributes from the streaming data.
2. Perform fault prediction based on modules built by LSTM. If the prediction result is 0, the hard disk is working properly, but a prediction result of 1 indicates imminent damage and triggers a warning message.

4   Conclusion and Future Works

With the emergence of cloud computing and related online services, many large enterprises are increasingly providing services to users through large data centers. Data center maintenance teams must ensure server reliability in order to provide stable, continuous cloud services. In a server, the hard disk is a component particularly prone to failure, thus effective hard disk maintenance can greatly contribute to overall server performance.

The proposed system collects device information for analysis, allowing for the possibility of predicting hard disk failure running the LSTM algorithm on Apache Spark. The system consists of two stages: modeling and real-time prediction. In the modeling stage, historical data is used to generate the LSTM module. Test data is used to verify the ability of the established prediction module to predict hard disk failures. In the real-time prediction stage, data collected from the device in real time is run through the trained prediction module to predict hard disk failure in real-time.

The proposed system allows IT teams to quickly and effectively monitor hard drive condition. Deep learning methods are applied to provide a new kind of insight into data, allowing for improved efficiency in the maintenance of large storage systems. Accurate prediction of device failure reduces the need for scheduled maintenance and ensures that each device is in service for its entire useful life time.

This study proposes an innovative approach to monitoring hard disk failures. The concepts developed extend beyond descriptive statistics and charts. The implementation uses SMART statistics to predict imminent hard disk failure. However, hard disks from various vendors will have different SMART statistics. While each manufacturer follows general guidelines in defining data attributes, the meaning of the SMART values ultimately depends on the definition provided by the hard disk manufacturer. Therefore, this proposed system provides a predictive model for hard disk manufacturers, but the predictive model cannot be applied to hard disks from all other manufacturers. Thus, when predicting and analyzing performance of hard disks from various manufacturers, one must use historical data specific to each disk to establish a new prediction module.

The Internet of Things (IoT) concept uses the Internet as a means of linking physical devices. In the future, IoT will raise many opportunities for improving productivity and efficiency. The cloud analytics platform could be used to collect data from multiple sources and multiple devices, providing significant advantages to managing large numbers of devices in a wide range of domains.

In this research, the BackBlaze data of the hard disk is used to train the module that can predict the hard disk failure based on the SMART data and issue alert when there is a possibility of failure. In order to improve the results of the prediction, we suggest developing a complete solution, which can predict the timing of the hard disk failure more accurately. Furthermore, it can inform the user when the monitored hard disk is about to be failed and show the remaining lifetime, helping the user understand the status of the hard disk more clearly.

References

1. Andy, K.: Backblaze Hard Drive Stats for 2017 (2017). https://www.backblaze.com/blog/hard-drive-stats-for-2017/
2. Canizo, M., Onieva, E., Conde, A., Charramendieta, S., Trujillo, S.: Real-time predictive maintenance for wind turbines using big data frameworks. In: 2017 IEEE International Conference on Prognostics and Health Management (ICPHM), pp. 70–77. IEEE, June 2017
3. Zhao, P., Kurihara, M., Tanaka, J., Noda, T., Chikuma, S., Suzuki, T.: Advanced correlation-based anomaly detection method for predictive maintenance. In: 2017 IEEE International Conference on Prognostics and Health Management (ICPHM), pp. 78–83. IEEE, June 2017
4. Samuel, A.L.: Some studies in machine learning using the game of checkers. IBM J. Res. Dev. 3(3), 210–229 (1959)
5. Brownlee, J.: Discover Feature Engineering, How to Engineer Features and How to Get Good at It (2014). https://machinelearningmastery.com/discover-feature-engineering-how-to-engineer-features-and-how-to-get-good-at-it/
6. Giles, C.L., Lawrence, S., Tsoi, A.-C.: Rule inference for financial prediction using recurrent neural networks. In: IEEE Conference on Computational Intelligence for Financial Engineering, p. 253. IEEE Press (1997)
7. Hochreiter, S., Schmidhuber, J.: Long short-term memory. Neural Comput. 9(8), 1735–1780 (1997)
8. Graves, A., Schmidhuber, J.: Framewise phoneme classification with bidirectional LSTM and other neural network architectures. Neural Networks 18(5–6), 602–610 (2005)
9. Pinheiro, E., Weber, W.-D., Barroso, L.A.: Failure Trends in a Large Disk Drive Population (2007)

Identifying Touristic Interest Using Big Data Techniques

Maritzol Tenemaza[1,2(✉)], Loza-Aguirre Edison[1], Myriam Peñafiel[1,2],
Zaldumbide Juan[1,2], Angelica de Antonio[2], and Jaíme Ramirez[2]

[1] Departamento de Informática y Computación, Escuela Politécnica Nacional,
Quito, Ecuador
{maritzol.tenemaza, edison.loza, myriam.penafiel,
juan.zaldumbide}@epn.edu.ec
[2] Laboratorio ETSI, Universidad Politécnica de Madrid, Madrid, Spain
{angelica, jramirez}@fi.upm.es

Abstract. The objective of this paper is to identify the most visited places through a sentiment analysis of the tweets posted by people who visited a specific region of a city. The analyzed data were related to preferences and opinions about tourist places. This paper outlines an architectural framework and a methodology to collect and analysis big data from twitter platform.

Keywords: Big data · User's interest · Sentiment analysis · Harvesting

1 Introduction

Big Data techniques are widely used in data harvesting studies. The amount of data traveling on the Internet today is large, complex and interesting. Big data is the way that information is handled. The processing of large quantities of data is complex, nevertheless, there are many predictive analytics tools that control data volume, velocity and variety. The value of data or quality and veracity or consistence of data are additional issues for a big data approach [1].

Twitter contains massive human – information. Nowadays, Twitter has 350 million users geographically distributed in all world. A twitter user has little geospatial information, because the users disable the user's location in their smartphone. Twitter user tracking by associating the longitude and latitude. Microblogging today has become a very popular communication tool among Internet users. Millions of messages are appearing daily in Twitter. The users share opinions on a variety of topics and discuss current issues. Microblogging Twitter become valuable sources of people's opinions of sentiment analysis [2].

In this paper, we propose the identification and validation of the most popular tourist places in a city by using Big data Techniques and Twitter as the data source. In our case, our interest in detected the best places to visit in a specific city. We use Microblogging of twitter for the following reasons: In Microblogging platforms, the people express their opinion and sentiments. This site is constantly updated in real time and grows moment by moment. The tourist's audience tweet in regular form,

© Springer Nature Switzerland AG 2020
T. Ahram (Ed.): AHFE 2019, AISC 965, pp. 169–178, 2020.
https://doi.org/10.1007/978-3-030-20454-9_17

this audience is representative; it is possible to collect information of individual tourists, familial tourists, and group tourists. The tweets contain positive, negative and neutral sentiments. The geo-location Twitter users are possible to detect.

Our proposal could be applied to any city, furthermore, in this article we collected data from New York, Paris, and London. We harvested and analyzed more than 16 million tweets to find better places to visit in these cities. These results are important because any recommendation system required information of the best places previously identified by other users.

The remainder of this paper is organized as follows: In Sect. 2, an overview of different twitter analysis is mentioned. Next, in Sect. 3 the data recollection, framework proposal, analysis, and results are described. Third, in Sect. 4 the results are analyzed. Finally, conclusions and future work are discussed in Sect. 5.

2   Literature Review

The growth of online environments has made the issue of information search and selection increasingly cumbersome [3]. The recent explosion of digital data is so important because using big data, managers can measure, and hence know, radically more about their business, and directly translate that knowledge into improved decision-making and performance [4]. As of 2012, about 2.5 Exabyte of data are created each day. More data cross the internet every second than was stored in the entire internet 20 years ago. This gives companies an opportunity to work with many petabytes of data in a single data set.

However, for some companies, Velocity is more important than volume. Real Time o nearly real time information makes it possible for a company to be much more agile. Additionally, big data takes the form of messages, updates and images posted to social networks [4]. Thus, variety is another characteristic to consider when we discuss Big Data.

Twitter is a popular microblogging service where users create status messages called tweets. Twitter is mainly characterized by social functions [5] These tweets sometimes express opinions about different topics [6]. Millions of people are using social network sites to express their emotions, opinions and disclose about daily lives. However, people write anything such as social activities or any comment on products. Through the online communities provide, an interactive forum where consumers inform and influence others. Moreover, social media provides an opportunity for business that giving a platform to connect with their customers such as social media for connecting with the customer's perspective of products and services [7]. Microblogging websites have evolved to become source of varied of information on which people post real time messages about their opinions on a variety of topics discuss current issues, complain and express positive sentiment for products they use a daily life. In fact, companies manufacturing such products have started to poll these microblogs to get a sense of general sentiment for their product [8].

The amount of information about travel destinations and their associated resources such as accommodations, restaurants, museums or events among others is commonly searched for tourists in order to plan a trip [9]. Additionally, tourists visiting urban destinations require identifiers the most interesting attractions [10]. For this reason, we observe the opportunity of analyses a data set based in microblogging Twitter, where the user's express opinions of their visit specific cities. We will analyze the sentiment expressed in a tweet the objective will be determining the most interesting places evaluated by the tourist. This information we called the opinion of other people. These results will be important for the tourism enterprises.

3 Method

3.1 Architecture

To use the Twitter API, a virtual machine was implemented. Elasticsearch, Kibana, Cerebro as servers have been used, additionally scripts Phyton were necessary to apply the harvesting architecture. The virtual machine was defined in a web server (Fig. 1).

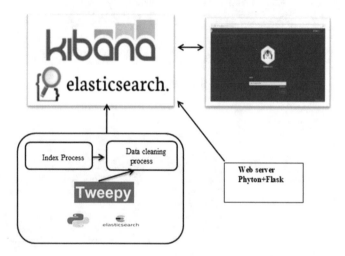

Fig. 1. Big data architecture

3.2 Methodology

For a recommender based in content, it is necessary the analysis of other tourist interest. Thus a Harvesting Methodology is applied. It defined: Data structuration, Data collection, Sentiment detection, Sentiment classification and the Presentation of results.

3.3 Data Structuration

The objective is to identify and map the attributes of Twitter that Elastic Search needs to collect. The tweets are collected in JSON format. They are send later to a NO-SQL database.

3.4 Data Collection

16 million of tweets were collected, by one month. The not structured data was structured. It was necessary to ensure that data is correct and representative. Tweepy library collected the information.

Data were collected in Paris, London and New York. Each city was segmented, Fig. 2 shown the segmentation on Paris, Fig. 3 on New York and Fig. 4 on London.

Fig. 2. Segmentation on Paris

Fig. 3. Segmentation on New York

Fig. 4. Segmentation on London

The data collected from twitter is necessary transform to format JSON. Then, it was stored in Elastic Search.

3.5   Sentiment Detection

The sentiment detection of a tweet is based in the analysis of the emotional charge makes it possible to distinguish the polarity (positive, negative, neutral), intensity (positive, negative) and emotion (happy, sad and others).

The sentiment analysis is detected by Text Blob library of Phyton, for that purpose is necessary import the code to the Text Blob, that is observed in Fig. 5.

```
import tweepy
import json
from tweepy.streaming import StreamListener
from tweepy import OAuthHandler
from tweepy import Stream
from textblob import TextBlob
from elasticsearch import Elasticsearch
# impor er
import re
# import twitter keys and tokens
from configLnl import *
```

Fig. 5. Importation of tweets to the Text Blob

In Fig. 6 we observe the code to obtain the polarity of every tweet and writing the polarity positive, negative o neutral.

```
print ("tweet capture -ID"+ str(dict_data["id"]))

# output sentiment polarity
print (tweet.sentiment.polarity)
```

Fig. 6. Code for tweet sentiment detection

3.6   Sentiments Classification

A learning supervised algorithm was applied, the machine was trained. The polarity (positive, negative, neutral) and the subjectivity (objective/subjective) is getting from the learning algorithm.

The informal nature of twitter requires the tweet pre-processing for demand the need to correct the colloquial expressions of the texts. The actions are: (a) eliminate the

Uniform Resource Locator (URL) of message, it is observed in Fig. 7 (b) tokenizer the words of twitter, (c) delete the stop words, whitespace and lines breaks, (d) replace smileys with their corresponding categories: happy, sad, tongue, wink and others, (e) exclude terms belonging to certain morph syntactic categories that are not significant for the analysis of feelings.

For natural language pre-processing, we use Text Blob for detection and classification of feelings. The classification is observed in Fig. 8.

```
cleantext = re.sub(r'^https?:\/\/.*[\r\n]*','',dict_data["text"],flags=re.MULTILINE)

cleantext = re.sub(r'^http?:\/\/.*[\r\n]*','',cleantext,flags=re.MULTILINE)
```

Fig. 7. Clean text of twit

```
# determine if sentiment is positive, negative, or neutral
if tweet.sentiment.polarity < 0:
    sentiment = "negative"
elif tweet.sentiment.polarity == 0:
    sentiment = "neutral"
else:
    sentiment = "positive"
```

Fig. 8. Sentiments classification

Because not all tweets have coordinates. It was necessary to separate the storage of tweets that have coordinates (Fig. 9) and those that not have coordinates (Fig. 10).

```
# add text and sentiment info to elasticsearch
es.index(index="londres_coord1",
         doc_type="test-type",
         body={"author": dict_data["user"]["screen_name"],
    "date": dict_data["created_at"],
    "tweet": dict_data["text"],
    "text": cleantext       ,
    "polarity": tweet.sentiment.polarity,
    "subjectivity":tweet.sentiment.subjectivity,
    "sentiment":sentiment,
    "location":dict_data["user"]["location"],
    "coordinates":dict_data["coordinates"],
    "geo_enabled":dict_data["user"]["geo_enabled"]})

print(f"Ingreso En elasticSearch paris_coord1")
```

Fig. 9. Code for storage the tweets in elastic search

```
    else:
        print(f"Ingreso En elasticSearch newyork_all1")
        # add text and sentiment info to elasticsearch
        es.index(index="newyork_all1",
                doc_type="test-type",
                body={"author": dict_data["user"]["screen_name"],
            "date": dict_data["created_at"],
            "tweet": dict_data["text"],
            "text": cleantext          ,
            "polarity": tweet.sentiment.polarity,
            "subjectivity":tweet.sentiment.subjectivity,
            "sentiment":sentiment,
            "location":dict_data["user"]["location"]})
        return True
# on failure
```

Fig. 10. Code for storage the tweets without coordinates

4   Results

The general purpose of the analysis is to transform the data obtained from twitter into meaningful information. The first step to ending the process is together the segments in each city. It is observed in Fig. 11, this process is known as re-indexation (Fig. 12).

Fig. 11. London segments

```
{ -
    "cordlondres": { -
        "settings": { -
            "index": { -
                "creation_date": "1532207983503",
                "number_of_shards": "5",
                "number_of_replicas": "1",
                "uuid": "LtDMxc8xTPKuYVW0q1qD_w",
                "version": { -
                    "created": "5060999"
                },
                "provided_name": "cordlondres"
            }
        }
    }
}
```

Fig. 12. Query to configure the London index

4.1 Scenarios

The three scenarios are Paris, New York and London. The data was collected from 20 May 2018 to 12 August 2018. In total, 16'064,840 tweets was collected. For this study only was analyzed the tweets with coordinates (Table 1).

Table 1. Scenarios by city

City	Tweets with coordinates	Tweets without coordinates
Paris	95,696	982,116
New York	439,996	4,121,340
London	187,908	2,109,975
World	7,139,185	-

Figure 13 shown the number of tweets in New York, The 15 July 2018 there are the greatest number of tweets. Figure 14 shown the best places in New York. The red area represents 50% of positive tweets. And the orange area represents 26% of positive tweets. The other areas son referenced in the rest of tweets. 90% of tweets refer to Times Square, Fifth Avenue, 40 theaters that make up the Broadway circle. Others references are the World Trade Center among others.

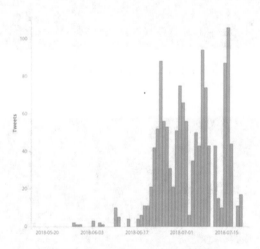

Fig. 13. Tweets in New York by date

Fig. 14. The best places in New York

The same process was applied to Paris and London. With a map of coordinates all the tourist places mentioned in tweets of Paris, London or New York were appreciated. The results are observed in Fig. 15 All these results are offered in a REST web service.

Fig. 15. Touristic places mentioned in tweets

5   Conclusions

Our results are coincident with the world barometer where the most visited city is New York and the most visited places in that city.

Our work had shown the potential of big data tools in the sentiment analysis of tweets. It is possible analyses emoticons, hashtags and others, it shows the potential of twitter information.

The big data – natural language processing tools used are useful and easily the processing of text, the polarity and translate of emoticons.

Microblogging data like twitter, on which users post real-time reactions to and opinions about specific places in different cities is the best material to define a tourism recommender.

Our future work will be developing a touristic recommender based in the other user's information based in analysis of twitter using Big Data tools.

References

1. Tole, A.A.: Big data challenges. Database Syst. J. **4**(3), 31–40 (2013)
2. Pak, A., Paroubek, P.: Twitter as a corpus for sentiment analysis and opinion mining. In: LREc (2010)
3. Gavalas, D., Konstantopoulos, C., Mastakas, K., Pantziou, G.: Mobile recommender systems in tourism. J. Netw. Comput. Appl. **39**, 319–333 (2014)
4. McAfee, A., Brynjolfsson, E., Davenport, T.H., Patil, D.J., Barton, D.: Big data: the management revolution. Harvard Bus. Rev. **90**(10), 60–68 (2012)
5. Wu, X., Zhu, X., Wu, G.Q., Ding, W.: Data mining with big data. IEEE Trans. Knowl. Data Eng. **26**(1), 97–107 (2014)
6. Go, A., Huang, L., Bhayani, R.: Twitter sentiment analysis. Entropy **17**, 252 (2009)
7. Sarlan, A., Nadam, C., Basri, S.: Twitter sentiment analysis. In: 2014 International Conference on Information Technology and Multimedia (ICIMU). IEEE (2014)
8. Agarwal, A., Xie, B., Vovsha, I., Rambow, O., Passonneau, R.: Sentiment analysis of twitter data. In: Proceedings of the Workshop on Languages in Social Media. Association for Computational Linguistics (2011)
9. Borràs, J., Moreno, A., Valls, A.: Intelligent tourism recommender systems: a survey. Expert Syst. Appl. **41**(16), 7370–7389 (2014)
10. Gavalas, D., Kasapakis, V., Konstantopoulos, C., Pantziou, G., Vathis, N., Zaroliagis, C.: The eCOMPASS multimodal tourist tour planner. Expert Syst. Appl. **42**(21), 7303–7316 (2015)

The Development of the Theory of Quality Assessment of the Information, Taking into Account Its Structural Component

Olga Popova[(✉)], Yury Shevtsov, Boris Popov, Vladimir Karandey,
and Vladimir Klyuchko

Kuban State Technological University, Krasnodar, Russia
popova_ob@mail.ru, pbk47@mail.ru,
{shud48, epp_kvy}@rambler.ru, kluchko@kubstu.ru

Abstract. Today is topical the automation of the production part of the search strategy. It is required to study the internal organization of information, which is presented not in a semantic way, but from the position of structuring information. Mathematic formulas reflecting the internal organization of information that affects the effectiveness of the choice of the method for solving the problem become actual. Modern quality indicators do not use the parameters of information structuring. But the structured information can forms the stable links and relationships between the procedural knowledge of the subject area used in modern computer systems to support scientific research, to organize effective selection of the most appropriate method for solving the current applied problem. The following scientific idea is proposed. A structured subject area has an optimal scope for making any right decision. Structured information can't be superfluous or incomplete, since it takes into account all ideal cases.

Keywords: Information · Entropy · Quality of information · Knowledge · Decision tree · Question and answer system binary tree

1 Introduction

At present, the quality of information is determined by such indicators as representativeness, content-richness, sufficiency, accessibility, relevance, timeliness, precision, validity, and stability. However, the structuredness of information indicator is not included among them, although it can show the nature of stable associations and relationships between procedural knowledge of a given domain, e.g. «optimization methods» that now has to be used in the modern computer systems for supporting the scientific research in order to organize efficient choice of the most suitable solution method for the current applied problem. The structured approach of depicting the information also allows meeting the needs of qualitative visualization of the domain. Currently, for determining the quality of information, they mainly use such variables as semantic capacity, semantic throughput capacity, and the information coefficient. The information coefficient can be considered the most useful for qualitative evaluation of information represented in a domain as of the present. The problem is the fact that

T. Ahram (Ed.): AHFE 2019, AISC 965, pp. 179–191, 2020.
https://doi.org/10.1007/978-3-030-20454-9_18

variables used in calculating this coefficient characterize the meaningful content of information (semantics) and the pragmatic content of information (pragmatics) while overlooking the structuredness of information completely. The scientific idea suggested by the authors puts forward the following fact as the basis: a well-structured domain has an optimum volume for making any correct decision. The information structured in the domain cannot be excessive as it is required for making a correct decision concerning the following problem which has its own set of conditions. It cannot be incomplete, too, as all the ideal cases have to be taken into account in the structure of the domain. The domain structure visualizes logical and graphic associations between procedural knowledge of the domain and helps analyze it, find new logical and graphic associations and obtain new procedural knowledge.

Within the contemporary cognitive approach [1], an object is described by: a set of the object parameters the meanings of which are given linguistically or numerically and a set of cause-effect relationships given over the set of all possible values of the parameters. Formally, the model is represented as an oriented graph – a cognitive map. The cognitive map is a model of expert's knowledge about processes in a dynamic situation with cause-effect relationships (W, F), where $W = |w_{ij}|$ – orgraph adjacency matrix, F – the set of dynamic situation factors, $F = \{f_i\}$, i, ..., n. The «hard» systems analysis is based on precise measurements and is oriented to optimizing the systems. For each factors, the set of linguistic values is determined, $X_i = \{x_{iq}\}$, and for modeling the dynamics of situation parameters change and obtaining the forecasts of development of the situation, a system of finite-difference equations is set as follows:

$$X(t+1) = WX(0), \qquad (1)$$

here X(0) is the vector of initial increment of values of the factors, t = 0;
X(t + 1) is the vector of change of the factors values at time points t = 1, ..., n;
W is the adjacency matrix.

The soft systems analysis (P. Checkland) is oriented to structuring the knowledge about a poorly defined situation in order to understand the main processes in the situation. Then, the decision-making task consists in analyzing the system state change dynamics which is represented by vectors of state at successive time points X(t), ∀t and interpretations thereof. An interpretation is the representation of the vector of state of the object obtained in the process of modeling within the interpreting system

$$\Psi : X(t) \rightarrow \vartheta, \qquad (2)$$

where ϑ is the interpreting system.

By interpreting system ϑ, the subjective conceptual system of the decision-making person is meant – the person's knowledge about the domain. The person's intellectual abilities Ψ – reasoning, generalization, and imagination, – are considered to be able to represent vector of state X(t) as a notion of domain ϑ in which the cognitive map is built. All these methods imply cognitive modeling in which cognitive processes within the human intellect are used that are generated by interpreting the results of modeling of a crude model of the reality in the form of a cognitive map into the subject's conceptual system. Meanwhile, exploratory research requires using the natural structuring – and in the human intellect, cognitive processes are generated by the processes of understanding and memorizing the knowledge belonging to a domain by means of structuring this knowledge. Cognitive maps being a well-known tool only help visualize the knowledge well in order to remember it.

This approach allows using Shannon's entropy formula in a completely different way, with the structural approach to be taken into account against reduction of uncertainty. Entropy is expected to decrease as the completely new property appears. This proves the hypothesis: a non-trivial method of solving a problem having a unique property reduces the uncertainty of the domain.

For carrying out the research, the object of the research was selected – the contemporary information system for supporting the scientific research, Optimel, in which one can efficiently select the most suitable method for solving the current applied problem [2, 3]. The subject of the research will be the optimization methods domain that features structuredness and visualization which is sufficient for conducting the research. Conclusions will be made using the rules and reasoning principles based on the empirical data about the object of the research. The basis for obtaining the data is the optimization methods domain that is to be studied experimentally. In order to explain any facts found, a hypothesis will be put forward. Shannon's model will be supplemented proceeding from the data obtained. Within the research, it is necessary to find the dependence between the contemporary indicators of the quality of information and the information structuredness extent.

2   Studying a Degenerated Structure

Let gradual structuring of the elements of set M be considered which are the solution methods for optimization problems having description and their own properties. First, the elements of set i have property $C_i(i) = C_0$

$$M = \{i \in I \mid C_i(i)\} = \{i \in I \mid C_0(i)\}. \tag{3}$$

Next, let several elements be successively isolated from set M. Each one will differ from other elements of the set in property $C_i(i)$, where $i = N, \ldots, 1$. For this, rule 1 is used which helps structure the set.

So, the following is obtained (see Figs. 1, 2, 3, 4 and 5):

$$M_1 = \{i \in I \mid C_1(i-1)\}, M_2 = \{i \in I \mid C_2(i-2)\}, M_3 = \{i \in I \mid C_3(i-3)\},$$
$$M_4 = \{i \in I \mid C_4(i-4)\}, \ldots, M_N = \{i \in I \mid C_N(1)\}. \tag{4}$$

As a result, a left-side degenerated tree is obtained. Even in case of such a structure, looking for a method from set M will be more efficient than in the case of using ready structure of the Prolog logical language.

Structural isolation i of the method can be considered an equally probable process in which the quantity of elements being structured changes at the increment of 1 from the maximum value N. Let Shannon's entropy formula H(AB) for this process be written down as:

$$H(AB) = H(A) + H(B) = -(1/i) * \log_2(1/i) - ((i-1)/i) * \log_2((i-1)/i), \quad (5)$$

where system A is solution method being structured (i);
system B is the quantity of the remaining solution methods (i);
system AB is the process of structuring;
i - the quantity of problem solving methods to be structured which is decreased at the increment of 1 in the process of the methods structuring, i = N, N − 1, N − 2, ..., 1.

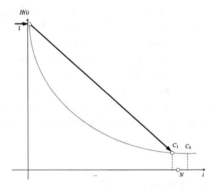

Fig. 1. Structural isolation of the tree root by property C_1.

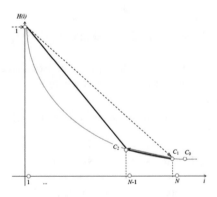

Fig. 2. Structural isolation of element N − 1 of set M by property C_2

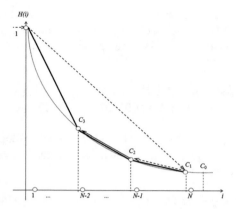

Fig. 3. Structural isolation of element $N - 2$ of set M by property C_3

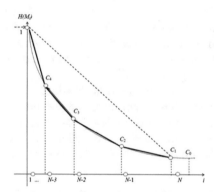

Fig. 4. Structural isolation of element $N - 3$ of set M by property C_4

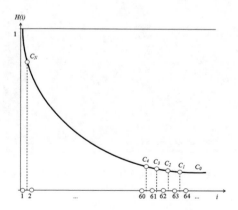

Fig. 5. Structural isolation of element 1 of set M by property C_N

Next, the data from the obtained structure of the binary question-answer system tree will be used. The structure was used in the Optimel software product [2]. First, the shared root is isolated, so i = N = 64. Now, group B includes 63 methods of solving the problems. This is why

$$H_{64}(AB) = -(1/64) * \log_2(1/64) - ((64-1)/64) * \log_2((64-1)/64) = \\ = 0,09375 + 0,02237 = 0,11612. \tag{6}$$

After isolating the method having the index of 63, group B will include 62 methods. Let the following be written down for structural isolation of the solution method having index 63:

$$H_{63}(AB) = -(1/63) * \log_2(1/63) - ((63-1)/63) * \log_2((63-1)/63) = \\ = 0,09488 + 0,02272 = 0,11759. \tag{7}$$

In the similar manner, Table 1 is filled in with data using formula (5) and dependence graph H(AB) is built (Fig. 6). It is clear from Fig. 6 that entropy of system AB consisting of i solution methods has the minimum value of 0,11 at position i = 64. Therefore, the higher the quantity of solution methods being structured, the lower the minimum value of uncertainty is. It should be reminded that the tree is left-side, degenerated and has the height of h = 63.

As a new method of solving the problem is added to set M, it will occupy the required position among structured elements i. The methods located on the right of the place where it was added will shift by one position to the right. Meanwhile, the minimum entropy value will decrease. Equation (5) have to be used, where i = 64 + 1 = 65. In the formulas, i = 64 + n if n methods are added to set M.

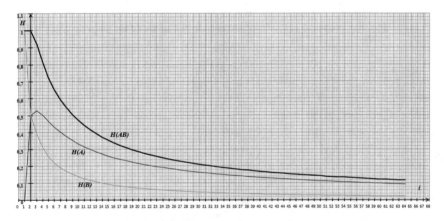

Fig. 6. Entropy of the system being in the process of structuring

3   Studying the Process of Structuring

When structuring a set, it is more efficient to obtain data structures that differ from the degenerated structures in the form. Such an approach helps obtain the optimal code. Let the calculation for the data structure be given – a binary question-answer system tree (see Figs. 7 and 8) which was obtained by O. Popova and B. Popov. Let the above Eq. (5) and Fig. 7 be used. Let Table 2 be filled in with calculated data according to which the system structuring process entropy (Fig. 9) is built. Let the resulting value of entropy of system ΣH and Ent(H) + 1 (Fig. 10) be built, too, as is the convention in calculating Hartley's measure.

Table 1. The value of Shannon entropy H(AB) of the system being structured.

i	H(A)	H(B)	H(AB)	i	H(A)	H(B)	H(AB)
2	0,5	0,5	1	34	0,149631	0,041802	0,191433
4		0,5	0,311278	0,811278	36	0,143609	0,039513	0,183122
6	0,430827	0,219195	0,650022	38	0,138103	0,037462	0,175565
8		0,375	0,168564	0,543564	40	0,133048	0,035613	0,168661
10	0,332193	0,136803	0,468996	42	0,128389	0,033938	0,162326
12	0,298747	0,11507	0,413817	44	0,124078	0,032413	0,156491
14	0,271954	0,099278	0,371232	46	0,120077	0,03102	0,151097
16	0,25	0,08729	0,33729	48	0,116353	0,029741	0,146094
18	0,231663	0,077881	0,309543	50	0,112877	0,028563	0,141441
20	0,216096	0,070301	0,286397	52	0,109624	0,027476	0,137099
22	0,202701	0,064064	0,266765	54	0,106572	0,026468	0,13304
24	0,19104	0,058842	0,249882	56	0,103703	0,025531	0,129234
26	0,180786	0,054407	0,235193	58	0,101	0,024658	0,125658
28	0,171691	0,050594	0,222285	60	0,098448	0,023843	0,122292
30	0,163563	0,047279	0,210842	62	0,096035	0,023081	0,119116
32	0,15625	0,044372	0,200622	64	0,09375	0,022365	0,116115

Table 2. The value of entropy of the system being structured at various levels of the tree

i	H_0	H_1	H_2	H_3	H_4	H_5	H_6	H_7	H_8	H_9	H_{10}	H_{11}	ΣH	entH
1	0,11	0,11	0,26	0,50									1,0	2
2	0,20	0,20	0,43	0,76	0,54	0,72	0,91						3,7	4
3	0,27	0,27	0,57	0,91	0,81	0,97	0,91						4,7	5
4	0,33	0,34	0,68	0,99	0,95	0,97							4,2	5
5	0,39	0,39	0,77	0,99	1	0,72							4,2	5
6	0,44	0,45	0,84	0,91	0,95								3,6	4
7	0,49	0,50	0,90	0,76	0,81	0,91							4,3	5

(continued)

Table 2. (*continued*)

i	H_0	H_1	H_2	H_3	H_4	H_5	H_6	H_7	H_8	H_9	H_{10}	H_{11}	ΣH	entH
8	0,54	0,54	0,94	0,50	0,54	0,91							4,0	5
9	0,58	0,59	0,97										2,1	3
10	0,62	0,63	0,99	0,39	0,72								3,3	4
11	0,66	0,66	1	0,61	0,97								3,9	4
12	0,69	0,70	0,99	0,77	0,97	0,91							5,0	6
13	0,72	0,73	0,97	0,89	0,72	0,91							4,9	5
14	0,75	0,76	0,94	0,96									3,4	4
15	0,78	0,79	0,90	0,99	0,54	0,72	0,91						5,6	6
16	0,81	0,81	0,84	0,99	0,81	0,97	0,91						6,1	7
17	0,83	0,84	0,77	0,96	0,95	0,97							5,3	6
18	0,85	0,86	0,68	0,89	1	0,72							5,0	6
19	0,87	0,88	0,57	0,77	0,95								4,0	5
20	0,89	0,90	0,43	0,61	0,81	0,91							4,5	5
21	0,91	0,91	0,26	0,39	0,54	0,91							3,9	4
22	0,92	0,93											1,8	2
23	0,94	0,94	0,16	0,19	0,81	0,91							3,9	4
24	0,95	0,95	0,28	0,32	1	0,91							4,4	5
25	0,96	0,96	0,37	0,43	0,81								3,5	4
26	0,97	0,97	0,46	0,52									2,9	3
27	0,98	0,98	0,53	0,60	0,21	0,91							4,2	5
28	0,98	0,99	0,60	0,67	0,35	0,91							4,5	5
29	0,99	0,99	0,65	0,73	0,46								3,8	4
30	0,99	0,99	0,71	0,78	0,56	0,22	0,91						5,2	6
31	0,99	0,99	0,75	0,83	0,65	0,38	0,91						5,5	6
32	1	0,99	0,80	0,87	0,72	0,50							4,9	5
33	0,99	0,99	0,83	0,90	0,78	0,60	0,24						5,3	6
34	0,99	0,99	0,87	0,93	0,83	0,69	0,41						5,7	6
35	0,99	0,99	0,90	0,95	0,88	0,76	0,54	0,26	0,35	0,65			7,3	8
36	0,98	0,98	0,92	0,97	0,91	0,82	0,65	0,43	0,56	0,91			8,1	9
37	0,98	0,97	0,94	0,98	0,94	0,87	0,73	0,57	0,72	1	0,81		9,5	10
38	0,97	0,96	0,96	0,99	0,97	0,91	0,81	0,68	0,83	0,91	1	0,91	10,9	11
39	0,96	0,95	0,97	1	0,98	0,95	0,87	0,77	0,91	0,65	0,81	0,91	10,7	11
40	0,95	0,94	0,98	0,99	0,99	0,97	0,91	0,84	0,97				8,5	9
41	0,94	0,93	0,99	0,98	1	0,99	0,95	0,90	0,99	0,50	0,81		10,0	11
42	0,92	0,91	0,99	0,97	0,99	0,99	0,97	0,94	0,99	0,76	1		10,5	11
43	0,91	0,90	0,99	0,95	0,98	0,99	0,99	0,97	0,97	0,91	0,81		10,4	11
44	0,89	0,88	0,99	0,93	0,97	0,99	1	0,99	0,91	0,99			9,5	10
45	0,87	0,86	0,98	0,90	0,94	0,97	0,99	1	0,83	0,99	0,72		10,1	11
46	0,85	0,84	0,97	0,87	0,91	0,95	0,97	0,99	0,72	0,91	0,97		10,0	11

(*continued*)

Table 2. (*continued*)

i	H_0	H_1	H_2	H_3	H_4	H_5	H_6	H_7	H_8	H_9	H_{10}	H_{11}	ΣH	entH
47	0,83	0,81	0,96	0,83	0,88	0,91	0,95	0,97	0,56	0,76	0,97	0,91	10,4	11
48	0,81	0,79	0,94	0,78	0,83	0,87	0,91	0,94	0,35	0,50	0,72	0,91	9,4	10
49	0,78	0,76	0,92	0,73	0,78	0,82	0,87	0,90					6,5	7
50	0,75	0,73	0,90	0,67	0,72	0,76	0,81	0,84	0,59				6,8	7
51	0,72	0,70	0,87	0,60	0,65	0,69	0,73	0,77	0,86				6,6	7
52	0,69	0,66	0,83	0,52	0,56	0,60	0,65	0,68	0,98	0,72	0,81	0,91	8,6	9
53	0,66	0,63	0,80	0,43	0,46	0,50	0,54	0,57	0,98	0,97	1	0,91	8,4	9
54	0,62	0,59	0,75	0,32	0,35	0,38	0,41	0,43	0,86	0,97	0,81		6,5	7
55	0,58	0,54	0,71	0,19	0,21	0,22	0,24	0,26	0,59	0,72			4,3	5
56	0,54	0,50	0,65										1,7	2
57	0,49	0,45	0,60	0,59	0,81								2,9	3
58	0,44	0,39	0,53	0,86	1	0,91							4,1	5
59	0,39	0,34	0,46	0,98	0,81	0,91							3,9	4
60	0,33	0,27	0,37	0,98									1,9	2
61	0,27	0,20	0,28	0,86	0,91								2,5	3
62	0,20	0,11	0,16	0,59	0,91								1,9	2
63	0,11												0,1	1
64													0	1

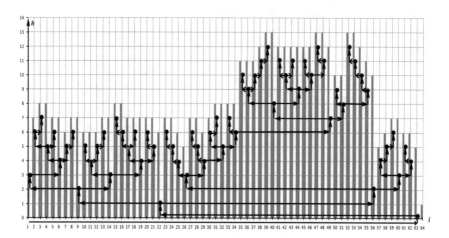

Fig. 7. The structure of binary question-answer system tree

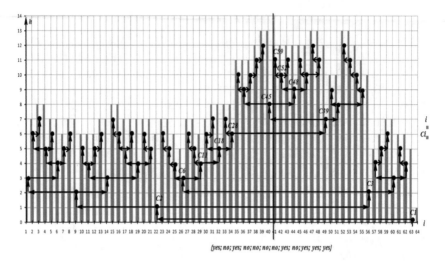

Fig. 8. Identification of the property for method M41 (M41 = {41 ∈ 64 | C1, C2, C3, C6, C11, C18, C28, C39, C45, C48, C52, C59})

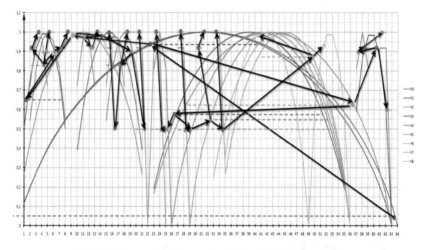

Fig. 9. Entropy of the process of Fig. 7 system structuring

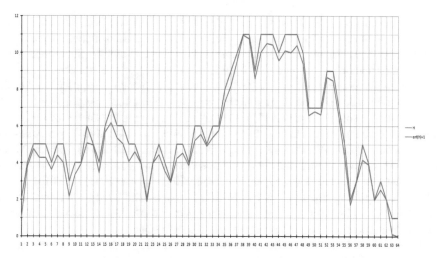

Fig. 10. Entropy dependence graph of system ΣH and $Ent(H) + 1$

4   Results and Discussion

Having compared Figs. 7 and 10, one can conclude that the outline of the tree repeats the shape of the entropy graph. Let it be considered that location height value h of element i on the tree for set M is Hartley's measure. In Fig. 7, height h is represented by the blue column of the histogram. Height h for property C_i is shown by the red column of the histogram. Therefore, Hartley's measure and Shannon's uncertainty measure coincide, with the structural isolation of each element i from set M being equally probable and not depending statistically. Hence the information capacity of the system consisting of set M is used up fully. Now M is structured and uncertainty of the situation is completely eliminated, as M includes all the known methods for solving optimization problems.

Therefore, the information of a structured domain equals entropy:

$$I = H. \tag{8}$$

In terms of the scientific and technical advance which implies new optimization problems emerging that require new solution methods to be found, the situation with the optimization methods domain is only resolved partially. This is why the new knowledge is the partial information which is difference between initial H_0 and final H_1 entropy:

$$I = H_0 - H_1. \tag{9}$$

Representing the domain information from the standpoint of its being structured is detailed by the classical definition of the notion «bit information» of Shannon's model in full: «… it equals the quantity of yes and no answers to reasonably posed questions using which the same information can be obtained» [4].

The contemporary Shannon's model cannot predict a list of possible states of the system in advance, for instance, emergence of a new solution method for an optimization problem. Popova and Popov suggest using logical and graphical representation of information in the domain. Infological structure of the domain knowledge obtained by the new algorithm of knowledge structuring which is approximated to the human intelligence helps not only find the strongest solution for the problem but also discover a new solution method [5] by selecting a new property of the method. The structured domain [6] contains information featuring undisputable importance and value because it is only ideal knowledge that is accumulated in the domain. When progressing along the tree, the most suitable one or the strongest problem solving method will be selected. The properties passed will contain explanation of the choice [7] made. Therefore, structuring the domain information and taking it into account in the calculations allows eliminating the formal aspect and considering such indicators of the information as its value, importance and content-richness.

O. Popova and B. Popov believe the sign side of messages should not be rejected in Shannon's model because a relevantly fitted sign system can improve structuring and help obtain efficient logical and graphical diagrams. For solving the above example, the sign system using 4 denotatums (Table 3) was used. The denotatums and the corresponding connotatum form the binary question-answer system tree which visually repeats the structural form of the way knowledge is represented by the human intellect and uses new logical and graphical associations.

Table 3. Sign system.

denotatum	Property C	Method i	↗ no	yes ↘
connotatum	a property represented in the question in the natural language form	name of the solution method in the natural language form	the problem has no such property, transition	the problem has such a property, transition

It can be concluded that structural representation of information and taking into account the structural constituent in indicators of the quality of information representation allow enhancing Shannon's model. It may also be spoken about the structured information of the domain being already represented in a qualitative manner. Now the efficiency of ways of representing the domain information can be compared from the standpoint of structuring it.

Acknowledgments. The work was carried out with the financial support provided by the Russian Foundation for Humanities within the research projects No. 16-03-00382 within the theme "Monitoring the research activity of educational institutions in the conditions of information society" of 18.02.2016.

References

1. Kulinich, A.A.: Situational, cognitive and semiotic approach to decision-making in organizations. Open Educ. **20**(6), 9–16 (2016)
2. Popova, O., Popov, B., Romanov, D., Evseeva, M.: Optimel: software for selecting the optimal method. SoftwareX **6C**, 231–236 (2017). http://www.sciencedirect.com/science/article/pii/S2352711017300316?via%3Dihub
3. Optimel smart selection system. RF patent No. 2564641, 27 May 2014. Bulletin No. 28, 21 p. Accessed 10 Oct 2015
4. Chapter 2. The quantity of information. http://izi.vlsu.ru/teach/books/002/le2/page2.html
5. Popova, O., Shevtsov, Y., Popov, B., Karandey, V., Klyuchko, V.: Theoretical propositions and practical implementation of the formalization of structured knowledge of the subject area for exploratory research. In: Karwowski, W., Ahram, T. (eds.) Advances in Intelligent Systems and Computing, IHSI 2018, vol. 722, pp. 432–437. Springer, Dubai (2018)
6. Popova, O., Shevtsov, Y., Popov, B., Karandey, V., Klyuchko, V., Gerashchenko, A.: Entropy and algorithm of the decision tree for approximated natural intelligence. In: Karwowski, W., Ahram, T. (eds.) Advances in Intelligent Systems and Computing, AHFE 2018, vol. 787, pp. 310–321. Springer, Orlando (2018)
7. Popova, O., Shevtsov, Y., Popov, B., Karandey, V., Klyuchko, V.: Studying an element of the information search system – the choice process approximated to the natural intelligence. Advances in Intelligent Systems and Computing, IntelliSys 2018, vol. 868, pp. 1150–1168. Springer, London (2018)

A Meta-Language Approach for Machine Learning

Nicholas Caporusso$^{(\boxtimes)}$, Trent Helms, and Peng Zhang

Fort Hays State University, 600 Park Street, Hays 67601, USA
n_caporusso@fhsu.edu,
{tehelms,p_zhang15_sia.se}@mail.fhsu.edu

Abstract. In the last decade, machine learning has increasingly been utilized for solving various types of problems in different domains, such as, manufacturing finance, and healthcare. However, designing and fine-tuning algorithms require extensive expertise in artificial intelligence. Although many software packages wrap the complexity of machine learning and simplify their use, programming skills are still needed for operating algorithms and interpreting their results. Additionally, as machine learning experts and non-technical users have different backgrounds and skills, they experience issues in exchanging information about requirements, features, and structure of input and output data.

This paper introduces a meta-language based on the Goal-Question-Metric paradigm to facilitate the design of machine learning algorithms and promote end-user development. The proposed methodology was initially developed to formalize the relationship between conceptual goals, operational questions, and quantitative metrics, so that measurable items can help quantify qualitative goals. Conversely, in our work, we apply it to machine learning with a two-fold objective: (1) empower non-technical users to operate artificial intelligence systems, and (2) provide all the stakeholders, such as, programmers and domain experts, with a modeling language.

Keywords: End-user development · Artificial intelligence · Machine learning · Neural networks · Meta-design · Goal question metric

1 Introduction

In the recent years, artificial intelligence (AI) systems [1] ended a long incubation phase and became mainstream technology. Nowadays, different types of machine learning (ML) algorithms are utilized for a variety of tasks (e.g., computer vision, natural language processing, and audio and speech recognition) in several different applications in diverse contexts and domains (e.g., robotics, manufacturing, and healthcare) [2]. Indeed, the design and implementation of ML algorithms, and specifically, neural networks (NNs), involve a number of challenges, such as, selecting the most relevant features and creating a representative model, identifying the type of task (i.e., supervised or unsupervised learning), choosing which family of algorithm to use (e.g., logistic regression, Support Vector Machines, or Artificial Neural Networks), selecting the most effective approach (e.g., recurrent or feed forward NNs), configuring

© Springer Nature Switzerland AG 2020
T. Ahram (Ed.): AHFE 2019, AISC 965, pp. 192–201, 2020.
https://doi.org/10.1007/978-3-030-20454-9_19

parameters (e.g., number of input and hidden layers), and fine-tuning the system for increasing its performance (e.g., introducing modifications, pre-processing input, or using hybrid approaches) [1]. Consequently, the overwhelming complexity of ML restrained its progress, in the last decades. To this end, communities in the field of AI joined their efforts and created software libraries that package ML systems into ready-to-use toolkits. These, in turn, enable a larger spectrum of professionals to choose among many available algorithms and to conveniently use them by simply customizing their behavior. As a result, developers do not have to deal with the complexity and risks of programming a system from scratch. Examples include Tensorflow [3], Apache MXnet [4], Microsoft Cognitive Toolkit (CNTK) [5], Scikit-learn [6], Pytorch [7], and many other toolkits that empower developers to operate ML systems and incorporate them in projects. As several software libraries are released under Open Source licenses, they foster the growing adoption of ML algorithms and, simultaneously, contribute to addressing current open issues and to improving performance. Moreover, the Open Source community, as well as organizations, released additional libraries. Nowadays, packages (e.g., sample datasets, visual or command-line tools, software automatic feature selection) can be layered on top of ML engines to achieve higher levels of abstraction and streamline the most tedious operations. In addition, by hiding complexity, toolkits render ML systems accessible to less-expert users, increase their awareness on AI, and promote their willingness explore the field [8].

Nevertheless, understanding and operating ML libraries still require some degree of programming literacy, which prevents many individuals who lack coding skills from being able to approach the use of machine learning algorithms and even understand their dynamics. Also, this is due to several factors that contribute to rendering ML obscure [9], such as, the inherent properties of the hidden layers of a NN. As a consequence, the intrinsic and intentional opacity of algorithms and requirements in terms of technical literacy increase the current divide between programmers who are expert about AI but lack domain knowledge [10] and customers who should not be required to be proficient in ML and yet, are not provided with any actionable framework for approaching it. This, in turn, leads to unrealistic expectations, communication barriers, and slower adoption rate, especially among small businesses. As the number of organizations that demand or are offered ML-based solutions is increasing, facilitation tools are needed to benefit both domain experts and AI scientists. Regardless of the experience level and the availability of powerful tools, given the novelty of this field, there is a lack of meta-design methodologies that improve the ergonomic aspects of machine learning and enable end-user development (EUD) of AI systems.

In this paper, we focus on the socio-technical dimension of artificial intelligence, and we propose a meta-language and design framework that (1) supports individuals with a broader spectrum of expertise to approach and leverage AI, (2) facilitates communication and information exchange between stakeholders having different backgrounds and heterogeneous skills, and (3) provides users with an actionable language for representing the meta-design of a ML system. To this end, we adopt the Goal-Question-Metric (GQM) approach as a viable framework for supporting end-user development across multiple disciplines. We describe the model and we detail examples of its implementation.

2 Related Work

Artificial Intelligence has the objective of creating systems that enable computers to supplement or replace human intelligence in solving problems that require some degree of decision-making [11]. In the last decades, research led to algorithms that, in addition to processing data, learn from them, that is, they are able to modify their internal structure based on the information they are trained with. By doing this, they can cope with certain degree of ambiguity and variation, fill-in data gaps, and leverage existing knowledge to take decisions in unprecedented scenarios. As an example, neural networks are being utilized in the healthcare domain to automatically identify cancer [12], monitor eye conditions [13], detect movement disorders [14], or support physicians' decisions in rehabilitation programs [15]. Moreover, recent advances in AI produced deep learning systems that are able to autonomously design their internal structure based on the input data. As a result, they achieve higher accuracy and performance in tasks that involve unstructured information (e.g., text, sound, and images) [16], though they require longer training time or more powerful hardware resources.

Although several groups pursuit the development of a general AI, that is, a unique system that can deal with any type of task, most of the problems are solved with narrow algorithms, each focused at one specific aspect. To this end, toolkits are extremely useful, as organizations can access a collection of methods and use them as building blocks, on a case by case basis. Although well-curated packages are more functional and easier to use than pure algorithms, they are not fully self-operational [17] and their implementation requires highly-skilled AI scientists, though there are early attempts to create high-level programming languages that support end-users [18].

To this end, several methodologies for the operationalization of conceptual models are available. The goal-question-metric paradigm is a structured approach for translating high-level objectives into actionable, quantitative metrics [19]. Most of the work using the GQM method is related to software engineering [20, 21]: as an example, the authors of [22] proposed a solution based on GQM for classifying well-known design patterns using machine learning. However, their goals and metrics consist in a perfect bijection. Hence, there is little advantage in using a predictor and in approaching the problem with the use of machine learning. Conversely, in [23], the authors describe a framework for risk management: the GQM methodology is utilized to determine which metrics affect projects. To this end, they envision a high-level goal, identify questions that need to be answered in order to evaluate whether the objective is met, and specify metrics that help take a quantitative approach to addressing each of the questions. After identifying the key metrics, they use them as features to train an artificial neural network (ANN) to produce a regression function estimate and to establish the success probability of a project based on the comparison between the threshold value and the regression function. The authors of [24] utilize ML techniques to evaluate the quality of software metrics defined using a GQM approach. Nevertheless, the structure of GQM itself can be utilized as an interpretation framework for artificial reasoning and, specifically, for describing the structure and functioning of neural networks, though it can be applied to other families of AI algorithms as well, as detailed in the next Section.

3   GQM as a Meta-Language for Machine Learning

GQM was initially designed as a language for translating business goals into questions and, ultimately, into metrics, in a top-down fashion. As a result, it can be utilized as a tool for supporting the top management in taking business decisions and developing assessment plans: specifying questions that enable middle managers to determine whether certain goals are met helps teams focus on the most important aspects, that is, improving the metrics that are meaningful to reaching their objectives. Simultaneously, data collected from low-level processes can be interpreted by middle managers, who can report to the executives in a bottom-up fashion. This type of hierarchical structure can be represented as a pyramid of connections: the graph resulting from a GQM model and its closed-loop process are shown in Fig. 1.

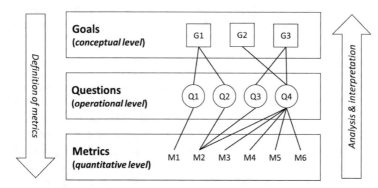

Fig. 1. An example of the logical structure of GQM, adapted from [25]: conceptual goals are specified into questions that can be answered by evaluating quantitative elementary or aggregated metrics. The model can be approached both in a top-down and in a bottom-up fashion to close the information loop between the conceptual, operational, and quantitative abstraction levels.

Although the GQM method is primarily considered as a top-down approach to problem solving, it can be utilized in a bottom-up fashion to identify the questions and goals that can be addressed using data and metrics that are already available: the simplicity of this model makes it very powerful, because it provides a versatile, bi-directional generative process for (1) quantifying high-level concepts and define which metrics are needed and, simultaneously, (2) for analyzing available measurements to generate hypotheses and theories. Furthermore, it results in a closed-loop system: information from goals and metrics can be utilized to evaluate whether the structure consistent and, in case, to make changes that can be reflected in the process.

Indeed, current applications of the GQM methodology rely on human intelligence and domain knowledge for processing information. However, its logic is similar to a neural network: a set of quantitative features (metrics) collected from an input layer (quantitative level) are fed into an internal layer (operational level) in which nodes (question) enable processing data and generating outputs (goals). Both in GQM and in NNs, metrics (features), either in elementary or aggregated form, serve as relevant predictors for the goal (output). The specificity of the intelligent system relies in the operational level: humans use domain knowledge for specifying the connections that link the conceptual and the quantitative levels (and vice versa) and build the GQM graph, whereas the hidden layer of a neural network incorporates methods, such as, activation function and backpropagation, that introduce non-linearity and enable weighing nodes and pruning connections. The analogy applies to most families and types of ML algorithms. The main difference between the goal-question-metric approach and a neural network is in that the former enables specifying multiple goals, whereas the outputs of a NN typically belong a homogeneous set.

In our work, we propose the GQM approach as a conceptual framework for the operationalization of systems and processes and, specifically, as a meta-language for designing applications based on machine learning. Specifically, we aim at leveraging it as a user-friendly and algorithm-agnostic modeling instrument that can abstract the lower-level components of a ML system and their technical implementation, so that AI experts and non-technical users can utilize it as a meta-design and collaboration tool.

3.1   Meta-Language

In its traditional use, GQM serves two purposes in representing systems and processes: as a top-down modeling instrument, its objective is helping identify measurable elements (i.e., metrics); on the other hand, after the quantitative level has been structured, it is a bottom-up assessment tool that uses data to measure qualitative goals. Although the state of the art primarily considered goal-question-metric for its applied outcome, its conceptual design is extremely valuable as a meta-language. As with other modeling languages utilized in software engineering, such as, the Unified Modeling Language (UML) [26], it can be applied to achieve a formal representation of a system. However, GQM is especially relevant to the ML domain from a system design standpoint: as its elements and graph-like structure mimic the structure and functioning of artificial intelligence algorithms, from machine learning to deep learning, they can be utilized as components of a meta-language for modeling the relationship between input features and output goals.

Self-explaining, Cross-Domain Definitions. The use of straightforward vocabulary, such as, metrics and goals, makes the language easy to understand by non-technical users, as the terms are already utilized in business rules across different backgrounds. Moreover, abstracting from the concepts of inputs and outputs removes any potential suggestion of an implicit direction, enables approaching the model in a top-down as well as in a bottom-up fashion, and supports its use for data-driven specification and for conceptual description and analysis.

Design Process as a Meta-representation of How ML Works. As discussed earlier, the fundamental components of GQM especially mimic the structure of a neural network. Specifically, defining the conceptual model with the goal-question-metric approach involves a process in which a human agent with sufficient expertise of a domain defines the layers and nodes and trains the resulting network: questions, which represent nodes in the hidden layer, leverage explicit and tacit knowledge and facilitate connecting input features and outputs. As human experience enables to directly prune nodes and connections that have low relevance, the graph resulting from GQM modeling directly translates into a network in which all links are significant and have the same weight. Although the similarity with NNs is straightforward, the goal-question-metric approach can similarly be applied to any ML algorithm.

Actionable by All the Stakeholders. On a merely conceptual level, the GQM meta-language, as other types of modeling tools, supports communication and exchange of information between domain experts and engineers. In addition, differently from instruments that serve a representational function only, structures obtained using the GQM meta-language are immediately actionable: the defined metrics can directly be utilized, either for helping humans measure goals or as an input dataset for a machine learning algorithm. Therefore, non-technical users can describe objectives, formalize domain knowledge, and explain the information needed for the assessment of goals. Conversely, AI experts can utilize the quantitative level as a model of the dataset; questions in the operational layer can become criteria for deciding the family and type of algorithm, whereas goals suggest the overall complexity of the problem and help refine implementation choices.

Interoperable and Robust to Iterations. Although the GQM meta-language can be utilized as a shared collaboration tool by any stakeholder, it supports separation of concerns and removes any interdependence between the model and its implementation. As a result, decisions regarding the configuration of a specific application do not affect how the graph structure is defined, and vice versa, changes in the GQM diagram can be treated as iterations and managed using a versioning system. This, in turn, enables reflecting them at an application level. Furthermore, the GQM meta-language is algorithm-agnostic: as changes in the machine learning engine do not affect the model, they can be implemented without impacting non-technical end users. Consequently, it is an interoperable, reusable, scalable, and suitable tool for designing systems that cope with high degree innovation and for domains that are exposed to intrinsic and extrinsic change factors.

Extensible. The goal-question-metric meta-language supports adding features and components that can improve its expressiveness. For instance, metrics and questions in a GQM graph can be associated with different weights, so that the structure can include a specification of the importance of each element, regardless of whether their relevance is decided by a human agent or calculated by a machine. This, in turn, can be utilized by non-technical stakeholders to be more specific in assessing the goal, or by AI experts, who can take weights into consideration when designing the network (e.g., for training purposes and for fine-tuning it). Moreover, software implementations of the GQM meta-language could interact with packages that incorporate ML algorithms, data visualization libraries, and performance analysis tools.

3.2 Domain-Knowledge Sharing Tool

One of the crucial activities in the development of software, especially if they incorporate some form of AI, is the collection and analysis of system requirements. As an example, numerous applications use artificial NNs in medical diagnosis, though interaction between stakeholders is often inefficient because health professionals and AI scientists approach data collection differently, structure information in dissimilar fashion, and present concepts using their own jargons. This is especially relevant to the outcome of a ML-based system: not being able to capture important domain features is among the main reason of poor performance.

To this end, GQM can be utilized as a tool that facilitates communication between ML-experts and non-technical users. The methodology inherently gathers information and structures it in an easy-to-understand and non-ambiguous written format that supports knowledge transfer across domains and among stakeholders. In addition, it can be utilized to elicit and formalize tacit knowledge, that is, information that would otherwise be difficult to collect or transfer using traditional means (e.g., automatic knowledge based on acquired expertise). Simultaneously, this approach results in better project documentation and helps track, revise, and improve software requirements and specifications over project iterations. In this context, the GQM meta-language can help engineer a lean participatory design and development process, increase communication and awareness among all stakeholders, and produce project documentation for future use. Indeed, other methodologies and meta-languages (e.g., UML) can be utilized for this purpose. However, in comparison, the GQM meta-language can especially capture the key items of a system based on machine learning, and it has the unique advantage of incorporating the conceptual and operationalization layers in a single model that completely describes the system and its conceptual, operational, and quantitative abstraction levels, while remaining independent from a specific implementation.

Incorporating the three layers in a single GQM meta-model results in a self-explanatory, actionable instrument that encapsulates and describes a problem in a way that is useful for the community of domain experts as well as for ML developers. Consequently, it can be utilized to create shared repositories of reusable practices and processes in which users could publish a formal description of the problem together with datasets that represent the defined metrics. This, in turn, would result in a self-contained package that provides the ML community with the necessary information for developing or customizing algorithms and contributing solutions. Simultaneously, this could give an impulse and stimulate different approaches in the context of AI.

3.3   Feature Selection

Indeed, having too many irrelevant features in a dataset affects the accuracy of AI models and causes overfitting. Moreover, the more data available, the longer it takes to train the system, especially in the case of deep neural networks. Therefore, feature selection, that is, preparing the dataset to remove redundant and misleading information, is among the most crucial tasks in machine learning. Although there are tools that automate this activity, this step is often the most labor-intensive part of building an AI system for a two-fold reason: (1) most businesses collect data in formats that are not ready for being utilized in ML (e.g., scarce, scattered in multiple databases, or stored in unstructured and ambiguous formats); moreover, domain knowledge is required for distinguishing relevant features from less significant ones, depending on the objective. In this regard, the GQM meta-language inherently helps highlight the relationship between relevant metrics and goals and empowers domain experts to contribute to the process and make it more efficient. Furthermore, it provides all stakeholders with a formal protocol that describes the type of information that needs to be collected for each goal and the format required for each metric.

3.4 Network of Machine Learning Agents

As discussed earlier, GQM supports designing models that include multiple goals. This contrasts with most implementations of narrow AIs, which typically have one objective, only. Nevertheless, this feature is advantageous in that a single structure defined in the GQM meta-language can be utilized to represent the entire set of multifaceted aspects in the problem space: the resulting graph describes a network of ML agents that use a shared pool of metrics to achieve their individual objectives. By doing this, the meta-language provides a way to measure the complexity of the model and results in a better understanding of the algorithms that will be required for implementing a solution. Furthermore, it creates a single representation of the several datasets that are required for assessing the different objectives. This, in turn, optimizes consistency, efficiency, and data reuse. Simultaneously, an overview of the entire problem space might help stakeholders identify alternative assessment and implementation strategies.

3.5 End-User Development

The simple and intuitive structure of GQM provides an interpretation framework for ML and it can help non-technical users understand how AI algorithms work without overwhelming them with non-necessary implementation details. Moreover, the GQM meta-language is suitable for supporting end-user development and for empowering domain experts to become active contributors. The introduction of UML as a modeling tool for software resulted in systems that support end-user development, which, in turn, simplified programming and helped refine modeling tools. The GQM meta-language offers a similar opportunity for system-system co-evolution in the context of ML. In this regard, software using the GQM meta-language as a specification format could interact with ML toolkits to enable users to dynamically instantiate agents that implement a specific algorithm depending on the goal. This would result in the possibility of empowering non-technical users to design the GQM graph, select an option from the list of available agents, train it using datasets defined in accordance with the metrics, and compare its performance with other algorithms.

Finally, from a human-system co-evolution perspective, reviewing the structure of a GQM model based on the performance of ML algorithms can result in a self-reflection process that supports domain experts in improving the definition of their objectives, the questions that must be asked to meet goals, and the metrics that must be collected to answer the questions, thus making the overall process more efficient.

4 Conclusion

Artificial intelligence is expected to be one of the disciplines that will impact society the most, in the next years. However, implementing machine learning algorithms still requires sophisticated programming skills: though packages, libraries and toolkits are available, most of their potential is accessible to experts and developers.

In this paper, we proposed the goal-question-metric approach as a meta-language for modeling machine learning systems. Although GQM has primarily been utilized to evaluate software metrics, we demonstrated that it is especially suitable as a description language for the meta-design of ML applications. We detailed its advantages for non-technical users, that is, it wraps the complexity of the implementation of machine learning in an easy-to-access abstraction layer, simplifies the understanding and use of algorithms, and supports end-user development. Moreover, we detailed how it can be utilized by AI experts an interoperable, reusable, and scalable modeling tool. Being algorithm-agnostic, the GQM meta-language is independent from any technical implementation and, thus, can facilitate communication among all the stakeholders involved in the development of a ML-based system, regardless of their level of proficiency with AI or programming languages. Finally, we outlined the main differences with other modeling languages, such as UML, and we presented examples of the proposed methodology as a viable instrument for gathering requirements, for feature selection, and for dynamically reflecting changes to the model.

References

1. Nasrabadi, N.M.: Pattern recognition and machine learning. J. Electron. Imaging **16**(4), 049901 (2007)
2. Sebastiani, F.: Machine learning in automated text categorization. ACM Comput. Surv. (CSUR) **34**(1), 1–47 (2002)
3. Abadi, M., Barham, P., Chen, J., Chen, Z., Davis, A., Dean, J., Devin, M., Ghemawat, S., Irving, G., Isard, M., Kudlur, M.: TensorFlow: a system for large-scale machine learning. In: OSDI, vol. 16, pp. 265–283, November 2016
4. Chen, T., Li, M., Li, Y., Lin, M., Wang, N., Wang, M., Xiao, T., Xu, B., Zhang, C., Zhang, Z.: MXNet: a flexible and efficient machine learning library for heterogeneous distributed systems. arXiv preprint arXiv:1512.01274 (2015)
5. The Microsoft Cognition Toolkit (CNTK). https://cntk.ai
6. Pedregosa, F., Varoquaux, G., Gramfort, A., Michel, V., Thirion, B., Grisel, O., Blondel, M., Prettenhofer, P., Weiss, R., Dubourg, V., Vanderplas, J.: Scikit-learn: machine learning in Python. J. Mach. Learn. Res. **12**, 2825–2830 (2011)
7. Paszke, A., Gross, S., Chintala, S., Chanan, G., Yang, E., DeVito, Z., Lin, Z., Desmaison, A., Antiga, L., Lerer, A.: Automatic differentiation in PyTorch (2017)
8. Erickson, B.J., Korfiatis, P., Akkus, Z., Kline, T., Philbrick, K.: Toolkits and libraries for deep learning. J. Digit. Imaging **30**(4), 400–405 (2017)
9. Burrell, J.: How the machine 'thinks': understanding opacity in machine learning algorithms. Big Data Soc. **3**(1), 2053951715622512 (2016)
10. Bach, M.P., Zoroja, J., Vukšić, V.B.: Determinants of firms' digital divide: a review of recent research. Procedia Technol. **9**, 120–128 (2013)

11. Jordan, M.I., Mitchell, T.M.: Machine learning: trends, perspectives, and prospects. Science **349**(6245), 255–260 (2015)

12. De Pace, A., Galeandro, P., Trotta, G.F., Caporusso, N., Marino, F., Alberotanza, V., Scardapane, A.: Synthesis of a neural network classifier for hepatocellular carcinoma grading based on triphasic CT images. In: Recent Trends in Image Processing and Pattern Recognition: First International Conference, RTIP2R 2016, Bidar, India, 16–17 December 2016, Revised Selected Papers, vol. 709, p. 356. Springer, April 2017. https://doi.org/10. 1007/978-3-319-60483-1_13

13. Bevilacqua, V., Uva, A.E., Fiorentino, M., Trotta, G.F., Dimatteo, M., Nasca, E., Nocera, A. N., Cascarano, G.D., Brunetti, A., Caporusso, N., Pellicciari, R.: A comprehensive method for assessing the blepharospasm cases severity. In: International Conference on Recent Trends in Image Processing and Pattern Recognition, pp. 369–381. Springer, Singapore, December 2016. https://doi.org/10.1007/978-981-10-4859-3_33

14. Bevilacqua, V., Trotta, G.F., Loconsole, C., Brunetti, A., Caporusso, N., Bellantuono, G.M., De Feudis, I., Patruno, D., De Marco, D., Venneri, A., Di Vietro, M.G.: A RGB-D sensor based tool for assessment and rating of movement disorders. In: International Conference on Applied Human Factors and Ergonomics, pp. 110–118. Springer, Cham, July 2017. https:// doi.org/10.1007/978-3-319-60483-1_12

15. Bevilacqua, V., Trotta, G.F., Brunetti, A., Caporusso, N., Loconsole, C., Cascarano, G.D., Catino, F., Cozzoli, P., Delfine, G., Mastronardi, A., Di Candia, A.: A comprehensive approach for physical rehabilitation assessment in multiple sclerosis patients based on gait analysis. In: International Conference on Applied Human Factors and Ergonomics, pp. 119–128. Springer, Cham, July 2017. https://doi.org/10.1007/978-3-319-60483-1_13

16. LeCun, Y., Bengio, Y., Hinton, G.: Deep learning. Nature **521**(7553), 436 (2015)

17. Gerbert, P., Hecker, M., Steinhäuser, S., Ruwolt, P.: Putting artificial intelligence to work. BCG Henderson Institute, The Boston Consulting Group, Munich, Germany (2017). Accessed 22 Jan 2018

18. Almassy, N., Kohle, M., Schonbauer, F.: Condela-3: a language for neural networks. In: 1990 IJCNN International Joint Conference on Neural Networks, pp. 285–290. IEEE, June 1990

19. Basili, V.R.: Software modeling and measurement: the goal/question/metric paradigm (1992)

20. Caldiera, V.R.B.G., Rombach, H.D.: Goal question metric paradigm. Encycl. Softw. Eng. **1**, 528–532 (1994)

21. Fontana, F.A., Zanoni, M., Marino, A., Mantyla, M.V.: Code smell detection: towards a machine learning-based approach. In: 2013 29th IEEE International Conference on Software Maintenance (ICSM), pp. 396–399. IEEE, September 2013

22. Uchiyama, S., Washizaki, H., Fukazawa, Y., Kubo, A.: Design pattern detection using software metrics and machine learning. In: First International Workshop on Model-Driven Software Migration (MDSM 2011), p. 38, March 2011

23. Sarcià, S.A., Cantone, G., Basili, V.R.: A statistical neural network framework for risk management process. In: Proceedings of ICSOFT, Barcelona, SP (2007)

24. Werner, E., Grabowski, J., Neukirchen, H., Röttger, N., Waack, S., Zeiss, B.: TTCN-3 quality engineering: using learning techniques to evaluate metric sets. In: International SDL Forum, pp. 54–68. Springer, Heidelberg, September 2007

25. Gasson, S.: Analyzing key decision-points: problem partitioning in the analysis of tightly-coupled, distributed work-systems. Int. J. Inf. Technol. Syst. Approach (IJITSA) **5**(2), 57–83 (2012)

26. Rumbaugh, J., Booch, G., Jacobson, I.: The Unified Modeling Language Reference Manual. Addison Wesley, Boston (2017)

Cascading Convolutional Neural Network for Steel Surface Defect Detection

Chih-Yang Lin[1]([⊠]), Cheng-Hsun Chen[1], Ching-Yuan Yang[1],
Fityanul Akhyar[1], Chao-Yung Hsu[2], and Hui-Fuang Ng[3]

[1] Department of Electrical Engineering, Yuan-Ze University, Taoyuan, Taiwan
andrewlin@saturn.yzu.edu.tw
[2] Automation and Instrumentation System Development Section,
China Steel Corporation, Kaohsiung, Taiwan
[3] Department of Computer Science, FICT, Universiti Tunku Abdul Rahman,
Petaling Jaya, Malaysia

Abstract. Steel is the most important material in the world of engineering and construction. Modern steelmaking relies on computer vision technologies, like optical cameras to monitor the production and manufacturing processes, which helps companies improve product quality. In this paper, we propose a deep learning method to automatically detect defects on the steel surface. The architecture of our proposed system is separated into two parts. The first part uses a revised version of single shot multibox detector (SSD) model to learn possible defects. Then, deep residual network (ResNet) is used to classify three types of defects: Rust, Scar, and Sponge. The combination of these two models is investigated and discussed thoroughly in this paper. This work additionally employs a real industry dataset to confirm the feasibility of the proposed method and make sure it is applicable to real-world scenarios. The experimental results show that the proposed method can achieve higher precision and recall scores in steel surface defect detection.

Keywords: Fully convolutional networks · Defect detection · SSD · ResNet

1 Introduction

The billet is an upstream product of the rod and wire from Sinosteel. The process of producing billet from casting to production involves cooling, sand-blasting, rusting, inspection, grinding and heating, and finally rolling into strips. Billets are approximately 145 mm × 145 mm in size, and can be supplied to strip and wire factories for rolling into strip steel, wire rods and linear steel. However, inspection is necessary to ensure the quality of the product before it is sent out.

The surface temperature of billets reaches as high as 700 to 900° [1] in the production environment. These conditions make defect detection on billets difficult to achieve. Traditional billet defect detection methods are divided into visual inspection [2, 3] and magnetic particle inspection [4]. However, visual inspection is more cost and time efficient; therefore, we will only focus on visual inspection in this paper. The types

© Springer Nature Switzerland AG 2020
T. Ahram (Ed.): AHFE 2019, AISC 965, pp. 202–212, 2020.
https://doi.org/10.1007/978-3-030-20454-9_20

of defects found can indicate the cause of defect formation and be used to improve the steelmaking process, since different defects have different causes.

In this paper, we develop a billet defect detection technology based on convolutional neural network. We propose a hierarchical structure to defect defects with revised SSD and ResNet50 [5, 6]. The experimental results show the effectiveness of the proposed method.

2   Architecture Overview

2.1   Structure of SSD

With the rise of convolutional neural networks, many models have evolved, such as Faster RCNN [7], Mask RCNN [8], Single Shot Multibox Detector (SSD) [9], and You Only Look Once (YOLO) [10]. All of these models have object detection capabilities. Among them, we chose the SSD300 version as the basic model. The reasons that we selected SSD300 are as follow:

- Faster RCNN and Mask RCNN are two-stage methods, which means that the training process is performed in two steps. In contrast, SSD and YOLO are one-stage methods, which are more efficient.
- The detection speed is better than that of other models. According to the author's paper, the detection speed of SSD300 is 59 FPS (frames per second).
- The architecture of SSD300 is simpler than other models' architectures and easier to adjust.
- SSD has multi-scale predictions.

The original SSD300 contains anchor boxes that are a combination of horizontal and vertical rectangles as shown in Fig. 1(a). In this work, we only use three horizontal rectangles as shown in Fig. 1(b). When there are many prediction boxes on an object, as shown in Fig. 2(a), non-maximum suppression (NMS) in SSD can solve this problem as shown in Fig. 2(b).

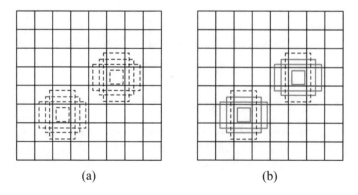

(a)   (b)

Fig. 1. (a) Original anchor box. (b) Customized anchor box.

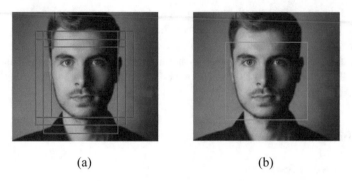

(a)   (b)

Fig. 2. (a) Bounding boxes before NMS. (b) Bounding box after NMS.

In Figs. 3–5, we present our revisions to the SSD architecture based on charac-teristics of collected defect images. Billet defects are mainly small ones. In Fig. 4, a 75 × 75 feature map is added to convolutional block 3 of the VGG16 layer, and the last two feature maps (3 × 3 and 1 × 1) are removed. In order to compare advantages and disadvantages of various SSD structures, the original SSD module (SSD300) and modified SSD module (revised-SSD300) will be trained. In addition, the revised-SSD300 will be extended to a revised-SSD600 with an input size of 600 × 600 as shown in Fig. 5. Therefore, in total, three models will be trained for comparison.

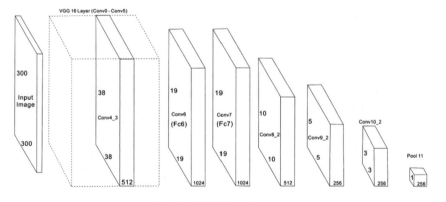

Fig. 3. SSD300 architecture.

2.2   Introduction of SENet and ResNet

In our main task, we need to detect two defects, called "sponge" and "scar" defects. The task of SSD is to determine whether the defects exist and where they are. We also added Squeeze-and-Excitation Net (SENet) [11] structure in our model to boost the

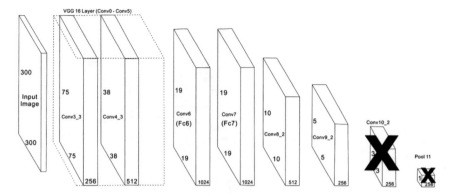

Fig. 4. Revised-SSD300 architecture.

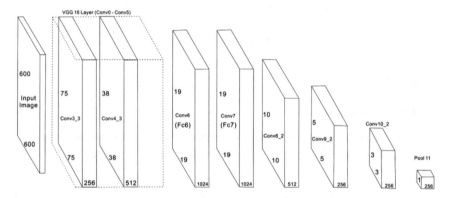

Fig. 5. Revised-SSDSE600 architecture.

results with adaptive weights for each feature map. SENet is not a complete network structure, but rather a small architecture in between convolution blocks. When SENet is applied, our method is called Revised-SSDSE, which is shown in Fig. 5.

Sometimes, another non-defect factor, called "rusty factor" as shown in Fig. 9(c), will be present in the dataset. The rusty factors, which are not defects, have various shapes and features and significantly affect our results. In order to detect rusty factors in the dataset, the 3 ∗ 3 and 1 ∗ 1 layers must be added back to the revised SSD network.

After determining the existence and location of defects, ResNet should identify the name of the defect. In this paper, we use ResNet50 [12] and classify three categories of defects as shown in Fig. 9. The defect from SSD will be resized to 224 × 224 to fit the input size for ResNet50. The combination of revised-SSDSE600 and ResNet50 forms the complete hierarchical structure as shown in Fig. 6.

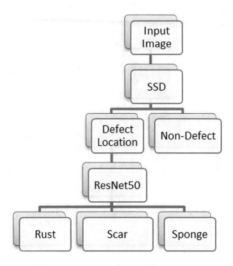

Fig. 6. Final hierarchical structure.

3 System Requirements

3.1 Hardware and Software

The hardware and software environment used in this paper is given in Table 1. The software part of the system includes Anaconda and GPU environment settings.

Table 1. Hardware and software environment.

Environment	Description
Operating System	Windows 10
Central Processing Unit (CPU)	Intel(R) Core(TM) i7-7700 CPU @ 3.60 GHz
Graphics processing unit (GPU)	NVIDIA GeForce GTX 1080 Ti 12G
CPU Memory	8 GB
Distribution	Anaconda v4.1.0
Programming Language	Python v3.5.4
Python Library	TensorFlow v1.8.0, Keras v2.1.6, Matplotlib v2.2.0, PIL v5.1.0
Editor	Visual Studio Community 2017
GPU Platform	NVIDIA CUDA 9.0
GPU Library	NVIDIA cuDNN 7.0

3.2   Data Annotation

We use LabelImg v1.6.0[1] tool to mark defect locations and non-defect classes in the dataset for the SSD model. LabelImg supports several operating system platforms, like Windows, Linux and Mac OS X. In this work, we use a Windows environment. After the labeling process is completed, the label result is saved in an XML format.

4   Experimental Results

The detection results are affected by camera types, illumination, number of defective samples, and other factors. The training process is performed as follows.

- Collect various defect samples.
- Mark the defect samples and generate corresponding XML files containing defect information.
- Train marked defect samples through the neural network structure and save the training results.

Table 2. Initial results of models.

Model	Sponge	ScarRemain
SSD 300	60%	85%
revised-SSD300	74%	95%
revised-SSD600	76%	95%

4.1   Initial Test

In the initial test, we prepared defect data with 464 Scar and 246 Sponge images in the dataset, 10% of which were validation and 90% of which were training data. The experimental results are shown in Table 2. The results in Fig. 7 show that the performance of the revised-SSD300 is similar to that of the revised-SSD600, but better than SSD300. There are too many redundant boxes when SSD300 is applied, as presented in Fig. 7b.

We test the daily images provided by the onsite database and used the following parameters as an accurate benchmark for calculating the system performance [13]:

- True Positive (TP).
- True Negative (TN).
- False Positive (FP).
- False Negative (FN).
- Precision (P) in Eq. (1).
- Recall (R) in Eq. (2).

---

[1] https://github.com/tzutalin/labelImg.

- F-Measure is a comprehensive evaluation index, which is used to understand whether two values of Precision and Recall are good, as shown in Eq. (3).

$$P = TP/(TP + FP). \tag{1}$$

$$R = TP/(TP + FN). \tag{2}$$

$$\text{F-Measure} = (2 \times P \times R)/(P + R). \tag{3}$$

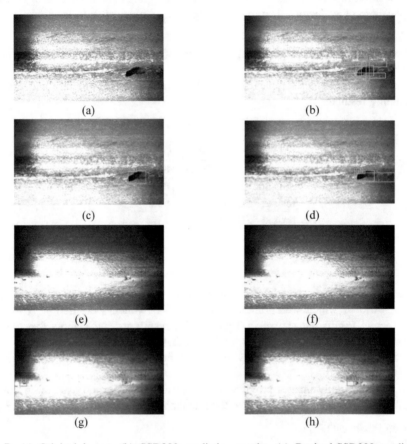

Fig. 7. (a) Original image. (b) SSD300 prediction results. (c) Revised-SSD300 prediction results. (d) SSD600 prediction results. (e) Original image. (f) Original image. (g) Revised-SSD300 prediction results. (h) SSD600 prediction results.

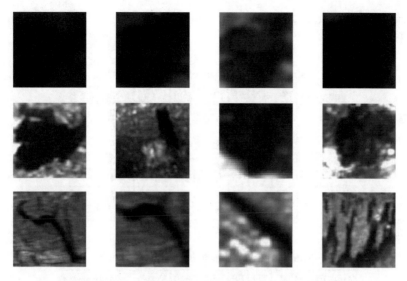

Fig. 8. Defect samples produced from SSD directly.

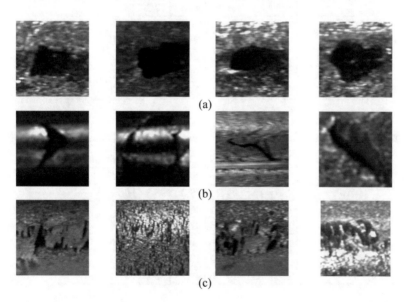

Fig. 9. Extension of defect range. (a) Scar. (b) Sponge. (c) Rust.

4.2 Final Test

According to Tables 3 and 4, after a seven-day training period, the highest precision and recall of the revised-SSD300 were 100% and 77.6%, respectively. The revised-SSD600 had a better recall due to its high-resolution images. However, the combination of the revised-SSDSE600 and ResNet50 achieved the highest precision and recall rates.

Table 3. Final test results of revised-SSD300.

Time (day)	TP	TN	FP	FN	P (%)	R (%)	F-Measure (%)
1	302	19	1	87	99.6	**77.6**	87.2
2	142	13	4	95	97.2	59.9	74.1
3	102	38	19	88	84.2	53.6	65.5
4	97	38	11	75	89.8	56.3	69.2
5	35	3	0	37	**100**	48.6	65.4
6	54	5	0	43	**100**	5.56	71.5
7	73	5	1	51	98.6	58.8	73.7

Table 4. Final test results of revised-SSD600.

Time (day)	TP	TN	FP	FN	P (%)	R (%)	F-Measure (%)
1	315	19	1	74	99.6	**80.9**	89.3
2	183	13	4	54	97.8	77.2	86.3
3	139	38	19	51	87.9	73.1	79.8
4	105	38	11	67	90.5	61.0	72.9
5	45	3	0	27	**100**	62.5	76.9
6	61	5	0	36	**100**	62.8	77.2
7	89	5	1	35	98.8	71.7	83.1

Table 5. Final test results of the combination of revised-SSDSE600 and ResNet50.

Time (day)	TP	TN	FP	FN	P (%)	R (%)	F-Measure (%)
1	353	19	1	36	99.7	90.7	95.0
2	226	13	4	11	98.2	95.3	96.7
3	179	38	19	11	90.4	94.2	92.2
4	138	38	11	34	92.6	80.2	85.9
5	66	3	0	6	**100**	91.6	95.6
6	95	5	0	2	**100**	**97.9**	98.9
7	118	5	1	6	99.1	95.1	97.1

Note that Fig. 10 shows that if we used defect bounding boxes directly from SSD for ResNet as shown in Fig. 8, the training process was hard to converge because the bounding boxes were too fitted to the defects. Therefore, we enlarged the range of the bounding boxes as shown in Fig. 9. After extending the bounding box, the training process could converge, which enabled the performance in Table 5 to be achieved.

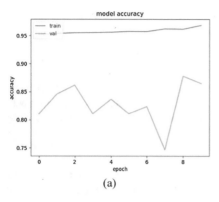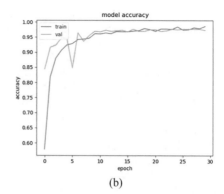

Fig. 10. Train results of revised-SSD600 with ResNet50. (a) With first samples of defect. (b) With second samples of defect.

5   Conclusions

In this paper, we design a hierarchical model to build a defect detection system for steel billets. We have modified the architecture of SSD by changing the sizes of feature maps and the sizes of anchor boxes to fit the shape of defects. The experimental results demonstrate the effectiveness of the proposed method. In further work, we will collect more rust defect images, because rusty types include many variations.

Acknowledgement. This work was supported by the Ministry of Science and Technology, Taiwan, under Grants MOST 106-2218-E-468-001, MOST 107-2221-E-155-048-MY3, and MOST 108-2634-F-008-001, and under Grants from China Steel Corporation RE106728 and RE107705.

References

1. Peacock, G.R.: Review of noncontact process temperature measurements in steel manufacturing. In: SPIE Conference on Thermosense XXI, pp. 171–189. SPIE, Florida (1999)
2. Yun, J.P., Choi, S., Kim, J.-W., Kim, S.W.: Automatic detection of cracks in raw steel block using Gabor filter optimized by univariate dynamic encoding algorithm for searches (uDEAS). NDT E Int. **42**, 389–397 (2009)
3. Duan, X., Duan, F., Han, F.: Study on surface defect vision detection system for steel plate based on virtual instrument technology. In: 2011 International Conference on Control, Automation and Systems Engineering (CASE), pp. 1–4. IEEE, Singapore (2011)
4. Balchin, N.C., Blunt, J.: Health and Safety in Welding and Allied Processes, 5th edn. Woodhead Publishing, Cambridge (2002)
5. Simonyan, K., Zisserman, A.: Very Deep Convolutional Networks for Large-Scale Image Recognition. CoRR abs/1409.1556 (2014)
6. Krizhevsky, A., Sutskever, I., Hinton, G.E.: ImageNet classification with deep convolutional neural networks. In: Advances in Neural Information Processing Systems 25 (NIPS 2012), pp. 1097–1105. Curran Associates, Inc., Nevada (2012)

 7. Ren, S., He, K., Girshick, R., Sun, J.: Faster R-CNN: towards real-time object detection with region proposal networks. In: Advances in Neural Information Processing Systems 28 (NIPS 2015), pp. 91–99. Curran Associates, Inc., Canada (2015)
 8. He, K., Gkioxari, G., Dollár, P., Girshick, R.B.: Mask R-CNN. In: 2017 IEEE International Conference on Computer Vision (ICCV), pp. 2980–2988. IEEE, Venice (2017)
 9. Liu, W., Anguelov, D., Erhan, D., Szegedy, C., Reed, S., Fu, C.Y., Berg, A.C.: SSD: single shot multibox detector. In: Leibe, B., Matas, J., Sebe, N., Welling, M. (eds.) Computer Vision – ECCV 2016. Lecture Notes in Computer Science, vol. 9905. Springer, Cham (2016)
10. Redmon, J., Kumar Divvala, B., Girshick, R., Farhadi, A.: You only look once: unified, real-time object detection. pp. 779–788. IEEE (2016)
11. Hu, J., Shen, L., Sun, G.: Squeeze-and-excitation networks. In: The IEEE Conference on Computer Vision and Pattern Recognition (CVPR), pp. 7132–7141. IEEE, Salt Lake City (2018)
12. He, K., Zhang, X., Ren, S., Sun, J.: Deep residual learning for image recognition. In: The IEEE Conference on Computer Vision and Pattern Recognition (CVPR), pp. 770–778. IEEE (2016)
13. Powers, D.M.W.: Evaluation: from Precision, Recall and F-measure to ROC, Informedness, Markedness and Correlation. 2017 (2007)

Machine Self-confidence in Autonomous Systems via Meta-analysis of Decision Processes

Brett Israelsen[(✉)], Nisar Ahmed, Eric Frew, Dale Lawrence, and Brian Argrow

University of Colorado Boulder, Boulder, CO 80309, USA
{brett.israelsen,nisar.ahmed}@colorado.edu
http://www.cohrint.info

Abstract. Algorithmic assurances assist human users in trusting advanced autonomous systems appropriately. This work explores one approach to creating assurances in which systems self-assess their decision-making capabilities, resulting in a 'self-confidence' measure. We present a framework for self-confidence assessment and reporting using meta-analysis factors, and then develop a new factor pertaining to 'solver quality' in the context of solving Markov decision processes (MDPs), which are widely used in autonomous systems. A novel method for computing solver quality self-confidence is derived, drawing inspiration from empirical hardness models. Numerical examples show our approach has desirable properties for enabling an MDP-based agent to self-assess its performance for a given task under different conditions. Experimental results for a simulated autonomous vehicle navigation problem show significantly improved delegated task performance outcomes in conditions where self-confidence reports are provided to users.

Keywords: Human-Machine systems · Artificial intelligence · Self-assessment

1 Introduction

Autonomous physical systems (APS) designed to perform complex tasks without human intervention for extended periods of time require at least one or more of the capabilities of an artificially intelligent physical agent (i.e. reasoning, knowledge representation, planning, learning, perception, control, communication) [1]. Yet, an APS always interacts with a human user in some way [2]. As these capabilities expand, APS also become more difficult for non-expert (and in some cases expert) users to understand [2]. For APS driven by AI and learning, predictions about behavior and performance limits can be hard to match to intended behaviors in noisy, untested, and 'out of band' scenarios. This raises questions about user trust in autonomous systems, i.e. a user's willingness to depend on an APS to carry out a delegated set of tasks, given the user's own (often limited and non-expert) understanding of the APS capabilities.

This work examines how APS can be designed to actively adjust users' expectations and understanding of APS capabilities which inform trust. Many different *algorithmic assurances* have been proposed to this end [1]. Process-based meta-analysis

© Springer Nature Switzerland AG 2020
T. Ahram (Ed.): AHFE 2019, AISC 965, pp. 213–223, 2020.
https://doi.org/10.1007/978-3-030-20454-9_21

techniques allow APS to self-qualify their capabilities and competency boundaries, by evaluating and reporting their degree of 'self-trust' or *self-confidence* for a task.

Several definitions and algorithmic approaches have been proposed recently which enable APS to automatically generate self-confidence scores in the context of different tasks and uncertainty-based capability assessments [1, 3]. This work restricts attention to the following definition of self-confidence, which captures the others to a large extent: *an agent's perceived ability to achieve assigned goals (within a defined region of autonomous behavior) after accounting for (1) uncertainties in its knowledge of the world, (2) uncertainties of its own state, and (3) uncertainties about its reasoning process and execution abilities.* However, a general *algorithmic* framework for computing and communicating self-confidence for general decision-making autonomy architectures and capabilities has yet to be established. Furthermore, it has not yet been confirmed experimentally whether (or in what contexts) human-APS interfaces that incorporate self-confidence reporting improve the ability of users to delegate of tasks within APS competency limits, compared to status quo interfaces that do not use such reporting. This paper provides results aimed at addressing both of these gaps.

In Sect. 2, we motivate the machine self-confidence idea and ground the concept for autonomous planning problems in the context of general Markov Decision Processes (MDPs), using a concrete uncertain navigation application example where tasks are delegated by a user. We review a factorization based framework for self-confidence assessment called Factorized Machine Self-Confidence (FaMSeC), and in Sect. 3 present a novel strategy for computing the so-called 'solver quality' factor of self-confidence, drawing inspiration from empirical hardness modeling literature [4]. Numerical examples for the UGV navigation problem under different MDP solver, parameter, and environment conditions, indicate that the self-confidence scores exhibit desired properties. Section 4 describes the setup and initial outcomes of experimental trials to investigate the effects of reporting self-confidence to users on simulated instances of the UGV navigation problem. The results show significantly improved delegated task performance outcomes in conditions where self-confidence feedback is provided to users vs. conditions where no self-confidence feedback is provided, providing favorable evidence for the efficacy of self-confidence reporting. Section 5 presents conclusions and avenues for future investigation.

2   Background and Preliminaries

MDP-based Planning. We present here an algorithmic framework for assessing machine self-confidence with respect to APS capabilities that can be modeled as Markov decision processes (MDPs). MDPs are composed of finite states and actions that capture nondeterminism in state transitions by modeling probability distributions $p(\cdot)$ from which the next state will be sampled. MDPs have well-established connections to other techniques for decision-making and learning under uncertainty, such as partially-observable MDPs (POMDPs) in limited observability environments and reinforcement learning with incomplete model information [8]. This provides a pathway to generalizing the notions and analyses for self-confidence assessments developed here for MDPs.

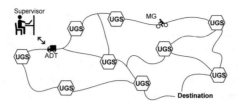

Fig. 1. ADT in road network evading MG with information from noisy UGS.

We consider generic MDP formulations of a task T delegated to an APS. In an MDP framing, the APS must find an optimal policy $\pi^* = u(x)$ for an MDP with state x and actions u, so that the value function $U = \mathbb{E}\left[\sum_{k=0}^{\infty} \gamma^i r(x_k, u_k)\right]$ is maximized for times $k = 0, \ldots, \infty$ – where $R(x_k, u_k)$ is larger (lower) for (un)favorable states and actions, $E[\cdot]$ is the expectation over all future states, and $\gamma \in (0, 1)$ is a future discount that can be tuned. Given u_k, x_k updates via a Markovian model $x_{k+1} \sim p(x_{k+1}|x_k, u_k)$, i.e. x_i is known at time k (no sensor noise), but state transitions $k \to k+1$ are not deterministic. The optimal state-action policy π^* is recovered from Bellman's equations via dynamic programming. In many practice, approximations π are needed to handle complex or uncertain dynamics (e.g. for reinforcement learning or very large state spaces) [5].

Donut Delivery Application. Consider a concrete grounding example called the 'Donut Delivery' problem (based on a 'VIP escort' scenario [4, 6]). As shown in Fig. 1, an autonomous donut delivery truck (ADT) navigates a road network in order to reach a delivery destination while avoiding a motorcycle gang (MG) that will steal the donuts if they cross paths with the delivery truck. The motorcycle gang's uncertain location is estimated via noisy observations from unattended ground sensors (UGS). The delivery truck selects a sequence of discrete actions: stay in place, turn left/right, go back, go straight. The ADT motion, UGS readings, and MG behavior are probabilistic and thus highly uncertain. As a result, the problems of decision-making and sensing are strongly coupled. Certain trajectories may allow the ADT to localize the MG before heading to the goal but incur a high time penalty. Other paths lead to rapid delivery, but since the MG can take multiple paths there can also be high MG location uncertainty, which in turn increases capture risk. A human supervisor monitoring the ADT does not have direct control of the ADT—but can (based on whatever limited amount of information is provided to the supervisor) provide 'go' or 'no go' commands to proceed with or abort the delivery operation before it commences.

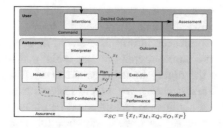

Fig. 2. Factorized Machine Self-Confidence (FaMSeC) information flow

The combined ADT and MG location states can be discretized to produce a discrete-time time-invariant Markov process. This leads to an initial state probability distribution and state transition matrix, which is a function of the ADT's actions. The reward $R(x_k, u_k) = R_k$ models preferences over the states, e.g. $R_k = -200$ if the ADT and MG are not co-located and the ADT is not yet at the goal; $R_k = -2000$ if the MG catches the ADT (i.e. they are co-located), and $R_k = +2000$ if the ADT reaches the goal and avoids capture. For simplicity, we treat the MG's state as fully observable (e.g. due to the availability of highly accurate UGS data), so that the ADT's planning problem leads directly to an MDP.

Self-confidence Factorization and Calculation. We briefly review and explore an algorithmic approach known as *Factorized Machine Self-confidence (FaMSeC)* for assessing and communicating machine self-confidence [7]. FaMSeC represents and computes self-confidence as a decomposable multi-factor function, which combines shorthand assessments of where/when operations and approximations that constitute a realized model-based autonomous decision-making system are expected to break down. Similar to the approach in [8], FaMSeC encodes metrics for correctness and quality of an autonomous system that an expert designer would use to account for variations in task, environment, system implementation, and context. The key innovation with respect to this previous work is that FaMSeC allows an APS to compute assessments of self-confidence in a completely automated manner, i.e. without the need for a human expert to specify how confident an APS ought to be under a specific set of a priori variations.

FaMSeC's self-confidence computing approach is depicted in Fig. 2. This uses a set of *self-confidence factors* (dashed lines) derived from core algorithmic decision-making components (white boxes in the 'Autonomy' block). The total self-confidence score and the score for each factor can be mapped onto an arbitrary scale, e.g. L_{xi} to U_{xi}, where L_{xi} indicates 'complete lack of confidence' (i.e. some aspect of task, environment, or context falls completely outside the system's competency boundaries), and U_{xi} indicates 'complete confidence' (i.e. all aspects of task, environment, and mission context are well within system's competency boundaries). These scales can vary and have different qualitative interpretations. The main idea is that these bounds in any case should provide a clear sense of 'confidence direction' (i.e. degree of self-trust).

Five self-confidence factors are considered by Ref. [7]: (1) x_I—*interpretation of user intent and task*: were the user's intentions properly translated into suitable tasks?; (2) x_M—*model and data validity*: are the agent's learned/assumed models and training data well-matched to the actual task?; (3) x_Q—*solver quality*: are approximations for finding decision-making policies appropriate for a given model?; (4) x_O—*expected outcome assessment*: does a particular policy lead to a desirable landscape of expected outcomes for states and rewards?; and (5) x_P—*past history and experiences*: how does the system's previous experiences inform assessment of competency for the current task at hand? These five factors provide only an initial and likely incomplete set of variables to consider in the highly limited context of MDPs. Furthermore, they are primarily aimed at self-assessment *prior* to the execution of a task, whereas it is conceivable that other self-confidence factors could be included to account for in situ and post hoc self-assessments. This work assumes for simplicity that the overall mapping from self-confidence factors to an "overall score" consists of a direct report of all or some fixed subset of the component factors, e.g. $x_{SC} = \{x_I, x_M, x_Q, x_O, x_P\}$.

We will use the Donut Delivery problem to examine two core issues: (i) how should these factors generally behave?, and (ii) how these factors be calculated?

To address (i), we may for instance consider the limiting behaviors and principal interactions that are expected for the various factors with respect to a given class of MDP solver, e.g. sampling-based Monte Carlo solvers for finding an approximately optimal policy π^* [9]. Figure 3 shows expected FaMSeC factor behaviors when using a hypothetical Monte Carlo solver, as conditions describing the task, environment, and solver realization are varied. The resulting notional factor scores are mapped onto an arbitrary finite range of -1 (total lack of confidence) to $+1$ (complete confidence). For instance, x_Q is expected to change as the number of samples used by the solver varies.

Scenario (Task/Environment/ System/Context Shift)	Expected Self-confidence Factor Behavior (Trends/Boundary Conditions)
ADT performing task for first time, R_k perfectly encodes supervisor's behavioral preferences	$x_I = 1, \quad x_P = 0$
R_k uncertain from supervisor input, system fared poorly on same road network in past	$x_I = 0, \quad x_P < 0$
Time-space resolution of dynamics model increases (decreases)	$x_M \to 1 \ (x_M \to -1)$
Number of samples used by sampling-based solver increases (decreases)	$x_Q \to 1 \ (x_Q \to -1)$
Additional task to retrieve valuable item at location A before reaching goal	x_O decreases as A gets further from goal

Fig. 3. Notional factors for Donut Delivery problem, with scaling to the interval $[-1, 1]$.

This thought exercise underscores the importance of accounting for complex inter-factor dependences while formulating approaches that address (ii). A simplifying assumption is to examine each factor in isolation along 'boundary conditions' where other factors do not change and do not contribute to overall self-confidence score. Reference [1] developed such an approach to compute x_O for MDPs, for boundary conditions $x_M = U_{xM}$ (perfect model of task), $x_I = U_{xI}$ (perfectly interpreted task and R_k reward), $x_Q = U_{xQ}$ (optimal π^* policy available), and $x_P = U_{xP}$ (previously encountered task). The resulting overall self-confidence score then depends only on x_O. This can be quantified as a measure of $R = p_\pi(R_\infty)$, the probability distribution of achievable cumulative reward values $R_\infty = \sum_{k=0}^{\infty} R_k$ achieved under policy π. Intuitively, R summarizes the landscape of possible outcomes for the APS if it were to apply the policy π to carry out its task, and thus provides useful information about the intrinsic difficulty or feasibility of a task by considering statistics beyond just the mean value of R (which π approximately maximizes via an MDP solver). For instance, if a large portion of R lies around very large negative values, then the task cannot be done without a high probability of encountering unfavorable outcomes (even with optimal decision-making).

Reference [7] compares several candidate measures of R, and recommends the logistically transformed upper partial moment/lower partial moment (UPM/LPM). This score quantifies the probability mass lying on either side of $\bar{R}_\infty$, which represents a user-defined minimally acceptable cumulative reward value (e.g. which can be translated from the maximum acceptable time to reach the goal). In the Donut Delivery problem, $\bar{R}_\infty$ corresponds to a user-specified maximum acceptable time to successfully reach the delivery destination. The distribution R can be empirically estimated from Monte Carlo sample simulations, by applying π to the assumed MDP state dynamics model. Figure 4 illustrates how this allows x_O to measure intrinsic task difficulty, and thus provide an uncertainty-driven signal to users to adjust APS performance expectations.

However, computation of other factors and inter-factor dependencies are not considered by Ref. [7]. Furthermore, studies with human users have not yet been done to validate and evaluate the impact of self-confidence reporting on task delegation. The remainder of the paper addresses both gaps.

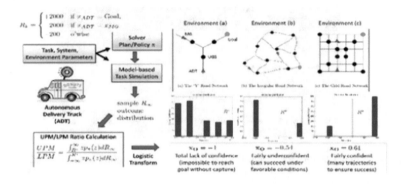

Fig. 4. x_O assessments for Donut Delivery problem in various environments.

3 Solver Quality Definition and Calculation

The insights above are extended to formally define and compute the solver quality FaMSeC factor x_Q. For APS planning, x_Q indicates how well a generic solver S for a planning problem will 'perform' on a given task T of class c_T. We restrict c_T here to the class of all applicable MDP problems, so that S is any type of MDP solver (although the ideas developed here are even more generalizable).

A formal definition for x_Q must make the notion of S's 'performance' precise to enable some form of quantitative evaluation. For MDPs, any particular task instance T has a corresponding optimal policy π^* which by definition leads to corresponding best achievable total reward. This suggests that a natural performance metric for assessing the competence of a generic solver S would be some quantitative comparison of the (approximate) policy $\hat{p} - \pi^*$. Such a 'strong' comparison can be done in many ways depending on the application context, e.g. by directly comparing actions taken by the policies in different states of interest, or by comparing the resulting expected total reward distributions or other artifacts of the different policies. But since π^* is usually unavailable (or else S would not be needed), x_Q should also allow for 'weak' comparison to other reference 'baseline' solvers that yield policies with performance as good as/very close to the true optimal π^*. This may require comparing two completely different types of solvers (e.g. a deterministic approximation vs. a Monte Carlo approximation), so x_Q should also ideally enable comparison across solver classes, as well as assessments of online/anytime solvers (for which complete state coverage and policy convergence may not be possible in large state/action spaces).

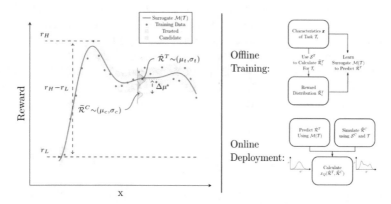

Fig. 5. (Left) key values for computing x_Q, where x is a 'feature of interest' for T or S; (right) offline training phase of surrogate M(T) and online calculation of x_Q.

At the same time, x_Q should account for the expected self-confidence metric trends and boundary conditions mentioned earlier in Fig. 3 for the Donut Delivery application. In particular, x_Q should be defined also to naturally account for the influence of both characteristics of S and features of T simultaneously. For instance, if the task T is not particularly complex or is characterized by a very small amount of uncertainty (e.g. so as to be nearly deterministic), then it is possible that a small sample size could suffice to closely approximate the optimal policy, in which case x_Q should indicate very high confidence. Moreover, while it is impossible to assess performance on *all* tasks in c_T, x_Q should be able to reflect S's performance on *any* task instance T in c_T, including previously unseen tasks (e.g. new road networks for the Donut Delivery problem).

A natural starting point for devising a quantitative assessment of x_Q according to these desiderata is to again examine how information about $R = p(R_\infty)$ (which is already used to form x_O) can be analyzed further. Note that x_O indirectly depends on x_Q, since reliable estimation of R requires knowing π^*. But in practice, an MDP-based APS will often employ an approximate policy $\hat{\pi}$ instead of the true optimal policy π^*. In turn, this leads to an estimate $p_{\hat{\pi}}(R_\infty)$ of the expected reward distribution which differs from the true $p_{\pi^*}(R_\infty)$ under the optimal π^*. We show next that, if a surrogate model $\hat{R}$ for R^T can be found when π^* is unknown but π^T is available (where $\pi^T \approx \pi^*$), then a quantitative comparison of the simulated R^C of a 'candidate' solver to the surrogate $\hat{R}$ leads to a suitable metric for x_Q (where x_Q is no longer considered 'perfect' as it was when deriving x_O).

Calculation. F Following the above discussion, a surrogate model M(T) can be learned, as shown in Fig. 5 (right), to predict the reward distribution of the 'trusted' reference solver S^T on task T (which ideally comes close to approximating the optimal π or converges to it with unlimited computing resources). The candidate solver S must then be evaluated online with respect to the trusted solver S^T. To do this, we compare the *predicted* performance of S^T to the simulated performance of solver S on task T, the latter of which is denoted by R. This approach is inspired by empirical hardness modeling techniques [10], which use surrogate models to predict the actual run time of NP-complete problem solvers. The key terms for computing x_Q are shown in Fig. 5 (left). The basic premise is to evaluate a scaled version of the 'distance' between the difference between the reward distributions of the trusted (T) and candidate (C) solvers, while accounting for the overall expected range of rewards of the trusted solver across multiple task instances. Reference [7] derives one possible formula for capturing this relationship,

$$x_Q = \frac{2}{1 + \exp(-q/5)}, \quad f = \frac{\Delta\mu}{r_H - r_L}, \quad q = \text{sign}(\Delta\mu)f^\alpha \sqrt{H^2(P,Q)} \quad (1)$$

where $H^2(P, Q)$ is the Hellinger distance between the R^T (P) and R^C (Q); $\Delta\mu$ is the difference in the means of these pdfs; r_H and r_L are largest and smallest R_∞ values observed in the S^T training set; and the α is a tuning parameter to adjust the level of interaction between f and H^2. The quantity x_Q ranges from $L_{xQ} = 0$ (total lack of confidence in S relative to S^T) to $U_{xQ} = 2$ (complete confidence in S relative to S^T), with the midpoint at 1 (equal confidence in S relative to S^T). Figure 6 shows the expected (uncertain) rewards for a trusted solver S^T and candidate solver S given some realization of a task variable/solver parameter. Shown in the table are the x_Q for varying conditions. At point B, the mean reward of S is lower than S^T's, and the corresponding variance is higher. This indicates that S provides a worse/less reliable policy compared to S^T, and so $x_Q < 1$. But when $r = 5$ (i.e. $r_H - r_L = 5$), the global reward range is 'large'), the candidate solver is now only slightly worse than the trusted solver, so $x_Q = 0.667$. When $r = 0.05$, S is clearly much less capable than S^T, and so $x_Q = 0.002$. At point C, the mean reward of S is now larger than that of S^T, but S's reward distribution has a larger variance. For $r = 5$, S is only marginally better than S^T, and so $x_Q = 1.095$. As the global reward range r decreases, the difference in capability between S and S^T increases, e.g. such that $x_Q = 1.995$ at $r = 0.005$.

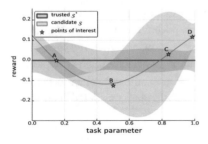

x_Q	Point of Interest			
	A	B	C	D
$x_Q(r=5.0)$	0.998	0.667	1.095	1.351
$x_Q(r=0.5)$	0.994	0.215	1.291	1.821
$x_Q(r=0.05)$	0.981	0.002	1.739	1.999
$x_Q(r=0.005)$	0.940	0.000	1.995	2.000

Fig. 6. x_Q assessments on synthetic $R = p(R_\infty)$ distributions for S^T and S.

4   Autonomous Donut Delivery Dispatch Experiment

An Amazon Mechanical Turk experiment was designed in order to investigate how x_Q and x_O effect user behavior, if at all. In the experiment participants acted as 'dispatch supervisors' for the ADT delivery problem described in Sect. 2. The hypothesis to be tested is that users who are presented with self-confidence metrics will perform better in the dispatch task, as measured by higher task scores. Of the two FaMSeC metrics discussed above (x_Q and x_O), each is either 'present' or 'absent'. Therefore the experiment is a $2(x_Q) \times 2(x_O)$ design, resulting in 4 conditions. Condition 1 is the 'control' where both x_Q and x_O are absent. In condition 2 x_Q is present, in condition 3 x_O is present, and in condition 4 both x_Q and x_O are present. A screenshot of a typical task is shown in Fig. 7 (in the figure both x_Q and x_O are displayed indicating that this is condition 4; other conditions had an identical layout with the applicable values missing/present). Based on feedback from a pilot study, the numerical values for x_Q and x_O were mapped to a 5 value Likert-type set of verbal descriptions ranging between 'very bad' and 'very good', to help users more easily grasp the scales when the two different metrics are placed next to each other. Participants were paid a base rate of $1.60 for completing the Human Intelligence Task (HIT), and were able to get a bonus of up to $1.00 based on their performance. Pilot data indicated that this would be equivalent to approximately $7–10/h. In practice the 'Turkers' were quite a bit faster than those in the pilot study, and earned around $13/h. Source code for the experiment is available on Github[1]. Generally, each participant must decide whether the autonomous delivery vehicle should attempt to make a delivery, or decline the delivery. A successful delivery attempt results in +1 point, failure in −1 point, and declining a delivery in −1/4 point. After being trained (and passing a quiz), participants are given a randomized sequence of 43 different delivery scenarios with varying maps. A pool of 500 possible road-networks were generated with varying N ($N = [8, 35]$) and varying transition probability (probability of traveling in the desired direction as opposed to a different direction). The experiment set of 43 tasks is a subset of the pool of 500 road-networks sampled in a pseudo random way over the x_Q/x_O space to avoid clustered samples.

Initial Results and Discussion. Some preliminary data comparing the difference in cumulative 'total score' is shown in Fig. 8. The total number of participants represented here is N = 255, with n = {63, 63, 64, 65} for conditions 1–4 respectively. A two-way analysis of variance was conducted on the influence of two independent variables (x_Q and x_O) on the participant's total score. Two levels were included for x_Q and x_O ('present' and 'absent'). All effects were statistically significant at the 0.05 significance level. The main effect of x_Q yielded an F ratio of $F(1,251) = 107.9$, $p < 0.001$, indicating a significant difference between x_Q 'present' ($M = -1.63, SD = 3.06$) and 'absent' ($M = -5.08, SD = 3.61$) conditions. The main effect of x_O yielded an F ratio of $F(1,251) = 124.5$, $p < 0.001$, indicating a significant difference between x_O 'present' ($M = -1.51, SD = 2.63$), and 'absent'

---

[1] https://github.com/COHRINT/SC_experiment.

$(M = -5.23, SD = 3.82)$ conditions. The interaction effect was also significant, $F(1,251) = 26.4$, $p < 0.001$. These results indicate that the participants were able to recognize limitations of the UDT and make more appropriate dispatch decisions when x_Q and x_O were present. However, post-hoc analysis is still needed to get a more detailed understanding of the differences. At the end of the experiment participants answered survey questions about their perception of the ADT, how much they trusted the system, and how capable they thought it was. Our hypothesis is that the presence of self-confidence metrics would also affect these responses, but these data are still being analyzed.

Fig. 7. Example screenshot from the Amazon Mechanical Turk experiment.

Arguably the 'Donut Delivery' task is too simple to be interesting. There are a few different reasons that it was chosen: (1) Using a more complex problem would introduce too many confounding factors that could limit the possibility of accurately observing effects of the FaMSeC metrics; (2) Having a simple task is desirable when trying to run an experiment a large, diverse, group of non-experts; (3) If support for the hypotheses were found in this experiment then more realistic, expensive, experiments could be run in the future.

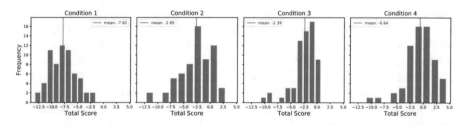

Fig. 8. Total score from each condition; red vertical lines indicate the sample mean.

5 Conclusions

We have investigated how Factorized Machine Self-Confidence (FaMSeC) factors can be implemented as *algorithmic assurances* to aid users of autonomous systems. Since these factors provide meta-cognitive self-assessments of different aspects of decision-making under uncertainty, reporting these factors can allow users to better task and use autonomous systems. Preliminary findings from a human participant study indicated that the reporting of two FaMSeC factors significantly improved users' ability to perform as a dispatch supervisors for the Donut Delivery task. Future work involves further analyzing the experimental results to identify effects of FaMSeC factors on user behavior and attitudes. Also, in condition 4 of the experiment the factors were presented separately; it would be interesting to investigate how those might be combined into a single summary metric. Such a combination might become even more critical for complex tasks or as other FaMSeC metrics are incorporated, since it may become difficult for users to process many different metrics simultaneously. Finally, future work will identify strategies for computing other FaMSeC factors for MDP-based autonomy, as well as explore other contexts for computing and using FaMSeC.

References

1. Israelsen, B.W., Ahmed, N.R.: "Dave…I can assure you …that it's going to be all right …" a definition, case for, and survey of algorithmic assurances in human-autonomy trust relationships. ACM Comput. Surv. **51**(6), 113:1–113:37 (2019)
2. Bradshaw, J.M., et al.: The seven deadly myths of "autonomous systems". IEEE Intell. Syst. **28**(3), 54–61 (2013)
3. Sweet, N., et al.: Towards self-confidence in autonomous systems. In: AIAA Infotech @ Aerospace, p. 1651 (2016)
4. Leyton-Brown, K., Nudelman, E., Shoham, Y.: Empirical hardness models: methodology and a case study on combinatorial auctions. J. ACM **56**(4), 22:1–22:52 (2009)
5. Kochenderfer, M.J.: Decision Making Under Uncertainty: Theory and Application. MIT Press, Cambridge (2015)
6. Humphrey, L.: Model checking UAV mission plans. In: AIAA Modeling and Simulation Technologies Conference. Guidance, Navigation, and Control and Co-located Conferences. American Institute of Aeronautics and Astronautics, August 2012
7. Aitken, M.: Assured human-autonomy interaction through machine self-confidence. M.S. thesis. University of Colorado at Boulder (2016)
8. Hutchins, A.R., et al.: Representing autonomous systems' self-confidence through competency boundaries. In: Proceedings of the Human Factors and Ergonomics Society Annual Meeting, vol. 59, pp. 279–283. Sage (2015)
9. Browne, C.B., et al.: A survey of Monte Carlo tree search methods. IEEE Trans. Comput. Intell. AI Games **4**(1), 1–43 (2012)
10. Israelsen, B.W., et al.: Factorized machine self-confidence for decision-making agents. arXiv:1810.06519 [cs.LG], October 2018

Research on Accuracy of Flower Recognition Application Based on Convolutional Neural Network

Jing-Hua Han$^{(\boxtimes)}$, Chen Jin, and Li-Sha Wu

Beijing Forestry University, Beijing, China
hanjing013@126.com

Abstract. Compared with traditional flower recognition methods, the existing flower recognition applications on the market use advanced deep learning technology to improve the accuracy of plant recognition and solve the problem of plant recognition. The article studied the five applications that users commonly use, comparing and analyzing their recognition accuracy, and finally putting forward the feasibility advice for further improvement of flower recognition applications. The method of sampling survey was adopted, this paper divides the garden flowers and wild flowers into different levels according to their common degrees. Each type of flower was shot from 5 different angles and scenes, and recognized by these five applications separately. The results showed that the rankings of the five applications evaluated were Hua Bangzhu, Hua Banlv, Xing Se, Microsoft's Flower Recognition, and Find Flower Recognition. At pre-sent, it is necessary to continuously improve from the aspects of technology, products and plant libraries.

Keywords: Convolutional neural network · Deep learning ·
Flower recognition · Accuracy

1 Introduction

Plant identification is one of the Botany research foundation technology, traditional plant identification method mainly through artificial sampling and contrast, inefficient and unable to popularize the public [1]. With the advent of Big data, the deep convolutional neural network has more complex network structure and the more powerful features learning ability and expression ability, because of its highly parallel processing ability, adaptive ability, more advantages of the fault-tolerant rate [2]. Developers will be introduced in the field of plant identification, greatly improve the efficiency of the plant identification [3]. Through machine learning method, the definition of image features of massive plant images is completed, and the features such as color, texture, shape and connected components are extracted for clustering analysis [4]. With this as the technical support, the application of flower recognition that is easy for users to use is developed, so as to solve the accuracy of plant recognition [5].

Deep learning concept put forward by Hinton and others in 2006, is a deep network structure of the artificial intelligence algorithm [6]. At present, there are many flower

© Springer Nature Switzerland AG 2020
T. Ahram (Ed.): AHFE 2019, AISC 965, pp. 224–232, 2020.
https://doi.org/10.1007/978-3-030-20454-9_22

recognition applications in the market, all of which use the deep learning technology based on convolutional neural network [7]. In this paper, the well-known flower recognition applications, such as Hua Bangzhu, Hua Banlv, Xing Se, Microsoft's Flower Recognition, and Find Flower Recognition, are selected to verify the recognition rate [8].

Find Flower Recognition can identify thousands of common flower plants. Hua Bangzhu and Hua Banlv are products developed by the same company, but the two products have different target groups. Hua Bangzhu is a platform for flower growers to communicate and interact. Hua Banlv has a powerful photo recognition function, is an intelligent flower identification sharps. At present, it has been able to identify nearly 3,000 genera and 5,000 species of wild and cultivated plants in China, almost covering common flowers and trees around. Microsoft flower recognition can currently identify 400 kinds of common flower varieties in urban greening and parks in China, and because of the development of the function of identifying common daily objects, it has a certain impact on the accuracy of flower recognition. Shape and color can currently identify more than 4,000 species of plants [9].

In this paper, through the study of the five applications with the most users, the analysis accuracy is compared and analyzed. The SPSS analysis software is used for data statistics and analysis, and finally the feasible suggestions for the further improvement of the flower identification application are proposed.

2   Materials and Methods

In this study, the sampling survey method was used to compare and analyze the recognition accuracy of these five flowers recognition applications. According to the common degree, the common, common and uncommon plants of garden plants and wild plants were investigated. 15 plants were selected as the research objects. The survey conducted in-depth excavation and analysis of user identification scenarios. Each plant randomly selected five different angles and different scenes for recognition (Fig. 1), and excluded interferences, all using clear flower maps of plants.

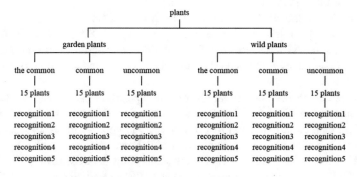

Fig. 1. Research content of flowers within the project

The five applications sequentially identify the selected plants, and score the accuracy according to the recognition results, and scores from 0 to 5 in the order of the answers (Fig. 2). Find Flower Recognition, Hua Bangzhu, and Microsoft's Flower Recognition all provide 5 answers, so the scores correspond to 0 points (no result), 1 point (recognition result 5), 2 points (recognition result 4), 3 points (recognition result 3), 4 points (recognition result 2), 5 points (recognition result 1); Hua Banlv provides 1 answer, so the score corresponds to 0 (no result), 5 points (have results); Xing Se provides 1 to 3 answers, so the scores correspond to 0 points (no result), 1 point (recognition result 3), 3 points (recognition result 2), and 5 points (recognition result 1). The total scores of garden plants and wild plants totaled 1125 points.

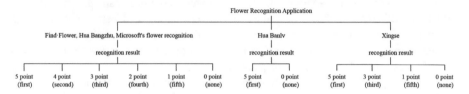

Fig. 2. The score of the five flower recognition applications

This study used these five applications to identify plants, and finally scored the recognition results, using SPSS analysis software for data statistics and analysis. The statistical methods involved include Reliability and Validity analysis, Nonparametric tests.

3   Results and Discussion

3.1   Reliability and Validity

This study used Cronbach's α coefficient to test the reliability of the survey. Usually the value of Cronbach's α coefficient is between 0 and 1. If it is greater than 0.70, it indicates that the survey has good index reliability.

The results of the analysis showed that the Cronbach's α coefficients of Find Flower, Hua Bangzhu, Microsoft's flower, Hua Banlv, and Xing Se were 0.967, 0.931, 0.958, 0.783, and 0.846, respectively. It shows that the experiment has high credibility and internal consistency for the investigation of each APP, and has very good reliability and stability.

In this study, the construct validity of the experiment was tested by KMO and Bartlett spherical test. The KMO value was greater than 0.70 and the Bartlett spherical test was less than 0.50, indicating that the survey had good validity. The results of the analysis showed that the KMO values of Find Flower Recognition, Hua Bangzhu, Microsoft's Flower Recognition, Hua Banlv, and Xing Se were 0.905, 0.871, 0.890, 0.729, and 0.819, all of which were greater than 0.70; the Bartlett spherical test had a p-value of 0.000. Rejecting the null hypothesis indicates that the experiment is effective.

3.2   Nonparametric Tests

The survey data does not follow the normal distribution, so this paper uses the non-parametric test Kruskal-Wallis H test and chart for comparative analysis. The essence of the Kruskal-Wallis H test is to test whether there is a significant difference in the overall recognition rate distribution among multiple applications and between multiple plant species under the premise of the same plant; the chart uses a histogram, a stacked map, a pie chart to The form of the statistical total score is comparatively analyzed, and the total score is high, indicating that the recognition rate is high; on the contrary, the recognition rate is low.

The Kruskal-Wallis H test of the nonparametric test was used to analyze the significant differences between the five applications (see Table 1); and the five applications were used to count the total scores of the different angles of each plant and plotted as columns (Fig. 3). It can be concluded that Hua Bangzhu, Hua Banlv, Xing Se and Find · Flower Recognition, Microsoft's Flower Recognition, and the asymptotic significance are all 0.000, rejecting the null hypothesis. Through the histogram, it can be seen that the first three are about 1.5 to 2 times of the latter two, which has obvious advantages in correct rate.

Table 1. Analysis of the difference significance in the application for recognizing five types of flowers

APP	Hua Bangzhu	Microsoft's Flower Recognition	Hua Banlv	Xing Se
Find · Flower Recognition	0.000	0.442	0.000	0.000
Hua Bangzhu		0.000	0.068	0.058
Microsoft's Flower Recognition			0.000	0.000
Ilua Banlv				0.434

In the research process, the recognition rate of the two applications of Hua Bangzhu and Hua Banlv is higher than that of the other three. On the one hand, there is a plant photo support of the Chinese plant image library, and on the other hand, the identification data generated by the user identification exchange is greatly improved. Its recognition rate. There is no significant difference between Xing Se and the top two, and it has a good recognition rate. It provides maps of plant attractions, and relies on user-provided plant pictures and supplements of expert identification to generate data, which improves the interaction between users and products. Achieving the goal of browsing the global selection of flower-seeking sites without leaving home, but because it will check the plant library through user feedback, the recognition rate has decreased. The lower total score was Find Flower Recognition and Microsoft's Flower Recognition, and the asymptotic p value of the two was 0.442, the difference was not significant. Among them, Microsoft's Flower Recognition relies on its own technological advantages to develop a proprietary identification model, and includes 400

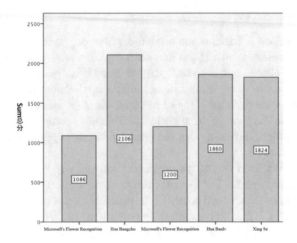

Fig. 3. Total score histogram of the five flower recognition applications

kinds of horticultural flowers in the system. The answer of this application provides higher credibility, thanks to its strong technical support, but its identification. The range is small, resulting in low recognition rate; the recognition rate of Find Flower Recognition is the lowest, the first is that its recognition range is small, the second is that its recognition model is not in training, the technology is not strong enough, resulting in low plant recognition rate. ($p > 0.05$, indicating that the difference is not significant; $0.01 < p < 0.05$ means significant difference; $p < 0.01$ means that the difference is extremely significant).

3.3   Plant Category Test

Through the comparative analysis of garden plants and wild plants, we can further understand the difference between the five applications for the overall recognition rate of plants (see Table 2).

Table 2. Analysis of difference significance in plant categories

Plant category	Overall	The common	Common	Uncommon
Garden vs wild	0.723	0.000	0.057	0.110

The recognition accuracy of these five applications for garden plants and wild plants is asymptotically p-value of 0.723, indicating that there is no significant difference. The pie chart shows that the six groups of data are close to equal (Fig. 4). But each of the gardens, the species of the species are slightly higher than the total score of the wild, and the recognition rates of the common plants in the garden and the common plants in the wild have obvious differences, and the phase difference value is nearly 400 points. The data shows that these five applications are better at the recognition rate of garden plants, depending on the majority of the user community's range of activities in the garden.

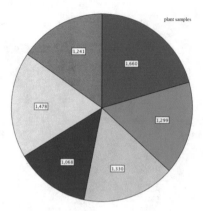

Fig. 4. Total score of plant samples

3.4   Internal Inspection of Flower Application

Through the identification and comparison of garden plants and wild plants in each application, the specific identification tendency of the application can be known (see Table 3).

Table 3. Analysis of difference between garden plants and wild plants in five flower recognition applications

APP	Find · Flower Recognition	Hua Bangzhu	Microsoft's Flower Recognition	Hua Banlv	Xing Se
Garden vs wild (overall)	0.000	0.893	0.000	0.486	0.320
Garden vs wild (the common)	0.001	0.489	0.014	0.389	0.199
Garden vs wild (common)	0.043	0.859	0.017	0.878	0.572
Garden vs wild (uncommon)	0.035	0.681	0.019	0.097	1.000

The results showed that the asymptotic Sig values of the garden plant recognition rate and the wild plant recognition rate of Find Flower Recognition and Microsoft's Flower Recognition application were both 0.000, less than the significance level of 0.05, and the cumulative score of the garden plants was found to be approximately It is twice as large as wild plants (Fig. 5).

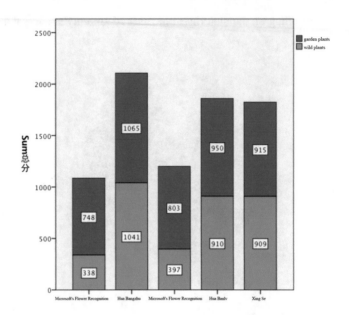

Fig. 5. Stacking of garden plants and wild plants of five flower recognition applications

At the same time, the asymptotic Sig values of common, common and uncommon categories of gardens and wild plants in these two applications are less than 0.05, indicating a very significant difference. The pie chart visually reflects the proportion of common, common and uncommon samples between garden plants and wild plants (Fig. 6). It can be found that the recognition rate of garden plants in these two applications is higher than that of wild plants. Degree level. Therefore, these two applications focus on the storage of garden plants, and the recognition rate has a lot of room for improvement.

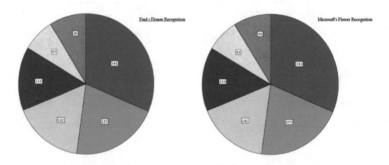

Fig. 6. Different levels of the total pie chart of Find · Flower and Microsoft's Flower Recognition

The asymptotic Sig values of the three applications of flower lord, flower companion and color are all greater than 0.05, indicating no significant difference. The total score of garden plants is slightly higher than the total score of wild plants, according to the common degree of plants, flowers Each common level of the helper's score is around 350, close to full marks; the scores of flower mate and color are around 300. These three applications should comprehensively develop the identification of garden plants and wild plants, and should continue to work hard to maximize the recognition rate.

4 Conclusion

Through the evaluation and comparison of the identification rate of five flower recognition applications, the paper analyzes that they should learn from each other's strengths, and improve from the aspects of technology, products, plant libraries, etc. Improve the accuracy of flower recognition and reduce the difficulty of flower recognition.

Artificial intelligence is not mature enough. To improve the recognition accuracy of literacy applications, new technologies and algorithms should be explored to improve existing identification models. For example, the extraction methods of image features such as color, texture, shape and connected components are more refined, and unique algorithms are developed for the field of plant identification. The existing ideas are extended to innovate existing machine training models to find deep learning based on convolutional neural networks. A new way out of technology.

The mature applications currently on the market are distributed in the fields of socialization, e-commerce, etc., and the application of literacy can learn from their advantages. It is difficult to maintain long-term data by relying on their own data. Only by letting users participate, it can save a lot of manpower by UGC operation. Material resources and greatly improve the accuracy of plant identification. Because of the large user base and the strong communication function, the flower lord has the highest recognition rate; the shape color provides a powerful plant navigation map because it provides users with the function of uploading while recognizing.

After investigation and data analysis, it is found that the application of literacy is not mature enough. The identifiable plant varieties are still at the level of domestic garden plants and common plants. It is difficult to identify specific varieties, and the recognition rate is not accurate enough. Knowledge-based applications should constantly improve their own recognition algorithms, increase plant stocks, especially wild plants, expand the breadth and depth of plant varieties, and achieve more accurate identification.

Acknowledgments. This work is supported by the Fundamental Research Funds for the Central Universities (2015ZCQ-YS-02) and Beijing Higher Education Young Elite Teacher Project (YETP0785).

References

1. Krizhevsky, A., Sutskever, I., Hinton, G.E.: ImageNet classification with deep convolutional neural networks. J. Adv. Neural Inf. Process. Syst. **25**, 1106–1114 (2012)
2. Hinton, G.E., Osindero, S., The, Y.: A fast learning algorithm for deep belief nets. J. Neural Comput. **18**, 1527–1554 (2006)
3. LeCun, Y., Bottou, L., Bengio, Y., et al.: Gradient-based learning applied to document recognition. J. Proc. IEEE **86**, 2278–2324 (1998)
4. Deng, J., Dong, W., Socher, R., et al.: ImageNet: a large-scale hierarchical image database. In: Computer Vision and Pattern Recognition, pp. 248–255. IEEE Press, Miami (2009)
5. LeCun, Y., Boser, B., Denker, J.S., et al.: Handwritten digit recognition with a back-propagation Network. In: Advances in Neural Information Processing Systems, pp. 396–404. Morgan Kaufmann Press, San Francisco (1990)
6. Hinton, G.E., Salakhutdinov, R.R.: Reducing the dimensionality of data with neural networks. J. Sci. **313**, 504–507 (2006)
7. Gardner, M.W., Dorling, S.R.: Artificial neural networks (the multilayer perceptron)-a review of applications in the atmospheric sciences. J. Atmos. Environ. **32**(14/15), 2627–2636 (1998)
8. Huang, Y.M., Lin, T., Cheng, S.C.: Effectiveness of a mobile plant learning system in a science curriculum in Taiwanese elementary education. J. Comput. Educ. **54**(1), 47–58 (2010)
9. Lee, S.H., Chan, C.S., Wilkin, P., et al.: Deep-plant: plant identification with convolutional neural networks. In: IEEE Image Processing (ICIP), pp. 452–456. IEEE Press (2015)

Mapping Digital Media Content

Stella Angova[(⊠)], Svetla Tsankova, Martin Ossikovski,
Maria Nikolova, and Ivan Valchanov

Department of Media and Public Communications, University of National
and World Economy, Osmi Dekemvri blvd, 1700 Sofia, Bulgaria
{sangova, s.tzankova, ossikovski, mnikolova,
valchanov}@unwe.bg

Abstract. The paper suggests that a variety of newly emerged/emerging journalistic forms have enriched online media content. The authors define these as innovative textual, audio, pictorial, and/or video techniques for presenting journalistic production. By examining digital content across both desktop and mobile-based platforms of media organisations, the authors identify 15 'hybrid' and 9 'genuine' new content forms and suggest a draft table-format map for their classification.

Keywords: Digital content · Online media · Content mapping

1 Introduction

Technological innovations have deeply changed the way media industry works. 'Cultural software' has become increasingly popular and the number of information and knowledge-sharing tools and services has been constantly growing [1]. Current studies see the resulting changes within the media ecosystem in this regard along various lines, e.g.: the influence of new technologies on society from a philosophical point of view [1–5]; on the media sector and the 'new newsroom' in particular [1, 6–13]; the essential features of new media [1, 7, 14]; the new types of journalism [15, 16]; the convergent culture [4]; media and journalism education [17]. Summarising those trends, Mitchelstein et al. [18] managed to identify five key research streams regarding 21-century online news production: the context of online news production; the processes of innovation in online journalism; online news production practices; professional and occupational matters of online news production; and the user as a content producer.

In this paper, we suggest that the introduction of cultural software in the field of media industry catalyses the creation and development of new forms of journalistic content. While leaving conventional media genre classifications aside, we suggest defining these 'new forms' as distinct and specific textual, audio, pictorial, and/or video techniques for presenting journalistic production. Essentially, we consider the use of such new forms equally a technological innovation, creative act, and business model.

© Springer Nature Switzerland AG 2020
T. Ahram (Ed.): AHFE 2019, AISC 965, pp. 233–238, 2020.
https://doi.org/10.1007/978-3-030-20454-9_23

2   Research Approach and Methodology

Whereas research interest regarding media content contexts, innovation in the sphere of media and journalism, and especially regarding innovative news production practices has been growing, the amount of available studies is insufficient, and the suggested identifications and classifications are often quite fragmented – or lacking altogether. Another problem is the notable lack of expert reports. The ones that we managed to trace were predominantly training-oriented materials focused on software used for the production of specific types of media content (e.g., timelines, story maps, infographics). We discovered few publications dealing with the features of new media content forms; however, a significant number of those were self-advertising materials of media outlets already using the new forms in question [19].

Bearing these lacunae in mind, for the current study we observed an analysed the mobile and desktop-based platforms of some traditional media brands (newspapers, radio and television stations), and of new media outlets without traditional analogues. We examined the mobile and desktop versions of the respective websites separately. All observations and analyses were completed between April 2017 and October 2018. For the sake of keeping a clearer focus, we decided to leave aside the social profiles of our media outlets available on platforms owned by other companies (such as Twitter, Instagram, Facebook). We used both qualitative and quantitative methods. The research went through the following phases:

Phase 1. Gathering of empirical data; selection and analysis of relevant academic and expert studies.
Phase 2. Observation and description of the content in research and news media platforms, including over 50 websites (media labs, major news providers, expert blogs, social media profiles).
Phase 3. Content analysis of expert and journalistic publications related to new media forms.
Phase 4. Case studies.
Phase 5. In-depth interviews with 10 Bulgarian media experts.

The mapping of new media content forms followed the gathering and analysis of a variety of empirical data: we collected over 600 information items coming from diverse traditional news and social media, both desktop and mobile-based platforms. We looked for answers to the following research questions: (1) what is the result of the journalistic uses of cultural software as regards media content? (2) What is the predominant type of new media narratives – textual, pictorial, or audiovisual? (3) What should be the logic behind the mapping of the new media narrative?

3   Findings

Regarding (1) *content*, our observations and analysis show that digital media forms can contain text, static visual imagery, dynamic visual imagery, audio, video, or a mix of textual and audiovisual content. Second, as regards the (2) *level of technological innovation*, these forms can be either 'hybrid' (i.e., innovative content transmitted via traditional technological means) or 'genuine new' forms (i.e., new types of content transmitted via new technological means). The intersection of these classification criteria (content, level of technological innovation) allows for laying out a tentative map of a variety of new journalistic forms.

Based on our empirical data analysis, we managed to identify and categorize 15 hybrid and 9 genuine new media forms (see Table 1 below). Further on, for each one of these we followed a uniform structure of subsequent in-depth analysis: (1) definition, (2) general description, (3) case studies, (4) available tools.

Table 1. Digital media content draft map (2018).

Type of content	Level of technological innovation	
	Hybrid forms	Genuine new forms
Text	Long read Notification content	(n/a)
Static imagery	Infographics	
	Vertical picture	Story map
	Square pictures	Timeline
	Slideshow	360-degree picture
	Satellite photo	3D picture
	Meme	Time lapse
	Drone photo	
Audio	3D radio	
	Podcast Audio slideshow	Visual radio
Video	Vertical video	360-degree video
	Video infographics	VR (virtual reality)
	Drone video	AR (augmented reality)

Expectedly, we identified a bigger number of audio, pictorial, and video media forms. We classified as 'hybrids' infographics, vertical pictures, square pictures, slideshows, satellite pictures, memes, drone pictures, 3D radio broadcasts, podcasts, audio slideshows, vertical videos, video infographics, and drone videos; respectively, story maps, timelines, 360-degree photos, 3D images, time lapses, visual radio broadcasts, 360-degree videos, virtual reality, and augmented reality fall into the 'genuine new forms' category:

This multitude of forms testifies well to the trend that access to technological innovation has enriched the journalistic practices for the production of media content. Media brands invest resources and effort into various ways of presenting information. However, the use of 'cultural software' does not appear to be an end in itself, but stands in line with the respective topic in order to provide for an accurate and improved presentation of the key content elements. We found substantial evidence that 360-degree videos, for example, allow for attention-catching presentations of otherwise uninteresting stories [20]; catching the exact meaning and the underlying emotion of news turns out to be a key function in this regard. According to professionals in the field, even raw 360-degree materials are far more compelling than conventional video footage [20]. As well, we noticed that linear and static video presentations do not attract new audiences due the low level of interactive content and their inability to put the user at the centre of action, i.e., to give them a 360-degree perspective. As well, the interaction of new technology and journalism adds substantial information value to the topic presented. Digital storytelling or the 'recreation' of a given event visualises factual content, and multimedia approaches change the way of perceiving information. Thus, the new forms of digital content allow for the creation of audiovisual simulations of alternative environments, where users can look and/or move in all directions. Technology has even made an additional further step, as evident from our descriptions of practices for the production of mixed-reality content: three-dimensional visualisations of real objects combined with augmented-reality objects presented physically in real-environment conditions. In sum, the key phrases describing this use of new technology in the process of media content production could be, e.g., content interaction, strengthening the realistic side of media content, added content value, mixed-reality content, new ways of content perception, stronger perceptions of physical space, content of higher information capacity, upgraded content, etc.

Our in-depth interviews helped us to verify these findings. Generally, our respondents think that technological changes have deep effects both on journalism in terms of professional qualification, as well as on the content, production, and ways of use of media products. The greatest challenges in this regard arise from the increased tempo and amount of work, the huge number of potential sources of information, the increasing variety of technical skills required for successful employment, the higher demand for video content, the need for ever better visualisations and multimedia products. Bulgarian media follow these general trends, albeit with some delay. As regards the quality of media content, the lag is more significant: visualisations are dominated by conventional photographs and videos, followed by infographics as a very popular presentation format; longreads, vertical, square, and 360-degree photos and videos are also well known to Bulgarian professionals. On the other hand, VR and AR are far less known and, respectively, rarely used. However, our interviewees emphasised that new technologies are merely a tool of journalism, the core values of which remain relatively stable, and their use cannot and does not guarantee the better quality of journalistic content.

4   Conclusions

Our findings confirm that there is a *steady growth in the number and range of practices* for the creation of media content in convergent media environment. Cultural software and technological innovations have changed the way of telling and presenting digital narratives. As a result, content-related journalistic practices have been re-defined to include the various ways of gathering information as well as the processes of creation, dissemination, and use of media products.

While a number of studies claim that video content has taken the lead and, accordingly, media companies should follow this general trend, our observations show that *many serious media outlets have not been 'infected' by the 'video virus'* to the extent of rejecting predominantly text-based forms. We found many examples of high-quality content with excellent multimedia presentations of the corresponding narrative. Established media brands, such as the *New York Times, Guardian, BBC, Telegraph, Washington Post, National Geographic, Wall Street Journal, CNN, Reuters, USA Today, Associated Press, Spiegel, San Francisco Chronicle*, and many others, demonstrate concepts for content which is both innovative and elite. Such brands indeed set up various basic norms regarding today's digital media formats – without neglecting the traditional principles of quality journalism at the same time.

The in-depth expert interviews confirmed our hypothesis that *new technologies deeply influence the media system as a whole*, especially as regards the impact of media texts, the growing potential of live broadcasting and data journalism, the enabling of convergence of content forms characteristic of different types of media. The interviewees also turned our attention to some emerging new content forms, such as satellite photos and time-lapses.

In general, text remains an important information vehicle, yet probably *steps aside as the core construction element* of media content. We ran across a number of cases of multimedia content where textual, pictorial, audio, and video elements could all function as autonomous vehicles of information. Multimedia stories remain valuable even when users can interact with one or few of their elements.

New media forms demonstrate a *higher engagement impact on the digital behaviour of audiences*. It is the digital behaviour of audiences that defines the strategic decisions regarding the specific profile and preferred types of media content. As a rule, today's audiences are active, and this has changed the understanding of how they can be involved in the media process. Users today employ a variety of devices to access online news, social networks, or corporate websites. The ways in which they find and interact with media products, as well as the technological means used for the creation and dissemination these products, are different – hence the differences in content designed for mobile and desktop-based platforms. In the case of mobile videos, for example, vertical formats have been trending due to the hardware specifics of mobile cameras and displays.

New media content *offers new types of experience to users*. Compared to traditional ones, new content forms allow for much higher degrees of empathy, as well as more realistic sensual perceptions. Technological innovations enable users to reach the inside core of the story, and even to 'replay' it in different reality formats. It is the story's context that determines the most appropriate forms for its media presentation.

New content forms *add value to media brands*. We consider content innovations to be an element of convergent business models. Content-based business models evolve around the idea of new content forms and rely on the changed paradigm of audiences as active participants in the media process. Content innovations contribute to the formation of loyal audiences around media brands.

We think our findings and draft map of new media content will be of benefit to media organisations, journalists, media experts, and university teachers and students in the field of media and journalism. As well, we look forward to discussing the classification criteria for digital media content mapping, the new content forms identified above, and the future of media production in general.

References

1. Manovich, L.: Cultural Software (2011). http://manovich.net/content/04-projects/070-cultural-software/67-article-2011.pdf
2. Castells, M.: The Rise of the Network Society. Wiley-Blackwell, Oxford (2009)
3. Manovich, L.: The Language of New Media. MIT Press, Cambridge (2001)
4. Jenkins, H.: Convergence Culture: Where Old and New Media Collide. New York University Press, New York (2006)
5. Dijk, J.: The Network Society. Sage, New York (2006)
6. Bradshaw, P.: The Online Journalism Handbook: Skills to Survive and Thrive in the Digital Age. Routledge, London (2011)
7. Deuze, M.: Media Work (Digital Media and Society). Polity, Cambridge (2013)
8. Deuze, M.: Managing Media Work. Sage, Thousand Oaks (2014)
9. Deuze, M.: Making Media: Production, Practices, and Professions. Amsterdam University Press, Amsterdam (2018)
10. Domingo, D.: Inventing online journalism: Development of the Internet as a news medium in four Catalan newsrooms. PhD dissertation. Universitat Autònoma de Barcelona, Bellaterra (2006)
11. Hall, J.: Online Journalism: A Critical Primer. Pluto Press, London (2001)
12. Pavlik, J.: Journalism and New Media. Columbia University Press, New York (2001)
13. Boczkowski, P.: The processes of adopting multimedia and interactivity in three online newsrooms. J. Commun. **54**, 197–213 (2004)
14. Lister, M., Dovey, J., Giddings, S., Grant, I., Kelly, K.: New media: A Critical Introduc-tion (second edition). Routledge, London (2009)
15. Deuze, M.: The professional identity of journalists in the context of convergence culture. Obs. J. **7**, 103–117 (2008)
16. Bull, A.: Multimedia Journalism. A Practical Guide. Routledge, London (2015)
17. Bardoel, J., Deuze, M.: Network journalism: converging competences of old and new media professionals. Aust. J. Rev. **23**, 91–103 (2001)
18. Mitchelstein, E., Boczkowski, P.: Between tradition and change: a review of recent research on online News production. Sage J. **10**, 562–586 (2009)
19. Radio 3 in binaural sound. BBC. http://www.bbc.co.uk/programmes/articles/29L27gMX0x5YZxkSbHchstD/radio-3-in-binaural-sound
20. Prat, C.: How 360 video can add value to journalism (2018). https://medium.com/journalism360/how-360-video-can-add-value-to-journalism-227c461b9aca

Social Media Competition for User Satisfaction: A Niche Analysis of Facebook, Instagram, YouTube, Pinterest, and Twitter

Sang-Hee Kweon[1]($\boxtimes$), Bo Young Kang[1], Liyao Ma[1], Wei Guo[1],
Zaimin Tian[1], Se-jin Kim[2], and Heaji Kweon[3]

[1] Sungkyunkwan University, Seoul, Korea
skweon@skku.edu, smileboyoung@gmail.com,
ayaolvo7@gmail.com, guoweikklk@gmail.com,
tzml786397607@gmail.com
[2] Korea University, Seoul, Korea
oosejinoo@naver.com
[3] Seoul National University, Seoul, Korea
hazzyjk@naver.com

Abstract. This paper explores five social media's media ecological location in the social user through niche analysis of five key SNS services (Facebook, Instagram, YouTube, Pinterest, and Twitter). Based on the results of 224 SNS user's questionnaire, factor analysis was carried out to extract five common factors of relationship, sociality, convenience, routine, and entertainment. The results of the niche analysis showed that Facebook had the widest niche in sociality (.627) and convenience (.636), and YouTube showed the widest niche in routine (.670) and entertainment (.615). For relationship (.520), Instagram had the widest niche. In terms of five factors, Facebook and YouTube have the greatest overlap in relationship (1.826) and sociality (2.696), while Pinterest and Twitter had the biggest overlap in routine (1.937); entertainment (2.263) and convenience (2.583). Besides, YouTube and Twitter had the most overlap. Facebook, Instagram, and YouTube had a competitive advantage over Pinterest in terms of all factors.

Keywords: Social media · Niche breadth · Niche overlap · Competitive superiority

1 Introduction

The technological progress of the media has come to a multimodal environment in which various media coexist. Especially, with the advent of social media such as Facebook, Twitter, and Instagram, their use and dependency are increasing, resulting in a different phase of the media environment.

This change in media convergence has increased the number of media in terms of quantity, and it has changed the structure of media in terms of quality. Today, along with the rapid development of the Internet, the paradigm of media environment and communication is changing in various ways and changing the usage pattern of users. Due to

© Springer Nature Switzerland AG 2020
T. Ahram (Ed.): AHFE 2019, AISC 965, pp. 239–249, 2020.
https://doi.org/10.1007/978-3-030-20454-9_24

the popularization of the Internet and the development of web technologies, network functions are expanding and communication and information sharing among people is accelerating due to the emergence and development of various social network services (SNS). Therefore, competition among media around service users is intensifying.

In this study, we use niche theory as the theoretical tool to analyze the user niche and compare the differences of the new media with similar characteristics for Facebook, Instagram, YouTube, Pinterest, and Twitter. Therefore, we try to analyze the competitive relationship of services with similar characteristics according to user utilization through niche theory.

Existing studies have paid attention to the usage motivation and adoption variables, but this study differs from such studies in that it focused on users' psychological factors such as utilization satisfaction and characteristics of social media and compared the competitive relationship between similar services.

2 Theories Background

2.1 Social Media Usage and Features

Social Media Usage. Social media is evolving into media in which various forms of content are created and shared by users. Social media is based on social networks. Social media has the characteristics of users' voluntarily participation, information sharing, and content creation. Competition has been gaining momentum as the fir0073t-generation social media, Facebook, and recently, Instagram, Snapchat, and other latecomers have experienced increases in usage rates. According to recent statistics from STATISTA [1], the number of social media users worldwide is projected to grow by 7.9% year-on-year to 2.46 billion in 2017, an annual average rate of 5.3%, and reaching 3 billion by 2021.

Social Media Features. In general, social media or SNS refers to the services that enable exchange and sharing of information and opinions on an online network. Boyd and Ellison [2] describe SNS as "a web-based service where public or semi-public profile information is created within a confined system, linked it with others, allowing the user to see the information created by others within the system". In short, they define the SNS in terms of its characteristics of the creation of information through voluntary participation of individuals, networking, and network. No [3] standardize SNS as a social media concept in the media domain, presenting networking and information sharing between people, self-expression and voluntary participation, and the possibility of expansion to new areas of SNS. Based on the Web 2.0 paradigm, social media has characteristics of participation, openness, dialogue, and community [4].

Since being created in 2004 Facebook has become a representative SNS which has the largest number of users after making itself available to use for all people over the age of 13 with an email address. One of the symbols of Facebook is the 'Like' button, which plays a big role in forming a consensus. Twitter is defined as social network service or microblogging service. But Seol [5] describes twitter not as a completely new SNS, but a new modified SNS that absorbs the strengths of existing blogs, SMS, messenger, and community, and is a media which strengthens 'user's choice', allowing users to filter only the stories that are relevant and the wanting to hear to communicate based on them.

Instagram' is a social network service that shares photos and videos more than text. The user is able to edit pictures according to taste and purpose, and it is easy to share it with others compared to SNS like Facebook. And YouTube is the UGC platform with the most users in the world. YouTube has evolved into a service that allows users to easily upload videos via the Internet and share them with others. Besides, Pinterest is a combination of pins used to fix objects on the wall and interest. It is an image-based social network service where images of individual interest are posted and are shared with friends in association with other SNS sites.

2.2   Applying the Niche Theory in the Usage Status of Social Media

Applied to the Niche Theory and Media's Niche Research. The niche theory began from the ecological theory that individual units apply to the given environment for their survival [6]. This theory pays attention to the "population-to-population competition" that occurs when a new population enters a limited space in which a particular population resides, as the new population also tries to consume the resources consumed by the existing population. Eventually, the theory explains the two groups form a complementary relationship and coexist or are replaced by a competitively superior population depending on the degree of competition.

The niche theory has also been applied to media research and has become a useful framework for analyzing the competition between media. Each media industry depends on limited resources, such as organisms in the environment. If multiple media depend on the same resources, the competition for survival among media becomes intense. In this process, media with superior resource acquisition replace inferior media, Strategic function differentiation creates a complementary relationship between the two media [7].

Applied to media studies, niche theory has become a useful analytical tool in studying the competition between media [8, p. 29]. Like organisms in the natural environment, each media survives on limited resources. When multiple media rely on the same resources, the competition for survival aggravates, through which process the superior media replace the inferior one or form a complementary relationship as a particular media diversifies its functions [7].

Also, niche theory presents "user satisfaction, satisfaction opportunity, user purchases, media use time, media content, and advertising costs" as the resources that media try to acquire through competition [8, p. 29]. That is, media compete with each other in order to secure user satisfaction, time and cost users pay, content to attract users, and advertising costs. Empirical studies analyzing media competition on the basis of niche theory have mainly been conducted for a share of advertising expenditure, media contents, and user satisfaction. User satisfaction is largely divided into 'acquired satisfaction' and 'satisfaction opportunity', where the former means satisfaction obtained by using specific media [8, p. 30]. On the other hand, the opportunity to meet means the convenience associated with the use of "whether media users can use specific media at specific places or times" [8, p. 31]. For user satisfaction, media competition can be explained by the concept of "functional alternative" proposed by the use and satisfaction theory [9, p. 175]. However, it is difficult to explain the competitive advantage that media with similar satisfaction levels have in each sub-dimension.

On the other hand, the theory of niche has the advantage of allowing the researcher to grasp the relative superiority among the comparative media in various sub-dimensions in the ability to satisfy users through the core concepts of niche breadth, niche overlap and competitive superiority [10, 11]. Therefore, in this study, we apply the niche theory to five social media of Facebook, Instagram, YouTube, Pinterest and Twitter.to ultimately determine the competitive advantage of each social media.

2.3   Niche Breadth, Niche Overlap and Competitive Superiority

Niche Breadth. The niche breadth refers to the relationship of resources and populations within a community [7]. In a spatial sense, the niche width is the size of a specific region formed along a specific resource dimension or axis [8]. Thus, the relative width is a measure of the population locality [12]. Figure 1 is a formula for calculating the niche width.

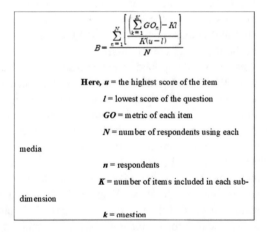

$$B = \frac{\sum_{n=1}^{N} \left[\frac{\left(\sum_{k=1}^{K} GO_n \right) - Kl}{K(u-l)} \right]}{N}$$

Here, u = the highest score of the item

l = lowest score of the question

GO = metric of each item

N = number of respondents using each

media

n = respondents

K = number of items included in each sub-

dimension

k = question

Fig. 1. Niche breadth calculation formula

Niche Overlap. Niche overlap refers to the degree of similarity among populations in resource utilization patterns, i.e., the extent to which they depend on a common resource [8, p. 37]. In other words, niche overlap is a concept that shows the degree of competition between media. As the niche overlap increases, the resources utilized by the two media become similar to each other, leading to a competitive relationship. Niche overlap is considered an indicator of the possibility to complement of a substitute; in user satisfaction, niche overlap would refer to the functional similarity between media objects of comparison [8, p. 80]. Figure 2 below is a formula for calculating the niche overlap.

$$O = \frac{\sum_{n=1}^{N} \sqrt{\sum_{k=1}^{K} \frac{(GO_i - GO_j)^2}{K}}}{N}$$

Here, i = media i

j = media j

GO = Achievement score of media i and media j in each item

N = the number of respondents using both media i and media j

n = respondents

K = number of items included in each sub-dimension

Fig. 2. Niche overlap calculation formulas

Competitive Superiority. Competitive superiority is a measure of the ability to identify which of the two media is in a superior position in resource utilization [8, p. 80]. The core concepts proposed by the niche theory are niche overlap and competitive superiority. This is an analytical tool that helps to grasp the rivalry between the media, helping to identify whether the comparative media is in a competitive or competitive substitution relationship through niche overlap and competitive advantage [13, p. 107]. Competitive superiority is calculated using the formula shown in Fig. 3 below.

$$S_{A>B} = \frac{\sum_{n=1}^{N} \sum_{k=1}^{K} m_{A>B}}{N}$$

$$S_{B>A} = \frac{\sum_{n=1}^{N} \sum_{k=1}^{K} m_{B>A}}{N}$$

Here, A = Media A

B = Media B

$mA > B$ = total score of the items whose B is higher than media A

$mB > A$ = the total score of the items whose A is higher than the media B

N = number of respondents using both media A and B

n = respondents

K = number of items included in each sub-dimension

k = question

Fig. 3. Competitive superiority calculation formulas

3   Research Problems and Methods

3.1   Research Problems

This study analyzes the competitive relationship between Facebook, Instagram, You-Tube, Pinterest and Twitter, and compares their competitiveness. We hope to focus on the niche theory to clarify the use conditions of Facebook, Instagram, YouTube, Pinterest, and Twitter, and analyze the difference.

In order to examine the competition among services in terms of user satisfaction by applying the formula of the niche theory, it is necessary to derive the common satisfaction factors for new media Facebook, Instagram, YouTube, Pinterest, and Twitter. Therefore, the following research problems were set up in this study.

Research Questions 1. What are the user's satisfaction factors in the use of Facebook, Instagram, YouTube, Pinterest, and Twitter?

Research Issues 2. What are the niche breadths, the niche overlap, and the competitive superiority in the use of Facebook, Instagram, YouTube, Pinterest, and Twitter?

Research Issues 3. What are the features of the competition between Facebook, Instagram, YouTube, Pinterest, and Twitter?

3.2   Method

Collecting Data. Survey was used as the data collection method for this study. From November 7th to November 14th, 2017, 10 questionnaires were distributed to all students of Sungkyunkwan University graduate school Communication Statistical Analysis course, and questionnaires were distributed to a total of 224 subjects. This study includes the social media of Facebook, Instagram, YouTube, Pinterest, and Twitter.

Factor Analysis and Reliability Verification for Each SNS. In order to use niche equation to investigate the competitiveness of Facebook, Instagram, YouTube, Pinterest and Twitter, we first extracted factors for each social media. Extracted factors were named relationship, sociality, convenience, routine, and entertainment according to previous existing studies on the characteristics of SNS.

The reliability coefficient for all factors is shows that, all, except for YouTube's routine and Facebook's recreation, have a satisfactory level of reliability with .6 directors.

4   Result

4.1   Niche Breadth Analysis

In this study, we applied the niche breadth formula to find the niche breadth of Facebook, Instagram, YouTube, Twitter, and Twitter for relationship, sociality, convenience, routine, and Entertainment. The results are shown in Table 1 below.

Table 1. Niche breadth of 5 SNS

Distinction	Relationship	Sociality	Convenience	Routine	Entertainment
Facebook	.478 (2)	.627 (1)	.636 (1)	.518 (2)	.559 (2)
Instagram	.520 (1)	.448 (3)	.606 (2)	.505 (3)	.553 (3)
YouTube	.452 (3)	.617 (2)	.587 (3)	.670 (1)	.615 (1)
Pinterest	.256 (5)	.254 (5)	.360 (5)	.340 (5)	.248 (5)
Twitter	.344 (4)	.310 (4)	.469 (4)	.470 (4)	.348 (4)

Note: 0 = maximum niche breadth, 1 = minimum niche breadth, and ( ) is the rank of the niche breadth among media by the factor.

Facebook showed a niche breadth from .478 to .636. Among the factors, the factors that showed the smallest breadth were convenience, followed by sociality, entertainment, and routine. Next, Instagram showed a niche breadth from .448 to .606. Just like Facebook, it showed the biggest niche breadth for convenience. YouTube showed a niche breadth from .452 to .670, and had relatively high niche breadth for all factors, of which routine was the highest. Pinterest showed a niche breadth from .248 to .360, and Twitter showed a small width from .310 to .470, showing relatively small niche breadths compared to other SNS.

Overall, Facebook has the biggest niche breadth, followed by YouTube, Instagram, Twitter, and Pinterest. For relationship, the Instagram showed the smallest breadth, Facebook showed the smallest in sociality and convenience, and YouTube showed the smallest breadth in daily and entertainment.

4.2 Niche Overlap Analysis

In the study of the media industry, the niche overlap is an indicator of the similarity between two media when it comes to resource utilization. The closer the value of niche overlap is to 0, the higher the similarity of resource utilization between the two media. Therefore, in this study, we applied a formula to find the niche overlap, and found the niche overlap between Facebook, Instagram, YouTube, Twitter, and Twitter in relationship, sociality, convenience, routine, and Entertainment as shown in Table 2 below.

In terms of respective factors, the niche overlap of Facebook and YouTube (1.826) was the highest for relationship, followed by the niche overlap of YouTube and Twitter (1.933). Conversely, it was found that the niche overlap of Facebook and Pinterest (2.730) and Instagram and Pinterest (2.956) were the lowest. In terms of Sociality, the niche overlap of Facebook and YouTube (2.696) was the highest, followed by Instagram and YouTube (2.821). The two groups with the lowest niche overlap are YouTube and Pinterest (3.641), Facebook and Pinterest (3.878). In terms of convenience, YouTube and Twitter (2.853) had the highest niche overlap, followed by Facebook and Instagram (2.756). On the other hand, Facebook and Pinterest (4.218), Instagram and Pinterest (4.263) showed the lowest niche overlap.

Table 2. Niche overlap by SNS factors

Distinction	Relationship	Sociality	Convenience	Routine	Entertainment
Facebook-Instagram	2.364 (6)	3.445 (7)	2.756 (2)	2.857 (4)	4.140 (9)
Facebook-YouTube	1.826 (1)	2.696 (1)	3.193 (5)	3.765 (10)	4.748 (10)
Facebook-Pinterest	2.730 (9)	3.878 (10)	4.218 (9)	3.758 (9)	2.846 (4)
Facebook-Twitter	1.953 (3)	3.182 (6)	2.961 (3)	3.671 (8)	3.180 (5)
Instagram-YouTube	2.288 (4)	2.821 (2)	3.710 (8)	3.140 (5)	3.432 (7)
Instagram-Pinterest	2.956 (10)	3.508 (8)	4.263 (10)	3.168 (7)	2.730 (3)
Instagram-Twitter	2.396 (7)	2.994 (4)	3.415 (7)	3.160 (6)	2.363 (2)
YouTube-Pinterest	2.687 (8)	3.641 (9)	2.975 (4)	2.627 (3)	3.270 (6)
YouTube-Twitter	1.933 (2)	3.163 (5)	2.583 (1)	2.392 (2)	3.914 (8)
Pinterest-Twitter	2.313 (5)	2.850 (3)	3.211 (6)	1.937 (1)	2.263 (1)

Note: 0 = maximum niche overlap, 6 = minimum niche overlap, ( ) is the rank of the niche overlap among media by the factor.

The two SNSs with the highest niche overlap in terms of routine were Pinterest and Twitter (1.937), and YouTube and Twitter (2.392) were second. On the other hand, Facebook and Pinterest (3.758), Facebook and YouTube (3.765) had the lowest rankings. Finally, in entertainment, we can see that the niche overlap of the Pinterest and Twitter (2.263) is the highest as usual, followed by the niche overlap of the Instagram and Twitter (2.363). Also, SNSs with the lowest number of duplicates are Facebook, Instagram (4.140), Facebook and YouTube (4.748). Overall, it is YouTube that showed the highest niche overlap among the five SNSs, while the SNS with lowest niche overlap was the Pinterest.

4.3   Competitive Superiority Analysis

Competitive superiority measures which SNSs are more competitive among the various SNSs that are in a competitive relationship with limited resources. In terms of satisfaction, competitive superiority includes the degree and direction of the competitive superiority of a particular SNS. For each sub-dimension, it sums up values where one SNS is greater than another SNS, the average of which is then compared. We also conducted a corresponding sample T-test to verify whether the mean of the two SNSs is really significant. Therefore, this study calculated the comparative superiority of the relationship, sociality, convenience, routine, and entertainment for five SNSs of Facebook, Instagram, YouTube, Pinterest and Twitter using the formula for calculating the superiority, as shown in Table 3 below.

Table 3. Competitive superiority by SNS factors

Distinction	Relationship		Sociality		Convenience		Routine		Entertainment	
	CS-value	t-value	CS-value	t-value	CS-value	t-value	CS-value	t-value	CS-value	t-value
FB > INS	1.520	−4.072***	3.596	6.548***	2.674	.971	2.161	−.286	2.243	−.412
FB < INS	2.850		1.330		2.302		2.262		2.417	
FB > YT	2.312	2.295*	2.837	1.467	3.313	5.077***	2.723	2.441*	1.888	−3.219**
FB < YT	1.601		2.265		1.522		1.740		3.277	
FB > PIN	3.237	11.100***	3.587	7.548***	4.318	14.810***	3.142	4.684***	3.749	9.313***
FB < PIN	.615		1.142		.499		1.303		.658	
FB > TT	2.510	4.030***	3.997	12.399***	3.308	4.668***	2.340	.056	3.284	5.481***
FB < TT	1.313		.673		1.623		2.320		1.205	
INS > YT	3.102	6.587***	1.011	−8.761***	3.140	5.909***	2.548	2.081*	1.848	−3.672***
INS < YT	1.119		3.772		1.227		1.721		3.234	
INS > PIN	3.472	10.933***	2.614	3.432***	4.184	14.363***	2.900	4.333***	3.859	14.721***
INS < PIN	.663		1.490		.500		1.227		.444	
INS > TT	3.095	6.549***	3.078	7 895***	3.198	4.630***	2.233	.310	3.384	7.544***
INS < TT	1.115		.876		1.569		2.123		1.069	
YT > PIN	3.029	9.241***	3.718	8.324***	3.773	11.739***	2.650	2.982**	4.372	16.569***
YT < PIN	.745		1.083		.634		1.353		.445	
YT > TT	2.274	2.630**	4.099	14.802***	2.689	2.674**	1.831	−1.255	3.957	9.825***
YT < TT	1.488		.515		1.769		2.328		.945	
TT > PIN	2.034	3.533***	1.226	−3.634***	3.077	8.273***	2.661	4.585***	2.348	5.889***
TT < PIN	1.117		2.254		.751		.962		.874	

*p < .05, **p < .01, ***p < .001

Note: CS = Competitive superiority, FB = Facebook, INS = Instagram, YT = YouTube, PIN = Pinterest, TT = Twitter

As shown in Table 3, the competitive superiority among the five SNSs is composed of 10 combinations. First, the competitive superiority of Facebook and Instagram is significant only in terms of relationship and sociality. Instagram is more competitive than Facebook for relationship, and Facebook is more competitive than Instagram for Sociality. Second, when looking at the competitive superiority of Facebook and You-Tube, all four but Sociality are significant. In terms of relationships, convenience, and routine, Facebook is more competitive than YouTube, and YouTube is in a competitive superiority in Entertainment. Third, in terms of the competitive superiority of Facebook and Pinterest, all of them showed meaningful results, and in all respects, Facebook has more competitive superiority than Pinterest. Fourth, for the competitive superiority of Facebook and Twitter, all factors except routine are significant, in all of which Facebook has a competitive superiority. Fifth, in terms of the competitive superiority of Instagram and YouTube, all aspects are significant. In terms of relationship, convenience, and routine, the Instagram has a competitive superiority, while YouTube is in a competitive superiority.in terms of sociality and entertainment.

Sixth, the competitive superiority of Instagram and Pinterest was also significant in all aspects, and in all five aspects the Instagram had a competitive superiority. Seventh, the competitive superiority of Instagram and Twitter are significant in the four aspects except for routine. In all four aspects, the Instagram is in a competitive superiority. Eighth, the competitive superiority of YouTube and Pinterest shows that the factors are significant in all aspects, and in all aspects, YouTube turned out to have competitive superiority. Ninth, the for YouTube and Twitter in terms of competitive superiority is significant in every aspect except for routine, and YouTube has a competitive superiority in all factors. Twelfth and last, the comparative superiority of Twitter and Pinterest shows a significant value in all respects. Twitter had more competitive superiority than Pinterest in all aspects except sociality.

In summary, Facebook, Instagram, and YouTube have a competitive superiority in relationship, sociality, convenience, routine, and entertainment compared to Pinterest, and have a competitive superiority with Facebook, Instagram, and YouTube, and Twitter except for in routine. In addition, Twitter has a competitive superiority compared to the Pinterest in all the factors except Sociality. Instagram shows meaningful results in measuring the superiority of relationship and sociality compared with Facebook, and Twitter derives meaningless results in comparison of superiority with all social media except Pinterest in routine

5   Conclusion

The technological progress of the media has come to a multimodal environment in which various media coexist. Especially, with the appearance of social media such as Facebook, Instagram, YouTube, Pinterest and Twitter, the media environment is facing another phase of usage and reliance on it increase. With the advent of multimedia environment, this study considers social media of Facebook, Instagram, YouTube, Pinterest, and Twitter are considered as the representative social media and conducted an empirical analysis using niche analysis to study competition between social media. Based on the results of the questionnaire, factor analysis was carried out to extract common factors of relationship, sociality, convenience, routine, and entertainment. Based on the results, we apply the formula of niche breadth, niche overlap, and competitive superiority to five social media and derive the following research results.

First, analyzing the niche breadth revealed that Facebook displayed the highest niche breadth in terms of sociality and convenience, followed by YouTube with the niche breadth of routine and entertainment. In relationship, Instagram showed the highest niche breadth. Also, setting aside sociality and convenience where Facebook showed the highest niche breadth, it had the second highest niche breadth in all other factors: relationship, routine, and entertainment. In addition, Twitter and Pinterest ranked 4th and 5th in each result. These results suggest that there is a difference according to the use of each social media. According to a survey by STATISTA (September 2017) [14], Facebook is now topping the world in terms of usage, Twitter usage is ranked 10th, and Pinterest is ranked 16th.

Second, we investigated the niche overlap of Facebook, Instagram, YouTube, Pinterest, and Twitter. In terms of relationship and sociality, Facebook and YouTube have the highest niche overlap, and Pinterest and Twitter are the most frequently overlapping for routine and entertainment, and YouTube and Twitter have the most overlap in convenience.

Third, a study of competitive superiority of Facebook, Instagram, YouTube, Pinterest, and Twitter revealed that Facebook, Instagram and YouTube have a competitive superiority in all aspects compared to Pinterest and when compared to Facebook,

Instagram, YouTube, and Twitter, has a competitive superiority in all aspects of aside from routine. Also, Twitter had a competitive superiority compared to the Pinterest in all the factors other than sociality.

Through the above studies, we have identified common satisfaction factors for the current social media of Facebook, Instagram, YouTube, Pinterest, and Twitter, and provided research results that can provide an understanding of the relationship between the superiority and the competitive superiority of the media, so this study was meaningful. On the other hand, the limitation of the research is that the differences in the number of users in the five representative social media is too big, which may raise doubts about the reliability of the questionnaire answers, especially with Pinterest, which have a very small of number of users, may have its results questioned.

References

1. "Number of Social Network Users Worldwide from 2010 to 2021" (2017)
2. Boyd, D., Ellison, N.: Social network sites: definition, history, and scholarship. J. Comput. Mediated Commun. **13**(1) (2007). Article 11
3. No STATISTA Young: A study on the motivation of SNS affecting flow e-business research. e-Bus. Stud. **16**(1), 287–304 (2015)
4. Song, K.J.: A study on political participation of the social network generation. Korean and World Politics **27**, 57–88 (2011)
5. Seol, J.-a.: The evolution and social impact of social media. In: The Korean Press Information Society Academic Conference, 12 (2009)
6. Dimmick, J., Patterson, S., Albarran, A.: Competition between the cable and broadcast Industries: a niche analysis. J. Media Econ. **5**(1), 13–30 (1992)
7. Dimmick, J., Rothenbuhler, E.: The theory of the niche: quantifying competition among media industries. J. Commun. **34**(1), 103–119 (1984)
8. Dimmick, W.J.: Media Competition and Coexistence: The Theory of the Niche. Lawrence Erlbaum Assoc Inc., Mahwah (2003)
9. Katz, E., Blumler, G.: Uses and gratifications research. Public Opin. Q. **37**(4), 509–523 (1974)
10. Lee, S.-Y.: A study on the utilization of mobile phones. J. Korean Press **47**(5), 87–114 (2003)
11. Feaster, J.C.: The repertoire niches of interpersonal media: competition and coexistence at the level of the individual. New Media Soc. **11**(6), 965–984 (2009)
12. Hellman, H., Soramaki, M.: Competition and content in the U.S. video market. J. Media Econ. **7**(1), 29–49 (1994)
13. No, K.-Y., Lee, M.-Y.: A study on the media competition of blogs: competitive analysis through the analysis of relational and information-oriented blogs. J. Korean Press **49**(3), 318–389 (2005)
14. STATISTA "Leading Social Networks Worldwide as of September 2017" (2017)

The Diffusion of News Applying Sentiment Analysis and Impact on Human Behavior Through Social Media

Myriam Peñafiel[1]($\boxtimes$), Rosa Navarrete[1], Maritzol Tenemaza[1],
Maria Vásquez[1], Diego Vásquez[2], and Sergio Luján-Mora[3]

[1] Escuela Politécnica Nacional, Quito, Ecuador
{myriam.penafiel, rosa.navarrete, maritzol.tenemaza,
maria.vasquez}@epn.edu.ec
[2] ESPE-UGT, Latacunga, Ecuador
ddvasquez@espe.edu.ec
[3] University of Alicante, Alicante, Spain
sergio.lujan@ua.es

Abstract. The Web is the largest source of information today, a group of these data is the news that is disseminated using Social Networks which are information that needs to be processed in order to know what is its main use in a way that contributes to understanding the impact of these media in the dissemination of news. To solve this problem, we propose the use of data mining techniques such as the Sentiment Analysis to validate the information that comes from social media. The objective of this research is to make a proposal of a method of Systematic Mapping that allows determining the state of the art related to the investigations of the diffusion of news applying Sentiment Analysis and impact on human behavior through Social Media. This initial research presented as a case study a time range until 2017 in research related to the news that uses Data mining techniques like to sentiment analysis for social media in major search engines.

Keywords: Data mining · Text analysis · Sentiment analysis · News ·
Social media · Human behavior · Systematic mapping

1 Introduction

The spread of data from the web and that come from digitalization, affordable technology, consumer hardware, social networks, communities, online media, cloud computing, etc., are known as "massive data" or Big data [1]. If you consider that every minute there are two million visits on YouTube, 1.7 million Facebook messages and hundreds of thousands of tweets occur per minute online [2].

Social media, social media (SM) are defined as a group of Internet-based applications that use ideological and technological foundations of Web 2.0, and that allow the creation and exchange of content generated by the user [3]. Also social media users like those who said "yes" to "Have you ever used a social networking site like Facebook, twitter or LinkedIn?" [4]. Therefore, the data obtained from these media

© Springer Nature Switzerland AG 2020
T. Ahram (Ed.): AHFE 2019, AISC 965, pp. 250–259, 2020.
https://doi.org/10.1007/978-3-030-20454-9_25

constitute a material relevant to the work of professionals who use this data as they become an active voice through the Web. The main sources of this data are those that come from twitter either in real time or through historical archives [5].

All these data that come from the web through all your means, need to be processed to become information. For this, new tools, methods, and techniques are needed each time to obtain knowledge from structured or unstructured textual data, in a precise, timely and clear way [6].

The use of Data Mining techniques allows obtaining the best of knowledge from large volumes of data generated by the development of the web (Big Data). One way to convert these unstructured texts into authentic information is through data mining techniques such as the Sentiment Analysis.

The Sentiment Analysis or Opinion Mining [7] is an emerging technique of Data Mining that applies Artificial Intelligence at different levels for the processing of these texts, classifying the opinion into a positive, negative or neutral feeling, obtaining values of polarity between [−1, 1], as a result of the processing of the data.

For example, in a study where Sentiment Analysis was applied to analyze the Brexit debate [8], an opinion such as "I do not agree, I am a migrant" would obtain a value of polarity equal to −1, an opinion like "Politicians can be divided into Brexit but people are united! Let's continue with that" they would obtain a polarity value equal to 1. On the other hand, if the opinion were "I'm not interested", it would obtain a polarity value equal to zero. Opinions are not always clear or obvious, for example, opinions such as: "We cannot have an open door for EU migrants between now and the end of the Brexit process. It is time to control the issue of immigration", "The British people have spoken and the answer is that we are out!", they require natural language processing which defines the analysis of feelings as a non-trivial process.

In order to solve problems like these, which was faced by the researchers who handle the extraction of information, this research is presented that defines a process for the realization of a systematic mapping about the use of Social Media for the transmission of news. As a case study, we consider the investigations that apply Sentiment Analysis to the news with the aim of analyzing the scope of application of these investigations to determine their fields of application, the data sources and the databases from which the greatest part of SM information.

2 Systematic Mapping

Knowing the state of the question about a specific area of knowledge is the first step to conducting an investigation. In this sense, there are two options for quality work focused on conducting a good literature review: systematic mapping and review of the literature itself.

An analysis of the question (state of the art), review of the literature, (literature review) are a means to identify, evaluate and interpret all available research relevant to a specific research question or thematic area, or phenomenon of interest [9]. This type of studies belongs to what is called secondary research whose main benefit is to identify what is known about the subject and therefore what remains to be investigated, the quality of the studies is evidenced if they apply the scientific method [10].

On the other hand, systematic mapping or scoping studies are designed to give an overview of a research area through the classification and counting of contributions in relation to the categories of that classification [11]. The objective is to look for the topics that have been covered, where they have been published, by whom and how the research has been carried out. The result of a mapping study is an inventory of the documents on the subject area, assigned to a classification. Therefore, a systematic mapping provides an overview of the scope of the subject and allows to discover the research gaps and trends [9].

The Table 1 describes the main differences between systematic mapping and systematic review according to [12]. The use of either of the two techniques depends on the level of depth that you want to achieve with the literature review, which is aligned with the objective of the proposed research. In any case, if you start with a systematic mapping and then you want to complement it until the systematic review, the mapping offers a way forward, as long as the systematic mapping offers clear, concrete evidence.

Table 1. Differences between systematic mapping and systematic review.

Process for	Mapping systematic	Revision systematic
Research question	General focused on obtaining a classification of research trends. The results can be quantity, type of studies, place, researchers, etc.	Specific, related to the results obtained in the studies Of the form: Is method A, (method, technique, technology, etc.) better than B why?
Search process	It is done by thematic area	Is done according to the research question
Requirements, search strategies	Less strict. The interest is the research trends	Very strict They should evaluate the completeness of the search, in addition, to be fair
Quality evaluation	Not required	With quality evidence based on the scientific method
Results	A number of related articles by thematic area	Articles that answer the specific research question

2.1 Process for a Systematic Mapping

According to Kitchenham [9] a Systematic Mapping is made up of three generic phases: planning of the review, carrying out of the review and reporting of the review.

2.2 Planning for Mapping: In This Phase, You Should Do

1. Identification of the need for a review. Review exhaustively and impartially if there are works already done.
2. Development of a review protocol. Specifies the methods that will be used to perform a systematic mapping, research question (optional), sources, criteria, etc.

2.3 Mapping: The Stages Associated with the Mapping Are

1. Identification of the investigation. Define an impartial search strategy.
2. Selection of primary studies. To evaluate its real relevance, it is intended to identify the primary studies that have direct evidence on the research question.
3. Evaluation of the quality of the study. Quality is the measure in which the study minimizes the bias and maximizes internal and external validity. One way to prove this is through evidence based on the scientific method. For example, Cohen's Kappa coefficient can be used [13].
4. Extraction and monitoring of data. This stage considers the extraction of data in defined forms that are piloted so that it contains all the necessary information.
5. Synthesis of data. It consists of collecting and summarizing the results of the primary sources.

These stages can be interactive according to the need of the process.

Mapping Reports. In this Phase, Quantitative Results are Presented in the Form of Clearly Identifiable Diagrams.

3 Results

To explain the systematic mapping, a case study has been selected whose objective is to carry out a mapping related to the application of Data Mining techniques such as the Sentiment Analysis to analyze the news that comes from SM. The objective of determining the number of studies that have been conducted on the subject, the sources of the news, the areas of application of Sentiment Analysis, the databases where these investigations have been conducted, and whether the information comes from the journalistic media.

Next, the mapping process will be applied to the selected case study using the State of the Art tool through Systematic Review (StArt) [14], which allows the process to be optimally developed, considering the phases and previously defined stages.

3.1 Phase 1: Planning for Mapping

The mapping protocol is established, which is recorded in the StArt tool. The revision protocol consisted in establishing the dates for the search from 2004 until 2017, in a search of articles that apply the Sentiment Analysis in the news, the search chain was "sentiment analysis news". Approximately 900 articles were found, If the number of items was minimum, it is considered that the topic is not mature, therefore there is not enough research in the area and a process of systematic mapping is not convenient. But it is the case.

Additionally, it is also necessary to determine the need for a review. For this purpose, an exhaustive search was made of systematic mappings in the subject in the scientific databases such as IEEE, Web of Science (WOS), ACM, Scopus, Google Scholar (GS). Results ordered by an author are detailed in the summary Table 2.

Table 2. Summary of literature review or systematic mapping of: "news of sentiment analysis in social media"

Num	Name	Results focus	Data base	Authors
1	Characterization of the use of social media in natural disasters: a systematic review	Disaster management	IEEE	(Abedin; Babar; Abbasi, 2014)
2	A survey of data mining techniques for social media analysis	Classification of the techniques used in feeling analysis in social media	GS	(Adedoyin-Olowe; Gaber; Stahl, 2013)
3	Approaches to Cross-Domain Sentiment Analysis: A Systematic Literature Review	Methods and techniques used in the analysis of feelings between domains	IEEE	(Al-Moslm; Omar; Abdullah; Albared, 2017)
4	Main concepts, state of the art and future research questions in sentiment analysis	Most used techniques for feelings analysis, supervised learning and without supervision a comparison	Scopus	(Appel; Chiclana; Carter, 2015)
5	Exploring the Ensemble of Classifiers for Sentimental	Identify techniques, classifiers, tools for analyzing feelings	ACM	(Athar; Butt; Anwar; Latif; Azam, 2017)
6	Opinion Spamming in Social Media: A Brief Systematic Review	Detection of opinion spamming (OSD)	WOS	(Baharim; Hamid, 2016)
7	Erratum to: Multilingual Sentiment Analysis: State of the Art and Independent Comparison of Techniques	Methods used and their applications	Scopus	(Dashtipour et al., 2016)
8	Supervised sentiment analysis in Czech social media	Supervised methods of automatic learning of feelings analysis, create a data set	Scopus	(Habernal; Ptáček; Steinberger, 2015)
9	Survey of visual sentiment prediction for social media analysis	Most used techniques for visual feelings analysis	ACM	(Ji; Cao; Zhou; Chen, 2016)
10	Sentiment Analysis: A Literature Rev	Methods used for the analysis of feelings	IEEE	(Nanli; Ping; Weiguo; Meng, 2012)
11	Text mining for market prediction: A systematic review	Text mining for market prediction	Scopus	(Nassirtoussi; Aghabozorgi; Wah; Ngo, 2014)

(continued)

Table 2. (*continued*)

Num	Name	Results focus	Data base	Authors
12	A review of the literature on applications of text mining in policy making	Use of text mining techniques for the policy formulation process	GS	(Ngai; Lee, 2016)
13	Sentiment Analysis: A Comprehensive Overview and the State of Art Research Challenges	Methods used for the analysis of sentiment, its applications and challenges to give an overview of the analysis of feelings Determines the influence of social media news in the field of investments	Scopus	(Ragini; Anand, 2016)
14	A survey on opinion mining and sentiment analysis: Tasks, approaches and applications	Comprehensive work on feelings analysis, table of available data sets for feelings analysis	Scopus	(Ravi; Ravi, 2015)
15	Opinion mining and analysis: A literature review	Analyze techniques, sites, type of social media source of opinion mining	IEEE	(Singh; Dubey, 2014)
16	Characterizing opinion mining: A systematic Mapping Study of the Portuguese Language	Methods and techniques of opinion mining (OM), Naive Bayes and Support Vector Machine were the main ones	WOS	(Souza; Vitório; Castro; Oliveira; Gusmão, 2016)
17	Social media and news sentiment analysis for advanced investment strategies Analysis - A Systematic Literature Review	Determines the influence of social media news in the field of investments	WOS	(Yang; Mo, 2016)

Note: References in this table were uploaded to a digital repository so as not to reload the article (https://goo.gl/n1ivWU)

Only two articles more related to the case study have been found in the review of the literature: that of [15] that presents a Systematic mapping in which it analyzes the Data Mining methods and techniques of Opinion Mining (OM). Determining as the most used methods for applications in SM is to Naive Bayes and Support Vector Machine as the main classifiers and to SentiLex-PT was the most used lexical resource.

There is also the research of [16] which conducts a survey to determine the classification of Sentiment Analysis techniques for social media and some tools, which are a starting point to apply the techniques of Sentiment Analysis. It is worth mentioning important contributions such as those of [17] that disclose corpus to be used in Sentiment Analysis. Additionally, it is important to note the absence of research on the subject in the Spanish language [18].

This research is justified by virtue of the fact that there is no work within the area of this research as it can be seen in the summary table that guides information professionals. About the current state of research in the field of data mining using sentiment analysis in the news that is generated through digital media, and that define about the areas of application, types of social media used, bases of source data, which are the research questions of the proposed study. Therefore, we move on to the next phase.

3.2 Phase 2: Mapping

Identification of the Investigation. For the identification of articles related to the theme, the four scientific databases that can accommodate the largest amount of research on the subject were taken into account: ACM, IEEE, Web of Science, Scopus, Google Scholar. The studies considered were extracted on several dates and the last group on June 1, 2017.

Evaluation of the Quality of the Study. Since this is a systematic mapping process, no additional criteria were considered as those that already consider the databases with which we worked. For a literature review it is essential to define additional quality criteria.

Selection of Primary Studies. From the database: Scopus, 430 articles were identified using the search string: "sentiment analysis" AND news.

From IEEE 160 articles were identified using the search string: "sentiment analysis news".

From Web of Science, 171 articles were identified with the search string: "Topic: (sentiment analysis) AND Topic: (news)".

From ACM, 318 articles were identified with the search string + sentiment + analysis + news.

With a total of 1079 articles, of which the tool eliminated 138 duplicates, thus achieving 941 articles. These results can be seen using the Start tool in Fig. 1.

Extraction and Monitoring of Data. In this stage, the articles are selected based on inclusion and exclusion criteria that were discussed by the researchers.

Data Synthesis. The synthesis process was carried out: from the 941 articles identified, the inclusion/extraction criteria were applied obtaining a total of 183 results. A one-to-one reading of the abstracts of the articles was carried out in such a way that 82 articles were obtained, and in this control 11 duplicate articles that were separated were detected. A matching evaluation of the keywords was also carried out, with the score obtained, the articles that had less than 15 points were eliminated giving a total of

59 accepted articles. These are the totally purified articles with which the final objective classification of this research was carried out.

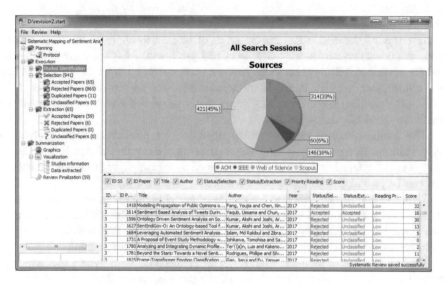

Fig. 1. Phase 2 of systematic mappings using StArt

3.3   Phase 3: Mapping Reports

Based on the proposed research questions, three classifiers were considered:

Means of Communication: Means of communication which was used as a source of information to perform the Sentiment Analysis. The following sources were included in this category: twitter, Facebook, blogs, online newspapers, online surveys, online news, and other digital opinion media.

Application Area: The sector where sentiment analysis was applied using news in this classifier was considered: politics, economy and finance, health, marketing, advertising, alerts and disaster prevention, entertainment and social.

Source of the News: Classified as formal journalism, 42% was obtained against 58% of informal news, which is precisely the reason for the lack of credibility on the part of information professionals.

To obtain these graphs, the results were exported to Excel, in order to make more explicit representations. The excel file of results of the mapping with its references is attached, which is available in the digital repository[1]

---

[1]   https://goo.gl/cmC7wY.

3.4   Conclusions and Discussions of Results

The big winner about the source from which the researchers take the news data to perform Sentiment Analysis is Twitter with 30% followed by news by digital means with 27%.

In terms of knowing what are the areas of interest to apply Sentiment Analysis using the news that comes from SM, is in first place with 31% economy and finance and this is understandable because they are the most interested in terms of revenues that this information can give them. Then there is the entertainment business with 22%, leaving the political sector in third place. Clearly see that the reasons that move the world are those of greatest interest.

The results obtained respond to the objectives of this research by applying a Systematic mapping, which is why we consider that the proposal can be applied to other areas of interest that should be investigated as education, considering the benefit that can be provided with this type of research in this field [6, 19].

This research has the inherent limitations of carrying out the process mechanically using a tool, that although it gives us quite accurate results, the relevant articles may also have been overlooked in the extraction phase given that the inclusion and exclusion criteria They may not use the searched keywords, for example.

Regarding the tool used although it facilitates the process of systematic mapping, since it considers the theoretical phases of the process, in the final stage of summarization, it does not allow detailed reports.

References

1. Gandomi, A., Murtaza, H.: Beyond the hype: big data concepts, methods, and analytics. Int. J. Inf. Manage. **35**(2), 137–144 (2015)
2. Warren-Payne, A.: 13 epic stats and facts from The State of Social webinar. https://www.clickz.com/13-epic-stats-and-facts-from-the-state-of-social-webinar/110510/
3. Kaplan, A.M., Haenlein, M.: Users of the world, unite! The challenges and opportunities of social media. Bus. Horiz. **53**(1), 59–68 (2010)
4. Cavazza, F.: Social media landscape 2017. https://fredcavazza.net/
5. Vinodhini, G., Chandrasekaran, R.-M.: Sentiment analysis and opinion mining: a survey. Int. J. Adv. Res. Comput. Sci. Softw. Eng. **2**(6), 282–292 (2012)
6. Peñafiel, M., Vásquez, S., Vásquez, D., Zaldumbide, J., Luján-Mora, S.: Data mining and opinion mining: a tool in educational context. In: Proceedings of the International Conference on Mathematics and Statistics (ICoMS2018), pp. 74–78. ACM (2018)
7. Liu, B.: Sentiment analysis and opinion mining. Synth. Lect. Hum. Lang. Technol. **5**(1), 1–167 (2012)
8. Del-Vicario, M., Zolloa, F., Caldarellia, G., Scalab, A., Quattrociocchia, W.: Mapping social dynamics on Facebook: The Brexit debate. Soc. Netw. **50**, 6–16 (2017)
9. Kitchenham, B.: Procedures for performing systematic reviews. Keele, UK, Keele Univ. **33**, 1–26 (2004)
10. Araujo-Alonso, M.: Las revisiones sistemáticas. Medwave. http://www.medwave.cl/link.cgi/Medwave/Series/mbe01/5220
11. Petersen, K., Vakkalanka, S., Kuzniarz, L.: Guidelines for conducting systematic mapping studies in software engineering: an update. Inf. Softw. Technol. **64**, 1–18 (2015)

12. Kitchenham, B., Budgen, D., Brereton, O.: Using mapping studies as the basis for further research–a participant-observer case study. Inf. Soft. Technol. **53**(6), 63–651 (2010)
13. Smeeton, N.-C.: Early history of the kappa statistic. Biometrics **41**, 795 (1985)
14. Marshall, C., Brereton, P., Kitchenham, B.: Tools to support systematic reviews in software engineering: a feature analysis. In: Proceedings of the 18th International Conference on Evaluation and Assessment in Software Engineering, pp. 1–13. ACM (2014)
15. Souza, E., Vitório, D., Castro, D., Oliveira, A., Gusmão, C.: Characterizing opinion mining: a systematic mapping study of the Portuguese language. In: International Conference on Computational Processing of the Portuguese Language, pp. 122–127 (2016)
16. Adedoyin-Olowe, M., Medhat-Gaber, M., Frederic S.: A survey of data mining techniques for social media analysis. arXiv preprint arXiv: 1312.4617 (2013)
17. Ravi, K., Ravi, V.: A survey on opinion mining and sentiment analysis: tasks, approaches and applications. Knowl. Based Syst. **89**, 1–46 (2015)
18. Martín-Valdivia, M.-T., Martínez-Cámara, E., Perea-Ortega, J.-M., Ureña-López, L.-A.: Sentiment polarity detection in Spanish reviews combining supervised and unsupervised approaches. Expert Syst. Appl. **40**(10), 3934–3942 (2013)
19. Peñafiel, M., Navarrete, R., Lujan-Mora, S., Zaldumbide, J.: Bridging the gaps between technology and engineering education. Int. J. Eng. Educ. **34**(5), 1479–1494 (2018)

Ensemble-Based Machine Learning Algorithms for Classifying Breast Tissue Based on Electrical Impedance Spectroscopy

Sam Matiur Rahman[1]([⊠]), Md Asraf Ali[1], Omar Altwijri[2],
Mahdi Alqahtani[2], Nasim Ahmed[3], and Nizam U. Ahamed[4]

[1] Department of Software Engineering, Daffodil International University,
Dhaka, Bangladesh
sammatiurrahman@yahoo.com
[2] Biomedical Technology Department, College of Applied Medical Sciences,
King Saud University, Riyadh, Saudi Arabia
[3] School of Engineering and Advanced Technology, Massey University,
Auckland, New Zealand
[4] Faculty of Manufacturing Engineering, University Malaysia Pahang,
Pahang, Malaysia

Abstract. The initial identification of breast cancer and the prediction of its category have become a requirement in cancer research because they can simplify the subsequent clinical management of patients. The application of artificial intelligence techniques (e.g., machine learning and deep learning) in medical science is becoming increasingly important for intelligently transforming all available information into valuable knowledge. Therefore, we aimed to classify six classes of freshly excised tissues from a set of electrical impedance measurement variables using five ensemble-based machine learning (ML) algorithms, namely, the random forest (RF), extremely randomized trees (ERT), decision tree (DT), gradient boosting tree (GBT) and AdaBoost (Adaptive Boosting) (ADB) algorithms, which can be subcategorized as bagging and boosting methods. In addition, the ranked order of the variables based on their importance differed across the ML algorithms. The results demonstrated that the three bagging ensemble ML algorithms, namely, RF ERT and DT, yielded better classification accuracies (78–86%) compared with the two boosting algorithms, GBT and ADB (60–75%). We hope that these our results would help improve the classification of breast tissue to allow the early prediction of cancer susceptibility.

Keywords: Breast tissue · Machine learning · Ensemble learning ·
Classification · Electrical impedance

1 Introduction

The density of breast tissue is one of the most important risk factors for the expansion of breast cancer, as has been widely recognized in the medical community [1, 2]. This type of cancer is the most common malignancy among women and the leading reason of

© Springer Nature Switzerland AG 2020
T. Ahram (Ed.): AHFE 2019, AISC 965, pp. 260–266, 2020.
https://doi.org/10.1007/978-3-030-20454-9_26

female death worldwide [3, 4]. Therefore, similarly to other forms of cancer, the early detection and prompt treatment of breast malignancy are crucial to the eradication of this disease. The electrical impedance of human breast tissue agrees with the electrical properties of human tissue, which include both resistance and capacitance, and it is well known that the electrical properties of biological tissues show notable differences depending on their structures and the frequency [5]. Thus, it is complex to identify its pattern deeply using traditional methods.

We are currently in a technological age, and thus, novel digital approaches for activity classification (e.g., smartphone health care apps and scientific software) must be as easy as possible to allow patients access to patient-centric care [6–9]. Machine learning (ML), which constitutes a subdivision of artificial intelligence that employs statistical, probabilistic and optimization techniques that allow computers to learn from previous examples and to detect very complex patterns from large, noisy or complex datasets, allows us to accomplish this goal [10]. Thus, ML can be incorporated in many modern medical applications that we often use in everyday life [11, 12]. Therefore, various state-of-the-art ML-based techniques can be used to better understand the complexities and pattern of breast tissues. A number of researchers have used different popular ML classification techniques, such as fuzzy C-means, K-nearest neighbour, naïve Bayes, support vector machine, multilayer perceptron, random forest and decision tree, to classify breast cancer [2, 13, 14].

Although these well-known ML algorithms are randomly used to classify breast tissue and cancer with good accuracy, we did not find any studies that compared the results from multiple ensemble-based ML classifiers, including bagging and boosting. It has been demonstrated that bagging and boosting are the two most easy-to-use ensemble-based non-linear machine learning models and are very powerful classification techniques because they can create a diverse ensemble of classifiers by manipulating the training data inputted to a "base" learning algorithm [15, 16]. In addition, ensemble-based ML classifiers are used as supervised learning methods in cancer research because these non-linear methods allow the development of predictive models with high accuracy, stability and easy clarification [17]. In addition, unlike linear models, these non-linear methods strategies allow easy mapping to non-linear relationships. Moreover, these ensemble-based ML classifiers have the heuristic to determine the significance of variable for predicting a class, which is also known as variable importance.

Therefore, the objective of this study was to investigate and compare the classification performances of five ensemble-based ML algorithms in classifying six breast tissues using nine electrical impedance spectroscopy variables. In addition, the importance of each variable in the classification margin was evaluated using three bagging ensemble ML approaches, namely, random forest (RF), extremely randomized trees (ERT) and decision tree (DT), and two boosting ensemble ML approaches, namely, gradient boosting tree (GBT) and adaptive boosting (AdaBoost, ADB).

2 Methods

We used the breast cancer datasets from the UCI Machine Learning Repository, which is freely accessible to academic users [18, 19]. These datasets comprise 120 impedance spectra collected from 64 female patients aged between 18 and 72 years. Each spectrum consisted of 12 impedance measurements obtained at different frequencies ranging from 488 Hz to 1 MHz. The data collection methods were previously described in details [20, 21]. The following nine impedance features were used for the classification: impedivity (ohm) at a frequency of 0 (IZ), phase angle at 500 kHz (PA500), high-frequency slope of the phase angle (HFS), impedance distance between spectral ends (DA), area under the spectrum (AREA), area normalized by the DA (A/DA), maximum of the spectrum (MAX IP), distance between I0 and the real part of the maximal frequency point (DR), and length of the spectral curve (P). Six classes of freshly excised tissues were studied using electrical impedance measurements (n represents the number of cases): carcinoma (n = 21), fibroadenoma (n = 15), mastopathy (n = 18), glandular (n = 16), connective (n = 14) and adipose (n = 22).

For classification, we used five nonlinear and ensemble machine learning algorithms: three bagging ensembles, namely, RF, ERT and DT, and two boosting ensembles, namely, GBT and ADB [15, 22]. The classification performances of all these classification algorithms were evaluated using the standalone Python programming language (version 3.6, www.python.org) [23]. The classification models were trained and cross-validated using the built-in Anaconda distribution of Python with notable packages, including matplotlib, numpy, scipy, and scikit-learn ("sklearn.ensemble.RandomForestClassifier") [24]. The K-fold cross-validation method using different training and validation data was used to ensure the correctness of the classification results [25]. Some of the important parameters of each ML algorithm were tuned manually because the performance of an algorithm is highly dependent on the choice of hyperparameters. Thus, these parameters were tuned until the highest accuracy was obtained, and the ranked order of variables based on variable importance was identified based on the highest accuracy.

3 Results

Table 1 presents the precision, recall, F-measure and accuracy values of each algorithm based on the context of classification. The results showed that RF and ERT exhibited improved performance, as demonstrated by higher values (>80%) for performance criteria. DT and GBT showed slightly decreased performance (75–79%) compared with two other ML algorithms. In contrast, ADB exhibited very low classification performance (<65%). In summary, the bagging ensemble algorithms showed better classification performance (>78%) compared with the boosting ensemble algorithms (<78%). A comparison of the accuracies (%) between all the ML algorithms is presented in Fig. 1.

Table 1. Performance of the classification tasks obtained with the ML algorithms

Ensemble classifier		Precision (%)	Recall (%)	F1-score (%)	Accuracy (%)
RF	Bagging	85.0	85.1	85.2	85.19
ERT		85.0	81.0	82.0	81.48
DT		79.0	78.0	78.0	78.10
GBT	Boosting	76.0	74.0	73.0	74.08
ADB		58.0	63.0	60.0	62.96

RF: random forest, ERT: extremely randomized trees, DT: decision tree, GBT: gradient boosting tree, ADB: adaptive boosting (AdaBoost).

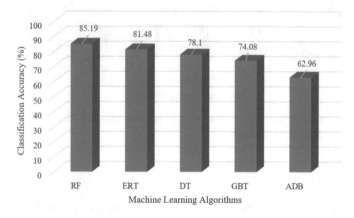

Fig. 1. Comparison of the accuracies of the five ensembled-based classifiers.

Values reflecting the importance of the variables in the five ensemble-based ML classifiers based on the classification margin are presented in Table 2 and Fig. 2. The rankings of the variables based on their importance with respect to the classification margin were different among all the classification methods. Specifically, one variable (IZ) was found to be the most importance, with values of 20%, 25% and 54%, in three ML algorithms, namely, RF, DT and ADB respectively. The P showed the highest importance values in the two other ML algorithms, ERT and GBT, with values of 21%

Table 2. Importance of variables in all the ML algorithms.

ML algorithms	Variable importance (%)								
	IZ	PA500	HFS	DA	Area	A/DA	Max IP	DR	P
RF	20	11	07	12	09	06	11	09	15
ERT	11	12	07	11	08	09	13	08	21
DT	25	17	02	14	05	00	11	15	11
GBT	07	13	11	14	03	08	13	04	27
ADB	54	14	02	04	14	02	00	04	06

The yellow-shaded values indicate the most important variables and 27%, respectively. In contrast, three variables, namely, HFS, A-DA and DR, were found to not be very important for the classification margin in all the ML algorithms. Two variables, namely, A-Da and Max IP, showed importance values of 0% in the DT and ADB classifiers, respectively.

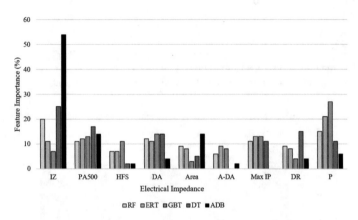

Fig. 2. Comparison of the importance of the different electrical impedance spectroscopy variables among the ML algorithms.

4   Discussion

The main objective of this study was to successfully analyse and compare the classification accuracies of the five most powerful ensemble and nonlinear-based ML techniques using a breast tissue database. The findings of the current study showed that three bagging ML algorithms, namely, RF, ERT and DT, yielded higher accuracies compared with the two boosting algorithms, GBT and ADB. In addition, two bagging ML algorithms, namely, RF and ERT, showed excellent classification performance in terms of precision, recall and accuracy, with values greater than 80% (Table 1). In contrast, one boosting ML algorithm, ADB, showed very low performance (<65%). Overall, the bagging approaches showed better performance compared with the boosting algorithms.

Breast cancer is one of the most common cancers that lead to death in women in both developed and undeveloped countries [2]. Therefore, researchers have attempted to develop approaches for improving breast cancer classification and thus establish improved and easier treatments. Because the reliable and accurate classification of breast tissue based on electrical impedance spectroscopy parameters is essential for the effective diagnosis and treatment of breast cancer, we determined the classification accuracies of different ensemble and nonlinear ML algorithms and evaluated the importance of the various parameters. Our results are consistent with those obtained in previous studies, which showed that bagging (RF, ERT and DT)-based ML techniques exhibit higher accuracy in terms of the classification margin [26, 27].

To our knowledge, this is the first study to compare classification performance to classify breast tissue based on different ensemble ML algorithms. To the best of our knowledge, this study constitutes the first comparison of the performances of different ensemble ML algorithms in the classification of breast tissue. We hope that our results will greatly contribute to cancer detection and classification. This finding is remarkably significant clinicians, doctors and specialists because it will aid the recognition of breast malignancy through the breast electrical impedance spectroscopy.

5 Conclusion

The early detection of breast cancer is important for reducing the mortality rate. The results of this study showed that all three tested bagging ML algorithms exhibit an improved classification accuracy compared with the two boosting algorithms. In addition, two electrical impedance variables, namely, IZ and P, were found to be more important for the classification margin. These results could help medical personnel better understand breast cancer-related information obtained from ensemble-based ML algorithms.

References

1. Oliver, A., Freixenet, J., Martí, R., Zwiggelaar, R. (eds.): A comparison of breast tissue classification techniques. In: International Conference on Medical Image Computing and Computer-Assisted Intervention. Springer (2006)
2. Oliver, A., Freixenet, J., Marti, R., Pont, J., Pérez, E., Denton, E.R., et al.: A novel breast tissue density classification methodology. IEEE Trans. Inf. Technol. Biomed. **12**(1), 55–65 (2008)
3. Ghasemzadeh, A., Azad, S.S., Esmaeili, E.: Breast cancer detection based on Gabor-wavelet transform and machine learning methods. Int. J. Mach. Learn. Cybern. **10**, 1–10 (2018)
4. Peto, R., Boreham, J., Clarke, M., Davies, C., Beral, V.: UK and USA breast cancer deaths down 25% in year 2000 at ages 20–69 years. Lancet **355**(9217), 1822 (2000)
5. Zou, Y., Guo, Z.: A review of electrical impedance techniques for breast cancer detection. Med. Eng. Phys. **25**(2), 79–90 (2003)
6. Moran, K., Bernal-Cárdenas, C., Curcio, M., Bonett, R., Poshyvanyk, D.: Machine learning-based prototyping of graphical user interfaces for mobile apps. arXiv preprint arXiv: 180202312 (2018)
7. Ahamed, N.U., Benson, L., Clermont, C., Osis, S.T., Ferber, R.: Fuzzy inference system-based recognition of slow, medium and fast running conditions using a triaxial accelerometer. Procedia Comput. Sci. **114**, 401–407 (2017)
8. Ahamed, N.U., Sundaraj, K., Badlishah Ahmad, M.R., Ali, M.A., Islam, M.A., Palaniappan, R.: Rehabilitation systems for physically disabled patients: a brief review of sensor-based computerised signal-monitoring systems. Biomed. Res. **24**, 370–376 (2013)
9. Hussain, A.T., Said, Z., Ahamed, N., Sundaraj, K., Hazry, D. (eds.): Real-time robot-human interaction by tracking hand movement & orientation based on morphology. In: 2011 IEEE International Conference on Signal and Image Processing Applications (ICSIPA). IEEE (2011)

10. Cruz, J.A., Wishart, D.S.: Applications of machine learning in cancer prediction and prognosis. Cancer Inform. **2**, 117693510600200030 (2006)
11. Palaniappan, R., Sundaraj, K., Ahamed, N.U.: Machine learning in lung sound analysis: a systematic review. Biocybern. Biomed. Eng. **33**(3), 129–135 (2013)
12. Nabi, F., Sundaraj, K., Kiang, L., Palaniappan, R., Sundaraj, S., Ahamed, N.U. (eds.): Artificial intelligence techniques used for wheeze sounds analysis. In: International Conference on Movement, Health and Exercise. Springer (2016)
13. Amrane, M., Oukid, S., Gagaoua, I., Ensari̇, T. (eds.): Breast cancer classification using machine learning. In: 2018 Electric Electronics, Computer Science, Biomedical Engineerings' Meeting (EBBT). IEEE (2018)
14. Polat, K., Güneş, S.: Breast cancer diagnosis using least square support vector machine. Digit. Signal Proc. **17**(4), 694–701 (2007)
15. Dietterich, T.G.: An experimental comparison of three methods for constructing ensembles of decision trees: bagging, boosting, and randomization. Mach. Learn. **40**(2), 139–157 (2000)
16. Bühlmann, P.: Bagging, boosting and ensemble methods. In: Handbook of Computational Statistics, pp. 985–1022. Springer, Berlin (2012)
17. Zhou, Z.-H., Jiang, Y., Yang, Y.-B., Chen, S.-F.: Lung cancer cell identification based on artificial neural network ensembles. Artif. Intell. Med. **24**(1), 25–36 (2002)
18. Da Silva, J.E., De Sá, J.M., Jossinet, J.: Classification of breast tissue by electrical impedance spectroscopy. Med. Biol. Eng. Comput. **38**(1), 26–30 (2000)
19. Dua, D., Karra, E.: UCI Machine Learning Repository. University of California, School of Information and Computer Science, Irvine (2017). http://archive.ics.uci.edu/ml
20. Jossinet, J.: Variability of impedivity in normal and pathological breast tissue. Med. Biol. Eng. Comput. **34**(5), 346–350 (1996)
21. Jossinet, J.: The impedivity of freshly excised human breast tissue. Physiol. Meas. **19**(1), 61 (1998)
22. Ge, G., Wong, G.W.: Classification of premalignant pancreatic cancer mass-spectrometry data using decision tree ensembles. BMC Bioinform. **9**(1), 275 (2008)
23. Pedregosa, F., Varoquaux, G., Gramfort, A., Michel, V., Thirion, B., Grisel, O., et al.: Scikit-learn: machine learning in Python. J. Mach. Learn. Res. **12**, 2825–2830 (2011)
24. Ahamed, N.U., Kobsar, D., Benson, L., Clermont, C., Osis, S.T., Ferber, R.: Subject-specific and group-based running pattern classification using a single wearable sensor. J. Biomech. **84**, 227–233 (2019)
25. Hothorn, T., Leisch, F., Zeileis, A., Hornik, K.: The design and analysis of benchmark experiments. J. Comput. Graph. Stat. **14**(3), 675–699 (2005)
26. Yu, Y., Chen, S., Wang, L.-S., Chen, W.-L., Guo, W.-J., Yan, H., et al.: Prediction of pancreatic cancer by serum biomarkers using surface-enhanced laser desorption/ionization-based decision tree classification. Oncology **68**(1), 79–86 (2005)
27. Ahamed, N.U., Kobsar, D., Benson, L., Clermont, C., Kohrs, R., Osis, S.T., et al.: Using wearable sensors to classify subject-specific running biomechanical gait patterns based on changes in environmental weather conditions. PLoS One **13**(9), e0203839 (2018). Epub 2018/09/19. https://doi.org/10.1371/journal.pone.0203839. PubMed PMID: 30226903

Cognitive Solutions in the Enterprise: A Case Study of UX Benefits and Challenges

Jon G. Temple[✉], Brenda J. Burkhart, Ed T. McFadden,
Claude J. Elie, and Felix Portnoy

IBM, Armonk, NY, USA
{jgtemple,bjburkha,ed.mcfadden,eliec,
fportno}@us.ibm.com

Abstract. The consumer market has witnessed a proliferation of cognitive solutions. This increase in consumer expectations for AI technology has led enterprise IT leaders to develop cognitive solutions to improve employee productivity, enhance marketing and sales insights, and make better data-driven decisions. As UX designers supporting the enterprise, we have been gaining experience working with cognitive solutions in multiple contexts, from sentiment analysis of people and news for sellers, cognitively-enhanced conflict resolution of conference calls, capability analysis of team performance, to various chatbots. We will discuss several different cognitive solutions that have been created for the enterprise and provide some recommendations and best practices.

Keywords: Cognitive solution · Chatbot · User experience

1 Introduction

Cognitive solutions are based upon artificial intelligence (AI) and machine learning. Cognitive systems are already changing and impacting our lives and the trend will continue. The consumer market has witnessed a proliferation of cognitive solutions, which have been used for everything from answering simple questions (e.g., "how tall is the Eiffel tower?"), helping out in the kitchen with recipes and timers, and changing thermostats and turning on lights, to playing interactive games like "20 questions". Cognitive solutions are often used to help users perform simple tasks, find information that is most relevant to them and to recommend products or solutions based upon intelligent analysis of domain knowledge. Not only are cognitive solutions being embraced by consumers, but they expect them to be easy to use, engaging, and useful [1, 2].

On the heels of its adoption in the consumer market, cognitive is finding root within the enterprise itself. In 2011, IBM dramatically announced its interest in cognitive solutions by demonstrating Watson, an AI system that defeated the legendary champions Brad Rutter and Ken Jennings on Jeopardy [3]. Since then, Watson has grown into a diverse line of marketable products capable of:

- Natural language classification, understanding and translation
- Machine learning

© Springer Nature Switzerland AG 2020
T. Ahram (Ed.): AHFE 2019, AISC 965, pp. 267–274, 2020.
https://doi.org/10.1007/978-3-030-20454-9_27

- Personality insights and tone analysis
- Data analysis, prediction, and decision support
- Visual recognition of images and patterns
- Conversational assistants (chatbots)

Enterprise IT leaders are embracing these new cognitive capabilities to improve employee productivity, enhance marketing and sales insights, and make better data-driven decisions.

As UX designers supporting the enterprise, we have been gaining experience working with cognitive solutions as AI has rolled out in multiple contexts. When deploying a cognitive solution, the enterprise setting poses a number of challenges that are quite different from the consumer space. As UX designers, we must ensure that the cognitive solution is well matched to the problem it is meant to solve, has a good user experience, and encourages adoption by users [4].

We will discuss several different case studies of cognitive solutions that have been created for the enterprise. Each solution has its unique benefits and challenges which we'll discuss in turn, from several chatbot implementations, to a sales tool that provides sentiment analysis of people and news for sellers, an application that prioritizes scheduled conference calls, and a tool that provides cognitive analysis of capabilities to help improve performance of sales teams. We will conclude with recommendations based upon our experiences and insights derived while designing the user experiences for these products.

2 Case Studies

2.1 Chatbots

Thanks to Alexa and Siri, people have become familiar with chatbot-style solutions and have developed high expectations around them. Any chatbot you create will need to reach or exceed the preset bar of expectations which can be a significant challenge, particularly for UX practitioners working with cognitive solutions in the enterprise environment due to the required investment in training to achieve the level of consumer-level chatbot solutions. The first case study represents the distillation of our experience working on multiple chatbot projects, including several single-purpose chatbots, a career development chatbot, an IT support chatbot, and a sales team performance chatbot. While each chatbot had its own unique challenges, they were many common threads. Based on this experience, we offer some lessons learned that should be applicable to any chatbot solution.

2.2 Too Many Chatbots

In efforts to move cognitive solutions into our enterprise, we have seen many groups develop their own chatbots specific to their product or service. These tend to be single-purpose chatbots that are trained in a single domain. For example, there could be a chatbot to advise you about which health plan to choose, or another to provide assistance completing a performance review. They have become relatively easy to build without a huge investment, and some vendors even sell chatbots that are already partially trained. This has led to a proliferation of chatbots [4], with multiple non-integrated knowledge bases that are incapable of taking advantage of potentially related or overlapping information, rules or analyses. With so many chatbots, finding the appropriate chatbot may be challenging, leading to such awkward workarounds as letting chatbots open other chatbots, or even chatbot directories.

Combine Knowledge Domains. Design a single chatbot system, if possible, and add new knowledge domains to it over time. Rather than creating additional single-purpose chatbots, acquire an additional knowledge domain that is compatible with a general purpose chatbot. Teaching users about new commands or capabilities for a single chatbot is far simpler than figuring out how tie the multitude independent chatbots together into a sort of quilt or creating a chatbot directory to help users navigate between them.

Reduce Cognitive Overhead. Making a chatbot reusable in multiple contexts enhances the ROI for a well-designed, easy to use chatbot. An added benefit is that users would not have the overhead of learning a new chatbot each time a new knowledge domain was plugged in. Finding a single product owner for this general purpose chatbot would also be of importance. This would require oversight among teams and team coordination as new domains are added.

2.3   What Can I Ask This Chatbot?

Improve Built in Education. Provide a way to communicate the domain of a chatbot. For example, Amazon sends out weekly announcements about what Alexa can now do this week and in some cases continually educates users on what it can do by suggesting things to ask. A simple way to do this is to have a very clear 'What can I ask?' function on the interface providing the user with an updated list with the most common and useful items for your bot that can be proactively summoned with clickable suggestions [5]. Another effective means is to promote the 'What can I ask?' when the bot does not understand the question or has a low confidence answer.

2.4   Competing Solutions: Search or Chatbot

Integrate Search with Intent Matching. If you have both search and a chatbot on your site, users may be confused as to which path will provide them with the best answers [4]. By combining these mechanisms, you eliminate a decision between two solutions that already overlap functionally and provide users with the best of both worlds. If a user performs a search, the search query is used to match to possible solution intents. If a match reaches a confidence threshold, you can present your matched high-confidence answer above the search results. If it isn't what the searcher is looking for, it fails gracefully to the search results. For example, we adopted this methodology on an IT Support site to great effect because it greatly manages expectations; users expecting just search results were delighted if the solution was displayed alongside the search results [6].

Consider User Needs. A common pitfall is adopting a technology just because it is new and cool. The first question to ask is, 'What problem and I trying to solve?' [7] If users are coming to your site and using search effectively to find what they want, you may be adding confusion by adding a new mechanism that may be less efficient/effective than what you have currently [4]. Is your domain small enough that a few easily scanned items in a FAQ would be more effective than adding the overhead of a chatbot? Is there a better way to provide information they need without resorting to putting critical information behind the chatbot curtain? Each project should consider these questions before deciding whether a chatbot is the right choice.

2.5 Do Not Waste the User's Time

Provide Direct Answers. "Let me search the web for you" is probably the most disappointing thing you can get from a chatbot short of a wrong answer. Users are expecting to get direct answers and the most successful chatbot interfaces provide them. Wherever possible don't link users off to where they can discover the answer but show it immediately [5]. For example, if a user asks a Learning chatbot, 'What courses do you recommend I take'? Instead of showing them a link where they can get a personalized list of recommendations, pull in the top two recommendations directly into the chat interface with a 'show all' link at the bottom that will take them to the full list if the top 2 weren't good enough. Direct answers to direct questions are a delightful experience.

Maintain Context. Maintain the user's context so that follow up questions aren't 'starting over'. Normal human conversation is heavily dependent on context, which reduces sentence and conceptual complexity by building upon earlier elements. Given the conversational nature of many chatbots, talking with a chatbot that cannot remember what you said in the prior sentence is like having a chat with a brain-damaged amnesiac who asks your name every time you enter the room.

Collect Feedback and Improve Your Chatbot Continuously. Collect user feedback on answers to improve accuracy but not at the cost of user experience. Never interrupt the user's flow to obtain feedback; keep in mind the user's primary purpose in using the chatbot is to learn or solve a problem, not to help you train your chatbot.

2.6 Sales and Customer Support Tool

In a different case study, we provided design support for an enterprise Sales and Customer Support tool that uses cognitive capabilities to provide intelligent support to sellers and includes a chatbot. The seller can access insights on an industry, a client company, clients and client executives. This tool also lets sellers receive sales guidance, get solution recommendations and access sales assets that can be tailored to their clients' needs.

There were many challenges that emerged as this project developed over time. Primarily, the amount of information about products, software, and services turned out to be very large and was constantly changing due to updates, improvements, and acquisitions. The same pattern was found for industries, companies, clients and executives, as well. This situation made keeping the knowledge base complete and up-to-date extremely difficult.

The original idea was to provide all cognitive insights through the chatbot and require users to ask questions for any information they sought. But users made it clear they wanted quick, direct access to available content rather than having to ask questions. Eventually, the chatbot interface was de-prioritized as the primary user interaction to the access relevant sales information and replaced with an interface to browse collections of products, services and sales assets. Similarly, the cognitive insights on clients and companies remained available, but not displayed prominently. At first, sellers thought these were "cool", but later they felt that the functions were not useful to their work.

Some Lessons Learned. Not all interactions are a good fit for a chatbot. Our sellers made it clear that this mode of interaction was an inefficient and clumsy way of getting at the volumes of information they needed. Another complicating factor was the inability of keeping the chatbot up to date with rapidly changing information, which undermined the informational value to the user and resulted in inconsistent answers to the same questions. All of this had a negative impact on how much the sellers felt they could trust the chatbot; consequently, it fell into disuse. When designing an application, start with the users' needs and find the solution that best meets them, not with a "cool" technology that you must retrofit to the users [7, 8].

2.7   Call Prioritization

An example of a lightweight cognitive solution in the enterprise is a mobile app used to join conference calls. The app extracts conference call information from users' calendars so they can join a conference call with a single tap. In an effort to help users prepare for their meetings, calendar conflicts are highlighted and prioritized for the user. Prioritization is currently based on an algorithm that considers more than 10 meeting characteristics including who is hosting and attending (management, VIPs, clients), the size of the meeting, and whether the meeting is just a phone call or includes a video conference link.

One design challenge was to decide how to highlight the meeting selected by the algorithm. Although the initial algorithm was based on a conjoint analysis of survey data that had employees choose between two meetings, it was unclear how accurate it would be in practice. For this reason, the feature was not emphasized beyond listing the "recommended" meeting first. This effectively allowed the user to benefit from the recommendation without the risk of user dissatisfaction from false positives.

In a future release, the app could build a personalized algorithm based on which meetings the user actually chooses to attend. This would essentially be a system that would learn how to weight the meeting criteria more closely to the user's prior selections. Recommendations with a high confidence could potentially be highlighted in a more prominent manner. Additional user behaviors could also be tracked to build personalized models to automate the resolution of conflicting meetings (e.g., request the less important meeting be rescheduled).

Some Lessons Learned. Using a conjoint analysis based on appropriate survey data [9] was a quick and easy way to create an initial model for this cognitive solution. It helped uncover general rules that are being used by users already to decide which meeting to attend when there are conflicts and established a foundation on which a more dynamic and personalized solution could be built in the future. Introducing this feature in a more passive way also allowed users to benefit from the recommendation without the negative effects that come with false positives.

2.8   Sales Team Performance Analysis

Another example of an enterprise cognitive solution is a tool that sales team managers can use to assess their overall team's performance and their individual team members' performance. The tool provides an objective assessment based upon available digital data. Cognitive models used 10 dimensions of successful team practitioners to assess how successful a team is in their client relationships, teaming skills, knowledge of offerings and industry knowledge. Examples of data sources to drive the assessments ranged from the value of sales made and planned, training taken, social network activity to personality insights. The goal was to identify any areas for improvement and make recommendations of how the team or individual could improve. Recommendations were derived using a predictive model to identify focus areas that would drive higher performance benefits for teams and individuals.

There were many design challenges encountered on this project. One challenge was the sensitivity and confidentiality of personal, performance, and monetary performance information used to inform the cognitive models. Cognitive personality insights were used as well, but how those were used to arrive at a rating on a dimension was not understood by the users nor explained in the user interface. Because users did not understand how the ratings were derived, they did not know what they could do to improve their ratings. Due to this lack of understanding of the data sources used to arrive at the ratings, the users did not trust the outcomes in the tool.

Another issue regarding data sources is that not all of them generalized to all team members. For example, social media sources such as the number of social connections or blog posts on a given topic were used, but not everyone chooses to use social media due to personal preference, workload or temperament.

Some lessons learned:

- Users will not want to use something that they do not trust. Build user trust through transparency by displaying enough information about the cognitive system so users can understand and interpret results.
- Also, build trust and set user expectations by showing the confidence level of information or do not use low confidence data at all.
- When the goal involves recommended improvements, provide actionable steps, along with the confidence level that the steps will be effective. For example, we provided links to learning resources that would improve team skills, along with how much improvement could be expected.

- Provide ways to safeguard confidential information [10], for example let users to opt-out for privacy reasons.
- Lastly, use social media data with care because it is not used equally by all people.

3 Overall Recommendations

Thanks to the commercial success of Alexa and Siri, people have become familiar with cognitive solutions and have developed high expectations for them. These expectations can present a substantial challenge for UX practitioners in the enterprise because it is rare that companies can invest in training AI systems to the extent of consumer-level solutions marketed to millions of people, and solutions may differ substantially in breadth, depth and versatility. To be successful, we recommend that cognitive solutions in the enterprise follow certain principles, including:

- Validate with users that a cognitive solution will be appropriate for the problems they must solve.
- User expectations about what you are trying to achieve must be aligned with the capabilities of the cognitive solution.
- Your system must be properly and sufficiently trained to address the problem you are trying to solve.
- Provide direct answers at a level of detail commensurate with the question.
- A cognitive solution should deliver only high confidence recommendations.
- Your system must provide graceful recovery when the cognitive system fails (e.g., makes a bad recommendation).
- A cognitive solution should include a loop so users can provide feedback on performance and the cognitive system can learn and improve.
- A cognitive system is only useful if people return to use it.
- Be transparent about how your cognitive system derives its recommendations; people must feel comfortable trusting its recommendations [11].

When these principles have been successfully followed, cognitive solutions can make a difference in the enterprise and improve employee productivity, support better informed decision-making, and increase work effectiveness.

References

1. Lau, T.: Why PBD systems fail: lessons learned for usable AI. In: CHI 2008 Workshop on Usable AI (2008)
2. Augello, A., Saccone, G., Gaglio, S., Pilato, G.: Humorist bot: bringing computational humour in a chat-bot system. In: International Conference on Complex, Intelligent and Software Intensive Systems (2008)
3. Markoff, J.: Computer wins on 'Jeopardy!': trivial, it's not. New York Times (2011)

4. Zamora, J.: I'm sorry, Dave, I'm afraid I can't do that: chatbot perception and expectations. In: Proceedings of the 5th International Conference on Human Agent Interaction, pp. 253–260. ACM, New York (2017)
5. Valerio, F.A.A., Guimaraes, T.G., Prates, R.O., Candello, H.: Here's what I can do: chatbots' strategies to convey their features to users. In: Proceedings of the XVI Brazilian Symposium on Human Factors in Computing Systems, Article No. 28. ACM, New York (2017)
6. Temple, J.G., Elie, C.J.: Beyond the chatbot: enhancing search with cognitive capabilities. In: Ahram, T.Z. (ed.) Advances in Artificial Intelligence, Software and Systems Engineering, pp. 283–290. Springer, Heidelberg (2018)
7. Wei, N.: So you want to be an AI designer? Interactions **24**(4), 44–49 (2017)
8. Følstad, A., Brandtzæg, P.B.: Chatbots: changing user needs and motivations. Interactions **25**(5), 38–43 (2018)
9. Green, P.E., Krieger, A.M., Wind, Y.: Thirty years of conjoint analysis: reflections and prospects. Interfaces **31**(3), 56–73 (2001)
10. Følstad, A., Brandtzæg, P.B.: Chatbots and the new world of HCI. Interactions **24**(4), 38–42 (2017)
11. Holmquist, L.E.: Intelligence on tap: artificial intelligence as a new design material. Interactions **24**(4), 28–33 (2017)

Academic Quality Management System Audit Using Artificial Intelligence Techniques

Rodolfo Bojorque$^{(\boxtimes)}$ and Fernando Pesántez-Avilés

Universidad Politécnica Salesiana, Cuenca, Ecuador
{rbojorque, fpesantez}@ups.edu.ec

Abstract. Quality management systems are a challenge for higher education centers. Nowadays, there are different management systems, for instance: quality, environmental, information security, etc. that can be applied over education centers, but to implement all of them is not a guarantee of education quality because the educational process is very complex. However, a few years ago the Quality Management Systems for higher education centers are taking importance especially in Europe and North America, although in Latin America is an unexplored field. Higher education centers quality is a very complex problem because it is difficult to measure the quality since there are a lot of academic processes as enrollment, matriculation, teaching-learning with a lot of stakeholders as students, teachers, authorities even society; in a lot of locations as campuses, buildings, laboratories with different resources. Each process generates a lot of records and documentation. This information has a varied nature and it is present at a structured and no-structured form. In this context, artificial intelligence techniques can help us to analyze and management knowledge. Our work presents a new approach to audit academic information with machine learning and information retrieval. In our experiments, we used information about syllabus, grades, assessments and online content from a Latin American University. We conclude that using artificial intelligence techniques minimize the decision support time, it allows full data analysis instead of a data sample and it finds out patterns never seen in the case study university.

Keywords: Quality Management Systems · Artificial intelligence · Audit techniques

1 Introduction

The quality concept is a cross-cutting theme that received and is receiving particular and continuous attention in many disciplines [1]. Quality has a lot of meanings since it depends on the context even the concept of quality has evolved over the years. For instance, [2] defines it as products and services quality of an organization is determined by the ability to satisfy customers, and by the expected and unforeseen impact on relevant stakeholders. Anyway, quality has subjective and objective concepts, it is a social construct concept [3].

In higher education context there are so many customers and stakeholders like students, industry, enterprises, even the society. 'Quality assurance' and 'quality enhancement' are two widely discussed topics, especially in higher education.

© Springer Nature Switzerland AG 2020
T. Ahram (Ed.): AHFE 2019, AISC 965, pp. 275–283, 2020.
https://doi.org/10.1007/978-3-030-20454-9_28

However, the centrality of discussion about quality assurance and enhancement is certainly based on the question "What is quality? [4]. The best way to assure the quality of products and services is to implement a Quality Management System (QMS), for instance, ISO 9000. Although, some authors refer to ISO 9000 standards as a 'straightjacket' because the translation of the standard when applied to educational institutions causes 'confusion and consternation' [5].

Quality management systems are a challenge for higher education centers. Nowadays, there are different management systems, for instance: quality, environmental, information security, etc. that can be applied over education centers, but to implement all of them is not a guarantee of education quality because the educational process is very complex. However, a few years ago the Quality Management Systems for higher education centers are taking importance especially in Europe and North America, although in Latin America is an unexplored field. Higher education centers quality is a very complex problem because it is difficult to measure the quality since there are a lot of academic processes as enrollment, matriculation, teaching-learning with a lot of stakeholders as students, teachers, authorities, even society; in a lot of locations as campuses, buildings, laboratories with different resources. Each process generates a lot of records and documentation. Nowadays people are living in a world crowded by so many information sources that it is impossible for us to absorb even a very small portion. This phenomenon is usually called "information overload". This information has a varied nature and it is present at a structured and no-structured form. In this context, artificial intelligence techniques can help us to analyze and management knowledge.

The remainder of this article is structured as follows: In section two, a review of quality management systems and artificial intelligence techniques as artificial neural networks and text mining are summarized. Section three details the proposed methodology and explains the different experiments and the kind of information used. Section four presents the results obtained using academic information, we contrast results with expert criteria. Finally, section five provides conclusions and recommendations.

2   Review

2.1   Quality Management Systems (QMS)

Several voices have been heard about the non-applicability at all of those management theories, especially because they derived from industry and had nothing to do with the higher education ethos [6, 7]. However, QMS and Total Quality Management (TQM) are widely used in higher education institutions. According to [8] a QMS includes activities through which the organization identifies its objectives and determines the processes and resources required to achieve the desired results. The QMS manages the processes that interact and the resources that are required to provide value and achieve results for relevant stakeholders. The QMS enables senior management to optimize the use of resources considering the consequences of their long and short-term decisions. A QMS provides the means to identify actions to address the expected and unintended consequences of the provision of products and services.

2.2 Artificial Intelligence Techniques

Artificial intelligence (AI) defined as a system's ability to correctly interpret external data, to learn from such data, and to use those learnings to achieve specific goals and tasks through flexible adaptation [9]. For [10] the analysis of data to model some aspect of the world. Inferences from these models are then used to predict and anticipate possible future events.

Artificial intelligence (AI) is an important technology that supports daily social life and economic activities. The market and business for AI technologies is evolving rapidly. In addition to speculation and increased media attention, many start-up companies and Internet giants are racing to acquire AI technologies in business investment [11].

AI is used for natural language processing, computer vision, robotic process automation, decision management, etc. For audit area, AI focuses on the financial detection fraud and intrusion detection systems to data networks through data mining techniques, neural networks, fuzzy logic and Bayesian networks [12–14]. All of them techniques are from machine learning, a sub-area of AI.

A machine learning definition is "the set of techniques and tools that allow computers to 'think' by creating mathematical algorithms based on accumulated data [15]". According to [16]: Broadly speaking, machine learning can be separated into two types of learning: supervised and unsupervised. In supervised learning, algorithms are developed based on labelled datasets. In this sense, the algorithms have been trained how to map from input to output by the provision of data with 'correct' values already assigned to them. This initial 'training' phase creates models of the world on which predictions can then be made in the second 'prediction' phase. Conversely, in unsupervised learning the algorithms are not trained and are instead left to find regularities in input data without any instructions as to what to look for. In both cases, it is the ability of the algorithms to change their output based on experience that gives machine learning its power. On the other hand, text mining is the process of deriving high-quality information from text [17]. Therefore, there are a lot of challenges about non-structured data that organizations have.

In this context, we combine text mining techniques and AI techniques from machine learning area for exploiting non-structured academic information. [18] proposes a model framework to explain educational fields where AI is applied according to Table 1. In addition, there are recent studies about AI techniques for the audit process [19–21]. However, academic quality management system audit is an unexplored field for AI.

Table 1. Application fields in education of AI.

Content and teaching methods	Assessment	Communication
Customize education content	Simplifying in MOOC assessment	Intelligent tutoring systems
Personalized learning	Identifying gaps in learning	
Educational robots		

3   Methodology

We apply two main AI techniques to address academic audit: (a) Text mining to transform text into real vectors. When the text is represented as a real vector we can use it as an input for other artificial intelligence techniques. In our case, we present an analysis of bibliography comments made by teachers over syllabus bibliography and (b) Artificial Neural Networks to complement previous analysis and to detect academic irregularities in the teaching-learning process planned by the syllabus, both issues are classification problems. These two analysis elements are a drawback for academic quality management system of Universidad Politécnica Salesiana since when the academic department used a sample the results are very poor, on the other hand, checking the whole universe of data is humanly impossible.

3.1   Text Mining Process

For text mining we use an academic dataset from Universidad Politécnica Salesiana, this corpus has 20,000 records about bibliography comments of teachers in the syllabus, all of them are classified by expert academics according to their expertise in different subjects. They used a binary classification, 'Good' for correct academic comments and 'Bad' for comments no adequate. The records belong to four previous academic periods. We used a 5-fold cross validation process to measure accuracy of the system.

Figure 1 shows the text mining process. We use three stages (a) stemming, (b) normalizing and (c) computing tf*idf.

The stemming stage is complex, in our case is as follow:

- Lower-casing: The entire input text is converted into lower case, so that capitalization is ignored.
- Stripping HTML: All HTML tags are removed from the input text. Many inputs often come with HTML formatting; we remove all the HTML tags, so that only the content remains.
- Word Stemming: Words are reduced to their stemmed form. For instance, 'research', 'researches', 'researched' and 'researching' are all replaced with 'research'. Sometimes, the Stemmer actually strips of additional characters from the end, so 'include', 'includes', 'included', and 'including' are all replaced with 'include'.

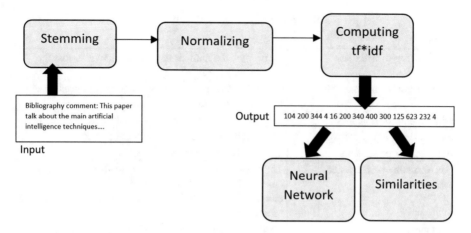

Fig. 1. Text mining process.

The normalizing stage consists in to do:

- Normalizing URLs: All URLs are replaced with the text 'httpaddr'.
- Normalizing Numbers: All numbers are replaced with the text 'number'.
- Normalizing cites: All cites used the IEEE citation format [number], they are replaced with the text 'cite'.
- Removal of non-words: Non-words and punctuation have been removed. All white spaces (tabs, newlines, spaces) have all been trimmed to a single space character.
- Removed word from stoplist: Our stoplist has a lot of very common words like the, as, a, an. They are removed.
- The result of this stage is a dictionary with all terms used in the comments of syllabus bibliography.

The computing tf*idf stage makes a vector (Term frequency – Inverse document frequency), in our case the documents are the comments and terms are the words in the dictionary. Finally, the tf*idf vector can be used by neural networks or to compute similarities measures.

3.2 Artificial Neural Networks Process

To apply artificial neural networks to predict course performance in regular/irregular categories, we used as an input a feature vector with information about syllabus according to Table 2, all features are normalizing using Max-Min norm. The output to train the network is the course status validate for experts, regrettable this evaluation occurs when the course is near to finish. The system aim is the early detection of course status.

The neural network architecture is presented in Fig. 2, using three layers with ten, five and three neurons produce the best accuracy for our experiments.

Table 2. Data input for neural network.

Feature	Description
F1: Quantity of activities	It specifies the number of learning activities that the teacher will use in his/her class
F2: Quantity of assessments	Regarding the number of assessments that teacher will give to his/her students
F3: Quantity of practices	It is the number of laboratory practices along the course
F4: Teaching hours per week	The number of hours for teaching per week
F5: Quantity of students	The number of students per class
F6: Academic performance average	The average of course's students grades about previous courses. If there are no grades (because is the first enrollment of the student in the university), we used high school grades
F7: Quantity of additional material	Readings and resources specified in the syllabus
F8: Professor's grade from the institutional evaluation	Institutional evaluation average of the three previous semesters

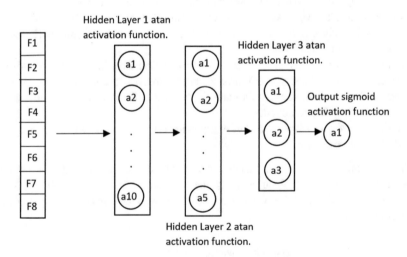

Fig. 2. Neural network architecture.

4   Results

To compute result we use quality metrics from confusion matrix for information retrieval, Table 3 shows how mapping results for precision, recall and fallout in Eqs. 1, 2 and 3.

Table 3. Confusion matrix for information retrieval.

	Classified as an academic result by the expert	Classified as not an academic result by the expert
Classified as an academic result by the machine	a	b
Classified as not an academic result by the machine	c	d

$$\text{precision} = \frac{a}{a+b} \tag{1}$$

$$\text{recall} = \frac{a}{a+c} \tag{2}$$

$$\text{fallout} = \frac{b}{b+d} \tag{3}$$

Table 4. Results about bibliography comments.

Precision	0.996
Recall	0.997
Fallout	0.521

We obtain a high precision for text analysis in bibliography comments. Table 4 shows results of precision, recall and fallout.

Table 5. Results about course status.

Precision	0.912
Recall	0.903
Fallout	0.709

In addition, course status predictions have a high precision. Table 5 shows results of precision, recall and fallout.

5 Conclusions

Using AI techniques for audit process can reduce the evidence verification time, we obtain high precision in predictions of academic information, comparing expert criteria and machine prediction. Our main contribution is a model to work with non-structured data in the audit process of Universidad Politécnica Salesiana. While this work analyzes only two academic processes, the model can be applying in any non-structured information.

The main challenge for quality management systems is optimizing auditor time and guide the effort in the monitoring process, in our case the experts, generally are university lecturers, they spend a lot of time checking and validating information. They prefer a system that can help them to detect academic irregularities and so to take advantage of the time.

References

1. Fiore, M., Contò, F.: The Quality Concept, Advances in Dairy Products. Wiley, New York (2017)
2. International Organization for Standardization: ISO 9001:2015 quality management systems-requirements (2015)
3. Pesántez, F.: Indicadores de gestión y calidad en la educación superior: un modelo de evaluación para la Universidad Politécnica Salesiana, Abya-Yala (2011)
4. Tanweer, M., Qadri, M.M.: Quality assurance in higher education: a framework for distance education. JDER J. Distance Educ. Res. 1(1), 6–24 (2016)
5. Kanji, G., Malek, A., Tambi, B.: Total quality management in UK higher education institutions. Total Qual. Manage. 10(1), 129–153 (1999)
6. Massy, W.F.: Honoring the Trust. Quality and Cost Containment in Higher Education. Anker Publishing, Bolton (2003)
7. Pratasavitskaya, H., Stensaker, B.: Quality management in higher education: towards a better understanding of an emerging field. Qual. Higher Educ. 16(1), 37–50 (2010)
8. International Organization for Standardization: ISO 9000:2015 quality management systems-fundamentals and vocabulary (2015)
9. Kaplan, A., Haenlein, M.: Siri, Siri, in my hand: who's the fairest in the land? On the interpretations, illustrations, and implications of artificial intelligence. Bus. Horiz. 62(1), 15–25 (2019)
10. Government Office for Science: Artificial intelligence: opportunities and implications for the future of decision making, 9 November 2016
11. Lu, H., Li, Y., Chen, M., Kim, H., Serikawa, S.: Brain intelligence: go beyond artificial intelligence. Mobile Netw. Appl. 32(2), 368–375 (2018)
12. Kirkos, E., Spathis, C., Manolopoulos, Y.: Data mining techniques for the detection of fraudulent financial statements. Expert Syst. Appl. 32(4), 995–1003 (2007)
13. Alrajeh, A., Loret, J.: Intrusion detection systems based on artificial intelligence techniques in wireless sensor networks. Int. J. Distrib. Sens. Netw. 9, 351047 (2013)
14. Idris, N.B., Shanmugam, B.: Artificial intelligence techniques applied to intrusion detection. In: 2005 Annual IEEE India Conference – Indicon, pp. 52–55 (2005)

15. Landau, D.: Artificial intelligence and machine learning: how computers learn. iQ, 17 August 2016. https://iq.intel.com/artificial-intelligence-and-machine-learning/. Accessed 7 Dec 2016
16. Information Commissioner's Office: Big data, artificial intelligence, machine learning and data protection (20170904 Version: 2.2) (2017). https://ico.org.uk/media/for-organisations/documents/2013559/big-data-ai-ml-anddata-protection.pdf. Accessed 1 Nov 2018
17. Berry, M., Castellanos, M.: Survey of text mining. Comput. Rev. 5(9), 548 (2004)
18. Chassignol, M., Khoroshavin, A., Klimova, A., Bilyatdinova, A.: Artificial Intelligence trends in education: a narrative overview. In: 7th International Young Scientists Conference on Computational Science, vol. 136, pp. 16–24 (2018)
19. Omoteso, K.: The application of artificial intelligence in auditing: looking back to the future. Expert Syst. Appl. 39(9), 8490–8495 (2012)
20. Zerbino, P., Aloini, D., Dulmin, R., Mininno, V.: Process-mining-enabled audit of information systems: methodology and an application. Expert Syst. Appl. 110, 80–92 (2018)
21. Holland, P., Rae, S., Taylor, P.: Why AI must be included in audits. KPMG (2018). https://assets.kpmg.com/content/dam/kpmg/uk/pdf/2018/06/why-ai-must-be-included-in-audits.PDF. Accessed 1 Dec 2018

Axonal Delay Controller for Spiking Neural Networks Based on FPGA

Mireya Zapata[1,2]($\boxtimes$), Jordi Madrenas[1]($\boxtimes$), Miroslava Zapata[3]($\boxtimes$), and Jorge Alvarez[2]($\boxtimes$)

[1] Department of Electronics Engineering, Universitat Politècnica de Catalunya, Jordi Girona, 1-3, edif. C4, 08034 Barcelona, Catalunya, Spain
mireyazapata@uti.edu.ec, jordi.madrenas@upc.edu
[2] Research Center of Mechatronics and Interactive Systems, Universidad Tecnológica Indoamérica, Machala y Sabanilla, Quito, Ecuador
jorgealvarez@uti.edu.ec
[3] Department of Electronics Engineering, University of the Armed Forces ESPE, Av. General Rumiñahui, Sangolquí, Ecuador
mazapata@espe.edu.ec

Abstract. In this paper, the implementation of a programmable Axonal Delay Controller (ADyC) mapped on a hardware Neural Processor (NP) FPGA-based is reported. It is possible to define axonal delays between 1 to 31 emulation cycles to global and local pre-synaptic spikes generated by NP, extending the temporal characteristics supported by this architecture. The prototype presented in this work contributes to the realism of the network, which mimics the temporal biological characteristics of spike propagation through the cortex. The contribution of temporal information is strongly related to the learning process. ADyC operation is transparent for the rest of the system and neither affects the remaining tasks executed by the NP nor the emulation time period. In addition, an example implemented on hardware of a neural oscillator with programmable delays configured for a set of neurons is presented in order to demonstrate full platform functionality and operability.

Keywords: Axonal delay · FPGA · Spiking Neural Networks

1 Introduction

Spiking Neural Networks (SNNs) process information with high biological realism and using spatio-temporal information in their calculations. SNNs model the neural activity based on a set of temporary pulses or spikes received through excitatory or inhibitory synapses, which may have different weights. The spikes are analyzed as a train of temporal signals, which frequency and timing contain processing information and neural perception [1]. It is considered that a spike is generated when the action potential exceeds the threshold voltage.

The main challenges of the emulation of nervous system functions are neural connectivity and temporal coding to transmit sensory information. The latter is used to control the speed at which the spikes travel through the axon. Thus, the network

© Springer Nature Switzerland AG 2020
T. Ahram (Ed.): AHFE 2019, AISC 965, pp. 284–292, 2020.
https://doi.org/10.1007/978-3-030-20454-9_29

dynamics is strongly linked to the synaptic and axon delays, and therefore to the spike-based learning processes among cortical areas [2–4]. It is important to mention that spikes travel through the neocortex relatively slowly. Transmission delay is reported in the literature in the range of 1 to 300 ms or more. [5]. Besides, the network dynamics and its stability are strongly related to the axonal time delay.

Our contribution is based on a multi-model Neural Processor architecture (NP). It has the capability to emulate neurons in real time, where a user can specify full synaptic connectivity and the neural algorithm to be executed by the neurons. More details will be described in the next section. Using NP architecture as a base, we present a compact axonal time delay controller (ADyC), this approach enables evolvable neural rate coding through a programmable path length for axon propagation time using a synchronous AER scheme previously implemented. The associated delay with each spike event support real-time modification. Besides, the network communication time is not affected due to the time delay value is not transmitted by the communication channel, as in other similar implementations as HICANN [6] where, the timestamp is contained in the transmitted packet. In this approach, a look-up table is used for the axonal delay values along with a delay sorter per neuron. It stores the arrival time and looks for a coincidence with the look-up table before re-transmitting the spike. Time information is forwarded to the channel until it is consumed, increasing the link payload. In our case, time information keeps local and is not transmmited. Others like [7] can implement only 4 axonal delay path per axon, meanwhile, we are able to implement axonal delays in all neurons in our system.

Section Neural Processor Architecture Overview describes the main characteristics of the platform. Section Materials and Methods presents the design that has been made for the ADyC. As an example, in Sect. 3 a neuron oscillator is presented as proof of concept along with Hardware test results in order to demonstrate full system operation. Finally in Sect. 4, conclusions and future work are presented.

1.1 Neural Processor Architecture Overview

The Neural Processor Architecture (NP) is a general purpose real-time FPGA-based approach [8, 9] that allows large-scale SNNs emulation. It is a hierarchical scalable platform, which is able to connect up to 127 NP in a ring topology in order to increase its scalability. Every NP is identified through an ID. Besides, it is possible to implement local and global spike in order to obtain hierarchical neural configurations.

As shown in Fig. 1, the NP consists of a 2D array of custom processing elements (PE) implemented with SIMD - type computing strategy [10]. In this way, every PE executes the same neural algorithm in parallel meanwhile synapses are processing sequentially. A neuron fires and a spike is generated when the voltage membrane exceeds the voltage threshold. The PE's are capable of performing arithmetic, Boolean and logical operations for the purpose to emulate the neural dynamic. It is possible to process the neural dynamics of more than one biological neuron by means of PE virtualization, i.e. time multiplexing.

It is responsibility of the Control Unit to synchronize the operations with the Communication Module and the flow control of data and instructions to and from the PE array.

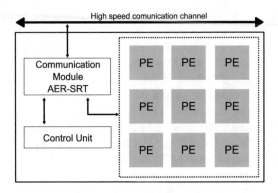

Fig. 1. Neural processor

Spike transmission is based on AER protocol over a Synchronous Serial Ring Topology (AER-SRT) [11]. It is a modified version of conventional asynchronous AER. In AER-SRT, whenever the neurons finish processing the neural algorithm, the generated spikes are encoded as address events, assigned to time slots and sent to the corresponding destination neurons through a broadcast serial bus. An address event is coded by 3 fields: virtualization, row and column PE position.

Figure 2 shows the 3 NP operation phases are shown: Initialization, Execution, and Distribution. Configuring the neural algorithm and synaptic and neural parameters are performed in the Initialization Phase. During the Execution Phase, neural and synaptic algorithms are computed. The generated spikes are delivered to the corresponding destination neuron throughout the Distribution Phase by means of AER-SRT protocol, which employs Aurora IP core [12] from Xilinx for establishing a high-speed serial communication.

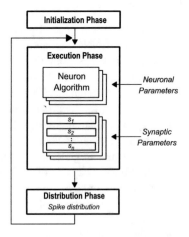

Fig. 2. Neural processor processing phases

It is remarkable to mention that an emulation cycle corresponds to the execution and distribution processing phases. In turn, the duration of both cycles represents a unit of time in biological terms, which is within the limits of real time. In NP, the signal eo_exec generated by the Control Unit controls the start and end of these phases.

2   Materials and Methods

2.1   Axonal Delay Controller

The Axonal Delay Controller (AyDC) allows to emulate axonal delays, i.e., a spike or group of spikes will not be transmitted immediately, but will reach their destination after a certain delay has elapsed, which in biological terms correspond to the axon length of the neural connection.

The AyDC is implemented as part of the AER-SRT Module as shown in Fig. 3. Through the shared bus the AER-SRT receives pre-synaptic spikes (sj_{nm}), which are temporarily stored in an Input FIFO in order to be processed in the execution phase. At the transmission side and during the distribution phase, the post-synaptic spikes (sj_{nm}) generated by the PE array are written directly to the Output FIFO in case it does not have any programmed delay (td = 0) waiting to be distributed by the communication link. Conversely, if sj_{nm} have assigned axonal delay, these are routed to the Delay Controller, where for each emulation cycle, a unit of time is subtracted (td-1). When td reaches zero, the sj_{nm} are placed in the Output FIFO so that they are transmitted in the next distribution phase.

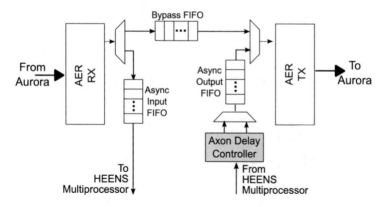

Fig. 3. AER-SRT Module with Axon Delay Controller

As shown in Fig. 4, the controller consists of two main blocks: RAM Time Delay and Delay FIFO. The RAM Time Delay is used to store the delay value (td) of the associated spike. The code 0 is reserved to spikes transmitted immediately after being generated i.e. td = 0.

During the distribution phase, the AER TX reads the Output FIFO, and dumps its contents into the Aurora Tx Side. Once the Output FIFO is empty, FIFO DELAY

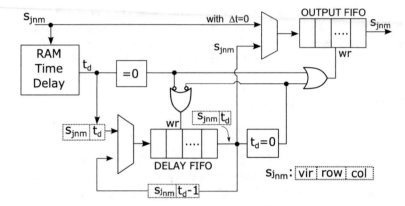

Fig. 4. Axonal Delay Controller. td: delay value; sj_{nm}: post-synaptic spikes

processing starts. This consists of verifying the delay value (td) of each of the stored spikes. If td $\neq$ 0, its value is decremented by 1 (td-1) and saved again into the FIFO DELAY. It operates as a circular buffer which is read and processed by each emulation cycle. A time unit is consumed per cycle. The opposite means that the sj_{nm} is ready for transmission and is sent to the OUTPUT FIFO. The processing of the controller ends when all the spikes stored on the FIFO DELAY have been examined.

3   Proof of Concept

3.1   Topology

As proof of concept, the topology shown in Fig. 5 was implemented, which is made up of an oscillator given by the four inner neurons connected to a combination of neurons with excitations and inhibitions that form a bandpass filter.

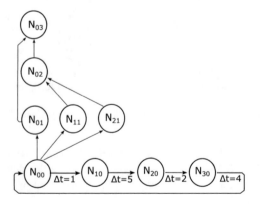

Fig. 5. Oscillator neural topology

The AER address events (11-bit) are defined by the virtualization level and the row and column PE position in the array where the spike is generated. In this application each PE emulates a neuron, no multiplexing is enabled. Table 1 shows the axonal delays assigned to every neuron in the topology.

Table 1. Assigned delay per neuron

Neuron	Virt	Row	Col	Delay
N00	0	0	0	1
N10	0	1	0	5
N20	0	2	0	2
N30	0	3	0	4
N01	0	0	1	0
N11	0	1	1	0
N21	0	2	1	0
N02	0	0	2	0
N03	0	0	3	0

3.2 Neural Algorithm

All the neurons of the network execute the same neural algorithm that corresponds to a Leaky Integrate and Fire (LIF) model [13] whose differential equation is solved with the Euler method. The implemented equation is depicted below:

$$V(t+1) = Vr + (V(t) - Vr).k_m + \sum w_i(t)S_j(t) + B(t) \tag{1}$$

$$V(t) > V_{th} => V(t) = Vr \tag{2}$$

- V(t): Membrane potential
- Vth: −55 mV: Threshold voltage (same to all neurons)
- Vr: −70 mV: Restitution voltage (same to all neurons)
- $k_m : e^{\frac{1}{\tau_m}}$: Decay constant ($\tau_m = 10$ ms)
- $w_i(t)$: Synaptic weight
- Sj: Post-synaptic spikes
- B(t): Background noise

In the topology presented, all the synaptic connections are excitatory although the system equally supports inhibitory synapses. The neurons emulated through the PE, contain a LFSR, which is used for generating background source of noise. The seeds were initialized with random values. Synaptic weighs and membrane voltages were also randomly assigned in a range of (0.1 mV $\leq w_i(t) \leq$ 1.5 mV) and (−65 mV V(t) $\leq$ 60 mV) respectively.

The neural algorithm is described by the assembly instructions of this architecture. The initial parameters of each neuron are defined in a text that contains all the information required to configure the application in the proposed prototype.

3.3 Results and Discussion

The ND architecture is described in VHDL language. Vivado [14] software from Xilinx is used for synthesizing, placement and routing. Implementation takes place in an FPGA Kintex-7 xc7k325tffg900-2 board.

AER-SRT protocol is used to configure the application and establish spike communication [15], operating at a serial bit rate of 1.125 Gbps. The clock frequency for data processing in this prototype is fclk = 125 MHz.

For the purpose of this work, a single board has been used, however, it is possible to scale the size of the network by connecting several nodes in a ring topology. Full architecture simulation is performed with QuestaSim [16] tool from Mentor Graphics.

Figure 6 represents the simulation of the delay controller signals in its main instances: RAM, Output and Delay FIFO. The eo_exec signal marks the beginning of the execution and distribution phase. The image illustrates the spike generated by the neuron N01 without a programmed delay (td = 0). It is transmitted at the same emulation cycle during the distribution phase to the target neuron N03.

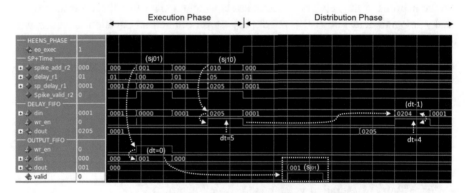

Fig. 6. Axon Delay Controller simulation; $s_j 01$: post-spike from axon N01 transmitted in the same emulation cycle; $s_j 10$ corresponds to the post-spike from axon N01 stored in the DELAY FIFO. It is processed during the Distribution Phase

The operation of the axonal delay controller in the spike generated by the neuron N10 with a delay of td = 5 is also observed. Since, td ≠ 0, is not transmitted in the distribution phase, instead, it is stored in the FIFO DELAY in order to be processed during the next distribution phases.

The Integrated Logic Analyzer (ILA) from Vivado was used to monitor the oscillator operation in neurons N00, N10, N20, N30 during the emulation cycles 566 to 578 in hardware (FPGA Kintex-7). As shown in Fig. 7, axonal delays established for this implementation are met.

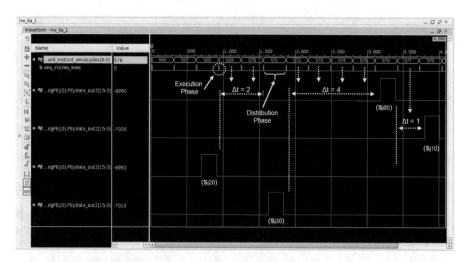

Fig. 7. Time simulation of **hardware implementation on FPGA - Kintex 7** using ILA tool. s_j20 represents the post-synaptic spike from neuron N20, it takes 2 emulation cycles before being transmitted to N30. s_j30 takes 4 cycles before reaching N00. Between N00 and N10, there is a delay of 1 cycle

4 Conclusions and Future Work

In the present work, we demonstrate the functionality of an Axonal Delay Controller implemented over a NP Architecture. This allows us to emulate spatio-temporal dynamics present in the biological neurons. The Delay Controller along with the AER-SRT based on synchronous AER is able to support hierarchical topologies and programmable path lengths for the axon propagation time in the range of 1 to 31 emulation cycles. Local and global spikes can be configured with axonal delays. Besides, the delay (td) storage on RAM allows real-time adaptation. RAM Time Delay and the DELAY FIFO width word determine the maximum propagation time that can be programmed, 5-bit are assigned to this field in this propose.

Longer axon propagation time may be obtained increasing the word-length in both RAM and FIFO. It is important to highlight that, in NP architecture dendritic and synaptic delays are feasible by software. Through the ADyC, delay adaptation is supported by NP, it can be accomplished in a single emulation step allowing learning delay.

Finally, as future work, we plan to implement cluster delays, i.e. delay at destination as a function of source chip ID.

Acknowledgments. This work has been partially funded by the Spanish Ministry of Science and Innovation and the European Social Fund (ESF) under Projects TEC2011-27047 and TEC2015-67278-R. Mireya Zapata held a scholarship from National Secretary of High Education, Science, Technology, and Innovation (SENESCYT) of the Ecuadorian government.

References

1. Ghosh-Dastidar, S., Adeli, H.: Third generation neural networks: spiking neural networks. In: Advances in Computational Intelligence, pp. 167–178. Springer, Berlin (2009). ISBN: 978-3-642-03156-4. https://doi.org/10.1007/978-3-642-03156-4_17
2. Brunel, N.: Dynamics of sparsely connected networks of excitatory and inhibitory spiking neurons. J. Comput. Neurosci. **8**(3), 183–208 (2000)
3. Roxin, A., Brunel, N., Hansel, D.: Role of delays in shaping spatiotemporal dynamics of neuronal activity in large networks. Phys. Rev. Lett. **94**(23), 238103 (2005)
4. Izhikevich, E.M., Gally, J.A., Edelman, G.M.: Spike-timing-dependent plasticity. Cereb Cortex **14**(8), 933–944 (2004)
5. Bringuier, V., Chavane, F., et al.: Horizontal propagation of visual activity in the synaptic integration field of area 17 neurons. Science **283**, 695–699 (1999)
6. Scholze, S., et al.: VLSI implementation of a 2.8 Gevent/s packet-based AER interface with routing and event sorting functionality. Front. Neurosci. **5**, 1–13 (2011). https://doi.org/10.3389/fnins.2011.00117
7. Wang, R.M., et al.: A mixed-signal implementation of a polychronous spiking neural network with delay adaptation. Front. Neurosci. 1–16 (2013). https://doi.org/10.3389/fnins.2014.00051
8. Sripad, A., et al.: SNAVA—a real-time multi-FPGA multi-model spiking neural network simulation architecture. Neural Netw. **97**, 28–45 (2018). https://doi.org/10.1016/j.neunet.2017.09.011
9. Zapata, M., Madrenas, J.: Synfire chain emulation by means of flexible SNN modeling on a SIMD multicore architecture. In: 25th International Conference on Artificial Neural Networks ICANN 2016, vol. 8681, pp. 222–229, September 2016. ISSN: 16113349. https://doi.org/10.1007/978-3-319-11179-7
10. Madrenas, J., Moreno, J.M.: Strategies in SIMD computing for complex neural bioinspired applications. In: Proceedings of 2009 NASA/ESA Conference on Adaptive Hardware and Systems, AHS, pp. 376–381 (2009). https://doi.org/10.1109/ahs.2009.31
11. Dorta, T., Zapata, M., Madrenas, J., et al.: AER-SRT: scalable spike distribution by means of synchronous serial ring topology address event representation. Neurocomputing **171**, 1684–1690 (2016). http://dx.doi.org/10.1016/j.neucom.2015.07.080
12. Xilinx: Aurora 8b/10 protocol specification, vol. SP002. https://www.xilinx.com/support/documentation/ip_documentation/aurora_8b10b_protocol_spec_sp002.pdf
13. Izhikevich, E.M.: Which model to use for cortical spiking neurons? IEEE Trans. Neural Netw. **15**(5), 1063–1070 (2004)
14. Xilinx Vivado. https://www.xilinx.com/products/design-tools/vivado.html
15. Zapata, M., Jadán, J., Madrenas, J.: Efficient configuration for a scalable spiking neural network platform by means of a synchronous address event representation bus. In: 2018 NASA/ESA Conference on Adaptative Hardware and System, pp. 241–248 (2018). https://doi.org/10.1109/AHS.2018.8541463
16. Questa Advanced Simulator. https://www.mentor.com/products/fv/questa/

Ubiquitous Fitting: Ontology-Based Dynamic Exercise Program Generation

Chuan-Jun Su$^{(\boxtimes)}$, Yi-Tzy Tang, Shi-Feng Huang, and Yi Li

135 Yuan-Tung Road, 32003 Chung-Li, Taiwan, R.O.C
{iecjsu, u6ytang, iesfhuang}@saturn.yzu.edu.tw,
s1068905@mail.yzu.edu.tw

Abstract. In order to reduce the incidence of disease and decrease the proportion of "sub-health", exercise regularly is one of the most important factors to solve these problems. Regular exercise has many positive effects on body's systems, while inappropriate forms of exercise can cause problems or even have adverse consequences for health. Therefore, this research aims to develop an ontology-driven knowledge-based system to dynamically generating personalized exercise programs. The generated plan exposing REST style web services, which can be accessed from any Internet-enabled device and deployed in cloud computing environments. To ensure the practicality of the generated exercise plans, encapsulated knowledge used as a basis for inference in the system is acquired from domain experts. Also, we integrate the system with wearable devices so that we can collect real-time data, for example, heart rate. In the future, break through the limitations of equipment, the accuracy and reliability can be promoted.

Keywords: Ontological knowledge base · Physical fitness ·
Personalized exercise program · Dynamically generate · Wearable device

1 Introduction

Due to the rapid growth of medical service, people focus not only on a cure but also on prevention of diseases. The best way to prevent disease is to keep ourselves healthy. The World Health Organization (WHO) gave the definition of health as follows:

> "Health is a state of complete physical, mental and social well-being and not merely the absence of disease or infirmity. [1]."

It shows that "health" not just talks about our physical body; we also have to concern our mind and the society. Take some common conditions for example, depression, insomnia, fatigue, poor memory, ache in muscles… and so on. These are the symptoms that will let us feel uncomfortable but won't make us sick, we can call this state "sub-health". It is a term that is being well-known in recent years. In brief, it is a state that between health and illness. Lots of people are trapped in this state. To escape from this state, personal health management have to be emphasized.

© Springer Nature Switzerland AG 2020
T. Ahram (Ed.): AHFE 2019, AISC 965, pp. 293–302, 2020.
https://doi.org/10.1007/978-3-030-20454-9_30

2 Related Works

2.1 Ontology-Based Systems

Ontology is an emerging technology in building knowledge-based information retrieval systems. It is used to conceptualize the information in human being understandable manner. Over the years, much importance was given to the systematic development of ontologies. As a result, many methods and strategies to build ontology in different domains came up in the field of ontological engineering [2].

2.2 Healthcare Systems

An ontology is a knowledge model that represents a set of concepts within a domain and the relationships among these concepts. Ontologies facilitate not only representation but also concept instantiation and instance-based reasoning within a domain. In the domain of healthcare, ontologies have been recognized as a key technology in helping to furnish the semantics required for deriving proper treatment through integrating clinical guidelines [3]. Riaño et al. [4] have used Protégé to develop an ontology for the domain of chronically ill patients requiring home care: the Case Profile Ontology which introduce an ontology for the care of chronically ill patients and implement two personalization processes and a decision support tool.

Each of a person has different physical conditions, and some of them may have diseases that will affect the choices of exercise type. In this research, we use ontology to exclude inappropriate type by their personal profile.

2.3 Personalized Physical Fitness

Su et al. [5] develop an ontology driven knowledge-based system for generating specifically designed exercise plan. This system (UFIT) also designed exercise plan based on: (1) the user profile; (2) the user's physical fitness; and (3) the user's health screening data. The research objective of UFIT is consistent with us. However, the results of UFIT are static, it cannot update dynamically. Thus, we will extend the structure of UFIT to complete our research.

2.4 Ontology and HL7 Standard

To cooperate with different systems and organizations, the interoperability is one of the most essential requirements for health care systems to reach the benefits promised by adopting HL7-based systems and Electronic Medical Records (EMRs). Slavov et al. [6] proposed an HL7-compliant data exchange software tool called Collaborative Data Network (CDN) aiming for clinical information sharing and querying. The clinical documents in CDN are modeled in compliant with HL7 v3 standard and encoded in eXtensive Markup Language (XML) format, which can be ultimately deployed in a cloud environment to support large-scale management and vast amounts of clinical data sharing.

As a key component of our system, the knowledge engine was built on top of problem-oriented medical record ontology "HL7-sample-plus-owl." defined by World Wide Web Consortium (W3C) [7]. According to W3C, the goal of "HL7-sample-plus-owl" is to define a minimal set of terms that connect representations from well-defined healthcare information and process models with more expressive foundational ontologies through the use of the criteria outlined in the traditional problem-oriented medical record structure. To ensure ubiquitous accessibility and wide area interoperability, we designed and developed this system, an HL7-compliant system driven by an ontology-based knowledge engine founding on HL7-sample-plus-owl. This system is capable of processing user health screening data and personal information from any HL7-enabled medical organization and subsequently generates personalized exercise plans.

3 Methodology

3.1 System Architecture

The following figure is the architecture of our system, it can be divided into 3 sections: (1) An ontological knowledge base that stores user profiles, wearable device data, physical fitness test data, health screening data, and exercise ontology; (2) Inference module consist SPIN rule and SPARQLmotion; (3) Environment and user interface to present and interact final exercise program. In this section will introduce each part in order (Fig. 1).

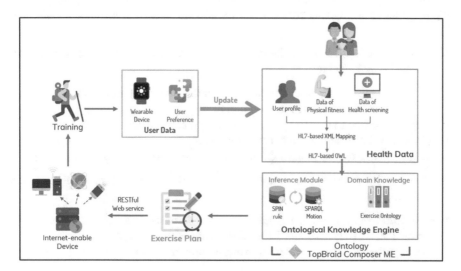

Fig. 1. System architecture

3.2   Ontological Knowledge Base

How to create an ontology is one of the most important things at the beginning of our research. Although there are various ways to build an ontological knowledge base, these methodologies encapsulate many common features. We follow a guide- Ontology Development 101, which is form by Noy and McGuinness [8]. There has a rule that they have especially mentioned:

> "There is no one correct way to model a domain— there are always viable alternatives. The best solution almost always depends on the application that you have in mind and the extensions that you anticipate."

As the rule said, our research objective is to develop an adaptive exercise program which will base on personal physical states. Thus, we should confine our domain to sports and physical fitness.

Here's the 7 steps:

Step 1. Determine the domain and scope of the ontology
Step 2. Consider reusing existing ontologies
Step 3. Enumerate important terms in the ontology
Step 4. Define the classes and the class hierarchy
Step 5. Define the properties of classes—slots
Step 6. Define the facets of the slots
Step 7. Create instances

To build a knowledge base that is close to knowledge of domain expert, we collect information from domain experts in MJ Health center and record the key point of how experts generate an exercise program. In this ontological knowledge base contain these types of data:

A. Exercise ontology: contains the exercise-related information acquired from domain professionals including:

- Goal of exercise (e.g., cardiopulmonary training, flexibility improvement)
- Type of exercise (e.g., jogging, swimming)
- Time of exercise (e.g., 10–15 min, 2 rounds, repeat ten times per round)
- Intensity of exercise (e.g., moderate, low, high)
- Frequency of exercise (e.g., 2–3 times/week, 3–4 times/week)

B. User profile ontology: contains personal information and physical test data including

- Basic information (e.g., name, sex, age, characteristics, preference, interest)
- Personal states (e.g., exercise habit, disabilities, impairments)
- User's preferences (e.g., preferred exercise, preferred time to exercise)

C. Health screening ontology: contains comprehensive health-screening information including

- Health-screening data (e.g., physiological data, triglyceride, cholesterol)
- Physical fitness test (e.g., grade 1, grade 2, grade 3, grade 4, and grade 5)

The example for generate a part of ontology will be presented in Fig. 2. It can clearly show the four component of building ontology, Class, Property, Instance, and Constraint.

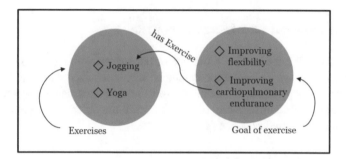

Fig. 2. Example of ontology

3.3   Inference Module

SIPN Rules and SPARQLMotion are two components of the Inference Module. The SPIN rules are pre-defined by developer. When we use SPARQL to query the information, the engine will find the information inside of the knowledge base. Finally, the knowledge engine will output the result by inference based on all SPIN rules, this process called SPARQLMotion.

3.4   RESTful Web Service

REST services provide an abstraction for publishing information and serve for communicating with Knowledge Engine and Database Server in this system. The REST services can be easily deployed via simple methods of Hypertext Transfer Protocol (HTTP) such as POST, GET, PUT and DELETE and produce discretely formatted responses (typically in XML or JSON with no dangling TCP connection). By applying REST services, this system is capable to offer ubiquitous access through virtually any Internet-enabled devices.

3.5   Wearable Device

Due to the growth of the wearable device become fully develop, there has various product can be chosen. Considering the objective of this research is only to access data from the wearable devices, rather than developing a product. We chose Apple watch as our integration target. There are some reasons why we select watch OS(Apple watch) instead of wear OS(Android watch):

- The Market share of watch OS growing more stable than wear OS.
- Watch OS integrated many functions in health kits also provide official tutorial for developer to capture health data.

4   Implementation and Usage Scenarios

In this part, we will introduce the prototype of our system with regard to our proposed usage scenarios. A system prototype embodied the generalized and functional design concepts for ontology-based application. And the scenarios can show the expected benefit of our system. Furthermore, we will compare the different between the manual way, and ontology-based exercise plans generation.

4.1   Knowledge Engine Development

As we mentioned above, knowledge engine is composed by SPIN rule and SPARQLMotion. The following will explain how we build the knowledge engine and what roles do these two components play.

- SPARQLMotion

SPARQLMotion is an RDF-based scripting language with a graphical notation to describe data processing pipelines. The basic idea of SPARQLMotion is that individual processing steps can be connected, so that the output of one step is used as input to the next. RDF graphs are the basic data structure that is passed between the steps, but named variables pointing to RDF nodes and XML documents can also be passed between steps. The behavior of each module is typically driven by SPARQL queries. The following graph presents the part of SPARQLMotion in our system, which arranged the work and order of each role like a script (Fig. 3).

Figure 5 shows the other script that describe the exercise plan which execute from the past script and update based on real-time heart rate.

- SPIN rule

SPIN combines concepts from object-oriented languages, query languages, and rule-based systems to describe object behavior on the web of data. One of the key ideas of SPIN is to link class definitions with SPARQL queries to capture rules and constraints that formalize the expected behavior of those classes. To do so, SPIN defines a light-weight collection of RDF properties.

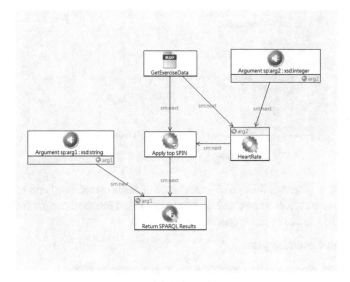

Fig. 3. Dynamic generation

The key point about Dynamic Exercise program generation is shown below (Figs. 4 and 5).

Fig. 4. CardioGrade1-3

These two figures are the five zones of exercise target heart rate, which have been converted to SPIN rule [9].

4.2 Demonstration of Scenarios

The objective of this research is to develop an Ontology-based Dynamic Exercise program generation in accordance with user profile, user's test data (the HL7-based

Fig. 5. CardioGrade4-5

data of physical fitness and the HL7-based data of health screening) and real-time heart rate to improve physical fitness and make the body less prone to common diseases. The following will introduce the scenarios.

- All types of exercise plan

    In this part we show the result of a complete exercise plan.

    Jessie has an annual health assessment including a fitness test. While traveling on business, she initiates the Ontology-based Dynamic Exercise program generation web service using her mobile phone, as illustrated in Figure. Ontology-based Dynamic Exercise program generation then extracts her profile and health data to generate a customized exercise plan, as shown in Figure. According to her health status, the system recommends three areas of exercise: cardiopulmonary, resistance, and stretching, but advises against certain exercise types, such as treadmill walking, because Jessie's medical history indicates she has been suffering from peripheral neuropathy, which can

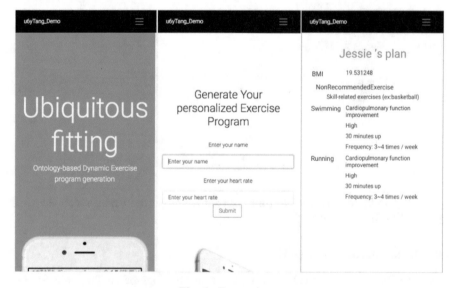

Fig. 6. Demo views

affect balance, placing her at greater risk of falling. Jessie selects the cardiopulmonary training. The recommended exercises associated with the selection are then generated, as depicted in Fig. 6.

- Dynamic generation (Take cardio exercise as example)

Because of the assessment will take place only once a year. The result of fitness tests is easily to be invalidated. Therefore, we add an element to make the system be able to update. This element is a real-time heart rate. It is relatively easy to get the data, also it has explicit definition to classify the level of exercise intensity.

Limited by current technology, we have to manual set up the interval of exercise so that can collect the heart rate during the exercise time.

5   Conclusion and Future Works

Exercise has a lot of benefit for people. However, inappropriate exercise will cause contrary effect. That's the reason why we build this system. To build an Ontology-based Dynamic Exercise program generation system which accordance with personal health data (physical fitness, health screening, real-time data from the wearable device) and profile (preference). Furthermore, the result which generated by our system can be accessed through any Internet-enabled device through the paradigm of RESTful web service.

However, monitoring people exercise is relatively simple to stimulation. It is worth to noting how to attract people to do exercise and do exercise regular. This is the most important also the most difficult part in health issue. After all, persevering exercise can really turn the population of sub-health to health also improve our body. Thus, it can be extended and research the issue about how to attract people to do exercise regularly.

In addition, the devices in our study are limited by current technology. We are able to provide only one input- heart rate to update the exercise program. If there's possible to collect data from any other exercise equipment, then we must be able to provide a more complete and accurate dynamic exercise program.

References

1. WHO. http://www.who.int/about/mission/en/
2. Hua Brenda, Y.: About sub-health, 11 February 2018. http://ezinearticles.com/?About-Sub-Health&id=5387585
3. Isern, D., SáNchez, D., Moreno, A.: Ontology-driven execution of clinical guidelines. Comput. Methods Programs Biomed. **107**(2), 122–139 (2012)
4. Riaño, D., Real, F., López-Vallverdú, J.A., Campana, F., Ercolani, S., Mecocci, P., et al.: An ontology-based personalization of health-care knowledge to support clinical decisions for chronically ill patients. J. Biomed. Inform. **45**(3), 429–446 (2012)
5. Su, C.-J., Chiang, C.-Y., Chih, M.-C.: Ontological knowledge engine and health screening data enabled ubiquitous personalized physical fitness (ufit). Sensors **14**(3), 4560–4584 (2014)

6. Slavov, V., Rao, P., Paturi, S., Swami, T.K., Barnes, M., Rao, D., et al.: A new tool for sharing and querying of clinical documents modeled using HL7 Version 3 standard. Comput. Methods Programs Biomed. **112**(3), 529–552 (2013)
7. W3C: A problem-oriented medical record ontology. http://www.w3.org/wiki/HCLS/POMROntology#A_Problem-Oriented_Medical_Record_Ontology. 11 Feb 2018
8. Noy, N.F., McGuinness, D.L.: Ontology development 101: a guide to creating your first ontology. Stanford knowledge systems laboratory technical report KSL-01-05 and Stanford medical informatics technical report SMI-2001-0880, Stanford, CA (2001)
9. American Heart Association: Target heart rates, 29 May 2018. https://healthyforgood.heart.org/move-more/articles/target-heart-rates

Innovation and Artificial Intelligence

Muhammad Sohaib Shakir[1(✉)], Farhat Mehmood[1], Zarina Bibi[1],
and Maliha Anjum[2]

[1] University of Lahore, Information Technology Department, Gujarat, Pakistan
hafizsohaib01@gmail.com
[2] University of Gujarat, Information Technology Department, Gujarat, Pakistan

Abstract. In this article, we endeavor to utilize the advanced workmanship in which computerized innovation remains in the focal point of its imaginative procedure – for instance of inventiveness to outline how innovativeness and AI can increase shared advantage from one another. Moreover, we discuss the distinctive effects of cognitive shifting and Intelligence on innovativeness. AIs create in information, they are likely going to be related with an impressive proportion of our endeavors, and we will extend ourselves with abilities to have the ability to make use of such extra knowledge, moving towards a cyborg situation, whereby the mental and physical spaces end up being dynamically and more clouded with our propelled understanding.

Keywords: Artificial Intelligence · Human-System · Cognition · Digital art

1 Introduction

Artificial intelligence (AI) is a term for reproduced knowledge in machines. These machines are modified to "think" like a human and copy the manner in which a man demonstrations. The perfect normal for computerized reasoning is its capacity to justify and take activities that have the most obvious opportunity concerning accomplishing a particular objective, in spite of the fact that the term can be connected to any machine that displays qualities related with a human personality, for example, learning and taking care of issues.

In the age of the fourth Industrial Revolution (4IR) (Xing and Marwala 2017), numerous nations (Shah et al. 2015; Ding and Li 2015) are defining out a larger objective of building/anchoring an "advancement driven" economy. As advancement underscores the usage of thoughts, inventiveness is normally viewed as the principal phase of development in which producing thoughts turns into the prevailing center (Tang and Werner 2017; Amabile 1996; Mumford and Gustafson 1988; Rank et al. 2004; West 2002). At the end of the day, if imagination is missing, advancement could be simply luckiness. we can think creativity like this way (Sanchez-Burks et al. 2015): "Creative people [are] able to connect experiences they've had and synthesize new things" [1].

In the last few years, we have seen a colossal ascent in the field of man-made brainpower, particularly in profound learning calculations. This has likewise prompted a colossal increment in the ability of machines to create content without anyone else,

© Springer Nature Switzerland AG 2020
T. Ahram (Ed.): AHFE 2019, AISC 965, pp. 303–307, 2020.
https://doi.org/10.1007/978-3-030-20454-9_31

in different structures, be it visual, sound related or message based. Nevertheless, a considerable measure of the interfacing of this AI with people remains principally advanced in frame, along these lines not partaking in our physical errands.

With the improvement of man-made reasoning, people are absolutely subject to machines. In couple of years back the undertaking that felt hard to finish have turned out to be significantly less demanding on account of quick improvement in this field, even than it assumes crucial job in day by day house hold work With a particular true objective to perceive practically performing human– fake collaboration structures in coordination's ex hazard for theory decision purposes, a multi-dimensional hypothetical framework is created [1, 2].

Man-made reasoning may do every one of those work where human physical body parts are major being used a precedent is that hand composing machine where machines make physically composed substance in continuation to human handwriting and move the human hand on the paper to form it out. As AI creates in its capacities, we acknowledge that an all the more firmly coupling of the data and yield procedure inside the space of physical assignments will be essential in making machine that incorporates substance and impacts you to make what the PC 'considers' through a magnet underneath the table the use of AI consistent [2].

2   Digital Art

The Mechanical improvement or unrest has a long history of affecting innovativeness. Take 'Gutenberg Revolution' (Kirschenbaum 2016; Winston 2005), the development of the printing press in the fifteenth century makes ready of large scale manufacturing of writings and pictures. With new correspondence, limit being empowered by this mechanical progression, the far reaching of material and scholarly trade winds up conceivable. These days, a large number of the working methodologies utilized by advanced artisans can be followed back to the good 'old days (between the 1960s) of the PC improvement. Since the rise of the World Wide Web during the 1990s, a differing assortment of chances were additionally opened for visual expressions with apparently boundless permutable measurements.

The explanations behind picking computerized craftsmanship as our talk stage are triple (Sefton-Green and Reiss 1999; Bentkowska-Kafel et al. 2005): right off the bat, it is such a typical practice for specialists, specifically youthful experts, to utilize an extensive variety of media expressions for inventive purposes, creating static/powerful pictures, and in addition controlling sound tracks and content contents; furthermore, advanced workmanship isn't a detached practice, isolated from different types of expressions. It is basically a technique that consolidates a wide range of interconnections with other craftsmanship practices together with other way of introductions and enquiries, showing that we are seeing and encountering another influx of imagination upheaval; last yet not the slightest, it merits seeing that a multitude of advanced craftsmen are currently working in various enterprises shoulder to bear with equipment and programming professionals at the cutting edge of development [1].

2.1 Techniques of Digital Art

The systems of computerized crafts man ship are utilized broadly by the prevailing press in commercials, and by movie producers to deliver visual impacts. Work area distributing has hugy affected the distributing scene, despite the fact that that is more identified with visual depiction. Both advanced and customary artisans utilize numerous wellsprings of electronic data and projects to make their work. Given the parallels among visual and melodic expressions, it is conceivable that general acknowledgment of the estimation of advanced visual workmanship will advance similarly as the expanded acknowledgment of electronically delivered music in the course of the most recent three decades (Fig. 1).

Fig. 1. Maurizio Bolognini, *Programmed Machines* (Nice, France, 1992–97). An installation at the intersection of *digital art* and conceptual art (computers are programmed to generate flows of random images, which nobody would see).

3 Cognitive Shifting and Intelligence on Creativity

The connection among imagination and official control has for some time been questionable. A few specialists see inventive reasoning as a defocused procedure with minimal official control contribution, though others guarantee that official control assumes an indispensable job in innovative reasoning. In this examination, we concentrated on one subcomponent of official control, intellectual moving, and analyzed its association with inventiveness by utilizing idle variable investigation and basic condition displaying. We likewise investigated whether this connection was interceded by knowledge. The outcomes demonstrated that: (a) subjective moving capacity had a positive association with imagination, yet just on the quantitative viewpoints (fluency and flexibility); (b) Intelligence had a positive association with both quantitative and subjective perspectives (inventiveness) of innovativeness, and its impact on subjective angle was more grounded than that on the quantitative viewpoint; (c) There was an interceding impact of insight on the connection between imagination psychological and moving [3].

Spatial cognition is an important branch of cognition science and very common in our daily life when people work together. As far as human-robot spatial discernment association, it would be emotional if the robot could collaborate with individuals like a human. Therefore, we outlined a wise robot, which could speak with individuals in regular dialect and think on the human's point of view in spatial discernment assignments, for example, get instruments and gather machines. The innovative thing is that the human-robot spatial cognition interaction system is established based on ACT-R (Adaptive Control of Thought-Rational) cognitive architecture [4].

There is in some quarters, concern about high– level machine insight and super keen AI coming up in a couple of decades, carrying with it significant dangers for humankind. In different quarters, these issues are overlooked or thought about science fiction. We needed to elucidate what the dissemination of conclusions really is, the thing that likelihood the best specialists as of now dole out to high– level machine insight coming up inside a specific time– outline, which dangers they see with that improvement, and how quick they see these creating [5].

4   Conclusion

The specialists imagine that super knowledge is probably going to arrive in a couple of decades and potentially terrible for mankind – this ought to be reason enough to do investigation into the conceivable effect of super insight before it is past the point of no return. We could likewise put this all the more unassumingly and still arrive at a disturbing end: We are aware of no convincing motivation to state that advancement in AI will come to a standstill (however, profound new experiences may be required) and we are aware of no convincing reason that super insightful frameworks will be useful for humankind. Thus, we should better examine the fate of super knowledge and the dangers it models for humanity. This study explored the relationship between cognitive shifting and creativity, the interceding impact of insight on this relationship, giving some new proof on the subject of whether inventive reasoning includes official control or not. It underpins the possibility that official control assumes a critical job in the age of original thoughts. Besides, past investigations on the connection between official control and innovativeness most centered on intellectual restraint, and this examination gives a more extensive point of view to this inquiry by considering subjective moving and the intervening impact of insight. It gives a model of how intellectual moving and knowledge influence the distinctive parts of innovativeness.

Acknowledgments. The heading should be treated as a 3rd level heading and should not be assigned a number.

References

1. Xing, B., Auckland Park: Creativity and artificial intelligence: a digital art perspective. papers. ssrn.com (2018)
2. Agrawal, H., Yamaoka, J., Kakehi, Y.: Artificial intelligence output via the human body. ACM (2018)
3. Pan, X., Yu, H.: Different effects of cognitive shifting and intelligence on creativity. J. Creative Behav. 1–18 (2016)
4. Mu, X.-L., Tian, Y., Tan, L.-F., Wang, C.-H.: The design of human-robot spatial cognition interaction system. IEEE (2016)
5. Müller, V.C., Bostrom, N.: Future Progress in Artificial Intelligence. Springer International Publishing, Switzerland (2016)

Software, Service and Systems Engineering

Service Model Based on Information Technology Outsourcing for the Reduction of Unfulfilled Orders in an SME of the Peruvian IT Sector

Renato Bobadilla[1(✉)], Alejandra Mendez[1(✉)], Gino Viacava[1(✉)],
Carlos Raymundo[2(✉)], and Javier M. Moguerza[3(✉)]

[1] Escuela de Ingeniería Industrial,
Universidad Peruana de Ciencias Aplicadas (UPC), Lima, Peru
{u201311160, u201311788, gino.viacava}@upc.edu.pe
[2] Dirección de Investigaciones,
Universidad Peruana de Ciencias Aplicadas (UPC), Lima, Peru
carlos.raymundo@upc.edu.pe
[3] Escuela Superior de Ingeniería Informática,
Universidad Rey Juan Carlos, Mostoles, Madrid, Spain
javier.moguerza@urjc.es

Abstract. In the current market, small- and medium-sized companies (SMEs) face losses due to poor process control. The core activities of information technology (IT) outsourcing service companies are to provide outsourcing services related to technology and information control, which is why it is crucial to work with standardized, efficient processes, to not affect the main process and resources involved. In this document, a case study of an SME is evaluated, related to a deficient billing process, which is not able to fulfill all of its orders. To solve the problem, we propose an IT outsourcing service model, based on the management of processes, knowledge, and change. After the model was validated, it was evidenced that it allowed the integration and finalization of the services provided by the company, increasing the monthly income by 80%.

Keywords: Process management · Service models based on IT outsourcing · Knowledge management · Change management · 5S management

1 Introduction

Worldwide, about 92% of industries invest in outsourcing services and 10 trillion dollars were invested in outsourcing services in 2017. In Latin America, the rate of outsourcing is the highest in Brazil. Peru is in fourth place in terms of the outsourcing rate. In Peru, 86% of companies outsource services, i.e., companies chose to focus on their key processes and outsource the rest of processes. This is done to maintain a good market position. Among the outsourcing services provided, 30% corresponds to information technology (IT) outsourcing, which comprises hiring a service that aims to meet and cover the technological and information needs of the company [1, 2].

© Springer Nature Switzerland AG 2020
T. Ahram (Ed.): AHFE 2019, AISC 965, pp. 311–321, 2020.
https://doi.org/10.1007/978-3-030-20454-9_32

Currently, 96.5% of companies in Peru are small- and medium-sized companies (SMEs), generating employment for around 8 million people. IT SMEs provide services to banks as well as retail and manufacturing companies, among others; however, they fail to complete and integrate processes that allow the total delivery of services. Thus, services cannot be presented to the customer for the corresponding payment. Every service company needs to receive income at specific moments.

The techniques, models, and processes of formal SMEs are well established, which might make services offered by these companies stand out. Models are established for the management of risks, human resources, technology and information, and recruitment of personnel. One of the limitations of the current service models is the centralization in the core developed by the company, without having a strategic plan that might incorporate service processes of IT SMEs.

Based on this background, we can state that the objective of this research project is to present a service model that includes all processes starting with quotes up to service invoicing. This will allow SMEs to complete services successfully and to meet existing deadlines. We propose a service model that integrates several engineering techniques. One of these is the management of processes through flow charts, SYPOC diagrams, procedures, BPMN diagrams, RACI matrix, and indicators, which allow the standardization of all its activities. In addition, knowledge management and change management were used to involve workers of the various IT SMEs [3, 4].

2 State of the Art

From the literature reviewed, it has been observed that several research projects on service management models based on IT outsourcing have been performed. However, all these models focus on the central process of an IT outsourcing company. We will mention some models in the following paragraphs.

2.1 Process Management

For process management, the organization is considered as an interrelated system of processes that jointly contribute to improving customer satisfaction [5]. Additionally, the implementation of process management in the organization will allow the following:

- Elimination of processes that do not add value
- Obtaining efficient results based on process performance
- Continuous improvement processes based on actual objectives [6]

Processes are represented using the SIPOC diagram, and improved visualization of component elements and flows of the main processes is achieved. In addition, process management is used to effectively find areas with insufficient information and/or unsatisfactory data and when it is applied, it is also possible to highlight the flow of high-level processes, creating the prerequisite for process improvement [7].

2.2   5S Technique

The 5S technique when implemented facilitates the improvement and maintenance of conditions of the organization, occupational safety and consequently total quality, productivity, competitiveness, and continuous improvement. The results of a case study are presented below. The implementation of the 5S technique at five factories reduced working time, unnecessary movements, and decreased inventory levels. It brought about an improvement in workflow and in the efficiency of the area [8].

2.3   Knowledge Management

The aim of knowledge management is to transfer knowledge from the place where it is generated to the location where it will be used. It entails the development of necessary skills in organizations to share knowledge and use it among their members and to make it endure over time. The implementation of knowledge management for the biofertilizers program in the municipality of Calixto García (Cuba) allowed the dissemination of information on the use and the need for soil improvement. Production in the municipality was also augmented. Additionally, a knowledge map was drawn up for the development of the program [9].

2.4   Change Management

Innovative companies are those that adapt best to environmental changes. They are also the most flexible ones and are best prepared to respond to market needs. However, to achieve this, it is necessary to prioritize the human factor, since it is essential for the implementation of changes in the organization as a key element of the system [10]. From the set of articles reviewed for the study, we can conclude that change management is a necessary tool for any organization that wants to implement a new work methodology or make a change that affects the operational level of the company. It is also required for the implementation of management techniques by processes and 5S.

3   Contribution

3.1   Rationale

From the literature reviewed, five models have been selected, which have different dimensions but are important to develop the new model. See Table 1.

Finally, the comparative analysis of the models can be applied because each one has a particular characteristic, focused on one or more dimensions, which also depends on the category and size of the company. Therefore, this enhances the proposal, since no model includes an integration of the dimensions mentioned above. Such models do not emphasize the combination of services with IT outsourcing through the integration of process management and 5S.

Table 1. Dimensions

Models	Dimensions				
	Human resources	Services	Technology and information	Change management	Knowledge management
Model 1		X	X		
Model 2			X		X
Model 3	X	X			
Model 4	X	X			
Model 5				X	
Bobadilla Castro, R. and Mendez Soto, A.	X	X	X	X	X

3.2   Proposed Model

According to the research that was performed for the development of the model and in relation to previous research, an opportunity for improvement was identified. Therefore, a service model based on IT outsourcing is proposed to integrate the aforementioned dimensions. In addition, it must be stated that the new model contributes to the development of new procedures and the registration and establishment of a new work method.

As you can see in Fig. 1, the proposal integrates management by processes and 5S techniques. They are also joined with other elements, such as change and knowledge management, for the correct operations of the model. Therefore, the conclusion is that the proposal is unique and it will contribute in a significant manner.

Fig. 1. General model

3.3   Components of the Proposed Model

This model contributed by incorporating the following techniques: knowledge management, change management, process management and 5S. Each of these techniques has stages, procedures, and formats for their correct implementation. In Fig. 2, the sequence of steps of techniques and the relationship between their procedures are provided.

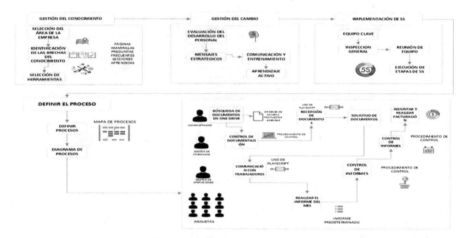

Fig. 2. Specific diagram

3.3.1   Knowledge Management
This technique is applied to evaluate the expertise of an organization and the importance of new knowledge. It is necessary to know the area and the personnel that will be involved in the management model. It will be divided into stages for correct implementation.

First stage: Selection of the organization area

Second stage: Identification of knowledge gaps

Third stage: Tool selection

3.3.2   Change Management
This technique is applied to evaluate and to create a contingency plan to deal with changes in work activities of the organization's personnel. It is divided into the following stages.

First stage: The activities performed by each worker should be evaluated and a comparison should be made with the activities modified for the process.

Second stage: A constant communication should be maintained, which allows the workers to express their doubts and queries about the process to be developed.

Third stage: Training should be provided on the change management strategy for the correct implementation of 5S and process management. A training procedure should be applied. See Fig. 3

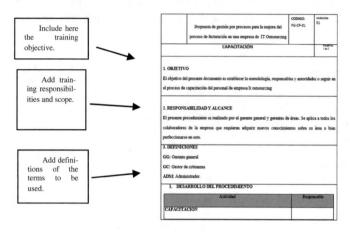

Fig. 3. Procedures

Fourth stage: Conduct a simulation of the development of new activities. This simulation will allow us to observe any changes in workers' behavior.

3.3.3   5S Technique

The 5S technique will be implemented to promote continuous improvement in organizations of the service industry. For proper implementation, the following stages must be followed:

First stage: A key team will be established and it will be in charge of adequately controlling the development.

Second stage: A general inspection of the chosen area of the organization must be made. A format will be used to analyze the current situation and a 0 to 10 score will be applied.

Third stage: Meetings of the work team should be held to discuss the budget, leaders of each activity, and presentation of inspection results. Therefore, a format to collect the information obtained during these meetings should be established.

Fourth stage: In this stage, the 5S technique steps should be implemented:

Seiton-Sorting: The work area should only have the necessary tools.

Seiton-Organization: It entails ordering the elements according to their importance to the organization.

Seiso-Cleaning: It refers to cleaning the work areas of the company.

Seiketsu-Standardization: It refers to regulating the work of the first three stages. For this stage, an improvement panel will be used to observe the changes. See Fig. 4.

Shitsuke-Discipline: It means to respect the previous stages and include them in the daily routine.

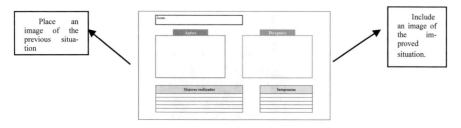

Fig. 4. Improvement control

3.3.4 Process Management

This technique will be applied to establish a sequence of process activities using graphical tools, procedures, and indicators. Like the previous techniques, they will be divided into steps. It is necessary to identify the inputs, outputs, and resources of the model. In addition, the critical factor of each process activity must be identified. Therefore, the use of a SIPOC diagram should be considered, which also shows the procedures that were established. See Fig. 5.

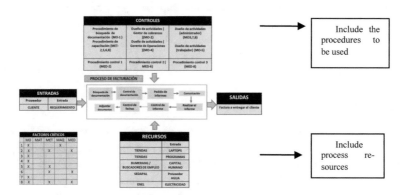

Fig. 5. SIPOC Diagram

3.4 Model Implementation Process

In the following flowchart, the application of the model can be observed starting with the evaluation of the process up to the simulation of the process with new activities. See Fig. 6.

3.5 General Indicators

- **Overtime:** Hours used to complete the tasks that were not done during the established time.
- **Personnel turnover index:** Percentage of people leaving an organization, without counting retired or dead employees.

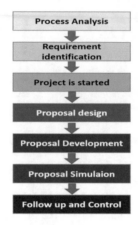

Fig. 6. Process view

- **% Unfulfilled Orders:** Percentage of orders not attended due to process inefficiencies.
- **Supply capacity:** Increase the capacity supplied by the company (increase services).

4   Validation

4.1   Case Study

The case study to be developed is an IT outsourcing company that provides human capital services to different financial entities, as well as retail and telephone companies. The organization does not have standardized processes to provide its services, that is, it does not complete the service development flow. An analysis was performed of its operational, support, and strategic processes to identify the area where the model should be applied. In this case, the case study will focus on the billing process area, since the services billed each month are not completely provided.

4.2   Current Diagnosis

Figure 7 shows that the company does not invoice approximately S/.30,000 each month.

4.3   Application of the Proposed Model (Simulation)

The validation process of the proposed model is shown in Fig. 8.

Fig. 7. Total non-billed services

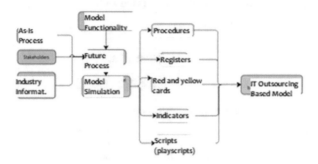

Fig. 8. Flowchart of the model

Table 2. Comparison of indicators

Indicator	Changes	Color Codes		Current Situation	Situation after improvements
		Red	Green		
Overtime	Decreased	< 12.5	= 12.5	38%	8%
Staff Turnover Index	Decreased	< 5%	= 5%	12%	0%
% Unfulfilled Orders	Decreased	< 100%	= 0%	40%	0%
Supply Capacity	Increased	< 10	10	6	11

To validate the current process, the process entities, attributes, and activities should be taken into account. The scenario of the proposed process adds the following improvement activities that are developed.

- Control 1
- Control 2
- Control 3
- Search for documents

Finally, based on the results of the proposed model, the conclusion is that there is an improvement in the billing process efficiency based on the following reasons:

Table 2 shows a comparison between current indicators and those proposed after the implementation of the model. As can be seen, a reduction in overtime, personnel turnover index and % of unfulfilled orders was achieved along with an increased supply capacity. Moreover, the monthly income increased to S/. 110,000, an increase of 80%. See Fig. 9.

Fig. 9. Monthly income

5 Conclusions

- The proposed model utilizes process management, 5S, knowledge management, and change management and can be applied to any specific area of a company area or other entities.
- The integration of Process Management, 5S, Knowledge Management and Change Management in our case study allowed us to increase the company's income by 70%.
- The number of unfulfilled orders was reduced by 100% (rejected invoices).
- Turnover of billing staff was reduced by 100%.
- A reduction of 30% of overtime was achieved in the billing process.

References

1. Mesias, M.: Outsourcing: 86% de empresas en Perú tercerizan servicios, Economía | Gestion. https://gestion.pe/economia/outsourcing-86-empresas-peru-tercerizan-servicios-232422. Accessed 29 Sept 2018
2. Gestión, D.: Outsourcing: ¿Cómo va la tercerización en Perú y el mundo? (2018)
3. Ferrer, M.: Raúl Marcelo Ferrer Dávalos 1(11), 102–114 (2015)

4. Torres Soler, L.C.: lctorress@gmail. co. (n.d.). http://search.ebscohost.com/login.aspx?
direct=true&db=aps&AN=131080953&lang=es
5. Rojas Pescio, H. G. grojas@ureus. c., & Roa Petrasic, V. A. veronica. roa@usach. c. (n.d.).
http://search.ebscohost.com/login.aspx?direct=true&db=aps&AN=131080954&lang=es
6. Baklizky, M., Fantinato, M., Sun, V., Thom, L.H., Hung, P.K.: Business process point
analysis: survey experiments. Bus Process Manage. J. **23**, 399–424 (2017)
7. Fleaca, B., Corocaescu, M.: Process map to create added value to customer based on quality
deployment function. In: 17th International Multidisciplinary Scientific GeoConfer-
ence SGEM 2017, 17 (53, SGEM2017 Conference Proceedings, ISBN 978-619-7408-10-
2/ISSN 1314-2704), pp. 657–664 (2017). https://doi.org/10.5593/sgem2017/53/S21.081
8. Jaca, C., Viles, E., Paipa-GaLeano, L., Santos, J., Mateo, R.: Learning 5S principles from
japanese best practitioners: case studies of five manufacturing companies. Int. J. Product.
Res. **52**, 4574–4586 (2014). https://doi.org/10.1080/00207543.2013.878481
9. Venkateswaran, S., Nahmens, I., Ikuma, L.: Improving healthcare warehouse operations
through 5S. IIE Trans. Healthc. Syst. Eng. **3**, 240–253 (2013). https://doi.org/10.1080/
19488300.2013.857371
10. Peña-Borrego, M.D., Rodríguez Fernández, R.M., Almaguer Pérez, N.A., Peña Rueda, Y.F.,
Infante, S.Z.: Gestión del conocimiento sobre biofertilizantes a nivel local: estudio de caso
municipio calixto garcía, cuba. Knowledge Management on the Use of Biofertilizers at Local
Level: Case Study in Calixto Garcia Municipality, Cuba. **39**, 41–50 (2018) http://search.
ebscohost.com/login.aspx?direct=true&db=fua&AN=130776134&lang=es&site=ehost-live

A Back-End Method Realizing the Ergonomic Advantage of Modularized Battery Systems: ICT Equipment, Electric Vehicles and Beyond

Victor K. Y. Chan[✉]

School of Business, Macao Polytechnic Institute, Rua de Luis Gonzaga Gomes,
Macao Special Administrative Region, China
vkychan@ipm.edu.mo

Abstract. Whereas the author's charging and discharging mechanism for modularized battery systems and the associated mathematical trade-off models proposed earlier minimize the human efforts in unloading/loading/(re)charging discharged modules, they per se cannot fully realize the potential ergonomic advantage of modularized battery systems in that considerable time is still necessitated for recharging the unloaded, discharged modules. Such time can apparently be obviated if unloaded, discharged modules are traded and swapped for some already fully charged modules at some battery swapping and charging stations. Battery module trading must be supported by a method to estimate the modules' different remnant energy storage capacities (hereinafter, capacities) so that such disparity can be financially offset by the trading parties. This article delineates such an estimation method, which comprises *Process 1* to calculate an indicator of a module's capacity and *Process 2* to estimate the module's capacity based on the indicator and some other parameters of the module.

Keywords: Battery ergonomics · Battery modularization ·
Remnant energy storage

1 Introduction

For decades, batteries have been deployed for powering electrical and electronic equipment. Batteries' performance (especially, if not exclusively, their charging speed [1], cycle life [2], power density [3], energy density [4], capital cost [5], and the capital cost of their charging infrastructures [6]) in powering mobile information and communication technologies (ICT) equipment (especially, mobile phones, tablet computers, etc.) and electrical vehicles is exceptionally topical in recent years given the currently overwhelming public focus on mobile e-commerce and environmentalism worldwide. Mobile ICT equipment enables mobile e-commerce whilst electrifying vehicles substantially alleviates air pollution and thus benefits the environment [7, 8].

In the same series of publications, the author presented a patented charging and discharging mechanism for modularized battery systems and proposed mathematical trade-off models to optimize the extent of modularization for the ultimate sake of the ergonomics of the battery modules' (re)charging in respect of minimizing the daily human efforts in unloading, loading, and/or (re)charging discharged modules [9, 10].

© Springer Nature Switzerland AG 2020
T. Ahram (Ed.): AHFE 2019, AISC 965, pp. 322–328, 2020.
https://doi.org/10.1007/978-3-030-20454-9_33

Nevertheless, the aforesaid mechanism per se cannot fully realize the potential ergonomic advantage of modularized battery systems. Whereas the mechanism minimizes human efforts in unloading, loading, and/or (re)charging the modules, considerable time is still necessitated for recharging the unloaded, discharged modules before reloading and enabling them to resume powering the equipment concerned. In contrast, an alternative practice of swapping the unloaded, discharged modules for some already fully charged modules enables virtually continuous powering of the equipment [11] but is beset by the extra cost of spare battery modules for replacing any discharged ones in the equipment concerned. This extra cost is partly avertable through infrastructures of battery swapping and charging stations (BSCSs) [11] around town if such BSCSs operate the business model in which battery users trade discharged modules with the BSCSs for fully charged ones, and the BSCSs recharge the discharged modules so "traded in" before "trading them out" subsequently to other battery users.

Unfortunately, battery module trading is in turn plagued by the disparity of the modules' remnant energy storage capacities (hereinafter, capacities) upon the modules having undergone different degrees of aging, been charged/discharged to different states of charge for different numbers of cycles and thus been degraded to different degrees. In fact, such a capacity of each module dictates how well the module powers equipment any further and thus the module's economic value [12]. Therefore, in any trading or swap transactions, estimation of such capacities of all modules involved is economically essential such that the disparity of the modules' capacities can be offset or compensated by monetary payments between the trading parties.

This article delineates a battery capacity estimation method, which is now covered by another patent and comprises the following two processes:

Process 1 calculates an indicator (i.e., index or proxy) of a battery module's capacity.

Process 2 estimates the capacity (as opposed to the indicator in *Process 1* above) of the module based on at least one of the parameters of the module below:

- the indicator in *Process 1* above,
- the self-discharge,
- the age since the first charge and/or discharge cycle,
- the previous average depth of discharge, and
- the number of charge and/or discharge cycles undergone so far.

The estimation in *Process 2* for a particular module is accomplishable by, for instance, statistical regression or neural networks given historical data about the capacities and the corresponding parameters in *Process 2* for modules of the same type.

This "back-end" method together with the aforesaid earlier mechanism is supposed to be able to fully realize the potential ergonomic advantage of modularized battery systems by minimizing both the human effort and the time in handling discharged battery modules.

2 A Battery Capacity Estimation Method

It is well empirically proven that the capacity of a battery (or a battery module) declines over its usage life with

- the decline growing alongside with the battery's degree of self-discharge and age [13],
- the rate of decline being affected by the battery's previous average depth of discharge [14], and
- the decline increasing with the battery's number of charge and/or discharge cycles undergone [15]

where the degree of self-discharge of any battery (or battery module) is the percentage decrease in the battery's (or battery module's) state of charge when the battery (or battery module) is open-circuited over a given time period upon the battery's (or battery module's) completion of a charge cycle, and the state of charge of any battery (or battery module) is the ratio of the amount of energy stored in the battery (or battery module) to the maximum amount of energy capable of being stored there.

Despite the various but almost equivalent possible ways to define a battery's (or battery module's) capacity, the capacity throughout this article refers to four battery capacity quantities below:

- the *remnant energy storage capacity* E_C, which is the battery's (or battery module's) maximum amount of energy capable of being stored there at any time or, more precisely, the maximum amount of energy capable of being stored there at any time and being released subsequently,
- the *remnant charge storage capacity* Q_C, which is the battery's (or battery module's) maximum quantity of charge capable of being stored there at any time or, more precisely, the maximum quantity of charge capable of being stored there at any time and being released subsequently,
- the *remnant lifetime total energy storage capacity* E_{CL}, which is the battery's (or battery module's) maximum total amount of energy capable of being stored there over the battery's (or battery module's) remaining usage life or, more precisely, the maximum total amount of energy capable of being stored there and being released subsequently over the battery's (or battery module's) remaining usage life, and
- the *remnant lifetime total charge storage capacity* Q_{CL}, which is the battery's (or battery module's) maximum total quantity of charge capable of being stored there over the battery's (or battery module's) remaining usage life or, more precisely, the maximum total quantity of charge capable of being stored there and being released subsequently over the battery's (or battery module's) remaining usage life.

In order to estimate these four capacity quantities, this article proposes a method composed of two processes as follows:

Process 1.
At a first point in time during any battery's (or battery module's) charge cycle, an initial state of charge S_1 is measured and recorded. Likewise, at a subsequent second point in time during the charge cycle, a final state of charge S_2 is measured and recorded.

Concurrently, the amount of energy E inputted into and the quantity of charge Q having flown within the battery (or battery module) between the first and second points in time are also measured and recorded. All these states of charge S_1 and S_2, energy input E, and charge flow Q are readily measurable. An indicator I_E of the battery's (or battery module's) E_C, which can later serve as a parameter for the estimation of E_C, and an analogous indicator I_Q of the battery's (or battery module's) Q_C, which can later serve as a parameter for the estimation of Q_C, are calculated as follows:

$$I_E = \frac{E}{S_2 - S_1} \tag{1}$$

$$I_Q = \frac{Q}{S_2 - S_1} \tag{2}$$

Process 2.
One or more of the following parameters of any battery (or battery module) are measured/calculated and recorded:

- the degree of self-discharge S_D,
- the time difference between the timestamp of the battery's (or battery module's) first charge or discharge cycle and the current time, i.e., the age A_G,
- the previous average depth of discharge D_P, and
- the number of charge and/or discharge cycles undergone C_D

where the depth of discharge of any battery (or battery module) is equal to one minus the battery's (or battery module's) state of charge.

Then, an estimate $\hat{E}_C$ of E_C for any battery (or battery module) in general is equated to a chosen function $f_{EC}(\cdot)$ of one or more parameters I_E, S_D, A_G, D_P, and C_D, an estimate $\hat{Q}_C$ of Q_C for any battery (or battery module) in general is equated to a chosen function $f_{QC}(\cdot)$ of one or more parameters I_Q, S_D, A_G, D_P, and C_D, an estimate $\hat{E}_{CL}$ of E_{CL} for any battery (or battery module) in general is equated to a chosen function $f_{ECL}(\cdot)$ of one or more parameters of either E_C or I_E, S_D, A_G, D_P, and C_D, and an estimate $\hat{Q}_{CL}$ of Q_{CL} for any battery (or battery module) in general is equated to a chosen function $f_{QCL}(\cdot)$ of one or more parameters of either Q_C or I_Q, S_D, A_G, D_P, and C_D. The chosen functions $f_{EC}(\cdot)$, $f_{QC}(\cdot)$, $f_{ECL}(\cdot)$, and $f_{QCL}(\cdot)$ here refer to mathematical functions chosen for the purpose of this *Process 2*, each function having some constant[1] parameter(s) (other than E_C, I_E, Q_C, I_Q, S_D, A_G, D_P, and C_D) being unknown and to be estimated subsequently and once or all in this *Process 2*.

These unknown constant parameter(s) of functions $f_{EC}(\cdot)$, $f_{QC}(\cdot)$, $f_{ECL}(\cdot)$, and $f_{QCL}(\cdot)$ are then estimated so that these functions can ultimately be leveraged to compute the estimates $\hat{E}_C$, $\hat{Q}_C$, $\hat{E}_{CL}$, and $\hat{Q}_{CL}$ of E_C, Q_C, E_{CL}, and Q_{CL} respectively for any particular

---

[1] More precisely, these "constant" parameter(s) are only relatively constant vis-à-vis E_C, I_E, Q_C, I_Q, S_D, A_G, D_P, and C_D and can be adaptive in nature and subject to update depending on the prediction model to be adopted.

battery (or battery module). The estimation of the unknown constant parameter(s) of, for example, $f_{EC}(\cdot)$ begins with the considerably large-scale data collection for past, known cases or observations 1, 2, …, n of one or more (most likely) batteries (or battery modules) at one or more (most likely) points in time, the data covering E_C and at least one of the parameters I_E, S_D, A_G, D_P, and C_D for each battery (or battery module) at each point in time. Then, a prediction model may come into play. Taking the example of employing statistical regression (among other plausible prediction models) as the prediction model for E_C and assuming without loss of generality that all the parameters I_E, S_D, A_G, D_P, and C_D (as opposed to a subset thereof) are included in the data collection, the estimate $\hat{E}_{C,i}$ of E_C for a particular case i is given by:

$$\hat{E}_{C,i} = f_{EC}\left(I_{E,i}, S_{D,i}, A_{G,i}, D_{P,i}, C_{D,i}; \beta_1, \beta_2, \ldots, \beta_p\right) \tag{3}$$

where $I_{E,i}, S_{D,i}, A_{G,i}, D_{P,i}$, and $C_{D,i}$ are respectively the values of I_E, S_D, A_G, D_P, and C_D for the particular case i, β_1, β_2, …, β_p are unknown regression coefficients corresponding to the aforementioned unknown constant parameters of $f_{EC}(\cdot)$, which are real numbers, and p is a natural number. Then, an optimization algorithm is applied to minimize $\sum_{i=1}^{n} g\left(\left|E_{C,i} - \hat{E}_{C,i}\right|\right)$ so as to obtain the optimal estimates $\hat{\beta}_1, \hat{\beta}_2, \ldots, \hat{\beta}_p$ of β_1, β_2, …, β_p where $E_{C,i}$ is the value of E_C for the particular case i, g: $\mathbb{R} \rightarrow \mathbb{R}$ is a monotonically increasing function, and $\mathbb{R}$ is the real number set. The aforesaid optimization algorithm can be traditional calculus, genetic algorithm, or otherwise. By now, the unknown constant parameters β_1, β_2, …, β_p of $f_{EC}(\cdot)$ can assume these known optimal estimates $\hat{\beta}_1, \hat{\beta}_2, \ldots, \hat{\beta}_p$, and the corresponding prediction model becomes:

$$\hat{E}_{C,i} = f_{EC}\left(I_{E,i}, S_{D,i}, A_{G,i}, D_{P,i}, C_{D,i}; \hat{\beta}_1, \hat{\beta}_2, \ldots, \hat{\beta}_p\right) \tag{4}$$

Please note that this part of this *Process 2* to estimate the unknown constant parameter (s) of $f_{EC}(\cdot)$ needs to be performed only once to arrive at (4). Once obtained, (4) can readily (and probably repeatedly) be utilized to compute the estimate $\hat{E}_C$ of E_C for any particular battery (or battery module) in general as detailed in other parts of this *Process 2*.

For any particular battery (or battery module) i_1 at hand, the (best) estimate $\hat{E}_{C,i_1}$ of the battery's (or battery module's) *remnant energy storage capacity* E_{C,i_1} is given by

$$\hat{E}_{C,i_1} = f_{EC}\left(I_{E,i_1}, S_{D,i_1}, A_{G,i_1}, D_{P,i_1}, C_{D,i_1}; \hat{\beta}_1, \hat{\beta}_2, \ldots, \hat{\beta}_p\right) \tag{5}$$

upon substituting $I_{E,i_1}, S_{D,i_1}, A_{G,i_1}, D_{P,i_1}$, and C_{D,i_1} into $I_{E,i}, S_{D,i}, A_{G,i}, D_{P,i}$, and $C_{D,i}$ of (4).

The unknown constant parameters of the remaining functions $f_{QC}(\cdot)$, $f_{ECL}(\cdot)$, and $f_{QCL}(\cdot)$ can analogously be estimated as can the corresponding prediction models for the battery's *remnant charge storage capacity* Q_C, *remnant lifetime total energy storage capacity* E_{CL}, and *remnant lifetime total charge storage capacity* Q_{CL}. It is noteworthy that the prediction model for Q_C has I_Q, S_D, A_G, D_P, and C_D as the parameters of $f_{QC}(\cdot)$, the prediction model for E_{CL} has either E_C or I_E, S_D, A_G, D_P, and C_D as the parameters of $f_{ECL}(\cdot)$, and the prediction model for Q_{CL} has either Q_C or I_Q, S_D, A_G, D_P, and C_D as the parameters of $f_{QCL}(\cdot)$. Prediction models other than statistical regression can also be alternatively adopted upon some adaptation to the above way of estimation.

3 Discussion

This article proposes a battery capacity estimation method particularly to estimate four battery capacity quantities, namely the *remnant energy storage capacity* E_C, the *remnant charge storage capacity* Q_C, the *remnant lifetime total energy storage capacity* E_{CL}, and the *remnant lifetime total charge storage capacity* Q_{CL} for any particular battery (or battery module). The method consists of two processes, *Process 1* being to calculate an indicator I_E of any battery's (battery module's) E_C and another indicator I_Q of the battery's (battery module's) Q_C according to (1) and (2), and *Process 2* being to estimate E_C, Q_C, E_{CL}, and Q_{CL} for any particular battery (or battery module) based on the values of some subsets of parameters I_E, I_Q, and others for the battery (or battery module). Such estimation can make use of various prediction models relating the values of E_C, Q_C, E_{CL}, Q_{CL}, to the aforesaid parameters in a large number of past, known cases or observations of one or more batteries (or battery modules) at one or more points in time. This article demonstrates a prediction model adopting statistical regression to estimate *remnant energy storage capacity* E_C, ending up with (4) to estimate $\hat{E}_{C,i}$ for any particular battery (or battery module) i in general. Prediction models for Q_C, E_{CL}, and Q_{CL} and prediction models adopting anything other than statistical regression should work similarly.

This battery capacity estimation method enables real-world, commercial trading of batteries (or battery modules) in that only with such a back-end method can disparity in the capacities as estimated as such between the batteries (or battery modules) being traded be financially offset by the trading parties through, for example, monetary payments. Most importantly, such a back-end method can support the realization of the potential ergonomic advantage of author's and a co-inventor's patented charging and discharging mechanism for modularized battery systems.

The only major outstanding step towards the above end is to collect data in the aforementioned large number of past, known cases or observations of one or more batteries (or battery modules) at one or more points in time, which is in essence practicable only through the implementation of an automated system to autonomously measure/calculate and record E_C, Q_C, E_{CL}, Q_{CL} and such parameters as I_E, I_Q, S_D, A_G, D_P, and C_D during the batteries' (or battery modules') normal charge and discharge cycles. The author is planning for such a step of implementation and subsequent automated data collection and is hopeful about reporting in further publications the progress in the foreseeable future.

References

1. Hsieh, G.-C., Chen, L.-R., Huang, K.-S.: Fuzzy-controlled Li-ion battery charge system with active state-of-charge controller. IEEE Trans. Ind. Electron. **48**, 585–593 (2001)

2. Omar, N., Monem, M.A., Firouz, Y., Salminen, J., Smekens, J., Hegazy, O., Gaulous, H., Mulder, G., Van den Bossche, P., Coosemans, T.: Lithium iron phosphate based battery – assessment of the aging parameters and development of cycle life model. Appl. Energy **113**, 1575–1585 (2014)
3. Shousha, M., McRae, T., Prodić, A., Marten, V., Milios, J.: Design and implementation of high power density assisting step-up converter with integrated battery balancing feature. IEEE J. Emerg. Sel. Top. Power Electron. **5**, 1068–1077 (2017)
4. Liu, Q.-C., Liu, T., Liu, D.-P., Li, Z.-J., Zhang, X.-B., Zhang, Y.: A flexible and wearable lithium-oxygen battery with record energy density achieved by the interlaced architecture inspired by Bamboo slips. Adv. Mater. **28**, 8413–8418 (2016)
5. Wood III, D.L., Li, J., Daniel, C.: Prospects for reducing the processing cost of lithium ion batteries. J. Power Sources **275**, 234–242 (2015)
6. Yilmaz, M., Krein, P.T.: Review of battery charger topologies, charging power levels, and infrastructure for plug-in electric and hybrid vehicles. IEEE Trans. Power Electron. **28**, 2151–2169 (2013)
7. Fichter, K.: E-commerce: sorting out the environmental consequences. J. Ind. Ecol. **6**, 25–41 (2003)
8. Shiau, C.-S.N., Samaras, C., Hauffe, R., Michalek, J.J.: Impact of battery weight and charging patterns on the economic and environmental benefits of plug-in hybrid vehicles. Energy Policy **37**, 2653–2663 (2009)
9. Chan, V.K.Y.: The modeling of technological trade-off in battery system design based on an ergonomic and low-cost alternative battery technology. In: Ahram, T., Karwowski, W. (eds.) Advances in Human Factors, Software, and Systems Engineering, Advances in Intelligent Systems and Computing, vol. 598, pp. 122–130. Springer, Cham (2018)
10. Chan, V.K.Y.: A generalized ergonomic trade-off model for modularized battery systems particularly for ICT equipment. In: Abram, T.Z. (ed.) Advances in Artificial Intelligence, Software and Systems Engineering, Advances in Intelligent Systems and Computing, vol. 787, pp. 523–534. Springer, Cham (2019)
11. Sun, B., Tan, X., Tsang, D.H.K.: Optimal charging operation of battery swapping and charging stations with QoS guarantee. IEEE Trans. Smart Grid **9**, 4689–4701 (2018)
12. Neubauer, J., Pesaran, A., Williams, B., Ferry, M., Eyer, J.: A techno-economic analysis of PEV battery second use: repurposed-battery selling price and commercial and industrial end-user value. SAE Technical paper, SAE 2012 World Congress & Exhibition (2012)
13. Peterson, S.B., Michalek, J.J.: Cost-effectiveness of plug-in hybrid electric vehicle battery capacity and charging infrastructure investment for reducing US gasoline consumption. Energy Policy **52**, 429–438 (2013)
14. Fernández, I.J., Calvillo, C.F., Sánchez-Miralles, A., Boal, J.: Capacity fade and aging models for electric batteries and optimal charging strategy for electric vehicles. Energy **60**, 35–43 (2013)
15. Han, X., Ouyang, M., Lu, L., Li, J., Zheng, Y., Li, Z.: A comparative study of commercial lithium ion battery cycle life in electrical vehicle: aging mechanism identification. J. Power Sources **251**, 38–54 (2014)

Conceptualizing a Sharing Economy Service Through an Information Design Approach

Tingyi S. Lin[(✉)]

Design Department, National Taiwan University of Science and Technology, 43,
Keelung Rd., 106, Taipei City, Taiwan
tingyi.desk@gmail.com

Abstract. With the rapid change in technology and the popularity of online-
offline activities, the applications of Internet of Things allows various objects,
things and services to connected to each other anytime, anywhere. The efficiency and the effectiveness of information enhances both production and
management. Better information design motivates users and enriches their
experience. This study aims to build a sharing economy service network through
a visual information approach. Through the processes of inventory, thinking,
planning and building, the research team searches for the touch points, gaps and
opportunities throughout the user's journey in order to understand the current
situation and design a new service. The relationship between stakeholders and
visitors/users is visualized during the analytical phase. Visual information
viewpoints are applied for design development and for preparing service guild
kits.

Keywords: Sharing economy · Visual information · IoT · Service

1 Introduction

The concept of sharing economy is enabling existing resources to be effectively used
and reused. By turning the ownership from "right to use" to "use it right", we can
synergize a new value for best results through a designed connectivity. Although the
concept of sharing economy has its long history that relates to the exchange of physical
goods and services, the term began to appear in early 2000s as an emerging business
structure due to the Great Recession [1]. PR Newswire in its CISION column (2014)
points out that Harvard Law School Professor Lawrence Lessig first used the term
'sharing economy' in 2008 [2]. He was describing the idea regarding to transactions
'for lending and borrowing rather than purchase and ownership'. By 2011, TIME
magazine name the term 'collaborative consumption' to be one of ten ideas that would
change the world [3]. Throughout those years, the rapid change in technology and the
popularity of online-offline activities are continuously amplifying possibilities and
opportunities in creating new business models. The application of Internet of Things
allows for various objects, things and services to be able to connect to one another
anytime, anywhere.

The network of Internet of Things often relies on visualized information to introduce, to communicate, and persuade users. The efficiency and the effectiveness of

© Springer Nature Switzerland AG 2020
T. Ahram (Ed.): AHFE 2019, AISC 965, pp. 329–335, 2020.
https://doi.org/10.1007/978-3-030-20454-9_34

information enhances both production and management. A better information design increases the users' motivation and enriches their experience. This study aims to build a sharing economy service network through a visual information approach. In order to construct innovative service, this paper documents the way to integrate an information design method to analyze and to optimize the processes.

2   Visual Thinking and Planning

Visual optimization plays a large role in this information era to make sure information is seen and understood. It is in great demand due to much larger data we create and information we receive on a daily basis New technology is taking us into an information era where we can receive information anytime, anywhere. With updated technology, we can also represent and perceive the living world through images, screens and scripts two- or three- dimensionally. Visual optimization here means not only paying attention to the ease of receiving, but also considering the ease of understanding—such as the ways in which the information is presented, which media to use, how to tell the story and how users can interact with it. The "inscribed surfaces" of two-dimensional visual representations remain "major principles of modern technology" [4], and the skill of visual thinking allows us to discover tangible and intangible objects, reasonable and unreasonable situations as well as interconnectivity and relationships. Patterns, relationships and solutions can be shown through an information design approach, with which the connection of existing iconicity, visuality, spatiality and graphism offers visual clues and signals.

The way users access information and dependencies is according to (1) the goals that users are seeking, (2) which reading mode users are in, (3) what kind of information provider it is and (4) how the content is prepared to be told. Through the processes of inventory, thinking, planning and building, the research team searches for the touch points, gaps and opportunities throughout the user's journey in order to understand current situations and design a new service. Tests and revisions are underway when submitting this paper. In order to provide transparent and understandable information, to guide and to encourage users' participation, to enrich their experiences and to assist them in making their decision right, the ultimate goal of this experiment is to build a sharing economy service and network for small-scale farmers in Puli Township through a visual information approach.

Few sharing economic business model such as Airbnb and Uber have successfully expanded in more than sixty countries around the world. Such new models are not only creating fresh commercial outputs, but are also reducing the idleness of resources. Moreover, such innovated service and/or business are providing more choices for users by the exchange of resources between individuals, business to business (B2B), as well as business to customers (B2C). Based on the process of resource circulation, Botsman and Rogers identified three kinds of sharing economies including Productive-Service Systems, Redistribution Markets, and Collaborative Lifestyles [5]. Information technology, social media, social commerce, urban lifestyle and rising costs are the driving forces behind most ever-growing startups. A new type of business concept is rising from the idea of 'collaborative consumption' with which to create mutual benefits to stakeholders through the power of the Internet of Things [3, 6]. This trend will increase circulation, reactivate the idleness and create new market opportunities (Fig. 1).

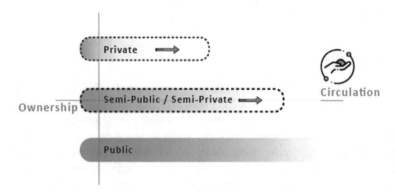

Fig. 1. The trend of sharing economy will increase circulation, reactivate the idleness and create new market opportunities through the concept of 'collaborative consumption'.

3 The Experiment

The successful application of technology creates collaboration between humans and machines. The idea of 'collaborative consumption' creates mutual benefits to participants and stakeholders through the power of Internet of Things. This new type of business concept facilitates life onward. This experiment is to energize small-scale agriculture not only to renovate the business but also to reserve their goodwill in natural farming. Reflecting upon previous experience in boosting market venues by connecting on and off-line services, the research team plans to develop a platform to connect seven farmers in Puli Township, Nantou County, Taiwan. Located in the mountainous center of Taiwan, Puli earns its reputation from mild weather and clean water. With its basin landscape protecting it from typhoons, the popular tourist spots include a brewery, paper factory and monastery. Adopting modern agricultural technology, Puli is also famous for growing high value crops such as flowers, mushrooms, sugarcane, white radish, passion fruit, water bamboo and others. The platform is established under a core value of sharing. Its purpose is to connect member farmers for discussions, exchanging resources and sharing ideas. Later, it will reach out to visitors for introduction, explanation, and promotion. With an innovative and visual thinking viewpoint, this collaborative consumption model will not only seek the common good between current farmer members but also create a welcoming environment for young people's willingness to return home.

Methods. In order to build a sharing economy service network through a visual information approach, this paper documents the way to comply an information design method to analyze and to optimize the processes – (1) explore current online and offline interactivities by observation and semi-structured interview; (2) define touchpoints and breakthroughs from the current relationship structure by mind-mapping and brainstorming; and (3) construct a new service module to link multiple entities through prototyping and in-group discussion. The module will be tested in the next stage. Further revision will fulfill defined touchpoints and breakthroughs, so as to strengthen the relationship and collaboration between stakeholders.

4   Results and Analysis

Through the processes of inventory, thinking, planning and building, the research team searches for the touch points, gaps and opportunities throughout the user's journey in order to understand the current situation and design a new service. Semi-structured interviews were conducted in the inventory stage to understand seven member farms' current production and service, their needs, expectations, opportunities and potentials. Figure 2 shows the initial platform structure that was deployed by understanding the surrounding environment and knowing stakeholder's demands. A short introduction will take users to a main page with graphics and a daily message will provide useful information in a brief amount of time. Updated networking activities and technological information that will be shared between member farmers will energize and connect them together for co-creation and collaboration. More detailed and individual information regarding each farm (such as introduction, know-how, tillage skill, workshop agenda…etc.) are structured under each individual session. However, the workshop announcement, the agenda and the registration links are connected between the main page and individual session. This arrangement brings a welcoming feel for participation and makes it easy to access.

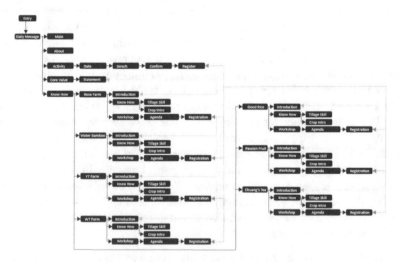

Fig. 2. An initial platform structure was deployed by understanding the surrounding environment and knowing stakeholders' demands by observation and semi-structured interviews.

Fig. 3. Brainstorming session is to collect feedback, reaction, and opinions for defining touch points, finding gaps and searching for opportunities within the initial platform structure.

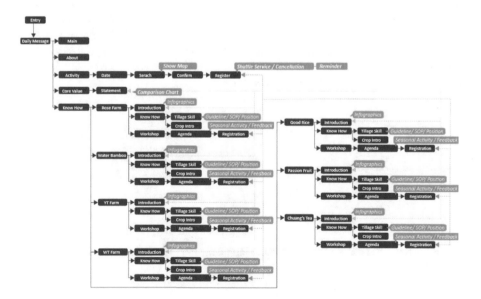

Fig. 4. More considerations need to be added on. Through the visual analysis and series of visual thinking process, solutions were raised to fulfill those considerations.

In order to define touch points, to find gaps and to search for opportunities within the initial platform structure, the research team conducted brainstorming sessions to collect feedback, reaction, and opinions from three designers who are in information design and service design fields (Fig. 3). Data was documented and analyzed via the thinking-planning-building processes. Several considerations and solutions were raised through this visual information approach—(1) show a map to assist users in searching and knowing locations and the relationship between them; (2) allow opportunities to change or to cancel the reservation, (3) provide shuttle service information, (4) send mobile and/or email reminders, (5) provide guideline and standard operating procedure to explain tillage skills, and (6) provide infographics and comparison charts to explain statement and each farm site's introduction instead of merely textual description (Fig. 4).

5   Conclusion

Recession is now a global issue. In order to cope with the result of insufficient demand, reusing and rearranging useful resources and surplus materials by exchange will be the next wave. It is not merely implementing the idea of 'recycling' but rather to 'discover' and 'recreate' new values from the old. An accessible model of a sharing economy concept will rekindle the hope in redefining and recounting the useful resources. With these sharing and exchanging efforts, participants and stakeholders benefit from each other. The altruistic system encourages selfless concern for the well-being of others. This stage is connecting farm to farm for a collaborating system. The relationship between stakeholders and visitors/users is visualized during the analytical phase. Visual information viewpoints are applied for design development and preparing service guild kits. The ultimate goal of this project is to build a service model in view of a shared-economy – so as to implement two-sided service model at a later stage for consumer outreach (Fig. 5). With the maturity of technology, sharing economy becomes one of the irresistible trends for future markets to grow.

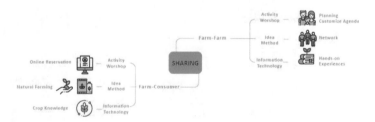

Fig. 5. A two-sided service model – farm to farm and farm to consumer – in view of a shared-economy.

Acknowledgments. Thankful for the research received supports from the Ministry of Science & Technology, ROC, Taiwan, project MOST 107-2410-H-011-020. The appreciation also extends to VIDlab project assistant, Hsing-Ya Tang, who helped up in data collection, mapping for analysis, and discussion together for solutions.

References

1. Stephany, A.: The Business of Sharing: Making It in the New Sharing Economy. Palgrave Macmillan, New York (2015)
2. PR Newswire: Honestay is the origin of sharing economy. https://www.prnewswire.com/news-releases/homestay-is-the-origin-of-sharing-economy-249415791.html. Accessed 27 Jan 2019
3. Walsh, B.: Today's smart choice: don't own share. In: 10 ideas that will change the world, TIME Magzine, March 17, 2011
4. Krämer, S.: Trace, writing, diagram: reflections on spatiality, intuition, graphical practices and thinking. In: András, B., Nyíri, K. (eds.) The Power of the Image: Emotion, Expression, Explanation, p. 6. Peter Lang, Frankfurt am Main and New York (2014)
5. Botsman, R., Rogers, R.: What's Mine is Yours: The Rise of Collaborative Consumption. Harper Collins Publishers, New York (2010)
6. Zoref, L.: Mindsharing: The Art of Crowdsourcing Everything. Portfolio Publishing, New York (2015)

Community Slow Transportation Service System Based on Driverless Vehicle

Jintian Shi[1](✉) and Rensi Zheng[2]

[1] College of Architecture and Urban Planning, Tongji University,
No. 1239 Siping Road, Shanghai, China
shijintian1017@126.com
[2] Tongji University, No. 281 Fuxin Road, Shanghai, China
zrensi@163.com

Abstract. Currently most of the urban transportation systems are planned to be car-centered. Yet very little is known about the research on slow transportation (also called non-motorized transportation), even though it takes account for more than 50% of the whole urban transportation volume and creates more interactions with passengers than urban fast transportation. Residential community is a typical representative of slow transportation environments, and its mobility should also be valued. As one of the biggest future trend, driverless vehicle will have a great impact on the field of transportation and it was also found to be more likely to perform well at slow transportation environments. So how to build up a community slow transportation service system based on driverless vehicle is the key purpose of this research. In this paper, we will introduce the general design process of the service system and represent some simple design results. Meanwhile, we have concluded some design summary and designing approach: (1) three main design types of community slow transportation service, (2) a Kano demand model-based service optimization tool, and (3) the slow transportation service design architecture.

Keywords: Slow transportation · Driverless vehicle · Service system · Residential community

1 Introduction

Slow transportation, also named non-motorized transportation, mainly refers to a means of transportation with a line speed of no more than 15 km/h [1], including walking, non-motorized transportation and vehicles with low-speed. Slow transportation takes account for more than 50% of the whole urban traffic volume at some Chinese metropolis [2], and its development goes beyond transportation itself, furthermore, has deep connections with energy crisis, environment pollution, social equity, and life style promotion [3]. Similar with the proposition of Slow Travel (an emerging conceptual framework offering an alternative to car travel, where people travel to destinations more slowly overland, stay longer and travel less [4]), slow transportation also concerns about locality, ecology, experience of mobility, and quality of life. Residential community is

© Springer Nature Switzerland AG 2020
T. Ahram (Ed.): AHFE 2019, AISC 965, pp. 336–346, 2020.
https://doi.org/10.1007/978-3-030-20454-9_35

one of the typical urban slow transportation areas [5], to achieve the above concerns, it remains to be innovated in combination with various emerging trends.

Diverse trends and developments may act as impetus to improve slow transportation, such as technological advance, servitization and sharing, social generations, scarcity of resources, and transportation paradigm change. The existing literature suggests that the technological intervention in the development of slow transportation is insufficient. As an inevitable technical trend, manned-vehicles will eventually exit the stage of history in the next few decades, through gradually being replaced by driverless vehicles [6]. It was also found that driverless vehicles may have better performances on service providing at slow transportation environments, such as residential communities, industrial parks, campuses, parking lots, airports [7]. But most of the current slow transportation solutions don't promote this trend and not likely to suffice to meet the requirements of future mobility [8], and how to make driverless vehicle serve better the residential communities also requires a proper designing way.

Represented by driverless technology, some digital innovation in slow transportation tends to concentrate on relatively isolated developments [9], lacking a comprehensive way to integrate different transport aspects into components of an entity. Recently, the notion of MaaS (Mobility as a Service, a system where a comprehensive range of mobility services are provided to customers by mobility operators) has developed in the transportation sector since it was popularized by Heikkilä [8] While, there is still not much discussion on the way how technology can improve slow transportation in the design logic of service system.

Here, we use large residential community as a typical representative of slow transportation environments, to show that how to use service system design thinking to reorganize the transportation offerings, in combination with driverless vehicle as a technological tool. We also present three main design types of slow transportation service, a Kano demand model-based service optimization tool, and the design architecture of slow transportation service system at communities. This paper has made some new attempts and explorations in the following aspects.

Driverless Vehicle-Specific. The iteration and advancement of transportation tools not only provide technical support for many functions, but also change the way people interact with vehicles. The design principles and methods for traditional human-car interaction are not fully applicable to the interaction design and service design of driverless vehicles. And this article will base on the characteristics of driverless vehicles to carry out service innovation in communities.

Service-System Design Thinking. Slow transportation is not a new topic but slow transportation service system is. Some existing studies classify the slow transportation system designing objects from the aspects of space, activity type, and behavior [2, 10]. The service design thinking transforms the single transport aspects to a comprehensive system containing mobility offerings, stakeholders and touchpoints. If regarding transportation service as the final designing object of the overall system, the other objects are the touch points acting as service carrier, and connecting these touch points in a reasonable way can create a smoother transportation service for users.

2 The Redefinition of Community Slow Transportation Service

The transformation of society has led to changes in design paradigm. Currently, product-based manufacturing society is turning to a service-based economic society. In this situation, the essence, mechanism, content, form of slow transportation service design need to be re-explored and even redefined. In addition, users will also generate more diverse slow transportation needs. How to discover emerging future slow transportation needs, and which demands can evolve into slow transportation services based on driverless vehicle, it should begin with the classification and analysis of slow transportation activities in communities.

2.1 Three Main Design Types of Slow Transportation Service

Based on the travel scope and transportation time, urban travels could be classified into urban travel, community travel, neighborhood travel, and basic travel [11]. The discussion scope of community slow transportation is from community travel to basic travel, and with the traffic transfer between urban travel and community travel. In these four kinds of travels, there are hundreds of community transportation activities and we have searched quantities of related service cases. After a period of analysis, it is found that location, collaboration, and relation could act as three hidden impacting factors of most of the activities. And we also conclude three main design types of slow transportation service in communities (Fig. 1).

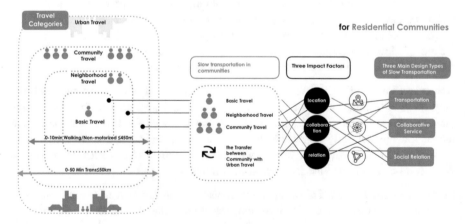

Fig. 1. Urban travel classification and the development procedure of three main design types of slow transportation in residential communities.

2.2 Demand Exploration Based on the Three Design Types

The above three main design types of slow transportation service may act as topics or tools to better explore, expand and discover possible related transportation demands.

Location-Based. The first type is location-based transportation demands and offerings. This kind of transportation tend to be direct point-to-point connection, and the uses always have specific destinations and certain requirements of time efficiency.

Collaboration-Based. The collaboration-based transportation demands and offerings focus on the cooperation between different objects, people with people, people with driverless vehicle, vehicle with vehicle. Taking vehicle with vehicle as an example, in this project, the driverless vehicle is designed to have four different assembled functional modules (Fig. 6). Not only the modules of one vehicle, but also different vehicles can combine or adjust at different using scenarios. This collaboration lying in vehicle with vehicle is helpful to providing more flexible transportation services.

Relation-Based. The third is relation-based demands and offerings, it focuses on discovering some offerings centered on the relations of people with people, people with vehicle, people with community, car with car, and car with community, in return these offerings will also strengthen the relations. Here we present a Human-Robot interaction based transportation offerings in this project. We regard the driverless vehicle as a robot system from the perspective of HRI (Human-Robot Interaction) [12], and there are two human-robot relations one is inside the vehicle, between passenger with the robot system. And the other is outside the vehicle, between pedestrians with the robot system. In the first relation, through smart human-robot interactions which initiated by the system but make users feel unintentional and natural, the robot systems get enough information, data or users' feed to update users' knowledge graph built in the previous, thus understand users more to provide more customized services. The relation between passenger and the robot is interdependent, passengers need the robot system understand themselves more for better services, and the robot system need passengers give it more information or data. In the outside pedestrian-robot relation, it also need effective mutual communication. Pedestrians need to know whether the robot system identifying them or other operation status to avoid being bumped or to take a ride, and the traveling vehicle also need to commute its status to pedestrians for removing people's fears. This relation claims building mutual-trust and effective design for communication mechanism and form.

Based on these three design types of slow transportation service, we eventually explore several possible demands and related directions, preparing for the next step of building a functional framework based on the demands.

3   Determining Service Content and Service Optimization Tool

In the last chapter, we get a certain number of slow transportation service scenario and possible directions, based on this, some service offerings could be explored, expanded and discovered. While how to evaluate which kind of transportation requirement is the most urgent one, which demand is just an optional "bonus", which need could be achievable by the service system, how to settle down a functional framework for the whole system, and how to transform needs to service. In this chapter, we will introduce a service optimization approach based on Kano demand Model.

3.1   Four Primary Slow Transportation Demands

Firstly, based on some possible functions and offerings of slow transportation in communities, we use Kano demand Model to classify and optimize the user satisfaction of the functions, thus determining the rough functional framework of the driverless vehicle based slow transportation service system design.

Kano demand model is a tool for classifying and prioritizing user requirements based on analyzing the impact of user needs on user satisfaction, reflecting the non-linear relationship between product performance and user satisfaction. According to the relationship between different types of quality characteristics and customer satisfaction, Kano demand model divides the quality characteristics of product services into five categories: must-be quality, one-dimensional quality, attractive quality, indifferent quality, and reverse quality [13]. Also, Kano demand model was found to be suitable for the building the functional framework in this project: Firstly, the service field based on driverless vehicles is still in the exploration stage. The boundary of the demands and subject should be clear. Kano demand could be helpful in defining the scope of demand. Secondly, Kano demand can help the service system pick out achievable needs through management-related research methods, avoiding meaningless needs.

We use Kano questionnaire, classification of service attributes and related tools to settle down a functional framework of slow transportation service system and figure out four main possible functions to be verified and evaluated (Fig. 2). Then four main functions are concluded as parcel delivery, commute, emergency/First-Aid service and garbage cleaning and collecting.

3.2   Slow Transportation Service Optimization Tool

Kano model focuses on the analysis users' satisfaction of a certain product or service, but for service system design, the user-centered thinking cannot solve all the problems. But in service system design, service providers and other stakeholders are also important components of the whole system, the high efficiency of the system is the purpose. So, service system design doesn't adopt the solution in exchange of satisfying service receivers' needs by remising service providers' interests. Taking account of the demands of both service providers and service receivers, providing high-performance services through effective system at as low as possible cost, the win-win design principle is one

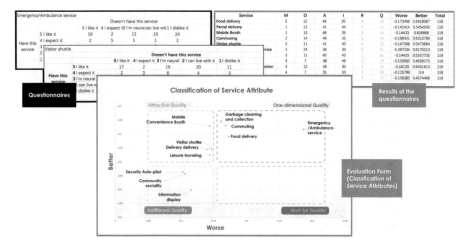

Fig. 2. The procedure of the development for rough functional framework of the slow transportation service system.

of the biggest difference from user-centered design with service system design. The functions or the services concluded from Kano model can highly represent the users' opinions, while from the perspective of the comprehensive system, we need a proper designing tool to evaluate these functions can be achievable to escalate to services at an appropriate cost.

It was found that, service time, system efficiency, and the social cohesion are three key factors of the feasibility analysis for the functions (or services-to be). Based on the previous research, we get a Kano model-based community slow transportation service optimization tool (Fig. 3).

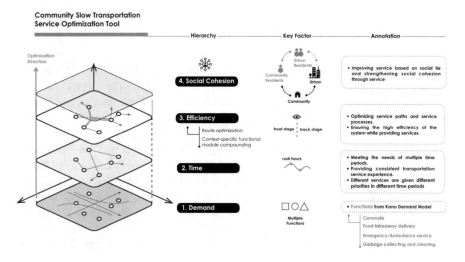

Fig. 3. Community slow transportation service optimization tool and its annotation

Demand. The first layer is the demand hierarchy representing the demands through Kano model.

Time. And the second hierarchy is the time hierarchy, it is to help analyze whether there is a large service time overlap or even conflict between various functions. In Fig. 4, if there is overlap, based on the current time interval, community characteristics, crowd characteristics, etc., it will give different priorities to different services, or remove some non-essential features.

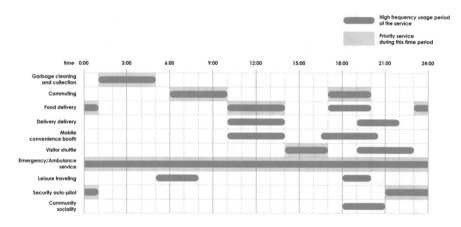

Fig. 4. Different services are given different priorities in different time periods.

Efficiency. The following layer is efficiency hierarchy. System efficiency can be improved from two aspects: route optimization and vehicle functional module compounding. (1) Route optimization. for users who have the same service requirements at the same time, improve service efficiency by optimizing the path or allowing multiple users to enjoy the service at the same time. For example, in the early commute rush hours, multiple users have reserved a commute service, and the system can provide the commuting service of the same or similar route for the users sharing the same time and geographical location, reducing the repetitive path. (2) Context-specific functional module compounding. In this project, the driverless vehicle is designed to have four different functional modules, and each module can be combined. In a certain scenario, different driverless vehicle functional modules could be altered or combined based on user/scenario/system (service priority) identification and analysis. For example, when user A booked the riding service going home in advance, the system found that user his package had been kept in the package cabinet of the vehicle modules for more than ten hours, also found his neighbor had a package to be taken in the cabinet as well (User identification) and the neighbor was at home, because she just ordered a bottle of juice from the "mobile convenience booth". That time was not a peak time of the off-duty, and based on the reservation data of the commuting service, knowing that there was not many commuting or riding demands (Scenario identification). Therefore, the driverless vehicle, under the help of system operator, altered the combination of two commuting modules for the commuting-season one hour before, and transformed into a combination of commuting module with a package cabinet (food takeaway delivery) module (Service priority identification). At the appointed time and at the community gate, the driverless vehicle picked user A up and sent him home with his package, then delivered the other two packages to his neighbors' home.

Social Cohesion. In the provision of these transportation services by driverless vehicles, it should be based on enhancing the social cohesion between different community residents, between community residents and urban residents, between communities and cities, enhancing relationships through services, and optimizing services or enriching services based on social networks.

Through this service optimization tool, from perspective of feasibility, we can have learned which transportation requirements could be advanced for services. From perspective of operability, it is instructive to know from which aspects to optimize the services.

4 Community Slow Transportation Service Design Architecture

In the previous work, we studied the advantages of driverless vehicles in a slow transportation environment and found that many technology-oriented thinking can also reverse the service design improvement. Community slow transportation activities were also studied and summarized into three main design types. Based on these three design types and Kano model, we also summarize the service optimization tool for the community slow transportation, as well as some specific optimization approaches. Based on these research results, we have come up with a community slow transportation service system design architecture in three-dimensions.

Just like Fig. 5 showing, the whole design architecture is divided into five hierarchies, namely, the functional, the structural, the path, the design, and the interaction hierarchy, and these layers are gradually advanced, where in the design layer and the interaction layer can be iterated with each other.

(1) **Functional Layout.** functional layout is performed on different functional partitions in the three-dimensional space of the community. Based on the defined functional framework, if the function to be designed already exists in the existing offerings of the community, it is called the stock function. If some of the features that are being designed are new and are planned to carry services on driverless vehicle as a mobile carrier, this is an incremental function. These stock and incremental functional areas or mobile functional points (driverless vehicles) should be placed on the community map.

(2) **Travel Structure.** The second is the travel structure hierarchy, which mainly refers to location-based, collaboration-based, relation-based transportation activities in communities.

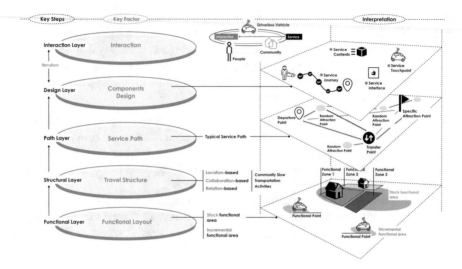

Fig. 5. The community slow transportation service design architecture in three-dimensional containing five different layers.

(3) **Service Path.** First, some typical users (personas) need to be analyzed, and then mark typical places in the community, such as departure points, connection(-transfer) points, specific attraction points, random attraction points, etc. So, we can derive their service paths in the community based on typical users and typical places, then superimposing paths on three dimensions.

(4) **Components Design.** At this stage, some contents need to be cleared: the service content, the service procedure, service interface, the service touchpoints and their location, etc.

(5) **Interaction.** The interaction here focuses on the interactions between people with people, people with vehicle, people with the environment. The interaction will optimize and improve the services, and the services will also better support the interaction. The fourth and fifth hierarchy can be iterated with each other to improve the overall system design.

5 Several Typical Community Slow Transportation Services

This design project aims at providing both significant also promising transportation services for the community residents, based on the characteristics of driverless vehicles. As Fig. 6 indicates, there are four typical services, takeaway food delivery, commute, emergency/First-Aid guidance, and garbage collection and cleaning, being elaborated in the further research and design. For these services, the form of the driverless vehicle is also designed to have four different functional modules, and different modules are capable to be combined or disassembled. In different using scenarios, different functional modules can be reasonably combined based on the current number of users and the degree of demand for a certain service.

Fig. 6. Preliminary concept design of the four functional modules for the driverless vehicle.

In the next step, we hope to continue an in-depth research and detailed design of the driverless vehicle-based community slow transportation service system. The research on the corresponding interaction between human and driverless vehicle, interaction between driverless vehicle and vehicle, the human-robot interface, and the interaction mechanism and design between driverless vehicles and pedestrians. After completing this part of the work, we hope to start the research on other typical urban slow transportation environments, such as industrial parks and airports, to discover how to take advantage of driverless vehicle-based service system to benefit more potential stakeholders.

References

1. ScienceNet.cn. http://blog.sciencenet.cn/blog-435786-374226.html
2. Yun, M.-P., Yang, X.-G., Li, S.: A brief review of planning for ped and bike system. Urban Transp. China **7**, 57–59 (2009)
3. Lin, G.: From Berkeley to Davis: towards eco-city via non-motorized transportation. Urban Plann. Int. **27**(5), 90–95 (2012)
4. Dickinson, J., Lumsdon, L.: Slow Travel and Tourism. Routledge, London (2010)
5. Liu, S.: Study on the emotional design of city slow traffic environment—taking the Latitude 7 road in Lanzhou new area as example: Lanzhou Jiaotong University. MsD (2017)
6. Lipson, H., Kurman, M.: Driverless: Intelligent Cars and the Road Ahead. MIT Press, Cambridge (2016)
7. Shi, J., Sun, X.: Driverless vehicle-based urban slow transportation service platform. In: International Conference on Cross-Cultural Design (2018)
8. Heikkilä, S.: Mobility as a service: a proposal for action for the public administration. Case Helsinki (2014)

9. Turetken, O., Grefen, P., Gilsing, R., et al.: Service-dominant business model design for digital innovation in smart mobility. Bus. Inf. Syst. Eng. **61**, 1–21 (2018)
10. Tian, X.: Strategies for designing urban slow traffic system. J. Transp. Inf. Saf. (5), 81–84 (2010)
11. Zhang, J., Li, M.: Research on the mode of the aged living in metropolitan community based on the concept of "1 hour traffic circle". City House **23**(1), 51–55 (2016)
12. Goodrich, M.A., Schultz, A.C.: Human–robot interaction: a survey. Found. Trends Hum. Comput. Interact. **1**(3), 203–275 (2008)
13. Sauerwein, E., Bailom, F., Matzler, K., et al.: The Kano model: how to delight your customers. Int. Work. Semin. Prod. Econ. Innsbr. **1**(4), 313–327 (1996)

Online Service Quality Measurement Utilizing Psychophysiological Responses

Peixian Lu, Lisha Li, and Liang Ma[⊠]

Department of Industrial Engineering, Tsinghua University,
Beijing 100084, People's Republic of China
liangma@tsinghua.edu.cn

Abstract. This study aims to measure the online service quality in real time utilizing psychophysiological responses of customer experience. Instead of using questionnaires, the psychophysiological responses can reflect the service quality in real time. In the experiment, we designed a searching task on the "Xiaomi" website. We measured the objective experience of the searching task including mental workload and emotional experience of customers by measuring their EEG and EDA, respectively. During the experiment, we used Think-A-Loud to obtain the subjective experience of customers. Eye tracker was used to determine the position of the special point they concerning about. The consistent analysis of the data of psychophysiological responses and the data of Think-A-Loud showed a high consistency. The result showed that the psychophysiological responses can be used to measure the user experience of using the service to reflect the service quality.

Keywords: Online service quality · Psychophysiological response · Customer experience

1 Introduction

Service quality plays a central role in the analysis of competitiveness of service industries (e.g. public school and universities, hospitals, and tourism agencies) and private economic companies (e.g. restaurant, outlet stores) [1]. Similar can be said for the online service quality. As e-commerce proliferates, companies are increasingly turning to the Internet to market products and services. However, the effectiveness of such online commerce systems depends on the degree of comfort that customers feel with the technology-based interactions between the customers and companies [2]. So, the online service quality measurement is very important to the customers and service supplier.

This study aims to measure the online service quality in real time utilizing psychophysiological responses from customer experience. For the traditional service quality measurement, according to our survey, several models and methodologies can be employed (e.g. the Total SQ model, Expectations-Disconfirmation model and the SERVQUAL model) [3–7]. When it comes to implement, these theories will need to be transferred into different questionnaires, mainly aiming to measure five aspects of service quality: reliability, tangibility, responsiveness, assurance and empathy [8]. This kind of

© Springer Nature Switzerland AG 2020
T. Ahram (Ed.): AHFE 2019, AISC 965, pp. 347–352, 2020.
https://doi.org/10.1007/978-3-030-20454-9_36

questionnaires measure service quality based on customers' previous service experiences, rather than the feeling of using the service in real time, which may make customers miss some details for quality evaluation [9]. Therefore, assessing service quality of the customers' perspective in real time should be beneficial to overall service quality measurement. If we want to measure the service quality in real time, we should use other method instead of the questionnaires. In this study, we used customers' psychophysiological responses when they are using online service to measure the real-time service quality. As we know, service can be regarded as a kind of special product [10]. As a result, the service quality can be reflected by the user experience of using the service. According to the evaluation model of Mahlke [11], user experience can be assessed from two aspects: mental workload experience and emotional experience. Hence, we can evaluate the service quality by measuring the mental workload experience and emotional experience of customers utilizing their psychophysiological responses.

2 Method

In this study, we recruited 20 participants including 12 males and 8 females. They are all college students aged 18–31. They were required to sign consent forms and read the experiment contents. After this, we helped them to wear the equipment and encouraged them to use Think-A-Loud method to express their feeling in real time during the experiment.

In the experiment, we designed a searching task on the "Xiaomi" website. First, the participants were required to keep tranquillization status by listening to soft music and watching scenery pictures for 4 min before the searching task. Then, the participants were required to find the "customer service interface" webpage in five minutes with a start of the home page of the website. During the searching task, they were reminded to use the Think-A-Loud method to express their operation, emotion and feelings and other thoughts.

We measured the psychophysiological responses during the task including mental workload and emotional experience of customers by recording their EEG and EDA, respectively (see Fig. 1). The EDA was measured by the ErgoLAB polygragh with a sampling frequency of 4 Hz, a transmission distance of 100 m and a weight of 20 g. The EEG was measured and transferred into attention level by the NeuroMedia hardware and software with high robust and noise cancel capacity. During the experiment, we used Think-A-Loud method to obtain the subjective experience of customers, which was used to confirm the validity of the psychophysiological measurement. Eye tracker was used to determine the position of the special concerning point of the participants. The fixation path was shown and recorded in real time during the experiment.

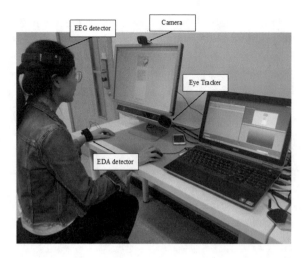

Fig. 1. The experiment scene and equipment used in this experiment.

3   Result and Discussion

3.1   Data Analysis

Three participants were interrupted unexpectedly during the experiment, and data of the left 17 participants can be used as reliable results. For each participant with a reliable result, we obtained the EEG data and EDA data of each webpage. We transferred EEG data into the attention level to represent the mental workload. If the attention level is increasing, it means the mental workload is raising. Figure 2 shows the EDA and attention level (transferred from EEG data) in each webpage for one participant.

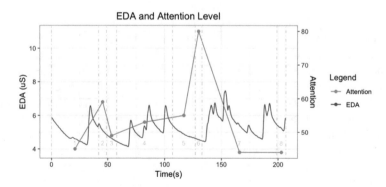

Fig. 2. The EDA and attention level measured on each webpage from one participant. The gray numbers in the figure is the code name of each webpage.

To analyze the attention level of all participants, we obtained the mean attention level of 17 participants on each webpage (see Fig. 3). Results showed that, for the mental workload, their attention level is very high on the webpage 3 (see Fig. 4a). Besides, they also paid much attention on the webpage 4 (see Fig. 4b).

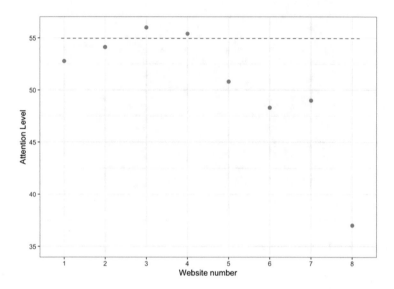

Fig. 3. The mean attention level of 17 participants on each webpage. This figure showed that the mean attention level of the participants is very high on the webpage 3 and webpage 4.

For the emotional experience, we also obtained the mean increased percentage of EDA of 17 participants on each webpage comparing with the tranquillization status (see Fig. 5). We can find that their emotional arousal level become very high on webpage 4 and 5.

According to the results of Think-A-Loud of each participant, we found that, for the attention level, they felt confused on the webpage 3, because there are so many words to read and understand. We also found that they spent too much time staring the words on the webpage 3 according to the gaze data. Besides, they also paid much attention on the webpage 4, because they wondered whether they should login in. These results are consisted with the result of the attention level measured by EEG (see Fig. 3).

For the emotional experience, according to the results of Think-A-Loud of each participant, we found that they felt disgusted on the webpage 4 and 5 (see Fig. 4b), because they think logging in is too bothering. These results are consisted with the result of the EDA level which reflecting their emotional arousal level (see Fig. 5).

From the analysis above, we can find that the psychophysiological responses measured are consisted with the results of Think-A-Loud. This result shows that the psychophysiological responses can be used to measure the user experience of using the service to reflect the service quality. The gaze data measured by eye-tracker can be used to orientate the trouble point of customers as supplementary.

Fig. 4. Webpages on which participants paid much attention. a. Webpage 3. b. Webpage 4.

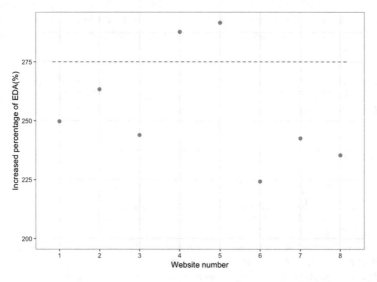

Fig. 5. The mean increased percentage of EDA of 17 participants on each webpage comparing with the tranquillization status. This figure showed that their emotional arousal level become very high on webpage 4 and 5.

3.2 Further Implication of Our Method

This study implies that we can measure the service quality in real time by analyzing the real-time psychophysiological responses of customers when they are experiencing the service, and eye-tracking can be used to determine the concerning point of the service quality. This method can be used for online service quality measurement that concerning more about the details. Furthermore, considering online service as a form of product, this study can be implanted into product development cycle, for this method can find details and pain points that have room for improvement, and thus can contribute to current product improvement or upgrade, or new product development. With the development of the wearable devices, recording users' the psychophysiological responses would be easier and simpler, and online service quality measurement utilizing psychophysiological responses would be easier and possibly, more accurate, as psychophysiological responses would not have to be collected in a laboratory environment, but in the real service encounter instead [12]. Our aims for further study would be finding a more systematic relationship between users' psychophysiological responses and online service quality, and apply wearable devices to collect psychophysiological responses in a non-laboratory environment.

4 Conclusion

This study measured the online service quality in real time utilizing psychophysiological responses of customer experience. This method can be used to measure the service quality in real time. Moreover, it can be used to find the pain point of the service. It is suitable if the service has many details to be evaluated. The measurement used in this study may be more useful with the development of wearable devices in the future.

Acknowledgments. This study is supported by the Natural Science Foundation of China (project number: 71471095) and the Ministry of Science and Technology of the People's Republic of China (project number: 2017YFC0820200 and 2016IM010200).

References

1. Ciavolino, E., Calcagnì, A.: Generalized cross entropy method for analysing the SERVQUAL model. J. Appl. Stat. **42**(3), 520–534 (2015)
2. Zhang, X., Prybutok, V.R.: A consumer perspective of e-service quality. IEEE Trans. Eng. Manage. **52**(4), 461–477 (2005)
3. Hwang, L.J.J., Eves, A., Desombre, T.: Gap analysis of patient meal service perceptions. Int. J. Health Care Qual. Assur. Inc. Leadsh. Health Serv. **16**(2–3), 143 (2003)
4. Ladhari, R.: Alternative measures of service quality: a review. J. Serv. Theory Pract. **18**(1), 65–86 (2008)
5. Lam, T., Hanqinqiu, Z.: Service quality of travel agents: the case of travel agents in Hong Kong. Tour. Manage. **20**(20), 341–349 (1999)
6. Seth, N., Deshmukh, S.G., Vrat, P.: Service quality models: a review. Int. J. Qual. Reliab. Manag. **22**(9), 913–949 (2005)
7. Dyke, T.P.V., Prybutok, K.V.R.: Measuring information systems service quality: concerns on the use of the SERVQUAL questionnaire. MIS Q. **21**(2), 195–208 (1997)
8. Parasuraman, A., Zeithaml, V.A., Berry, L.L.: Refinement and reassessment of the SERVQUAL scale. J. Retail. **67**(4), 420–450 (1991)
9. Jimenez-Molina, A., Retamal, C., Lira, H.: Using psychophysiological sensors to assess mental workload during web browsing. Sens. (Switz.) **18**(2), 1–26 (2018). https://doi.org/10.3390/s18020458
10. Vargo, S.L.: The four service marketing myths: remnants of a goods-based, manufacturing model. J. Serv. Res. **6**(4), 324 (2004)
11. Mahlke, S.: Studying affect and emotions as important parts of the user experience. Position Paper at the Workshop on 'The Role of Emotion in Human-Computer Interaction' (2005)
12. Tsao, L., Li, L., Ma, L.: Human work and status evaluation based on wearable sensors in human factors and ergonomics: a review. IEEE Trans. Hum. Mach. Syst. **49**, 72–84 (2018)

Cubo: Communication System for Children with Autism Spectrum Disorders

Sofia Espadinha Martins[1,2(✉)] and Paulo Maldonado[1,2,3,4]

[1] Centro de História de Arte e Investigação Artística (CHAIA),
Universidade de Évora, Palácio do Vimioso, Largo Marquês de Marialva 8,
7000-809 Évora, Portugal
sofiaespadinha@hotmail.com,
paulomaldonado@inspaedia.com
[2] Departamento de Artes Visuais e Design, Antiga Fábrica dos Leões,
Universidade de Évora, Escola de Artes, Estrada dos Leões,
7000-208 Évora, Portugal
[3] Centro de Investigação em Território, Arquitetura e Design (CITAD),
Universidades Lusíada, Rua da Junqueira, 188-198, 1349-001 Lisbon, Portugal
[4] Faculdade de Arquitetura, Centro de Investigação em Arquitetura,
Urbanismo e Design (CIAUD), Universidade de Lisboa, Rua Sá Nogueira,
Polo Universitário, Alto da Ajuda, 1349-055 Lisbon, Portugal

Abstract. The current research, focusing on Social Innovation and Inclusive Design, seeks to understand and explore the interpersonal communication of children with autism spectrum disorders. The main objective is to contribute to the development of the cognitive and social skills of these children, improving and facilitating their difficulties in three domains: verbal language and communication; interpersonal relationships and in the field of thought and behavior. In order to help solving these problems, Cubo has been developing, an innovative system of universal and inclusive communication, composed of a new universal and alternative/augmentative alphabet and digital object that promotes autonomy, social integration, personal development and interpersonal relationships. This first paper aims to inform and promote the discussion of the process and results of the ongoing research, describing the first phases of the design process itinerary and methods applied, as well as the description of the following phases that cause multidisciplinarity and co-creation.

Keywords: Autism spectrum disorder · Inclusive design ·
User Experience Design (UXD) · Digital experience · Interaction Design (IxD) ·
User Interface Design (UI) · User-centered design · Universal language

1 For Whom? The Context of Research

The social inclusion of minorities must now be a concern in the design of new solutions and design processes, and designers need to contemplate the needs and specific abilities of these users in the design of products and/or services, integrating them into the

T. Ahram (Ed.): AHFE 2019, AISC 965, pp. 353–365, 2020.
https://doi.org/10.1007/978-3-030-20454-9_37

society and contributing to the equal rights and opportunities of all citizens.[1] Design should seek to understand the context in order to give answers to the needs and desires of consumers/users and thus to be able to improve the world, in a holistic perspective of what is socially responsible design [1]. Being Design Thinking such a fundamental and indispensable capacity for the origin of new products and services that bring changes to the lives of people with specific needs, the growing emergence of projects in recent years that have as main objective the social integration of people with different mental and/or physical abilities is noticeable. We can say that Design is mainly at the service of people and their needs, placing the designer in a position of extreme importance in the construction of change of mentalities and the quality of life of society, as stated by David Berman: «Designers have an essential social responsibility because design is at the core of the world's largest challenges… and solutions. Designers create so much of the world we live in, the things we consume, and the expectations we seek to fulfill. They shape what we see, what we use, and what we waste. Designers have enormous power to influence how we engage our world, and how we envision our future» [2]. Based on these principles and good design practice in several areas (Inclusive Design, Participatory Design, User-Centered Design, UX, UI, IxD), this research seeks to understand how this discipline can contribute to the development of interpersonal communication in children with ASD[2] and to facilitate interaction/social integration in the various contexts in which the child lives (research question of our Master's Degree in Design that is underway). It is estimated that there are around 70 million people with autism spectrum disorders, or about 1% of the world's population, but although these numbers represent a small percentage, autism is a developmental disability which has grown the most in recent years [3–6].[3] Autism Spectrum Disorder are severe disorders of the child's neurodevelopment that persist throughout life, and can coexist with other pathologies, characterized by difficulties in the areas of communication, interaction and behavior [8]. The characteristics of this disorder cause clinically significant impairments in social and occupational functioning and compromise the day-to-day functioning of these children, of which we highlight [9]:

(i)   Limitations in communication and social interaction across multiple contexts (many individuals have language deficits, ranging from complete absence of speech to language delays, poor speech comprehension, echo speech or overly literal and pompous language);

(ii)   Deficit in social-emotional reciprocity and lack of awareness of feelings, affecting the ability to make friends;

(iii) Reduced sharing of interests, emotions or affection;

---

[1] «Thinking globally means recognising and celebrating human diversity. It means embracing difference, be it physical, intellectual, cultural, aspirational, or of lifestyle […] The design challenge is to include, and not to exclude unknowingly […] Since then, inclusivity has moved from the margins of design thinking to the mainstream» [1].

[2] (ASD) means Autism Spectrum Disorders.

[3] As an example, according to one of the last studies conducted (2014), in the USA, one in 59 children are diagnosed with ASD, affecting about 3 million people [3, 4, 7].

(iv) Abnormalities in eye contact and body language or deficits in the understanding and use of gestures;
(v) Difficulties in adjusting behavior in a way that suits the various social contexts;
(vi) Repetitive and restricted patterns of behavior;
(vii) Inflexible adherence to routines or ritualized patterns, resisting change;
(viii) Restricted and fixed interests that are abnormal in intensity or focus;
(ix) Deficits in non-verbal communication (gestures);
(x) Hyper or Hiporreactivity to sensorial stimuli.

Communication is an essential factor in the individual and social development of the child. The act of communicating, whether verbally or by gestures or expressions, is a universal necessity and an instrument of integration and social interaction that allows us to apprehend the world around us. According to Sim-Sim: «In the life of the child, communication, language and knowledge are three pillars of simultaneous development, with an eminently social and interactive penchant […] language is used in social context to express what we want, what we think and, not least, to initiate, maintain and control social interactions […] The communicative act is a dynamic, natural and spontaneous process that requires the interaction of at least two people in order to share needs, experiences, desires, feelings and ideas […] Interacting verbally, children learn about the physical, social and affective world» [10]. We can conclude that changes and disturbances in language and interpersonal communication can cause damage to the child's cognitive, emotional and social development. As mentioned earlier, some of the limitations of children with ASD are related to communication and verbal language, causing difficulties in social interaction, and consequently acceptance and integration in society. In this way, we justify the relevance of this study/project. The research focused on the need to create solutions that promote the interaction of these children with their family and friends, in an attempt to improve affective relationships through communication and understanding of the needs and desires of the child. Faced with this issue, it was decided to approach these two major domains, which complement each other: Domain of Social Interaction, in which it is intended to create moments of interaction with the other, making communication more intuitive and recurrent through the created system (Cubo); Domain of Communication and Language, by helping the child become familiar with verbal language and express needs, desires and feelings through a simplified visual alphabet/code, easy to understand. Thus, we can define as general objectives of research the contribution of Design to the development of cognitive and social capacities of children with ASD, improving and facilitating communication and interpersonal relationships, through a digital artifact that promotes interaction, integration, personal development and autonomy, and as specific objectives:

1. Help these children communicating and expressing needs, desires and feelings;
2. Create a visual alphabet/code, with the same visual identity language, simple and easy to perceive as a universal and inclusive alphabet that can be used not only in Autism Spectrum Disorders but also in other pathologies and which is recognized, interpreted and used by those who do not have these difficulties;
3. Develop a system that can be seen as a useful tool which can be used at home, in schools and in therapy sessions;
4. Promote affective relationships with family and friends;

5. Familiarize the child with verbal language;
6. Stimulate the understanding of everyday life and social contexts;
7. Integrate several functionalities into a single solution.

Therefore, the goal of this research is to build a communication system that targets children aged 5 to 12 years with difficulties and limitations in terms of communication and interaction. Autism spectrum disorder is a condition that also affects families, because parents face big challenges every day, resulting in a negative impact on the family and the need to resort to therapists and specialists. That's the reason why we consider it appropriate to include parents in the main target public of the project, and the remaining actors in the universe of children with ASD (other relatives, educators/teachers, specialists and therapists) are considered as a secondary target audience.

2 The Parallelism Between Alternative and Augmentative Communication and Universal Languages

As our main purpose is to overcome or minimize the limitation of communication and language in children with ASD, through the offer of a new alphabet/code that facilitates the construction of discourse and interaction with the other, we could not fail to aspire to universality as quality of this new offer, despite this form of communication being considered as an augmentative and alternative communication instrument.[4] Our vision for the future seeks to find a parallelism between alternative and augmentative communication and universal languages.[5] It is our desire and ambition that this new alphabet/code is understood not only by children with ASD, but also by all people who are not included in this context, promoting inclusiveness and equality. This desire extends to the possibility of being used in the therapy of other pathologies as a facilitator of the development of speech/communication and cognitive skills. Taking into account that universality is one of the objectives to be achieved, it is essential to establish a comparison with other languages that have inspired us and that have been a milestone in the development of communication as we know it today.[6] The will and need to communicate is intrinsic to the human beings from the earliest beginnings. Graphic images

---

[4] «An alternative augmentative communication system is an integrated set of techniques, aids, strategies, and capabilities that a person with communication constraints uses to communicate. Alternative and augmentative communication is considered to be any type of communication that replaces, amplifies or supplements speech»; «Alternative communication is a system that replaces speech and augmentative communication is a system that supports or complements speech» [11].

[5] When we refer to universal languages, we encompass all kinds of «functional languages», with utilities and information functions, world-wide recognized and understood, such as pictographic codes of signage and traffic signs and «graphic languages» [12] like the writings elaborated in antiquity. We also included within this spectrum the pictographic and ideographic language shared through the Internet - the Emojis. We have chosen to name this set of communicative languages as "universal languages", because it is more practical to reference them when necessary.

[6] «The expressive possibility of reciprocal understanding between the beings of a group, of a community, of a species is by antonomasia one of the most important conditions, from the very beginning of life, for survival» [13].

or signs have been used since Antiquity in order to materialize thought, such as «In Sumer, in 4000 B.C, there was a pictographic writing consisting of images that were not representations of things but "symbols" and whose articulation revealed the meaning, the discourse. This "pictographic writing" resulted from the combination (conventional, codified) of iconic figures which, more than pictograms, were ideograms, since they were based on figures that symbolized ideas and not figurative images» [12] (Fig. 1). Egyptian hieroglyphs are also an example of this type of writing (Fig. 1).

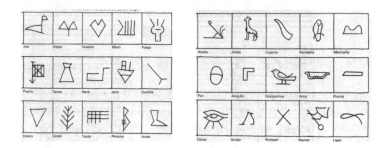

Fig. 1. Sumerian *(Left)* and egyptian *(Right)* cuneiforme writing [13].

Another example that we must take into consideration for the construction of a new universal and inclusive language is the signage. In these cases, the universality of pictograms and ease of perception are the characteristics that must be taken into account in order to provide user-friendly information instructions. In signage, «The crescent use of pictograms is determined by the language problem itself. Roads, railway networks, maritime and air lines extend far beyond national, linguistic and ethnic boundaries. An alphabetical polyglot description required of very oversized supports or boards, and the informative content would lose clarity» [13]. In the same way that signage is universally understood, also on traffic signs it is necessary to «find signs whose comprehension provides functional instructions practically accessible to everyone, instantly and unequivocally [...] for sure, they require for their comprehension of a learning process. However, when they have already incorporated into unconscious knowledge [...] the information they provide is immediate and spontaneous» [13]. The universality of the traffic signs is also due to its easy perception and memorization due to three factors: the form, the content and the color. These three characteristics determine specific rules and meanings. These signs are understood anywhere in the world and in any language. Bearing in mind the examples presented above, it is important to highlight the importance of Emojis, as pictographic and ideographic elements that are inspiring for what is intended, which is the development of the new alphabet/code. As far as emoji is concerned, it is important to emphasize the attribution of universal meanings (all people have the same immediate perception of the same emoji). This perception is transversal and independent of the country or culture.

According to Ai et al., «emojis do not have language barriers, making it possible to communicate among users from different countries. These advantages have attributed to the popularity of emojis all over the world, making them a "ubiquitous language" that bridges everyone» [14]. Despite the universality of the meanings of this language, the interpretation of each emoji or set of emoji may vary from individual to individual (interpretation will depend on the context of the conversation, the interpretation of the sender and the receiver), being able to express diverse ideas, feelings, emotions and actions [15]. Having said this, each of the examples mentioned above, carries to the project fundamental characteristics: the use of symbology to form a language of its own; the meaning attributed to colors, and the universality. In short, regardless of whether we use verbal language as a means of communication, other complementary alternatives to language may also be used, because «our civilization is and will be visual» [12].

3   An Alphabet for All – A Visual and Universal Language

Currently there are several augmentative and alternative communication systems and equipment that are used to improve or replace speech in individuals with difficulty or inability to express themselves through verbal language. These systems can be divided into two groups: systems without help and systems with help. For this research we are interested in talking about systems with help. These systems require external assistance through instruments, in which the user selects symbols by means of a platform, these graphic symbols being based on schematic drawings, with more or less structuring, and with different levels of symbolism [11]. Some of the sets of graphic symbols that we find today applied to several pathologies are: PIC, Rebus, Bliss, Sigsymbols, Picsyms, Oakland e PCS (Picture Communication Symbols),[7] the last set of symbols is used by many countries. These sets of graphic symbols, although used in different countries, cannot be considered universal in the perspective of what is intended with our proposal and objective, which is the transversal universality applied to a language that is intended for all, regardless of the capabilities, age, culture, nationality and education of each – the creation of a universal and inclusive library of symbols. Comparing the objectives of a new universal and inclusive alphabet/code, which we mentioned earlier in the context of the study/project (point 2), we highlight some features of these already developed sets of symbols, which in this perspective they do not confer on them the universality and coherence that we desire:

- If they are not accompanied by their description, they are not recognized and interpreted intuitively by people who do not have deficits in communication and language;
- They are not used outside the context and universe of the diseases to which they have been drawn (with rare exceptions), and are directed only at such persons;

---

[7] These graphic symbols were created in 1980 by speech therapist Roxanna Mayer-Johnson, who created an assistive technology company focused on augmentative communication products centered on these symbols. Later, this company was approached by DynaVox Systems [16] in 2004, and is now known by Boardmaker [17], which provides software that has a library of PCS symbols and tools that allow the construction of educational, recreational and augmentative and alternative communication resources.

- These sets are not available in their entirety free of charge to users because they are owned by various companies/brands and therefore cannot be used and shared without the permission and authorization of the company owning each set, restricting the spectrum of people using these symbols;
- Each set of symbols has a different and distinct visual identity, there is no consistency between them (even within each set, there are irregularities in the identity and visual language of each pictogram);
- There are different pictograms for a single meaning (even within each of the sets), there being no coherence;
- Some pictograms are too complex for their correct interpretation, either by the person with difficulties in verbal language, or by those who do not have these difficulties.

When we write a message (as in Messenger), we substitute our speech written by pictograms or ideograms as an alternative or as an addition to what we are trying to convey (as we have already given by example, Emojis - in which its direct meaning is understood a general form), it is our goal to create a universal alphabet/code that interconnects all people through universal communication and without barriers, so that children or adults with difficulty and incapacity in verbal language (as is the case of people with ASD), do not feel different and excluded, and are part of a whole that unites the differences of each one, be they cognitive or even cultural and national differences.

3.1   A New Alphabet/Code

In view of this main objective, in relation to one of the constituent parts of this research/project - the new alphabet/code to be developed, and taking into account the bibliographic review and analysis of the case studies mentioned above, it was necessary to outline specific way to characterize and achieve the desired solution:

- The alphabet/code should be universal and unique;
- The alphabet/code should include and integrate, not exclude in any circumstance;
- Each pictogram/ideogram that compose the alphabet/code must have only one assigned meaning (an object or idea, such as "apple" or "happy", is only represented by a single pictogram or ideogram) without different variants of the same meaning, for its easy understanding and memorization;
- This alphabet/code can be used in various contexts, both in the therapy of certain disorders or deficits in communication, and in conversation contexts on web platforms (mobile applications and websites). This point does not exclude other physical or digital media, where it may be feasible and beneficial to use this alphabet/code as communicative and informative form;
- The alphabet/code must have a unified and coherent language and visual identity;
- The alphabet/code should be grouped by categories and each category should be associated with a specific color (colors speed up the correct and fast location of the pictograms/ideograms favoring memorization and learning), similar to PCS graphic symbols referred to;
- The pictograms/ideograms should be simple and easy to memorize their meanings. Pictograms must also be monochromatic (less color = less abstraction);

- Each pictogram/ideogram must be understood in its entirety and instantaneously after the learning process and must be able to be associated with each other;
- The alphabet/code should be easily interpreted by all users, regardless of their cognitive abilities;
- The alphabet/code should be made available free to all users.

After defining the basic requirements for the development of the new universal and inclusive alphabet/code, an initial idea was designed, seeking to fulfill the defined characteristics (Fig. 2), in order to be validated and revised later. This alphabet/code will consist of pictograms and ideograms that represent objects, ideas, expressions, emotions, needs and even desires, which are associated with each other, forming phrases that facilitate discourse and interaction (Fig. 3). The alphabet/code is divided into categories (needs and desires, emotions, basic expressions - at an early stage these categories are most necessary to the autonomy of people with communication difficulties) and each category was assigned a color (blue, red and yellow, respectively).

Fig. 2. Initial sketch phase of the new alphabet/code. Source: Authors, 2019

Fig. 3. Sentence formation through the combination of pictograms/ideograms. Source: Authors, 2019

4   Research Methodology – Applied Methods and the Following Phases of the Design Process

During the Design process we have adopted a methodology through the application of non-interventionist and interventionist methods of a quantitative and qualitative nature, in order to create a graduated and effective project itinerary that will lead us to a solution that responds appropriately to the needs of the target audience and to the objectives and requirements of the project. We want to ensure, through this research journey, the effectiveness of the solution, generate empathy with all the stakeholders involved in the project. The design itinerary was designed based on the Design Thinking proposed by IDEO [18], because «Design Thinking gives you faith in your Creative abilities and a process to take action through when faced with a difficult challenge [...] It's a structured approach to generating and evolving ideas [...] It's a deeply human approach that relies on your ability to be intuitive, to interpret what you observe and to develop ideas that are emotionally meaningful to those you are designing for» [19]. Based on IDEO's itinerary, we divided the research process into several phases: (i) Understand and Interpret; (ii) Analyze and Explore; (iii) Idealize and (iv) Prototype and Validate, being an iterative process, in which there may be phases that coincide with each other. In order to respond to the challenges, we propose to respond, two methods have been defined, which form an integral part of these phases of the design process itinerary:

- Passive research: focuses on the concern to know the target audience and the context in which the study/project is inserted through the bibliographic review. At this stage of the research it was necessary to deepen some of the concepts underlying the challenge addressed, where the areas of multidisciplinary research were defined to be studied because they are relevant in the theoretical basis of the project and their interconnection, which will provide the construction of a real solution and focused on the needs of the user: ASD; Augmentative/Alternative Communication Systems; Theories of Child Development; Inclusive Design; Participatory Design; Digital Experience; User Experience; User Interface; Interaction Design; User-centered Design; Universal Languages; Semiotics and Signs. Within these areas of research, through the bibliographic review, are being studied, among others some authors such as: Piaget (1896–1980); Winnicott (1896-1971); Lev Vygostsky (1896–1934); Lorna Wing (1928–2014); Eric Schopler (1927–2006); Bill Mooggridge (1943–2012); Bill Buxton (1949); Alan Cooper (1952); Steve Krug (1950); Donald Norman (1935); John Maeda (1966); Roger Coleman (1943); Umberto Eco (1932–2016); Adrian Frutiger (1928–2015); Joan Costa (1926). In passive research, case studies have been analyzed, related or not related to the concepts and objectives of the project, in order to allow the identification of characteristics and functionalities that can be considered in the final solution of both the alphabet/code and the physical artifact (The Cubo).

- Active Research: in which contact with the reality of the study/project and with professionals from other scientific areas is essential, reinforcing the values of user-centered design and participatory design where co-creation and multidisciplinarity prove to be valuable tools for the development of an innovative, complete and effective solution. We should not understand the problem that we propose to solve in isolation. It is essential that users are actively involved throughout the process, so that they can contribute, in order to promote the discussion and debate about the project, contributing to a more rich and empathic final solution.[8] We divide active research into 3 phases: (i) Initial Phase: which corresponds to the deep knowledge of the target audience from initial exploratory conversations with therapists/psychologists and focus groups ("Focus Group" [21]) constituted by parents of children with ASD where the method "Diary Studies" [21] will be applied. The first session with the parents will aim to present the project in order to gather feedback and generate brainstorming ("Co Creation Session" [20]), with a view to the initial validation of the idea/project. At the end of the session, evaluation questionnaires will be applied; (ii) Intermediate phase: Initial validation of the alphabet/code will be carried out through two focus groups - one consisting of children with ASD with the accompaniment of a therapist or special education teacher and the other focus group with a set of people which do not have communication and language deficits – with the purpose of validating the first pictograms/ideograms for later revision. During this phase the methods "Prototyping" and "Evaluative Research" [21] will be applied; (iii) Final Phase: the alphabet/code will be validated through two focus groups, with the purpose of validating the final pictograms/ideograms and making a final evaluation of the results. The same methods from the previous phase will be used. All these sessions with focus groups will be conducted through schools. In the context of a master's dissertation, only the alphabet/code (with recourse to a prototype) will be validated, and the physical/digital object will be validated only from expert focus groups, in which the performance of the artifact's functionalities will be debated. During the implementation of the phases described above regarding the research methodology will be reviewed the various components of the future final solution. Subsequently, all components of the solution will be finalized, as well as their operation and the interconnection of all parts so that the Cubo System project can evolve outside the academic context.

5   Results and Discussion – Cubo System

After completing some of the stages of the project itinerary, we have the results of the initial phases of an investigation still open, the first proposition of alphabet/code (in which the whole project is based) and a hypothesis of solution of electronic/digital artifact – The Cubo, that constitutes the interface for children with deficits in interaction/communication (Fig. 4). The Cubo integrates the alphabet/code functionality to communicate and interact with others, which is expected to be effective in the interaction and integration of children with ASD in society. In summary, in addition to this central function, the Cubo incorporates many other features that will allow greater

---

[8] «The process is designed to get you to learn directly from people, open yourself up to a breadth of creative possibilities, and then zero in on what's most desirable, feasible, and viable for the people you're designing for. […] Empathy is the capacity to step into other people's shoes, to understand their lives, and start to solve problems from their perspectives. All you have to do is empathize, understand them, and bring them along with you in the design process. […] Only by listening, thinking, building and refining our way to an answer do we get something that will work for the people we're trying to serve» [20].

autonomy and personal development. The Cubo is an innovative system of universal and inclusive communication that consists of a physical object, the new alphabet/code and an App addressed specifically to the parents of these children. This functionality will be presented in the next paper to be submitted to AHFE 2020, following the new developments of the project.

So far, we have gathered very positive feedback from specialized professionals, who have been contributing to the development of the project, through exploratory talks, discussion of the solutions already conceived and their feasibility, ideal scenario for collecting inputs about characteristics of this disorder. Some topics of active research were also discussed, which resulted in the redefinition of the steps to be taken in the near future. After exploratory sessions, these experts showed great interest in the project, confirming the quality and relevance of the solutions so far projected, revealing that it will be a good way for the future. The next phases of the research process will be crucial for the development of the project and its success – we will go deep into the stage of active research, where we will promote contact with all stakeholders through the participation and involvement of schools and support centers. It is our desire that it be a highly collaborative process, generating added value, in the long term, for all users. It remains to be said that universality is an objective that will have to be achieved in the long term, since it will only be possible to test and validate the project, internationally, outside the academic context. In conclusion, although this study is directed at a specific neurodevelopmental disorder, we will emphasize and remember that we have as a goal that this solution will have, in the future, positive and significant repercussions not only in the therapy of this and other cognitive issues, but also that it is synonymous with the first steps towards a more conscious and humane society, and for a more inclusive future, where we can communicate in unison, because despite the differences, we are all equal.

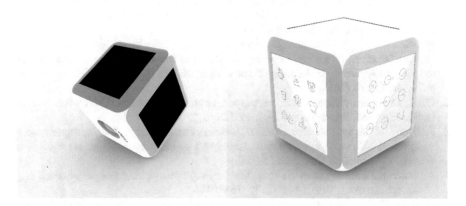

Fig. 4. Daily use tool that enables the use of the new alphabet/code and facilitates communication and interaction with others. Source: Authors, 2019

Acknowledgements. CHAIA – Centro de História de Arte e Investigação Artística, Universidade de Évora, Portugal – CHAIA/UÉ [2019] – Ref^a UID/EAT/00112/2013 Project financed by national founds from FCT/Fundação para a Ciência e Tecnologia. We thank to Dr. Rui Negrão, Clinical Specialist Psychologist at the SOERAD Hospital Unit in Torres Vedras (Portugal) and Professor Célia Sousa Cordinator of the Centro de Recursos para a Inclusão Digital (CRID) do Instituto Politécnico de Leiria (Portugal), and their valuable contributions.

References

1. Coleman, R., et al.: Design for Inclusivity: A Practical Guide to Accessible, Innovative and User-Centered Design. Routledge, New York (2007)
2. Berman, D.B.: Do Good: How Designers Can Change the World. New Riders, Berkeley (2009)
3. Autism Speaks. https://www.autismspeaks.org/autism-facts-and-figures
4. Autism Speaks 2017 Annual Report. http://www.autismspeaks.org/sites/default/files/2018-08/2017-annual-report_0.pdf
5. Autism Society. https://www.autism-society.org/what-is/facts-and-statistics/
6. World Health Organization. https://www.who.int/news-room/fact-sheets/detail/autism-spectrum-disorders
7. Center for Disease Control and Prevention. https://www.cdc.gov/mmwr/volumes/67/ss/ss6706a1.htm?s_cid=ss6706a1_w#suggestedcitation
8. Gonçalves, A., et al.: Unidades de Ensino estruturado para Alunos com Perturbações do Espectro do Autismo: normas orientadoras (Structured Learning Units for Students with Autism Spectrum Disorders: guidelines). Ministério da Educação: Direção-Geral da Inovação e de Desenvolvimento Curricular, Lisboa (2008)
9. APA, American Psichyatric Association – DM-5: Manual de Diagnóstico e Estatística das Perturbações Mentais. Climepsi, Lisboa (2014)
10. Sim-Sim, I., et al.: Linguagem e Comunicação no Jardim-de-infância: textos de apoio para educadores de infância (Language and Communication in kindergarten: support texts for early childhood educators). Ministério da Educação: Direção-Geral da Inovação e de Desenvolvimento Curricular, Lisboa (2008)
11. SOUSA: Célia – A comunicação aumentativa e as tecnologias de apoio (Augmentative Communication and Assistive Technologies). In: Ministério da Educação e Ciência, Portugal (ed.) A Acessibilidade de Recursos Educativos Digitais, pp. 51–63. Ministério da Educação e Ciência, Lisboa (2011)
12. Costa, J.: Design para os olhos: marca, cor, identidade, sinalética (Design for the Eyes: brand, color, identity, signage). Dinalivro, Lisboa (2011)
13. Frutiger, A.: Signos, símbolos, marcas, señales: elementos, morfologia, representación, significación (Signs, symbols, brands, signage: elements, morphology, representation). Editorial Gustavo Gili, Barcelona (2005)
14. Ai, W., et al.: Learning from the ubiquitous language: an empirical analysis of emoji usage of smartphone users. UBICOMP **16**, 12–16 (2016)
15. Parra, G.: Caracterización de los usos y funciones de los emojis em la comunicación mediada electrónicamente. Universidad de Extremadura, Badajoz (2017)
16. Tobii Dinavox. https://www.tobiidynavox.com
17. Boardmaker. https://goboardmaker.com
18. IDEO. https://www.ideo.com/eu

19. Fierst, K., Diefenthaler, A.: Design Thinking for Educators. Riverdale Country School–IDEO, New York, Palo Alto (2011)
20. IDEO. http://www.designkit.org/resources/1
21. Martin, B., Hanington, B.: Universal methods of design: 100 ways to research complex problems, develop innovative ideas and design effective solutions. Rockport Publishers, Beverly (2012)

Improve Your Task Analysis! Efficiency, Quality and Effectiveness Increase in Task Analysis Using a Software Tool

Vaishnavi Upadrasta$^{(\boxtimes)}$, Harald Kolrep, and Astrid Oehme

HFC Human-Factors-Consult GmbH,
Köpenicker Straße 325, Haus 40, 12555 Berlin, Germany
`{Upadrasta, Kolrep, Oehme}@human-factors-consult.de`

Abstract. As the field of Task Analysis (TA) is still fragmented and poorly understood by many, a software-tool, build by HFC Human-Factors-Consult GmbH, has been developed for easy and better TA. Purpose of the study is to evaluate the efficiency, quality and effectiveness of TA performed with the support of this software-tool. In the experiment, 36 participants conducted a total of two hierarchical-TA (HTA) on two given tasks, once using the new software-tool and the second using fundamental methods i.e. paper-&-pencil. The results indicated that the software-tool aided participants in producing good-quality analysis, provided support and guidance during the HTA-process and helped in maintaining a consistent performance-level in terms of both quality and effectiveness. The findings resulted in identifying the strengths and acknowledging the shortcomings of the new software-tool, thus providing a concrete direction for further improvements. Moreover, the study adds to literature by developing checklists to make this assessment, which in turn proposes components that characterizes a good TA.

Keywords: Method · Task analysis · Hierarchical Task Analysis · Software tool · Support software

1 Introduction

The origin of modern Task Analysis (TA) can be traced to the early 19th century with its foundations in the field of scientific management [1]. Since then, TA has been recognized to be one of the important tools of human factors, scientific discipline focusing on humans at work, especially with human's use of technology. It is required to solve problems that arise through computerization and automation. It attempts to apply knowledge about human ability and limitations to design machines, tools, tasks, and work environments to make them as productive, effective and safe as possible [2]. With digitalization and development of more complex technology, the demand for TA arose and today it plays a central role in Human-Machine Interaction, Human-Factors Engineering and Safety and Health Hazards [3].

TA is a broad and generic term that comprehends a large range of methods and techniques for, simply put, 'analyzing a particular task or tasks'. It aims at obtaining detailed information on actual or projected task performance as well as the organization

© Springer Nature Switzerland AG 2020
T. Ahram (Ed.): AHFE 2019, AISC 965, pp. 366–378, 2020.
https://doi.org/10.1007/978-3-030-20454-9_38

of this collected information in a suitable format [4]. It involves not only identifying but also predicting difficulties that are faced while the task is being performed. TA can be conducted in different stages of a project from preliminary to detailed design stages, to construction and development stage and finally to post development decomposition and evaluation stage. In other words, TA methods are used in predictive as well as in an evaluative manner [3]. There are multiple ways in conducting task analysis, but the general procedure includes identifying tasks or goals, collecting data, analyzing the collected data and representing the analyzed data [5].

2   Current Issues

TA is now routinely used in numerous settings and across various domains. Over a hundred TA methods and their permutations have been developed over time to suit specific needs of countless domains [6]. Human-factors practitioners and ergonomists are still developing new methods and approaches for a multitude of different purposes. Modern applications of TA include system and interface designing [6, 7], training [8] in conventional office environment [9] as well as non-conventional environment such as training in military and defense [10], Human-Error Identification and prediction [11], allocation of functions and system assessment [12], improving productivity [13], skills and knowledge acquisition [14], evaluating usability [15] and much more. This flexibility has proven to be advantageous, but it has also created a number of problems.

Felipe et al. [8] state 'the application of TA to training has become invaluable to the work place; however due to various methods of TA it has become difficult to train TA itself'. Even with decades of research, the process of TA has few guidelines [16] and the presented results follow no specific standardizations [1] or defined formats. Often, TA results have complex outputs and visualizations, which are difficult to understand. As a result, TA can be complicated and difficult to perform [7, 16], and is only understood by experts [6]. It can be argued that even experts find it difficult to understand and use TA. Often only the analyst or the group of analysts who perform the TA understand it. Besides, performing TA requires a specific set of skills to be learnt in time and with experience. Stanton [16], who stated that Hierarchical Task Analysis (HTA) requires craft skills that need to be learnt under expert guidance, supports this. Thus, TA is time consuming and expensive.

Moreover, the diversity of TA, limited guidelines and lack of standardization has resulted in very little guidance on the best practices in conducting TA and evaluating TA that have been performed by an analyst [17]. A review of TA literature has shown that there is indeed little research on what are considered as "good" characteristics of TA. In addition, considering the maturity of TA literature, it is also surprising that little effort has been made in developing software tools to simplify the different stages of the process of TA. A need for TA software tools and applications has been identified across literature [7, 10, 12, 13, 16]. Few tools have been developed over the years but lack in supporting the full breadth of TA. These tools have been identified to be either too generic and hence lack key aspects of TA or are developed for very specialized purposes and thus impose highly specific TA methods that are extremely complex [12].

In order to facilitate the process of TA in various life cycle stages, a software tool was built by HFC Human-Factors-Consult GmbH, Berlin. It supports various TA methods and techniques and uses Hierarchical Task Analysis (HTA) as its foundation. HTA is a goal-based analysis where task and sub-tasks are explained in terms of their goals [16]. The task is broken down into subtasks and operations, which are then graphically represented in a structured fashion. Plans are established to determine how these sub-goals are achieved.

The purpose of this study is to evaluate whether this software tool improves the performance of the TA conducted by the analyst. The objective of the current study is to verify the strengths and identify the weaknesses of the software tool. These insights could be useful to implement further improvements and provide design guidelines.

In this process, this study attempts in developing an evaluation framework in the form of a checklist to determine a "good" TA. The checklist comprises two major components: quality and effectiveness. These components help in systematically evaluating whether the analyst was able to successfully capture all the essential elements of the task in the best possible way. Moreover, efficiency of an analysis had also been considered to play a vital role in determining if the analysis has been well performed.

3   New Software Support

A new software tool is designed for human factors task analysis and human error identification to overcome the problems mentioned in Sect. 2. The software tool was built by HFC for internal usage. It was applied to examples and real used cases across healthcare, control room and aviation domains by human factors engineering specialists. In the last two years, HFC is adapting and testing the software tool within a German-government funded project 'KUKoMo' for Human-Robot Collaboration.

It was developed keeping several TA methods and techniques in mind in order to facilitate the empirical steps of TA procedure including data collection and visualization of this collected data. These methods and techniques include HTA, Applied Cognitive Task Analysis (ACTA), Human Error Templates, (HET) and Task Map Layering technique [18]. Features of the tool are also customizable so that it can be made suitable regardless of the domain.

The software tool supports the entire TA procedure and simplifies its application. It is designed not only to reduce effort and resources for conducting TA but also to provide non-expert analysts with some guidance. HTA is the starting step to model a task hierarchy, where task and subtask instead of the classical approach of goals and sub-goals are represented in the tree structure. This simplifies handling and understanding of the graphical representation, especially for non-human factors experts.

One key aspect of this tool is that it runs on mobile devices such as tablet PC (Fig. 1) enabling analysts to flexibly collect data and efficiently apply TA on the field as well as in the laboratory. The tool offers three forms of task model representation: table view, graphical task tree view (Fig. 2) and task map layering view. The table view and the tree view are continuously synchronized. These representations can be exported as a table (*.csv file) or image (screenshot of the graphical tree in PNG) and utilized for further analysis or documentation.

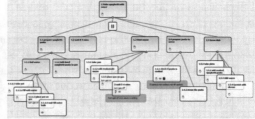

Fig. 1. HFC software tool on a tablet.

Fig. 2. Theoretical HTA tree view with HFC software tool.

The software tool provides seven different analytical and empirical steps to assist analysts in the procedure of TA, thus putting forward a handy, efficient and transparent way of TA. This proposed TA procedure consists of (1) Theoretical HTA, (2,3,4,5) Observation, Card Sorting, Interview and Top-Level Interview are built upon on ACTA, (6) Human Error Template and (7) Task Map Layering. For a detailed description of these steps, refer to Onnasch et al. [18]. All seven steps can be performed in a complete workflow, or subsets of steps can be chosen as per the purpose of the analysis. This is possible as each step is implemented as a separate tab in the user interface of the tool. Collection of data is supported in most of the steps and the created TA or task models can not only be transferred between the steps but also altered or reconstructed. Moreover, the tool offers standard templates for interviews and human error prediction.

4 Efficiency, Quality and Effectiveness Criteria

Over the years, the need for TA has been has been explicitly established across numerous domains. According to Stanton [16], the popularity of HTA along with its versatility and flexibility for conducting analysis and representing them has resulted in the development of practitioner's own adaptations and unique mutations, making it thus difficult to propose one right way of performing it. Various TA handbooks demonstrate procedures for conducting analysis, but there is limited literature describing the best practices.

In a panel discussion, TA experts [17] make an attempt to explore points of consensus with regards to the best practices in performing and evaluating cognitive task analysis. Viewpoints of the panel members range from establishing factors (for e.g. validity, reliability, generalizability, scope, completeness and usefulness) that play a role in the quality of a TA [19] to more generic but pragmatic criterion (for e.g. Klein [20] suggests that every well performed analysis would generate insights) for what counts as a good analysis. Efficiency may refer to the level of resources consumed in performing tasks [21] such as amount of time and costs. According to Diaper and Stanton [6], a good speed when conducting TA indicates good efficiency of the TA.

Effectiveness on the other hand, is defined as the extent to which a goal or a task is achieved [22]. In certain cases, it is easy to assess the effectiveness of a particular task by simply recognizing if the task was a success or a failure. In the case of TA however, recognizing if the conducted analysis is a success or not can be complicated because of its flexible nature in providing multiple interpretation. However, Diaper and Stanton [6] state that representations of the analysis are very important for the effectiveness.

A benchmark for good analysis is needed. In their paper, Felipe, Adams, Rogers, and Fisk [8] examined different types of training methods in order to select the best method to train novices in TA. For this purpose, they created a criterion consisting of five components (hierarchical representation, state high-level goal, state plan, state sub-goals, and state satisfaction criteria) to establish the performance of the conducted analysis by the novices.

Based on this available literature and some new ideas, two separate checklists were created to assess the quality and effectiveness of the performed TA for the current study. The following seven items/components were chosen to be used as a new quality checklist: state high-level goal, state plan, state sub-goal, use task categories, level of granularity, state optional steps, and state numbering system. The new effectiveness checklist includes five components to capture the overall success of the TA: completeness, representation, overall structure, overall performance, satisfaction level. Based on pre-defined scales for individual components for both checklists, master TA templates created for the respective task by TA experts for comparison purposes and analysts' self-evaluation, scores ranging from 1–3 are given. Sum of the individual component scores of both checklists contribute to the total quality and total effectiveness score. Furthermore, there is a potential to adapt these evaluation checklists to different needs and contexts across domains.

5   Method

The purpose of the paper is to investigate whether the TA software tool developed by HFC can aid in conducting better TA in terms of its quality and effectiveness as well as that this analysis will be performed efficiently. An experiment was designed where participants were required to perform HTA, on given tasks.

They were provided with two comparable online tasks, one where they were expected to perform TA with the help of the developed software tool and the other with the help of basic methods that included Excel table, Word document, and paper and pencil. For the purpose of this paper, these basic methods will henceforth be referred to as 'paper and pencil' (P&P) and the new developed software tool as 'software tool'. Participants were unfamiliar with both tasks but familiar with its domain, i.e. online shopping. The two online tasks were Task ABC (add a round coffee table to shopping cart) and Task XYZ (add a black sofa to shopping cart), one for each condition. Both tasks were divided into three parts where Task A and Task X provided a description on how to add a round coffee table or black sofa to the shopping cart. Task B and C and Task Y and Z were in the form of videos. These videos showed how the users actually performed the above-mentioned tasks.

The completed analyses (with the software tool and with P&P) collected from the participants was then evaluated on the grounds of its efficiency, quality and effectiveness. Standardized questionnaires and the two new checklists developed for the purpose of the current experiment (Sect. 4) were utilized to support the evaluation process.

Hypothesis. It was predicted that participants should perform better TA with the help of the software tool than with P&P. It was thus hypothesized that:

(1) The efficiency of the performed TA will be higher when using the software tool than P&P.
(2) The mean quality score for the performed TA will be better when using the software tool than P&P, and
(3) The mean effectiveness score for the performed TA will be better when using the software tool than using P&P.

More specific and detailed hypotheses were derived considering all dependent variables and their individual components.

Participants. The data of 36 participants, 15 females and 21 males, ages ranging from 19 to 53 ($M = 26.9$, $SD = 6.82$) were analyzed. 28 participants (78%) were students and the other eight participants (22%) were working professionals. The majority of the participants belonged to the fields of IT, engineering or psychology. All participants were required to either have practical knowledge on TA or at the least, theoretical knowledge on TA. The duration of the experiment was about 90 to 120 min and was performed in a quiet and well-lighted setting.

Design. The experimental design used is a randomized 2×2 repeated-measures design (Table 1). The independent variables comprised of two conditions and two tasks (the two different online shopping scenarios that were analyzed by the participants).

The two conditions included: Conducting HTA on the given task with the software tool (experimental condition) and conducting HTA on the given task with P&P (control condition). The experiment is primarily aimed to determine the main effect of the two conditions on the dependent variables. The second independent variable comprised of the two tasks, Task ABC and Task XYZ. However, this experiment is not aimed at analyzing the main effect of this variable on the dependent variables. Therefore, for this experiment it is assumed that the two tasks are comparable.

Table 1. Experimental design: randomized 2×2 repeated-measures design

Sample size (N) = 36	Task ABC	Task XYZ
Tool-first ($n = 18$)	Tool condition	P&P condition
P&P-first ($n = 18$)	P&P condition	Tool condition

372 V. Upadrasta et al.

The dependent variables included measures of efficiency, quality and effectiveness and its individual components. Efficiency was measured with the help of total time taken to finish the TA and the workload experienced in the process. Quality and effectiveness were evaluated using the developed checklist (mentioned in Sect. 4).

Materials. The following materials were utilized in the current experiment: (1) *Information on HTA* adapted from Stanton [16]. This gave the participants an idea on what was expected of them in this experiment. Graphical examples were used for better understanding. (2) HFC's *new software tool* installed on tablets and laptops and *tool support information* in the form of live tutorial and a handout containing specific information on menu and toolbar functions was provided. (3) *Step by step experimental instructions* of the experiment was provided. (4) After the completion of each condition, participants filled out the *after task questionnaires:* NASA-TLX[1][23] and a modified version of the SUS[2][21]. Minor adjustments were made to the SUS to suit this experiment.

Procedure. All participants received training on HTA at the beginning to ensure that all participants were on a comparable level of prior knowledge. They then had to perform HTA on both the tasks once using the software tool and once using P&P according to the respective sequences. Additional information on how to use the software tool along with some time to acquaint themselves with the tool was given right before the software tool condition. After completion of each task they filled out the after-task questionnaires. At the end they were debriefed and thanked for their participation. Each conducted analysis was then scored by an expert using the developed checklist and master templates of both tasks.

6 Results

The results were analyzed by calculating the mean difference between the two dependent means of the matched pairs using a repeated measures ANOVA.

Efficiency. Efficiency was assessed with two variables: time taken to complete the TA, and workload and its components. It was found that the total time was higher for the software tool condition ($M = 1602s$, $SD = 603$) than for the P&P condition ($M = 1106s$, $SD = 487$) and this difference was highly significant, $F(1, 34) = 30.40$, $p < .001$, $\eta = .472$. The mean of total workload score for the software tool condition ($M = 0.44$, $SD = 0.11$) was significantly higher than the mean of the P&P condition ($M = 0.39$, $SD = 0.12$), $F(1, 34) = 4.86, p < .05$, $\eta = .125$. These mean values indicate that the participants required less time and experienced less workload when they performed TA with P&P. Figures. 3 and 4 illustrate the interaction between the two conditions based on the sequence in which they were conducted for time taken and

---

[1] The NASA-TLX is a multidimensional, subjective rating procedure used to assess workload [23] which was developed by the human performance group at NASA Ames Research Center.

[2] The SUS System Usability Scale is a simple ten-item Likert scale that provides a global view of subjective assessments of usability.

workload experienced to complete the HTA. As Fig. 3 illustrates, there was a significant interaction between the sequences of the two conditions, $F(1, 34) = 7.87, p < .05$, $\eta = .188$. Tool-first participants took the longest ($M = 1768s, SD = 647s$) to complete the software tool condition but took the least amount of time ($M = 1020s, SD = 361s$) for P&P condition. P&P-first participants, for P&P condition took more time ($M = 1192s, SD = 585s$) than those in tool-first sequence. Whereas, for software tool condition they took less time ($M = 1436s, SD = 522s$) than those in tool-first sequence. In the case of workload, all participants in both sequences experienced significantly higher workload when they performed HTA with the help of the software tool ($M = 0.43, SD = 0.10; M = 0.45, SD = 0.12$) than P&P ($M = 0.38, SD = 0.11; M = 0.40, SD = 0.13$). Additionally, Fig. 4 shows that there was no significant interaction found, $F(1, 34) = 0.004, p = .95, \eta = .000 F(1, 34) = 0.004, p = .95, \eta = .000$.

Some of the individual components of the workload scale (NASA TLX) were then separately assessed. The results for mental demand and frustration components indicated that the mean scores were significantly higher for the software tool condition ($M = 0.60, SD = 0.26$), whereas lower for P&P condition ($M = 0.35, SD = 0.24$), $F(1, 34) = 19.02, p < .001, \eta = .359$. Participants' mean performance scores show that the participants performed better when they conducted the software tool condition ($M = 0.30, SD = 0.15$) than P&P condition ($M = 0.41, SD = 0.17$), $F(1, 34) = 15.79$, $p < .001, \eta = .317$. These results were found to be consistent for both sequences and there was no significant interaction between the sequences of the conditions, $F(1, 34) = 0.13, p = .72, \eta = .004$.

Usability. The mean SUS score of all participants for the software tool condition ($M = 55.10, SD = 15.00$) was significantly lower than that of the P&P condition ($M = 68.61, SD = 19.16$), $F(1, 34) = 10.59, p < .01, \eta = .237$, indicating that the software tool had a lower usability. After conducting further statistical analysis, it was observed that the total workload experienced by the participants for the software tool condition was significantly correlated to the usability of the software tool, $r = -.294$ $[-.594, .033], p = .04$.

Quality and Effectiveness. As hypothesized, the quality score (higher scores indicated lower quality) for the software condition ($M = 12.36, SD = 2.78$) was significantly lower than that for the P&P condition ($M = 13.61, SD = 2.61$), at the p-level = .001, $F(1, 34) = 14.01, p < .001, \eta = .292$. Likewise, effectiveness score was lower for the software tool condition ($M = 9.31, SD = 2.90$) than the P&P condition ($M = 10.00$, $SD = 3.06$), however, statistically this difference was not significant, $F(1, 34) = 3.70$, $p = .06, \eta = .098$. Figures 5 and 6 demonstrates the interaction between the two conditions based on the sequence in which they were conducted. For quality, participants belonging to both sequences received lower scores for the software tool condition ($M = 12.0, SD = 3.03; M = 12.72, SD = 2.54$) than for P&P condition ($M = 12.67$, $SD = 1.85; M = 14.56, SD = 2.96$), and no significant interaction was found between the sequences, $F(1, 34) = 3.05, p = .09, \eta = .82$. However, the difference between the score of the two conditions for the P&P-first participants was greater than that of tool-first participants. In the case of effectiveness, participants belonging to the P&P-first sequence received lower scores for the software tool condition ($M = 9.44, SD = 2.749$) than for the P&P condition ($M = 11.17, SD = 3.073$) whereas, participants in tool-first sequence received similar scores as participants in P&P-first sequence for the software tool condition (M = 9.17, SD = 3.111), but they obtained even lower scores for the P&P condition (M = 8.8.3, SD = 2.640). A significant interaction was found between the two sequences, $F(1,34)$ = 8.10, p = .01, η = .192.

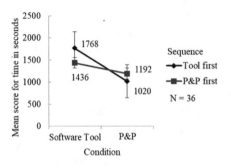

Fig. 3. Interaction diagram of the two sequences of the two conditions for time.

Fig. 4. Interaction diagram of the two sequences of the two conditions for total workload (lower score = lower workload).

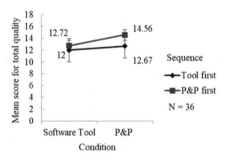
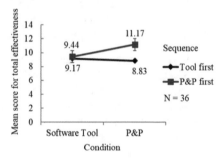

Fig. 5. Interaction diagram of the two sequences of the two conditions for total quality (lower score = better quality).

Fig. 6. Interaction diagram of the two sequences of the two conditions for total effectiveness (lower score = better effectiveness).

7   Discussion and Interpretation

Participants in both sequences took longer to complete their HTA with the software tool. This can be attributed to the learning curve effect. All participants had to concentrate on the online task at hand and needed to learn the new software tool. Tool-first participants took the longest to complete the software tool condition, as they had to simultaneously learn the software tool and analyze the online task for the first time i.e. their first condition. P&P-first participants required less time than tool-first participants did for the software tool condition, because they were able to familiarize themselves with the given online task in their first condition (P&P condition) which helped them to perform their second condition. Also, as they were acquainted with the task at hand, they were able to concentrate solely on learning and using the new tool. Thus, we conclude that the additional activity of learning and using the tool contributed to the added time. Even though tool-first participants took the longest to complete the software tool condition, they took the least amount of time for the P&P condition. As the software tool also aims at providing analysts with guidance for HTA procedure, the software tool is likely to have aided the participants in better understanding of HTA. This could be why tool-first participants performed much better for their second condition.

In the case of workload, all participants in both sequences experienced significantly higher workload when they performed HTA with the help of the software tool. As no interaction between the two sequences were found, the sequence of the conditions played no role in this outcome. Even individual component scores of workload, such as mental demand and frustration scores, were significantly higher for the software tool condition, which means performing the HTA with the software tool was more mentally demanding and highly frustrating for the participants. This clearly suggests that learning and using the software tool resulted in the participants experiencing higher workload. Yet, regardless of the high mental demand and frustration, participants' performance scores indicated that the participants in both sequences rated their performance better when they conducted HTA with the software tool than with P&P. This shows that participants performed better with the help of the software tool.

It was found that the software tool had low usability in comparison to P&P method. Also, the total workload experienced by the participants when they performed HTA with the software tool was significantly correlated to the software tool's usability score. This means that the higher the workload the participants experience, the lower was the usability score (SUS). This correlation suggests that the higher the workload when performing HTA during the software tool condition caused a lower evaluation result for the software tool's usability.

In the case of quality, as per the results, the overall quality of the performed HTA was significantly better using the software tool than using P&P. As displayed in Fig. 5, the means show that tool-first participants performed much better for their first condition using the software tool (software tool condition), but not as good (however relatively well) for their second condition with P&P (P&P condition). On the other hand, P&P-first participants performed their first condition (P&P condition) qualitatively least good but performed their second condition (software tool condition) with the help of the software tool much better. These mean scores indicate that the software tool enabled participants to produce better quality HTA i.e. the support received from the software tool was likely to have facilitated the tool-first participants to produce better quality HTA. Moreover, the guidance provided by the tool also increased their understanding of HTA, which they then utilized when performing HTA with P&P. P&P-first participants did not receive the support provided by the tool for their first condition. This showed in their poor scores. The quality of their HTA however increased tremendously when they performed HTA with the help of the software tool suggesting that the software tool aided in enhancing the quality of the participants' HTA.

For effectiveness, it can be stated that when HTA was conducted using the software tool, there was a tendency that participants were more effective in conducting HTA with the software tool than with P&P. As illustrated in Fig. 6, P&P-first participants did not perform their first condition (P&P condition) as well as their second condition (software tool condition). Tool-first participants received similar scores as P&P-first participants for the software tool condition, however, obtained even better scores for the P&P condition, i.e. they performed well in their first condition, but even better in their second condition. These scores can be interpreted in a similar way as the quality scores. P&P-first participants performed least effectively when they used P&P but showed tremendous improvement when they used the software tool. This suggests that the software tool supported the participants and enabled them to effectively conduct HTA. In addition, as tool-first participants performed their first condition similar to the second condition of P&P-first participants, it further confirms that the software tool not only facilitated HTA process but also supported the participants to maintain a consistent level and standard. Tool-first participants showed further improvement when they conducted HTA with P&P, thus demonstrating that the support and guidance received when using the software tool helped participants in their understanding of HTA and in turn their effective performance of HTA for the P&P condition. The presence of the sequence effect confirms that the software tool influenced the performance of the tool-first participants for their second condition (P&P condition).

8   Conclusion

The HFC software tool was designed to integrate the commonly used TA methods and techniques and to put forward a practical, convenient and acceptable software application that is suitable for TA processes. The software tool provides several benefits such as it helps analysts to reach better quality of their TA, maintaining a standardized and consistent level of performance in terms of quality and effectiveness of HTA and provides support and guidance which would help experts as well as enable novices to learn and understand the process of TA and enable them to produce good quality analysis. The results have also established some practical implementations of the software tool. It is useful for those who do not conduct TA on a daily basis, it allows visualization flexibility (graph view, table view, task map layering view) permitting analysts to remain at their comfort level, and one can use the tool in countless settings. Along with its strengths, there are however, weaknesses that make the software tool difficult and tedious to use. Higher time and workload and low usability indicate that the tool still needs improvement in terms of the interaction design to improve the usability. However, the tool has been demonstrated to give valuable guidance in performing the task analysis leading to better quality of the analysis.

Improvements and expansions on the tool are currently being tackled in the scope of KUKoMo project within Human-Robot Collaboration domain. New tool features have been recognized and included to better suit the domain requirements. A second evaluation, similar to the current research design, is currently being executed. Further tool evaluations will be implemented in the near future. It is also worthwhile to invest in the two checklists directing future research in establishing them as standardized guidelines of TA quality and effectiveness assessment measures.

Acknowledgement. The author is responsible for the content of this publication. We thank all participants for their time and effort and anonymous reviewers for their constructive comments. We would also like to show our gratitude to HMKW University of Applied Sciences for Media, Communication and Management, Berlin, for their support.

References

1. Annett, J., Stanton, N.A. (eds.): Task Analysis. CRC Press, London (2014)
2. Hollnagel, E.: Task analysis: why, what, and how. In: Handbook of Human Factors and Ergonomics, 3rd edn., pp. 371–383. Wiley, Hoboken (2006)
3. Stanton, N.A., Salmon, P.M., Rafferty, L.A., Walker, G.H., Baber, C., Jenkins, D.P.: Human Factors Methods: A Practical Guide for Engineering and Design. CRC Press, London (2017)
4. Bass, A., Aspinall, J., Walters, G., Stanton, N.: A software toolkit for hierarchical task analysis. Appl. Ergon. **26**(2), 147–151 (1995)
5. Annett, J., Duncan, K.D., Stammers, R.B., Gray, M.J.: Task Analysis. HMSO, London (1971)
6. Diaper, D., Stanton, N.A. (eds.): The Handbook of Task Analysis for Human-Computer Interaction. Lawrence Erlbaum Associates Inc. Publishers, Mahwah (2004)
7. Crystal, A., Ellington, B.: Task analysis and human-computer interaction: approaches, techniques, and levels of analysis. In: AMCIS 2004 Proceedings, 391, New York (2004)
8. Felipe, S.K., Adams, A.E., Rogers, W.A., Fisk, A.D.: Training novices on hierarchical task analysis. In: Proceedings of the Human Factors and Ergonomics Society Annual Meeting, vol. 54, no. 23, pp. 2005–2009. Sage, Los Angeles, California (2010)
9. Shepherd, A.: Hierarchial Task Analysis. CRC Press, London (2014)
10. Hone, G., Stanton, N.: HTA: the development and use of tools for hierarchical task analysis in the Armed Forces and elsewhere. Human Factors Integration Defence Technology Centre (2004)
11. Stanton, N.A., Salmon, P., Harris, D., Marshall, A., Demagalski, J., Young, M.S., Waldmann, T., Dekker, S.: Predicting pilot error: testing a new methodology and a multi-methods and analysts approach. Appl. Ergon. **40**(3), 464–471 (2009)
12. Hugo, J.: An integrated suite of tools to support human factors engineering. In: NPIC & HMIT 2015. Charlotte, NC (2015)
13. Kieffer, S., Batalas, N., Markopoulos, P.: Towards task analysis tool support. In: Proceedings of the 26th Australian Computer-Human Interaction Conference on Designing Futures: the Future of Design, pp. 59–68. ACM, New York (2014)
14. Dixon, K., Reed, Y., Reid, J.: Supporting teacher educator professional learning about assessment: insights from the design and use of a task analysis tool in a first-year BEd programme. S. Afr. J. High. Educ. **27**(5), 1099–1117 (2013)
15. Promann, M., Zhang, T.: Applying hierarchical task analysis method to discovery layer evaluation. Inf. Technol. Libr. **34**(1), 77–105 (2015)
16. Stanton, N.A.: Hierarchical task analysis: developments, applications, and extensions. Appl. Ergon. **37**(1), 55–79 (2006)

17. Roth, E.M., O'Hara, J., Bisantz, A., Endsley, M.R., Hoffman, R., Klein, G., Militello, L., Pfautz, J.D.: Discussion panel: how to recognize a "good" cognitive task analysis? In: HFES (ed.) Proceedings of the Human Factors and Ergonomics Society Annual Meeting, vol. 58, No. 1, pp. 320–324. Sage, Los Angeles, California (2014)
18. Onnasch, L., Bürglen, J., Tristram, S., Kolrep, H.: Break it down! a Practitioners' software tool for human factors task analysis and human error identification. In: Proceeding of 32nd Conference of European Association for Aviation Psychology – Thinking High AND Low: Cognition and Decision Making in Aviation. Cascais, Portugal (2016)
19. Bisantz, A.: Can (and should) i use these results? Six factors in CTA quality. In: Roth, E.M., et al.: Discussion Panel: How to Recognize a "Good" Cognitive Task Analysis? In: HFES, vol. 58, No. 1, pp. 320–324. Sage, Los Angeles, California (2014)
20. Klein, G.: Does a CTA study generate insights? In: Roth, E. M., et al.: Discussion Panel: How to Recognize a "Good" Cognitive Task Analysis? In: HFES, vol. 58, No. 1, pp. 320–324. Sage, Los Angeles, California (2014)
21. Brooke, J.: SUS-a quick and dirty usability scale. Usability Eval. Ind. **189**(194), 4–7 (1996)
22. Jordan, P.W.: An Introduction to Usability. CRC Press, London (1998)
23. Hart, S.G.: NASA Task load Index (TLX), vol. 1.0, Paper and pencil package. Ames Research Center, California (1986)

Applying the Ecological Interface Design Framework to a Proactively Controlled Network Routing System

Alexandros Eftychiou[(⊠)]

University College London, London, UK
A.eftyhiou@ucl.ac.uk

Abstract. The focus system of this study is the proactively controlled signal routing system used by BBC News Division to broadcast content around its global network. The Work Domain Analysis and Decision Ladder Analysis both associated with the Ecological Interface Design (EID) approach will be used to analyze the system. [7] states the EID approach can improve performance in a variety of domains based on the novel information requirements it uncovers, however we propose that certain alterations to how EID approaches proactively controlled systems would be beneficial. Specifically, when we conducted a work domain analysis, we focused on the object level of the model that specifies that system objects are normally arranged spatially on a display to reflect the actual layout of a system. We feel that explicit reconceptualization of how the objects are arranged is necessary when approaching proactively controlled displays. Specifically, we propose an addition in the form of a temporal arrangement of objects to the current guidelines for the spatial arrangement of objects. The reason for this is so that the trend-based element can be more explicitly stated in the work domain model. We also conducted a Decision Ladder analysis where the system state node of the framework which included the technical specification as well as time orientated usage profiles of resources. The analyses inform subsequent design of a time tunnel visualization to support proactive control of the BBC Broadcast Signal Routing System.

Keywords: Human factors · Systems engineering · Cognitive engineering · Work domain model · Decision ladder model · Proactive system control · Ecological interface design · Complex systems · Signal routing systems · Time tunnel displays

1 Introduction

1.1 A Proactively Controlled Signal Routing System

Complexity in systems arises from the interdependence of human and machine components that must interact to achieve the requirements of the system [6]. [7] states that compared to current design approaches the EID approach can uncover novel information in a wide variety of complex system application domains. This study aims to examine the above claim by attempting to model and uncover novel information requirements in the form of emergent properties that exist in a proactively controlled

© Springer Nature Switzerland AG 2020
T. Ahram (Ed.): AHFE 2019, AISC 965, pp. 379–388, 2020.
https://doi.org/10.1007/978-3-030-20454-9_39

network signal routing system. The focus system of this case study will be the BBC Master Control Room (MCR) at Broadcast House in London. In the following section we will provide a short background relating to the system. Broadcast control is a central hub that connects news crews out in the field to BBC studios. The control room operator controls the routing of signals between different locations in BBC global broadcast network. Routers are networking devices that forward packets of data between different networks, the BBC broadcast system utilizes multiple routers in interconnected networks allowing it to send and receive information internationally. The term source and destination will also be used throughout this case study. Points that create signals are described as sources while the locations the information arrives at are known as destinations. An example of a source could be a news crew out in the field reporting on a story while a destination might be a BBC news studio receiving the signal that could be in the form of video or just audio. MCR can be thought of as an intermediary, a central hub that is charged with connecting sources to destinations by opening all the right channels (routes) to allow the source to transmit to its destination. The task of connecting sources to destinations is known as Routing. During routing, signals are passed along various channel types that differ depending on the context of use. Proactive control is necessary to avoid resource conflict due to the finite resources controlled by MCR. This competition is further increased due to personal preferences that arise from individuals that request routes. The organizational structure within the broadcast control room involves a routing operator who will execute the routing of tasks that have been prepared by a managing operator. The managing operator who proactively controls the system must mentally integrate information from multiple sources to gain an overall view of the situation or "Big Picture" to plan and prepare routes that will not conflict with each other minimizing signal routing resource conflicts.

Critical information for the managing operator is scheduled information about future bookings for resources as well as historical information. Managing operators can scroll through three days' worth of future bookings, allowing them to anticipate future events. There are also resource bookings that are made last minute and thus are not registered on the system, in these cases the managing operator is expected to use historical information to determine trends and fill in the gaps in the schedule where last minute future bookings will arise. Broadcast Network Routing at the BBC typifies the complex sociotechnical domains that Cognitive Work Analysis is used to investigate.

1.2   The Time Tunnel Format

The time tunnel format display was adapted from the DURESS display, the Work Domain and Decision Ladder analyses described below will provide input to the development of display design concepts and the initial display prototypes. The time tunnel design is based on [1] where the Time Tunnel format was used to support proactive decision making and succeeded in providing superior support compared to a traditional line trend interface. We argue that based on [1] findings, that this format would be beneficial to the managing operators in the master control room. These interfaces provide the user with a representation of their ecology and the domain they are controlling as they present the relationships between variables, the goals, system constraints and the physical component states. The key data from the Work Domain Model will be mapped onto this display format in such a way to provide the operator with a way to perceive multiple levels of the model and how they are linked together. The temporal organization of the lowest levels of technical information on this visualization will permit operators to determine repeating usage patterns (trends) of source, destination and intermediate technical equipment over time and thus determine how they will be used in the future. Further the higher level of information flow will be visible as a vertical line through the interface in order to give operators an understanding how previous usage of technical equipment tied in with the overall ability of the system to perform its goals. This overview is of multiple levels of the system and their relationships now becomes external and can be used in conjunction with the operator mental model thus taking some of the mental load off the operator (Fig. 1).

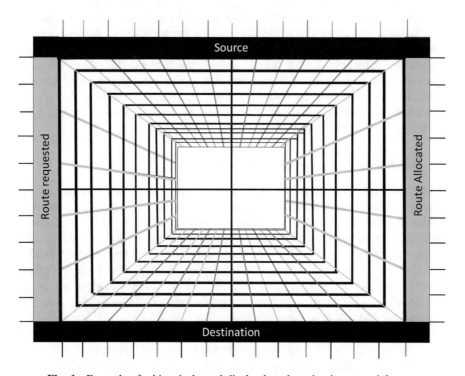

Fig. 1. Example of a historical trend display based on the time tunnel format.

2   Method

The case study involved an extensive ethnographic study using unstructured contextual interviews, performed with the managing operator of the control room to achieve two purposes, first to facilitate an exploration of the domain (exploratory interviews), and second to become familiar with specific aspects of the domain (familiarization interviews). The contextual enquiry included sessions where operators were required to think-aloud while carrying out tasks. The purpose of the think-aloud protocol sessions was to provide a deeper understanding relating to the workflow of the operators focusing primarily on the tools used to proactively control the Broadcast system. We analyzed the data using tools from the cognitive work analysis approach including the Work Domain Analysis (WDA) and the Decision Ladder Analysis or (ConTa). The outputs of the analysis were a Work Domain Model (WDM) of the broadcast control system and the complementing Decision Ladder Model (DLM).

2.1   Framing the System for Work Domain Analysis: Defining the Boundaries of the Analysis

We approached the modelling of this system by first defining the problem to be solved through use of the Work Domain Analysis (WDA). This assisted in identifying the boundaries of the system we were interested in modelling. Determining the boundaries of the system also allowed the identification of the technical equipment and their control surfaces currently used by operators in proactive control of the system as well as where the boundaries lie for this proactively controlled system. Determining the boundaries was a pragmatic consideration that would help to determine the focus system. The system boundaries are set at the sources and the destinations that operators are expected to communicate with and proactively control the flow of information through the system via the control room.

2.2   Defining the Nature of the Constraints

We identified technical constraints that govern the system, these include technical equipment at the source and the destination, in addition technical equipment also includes the routers and the circuits that are used to permit the flow of information between source and the destination. Technical constraints involve the correct matching of equipment at a source to circuits through which signals will be routed and finally to equipment that also matches these technical specifications at a source. Technical Constraints are hard constraints if not met will result in the system either not sending the information along the route or may damage the information while being sent down the route from the source to the destination.

Intentional constraints involve the constraints where the managing operator's intentions determine how he controls the system. These intentions arise out of ways in which resources are allocated in order to further improve the success of a routing task and get content to air. Intentional constraints revolve around helping ensure that personnel at a source and studio staff at a destination connect their equipment to the BBC broadcast routing system in as trouble free a manner as possible. An example of an intentional constraints is that content creating units always like to use the same resource repeatedly. The intentions of the staff who rely on the managing controller also affect the managing operator as the preferences of people at the source and the destination must also be considered to assist the flow of information through the linking of the route by the source or destination.

2.3   Structuring Data Collected Along the Work Domain Model

Structuring the data according to Means-End Relation Hierarchy used in the Work Domain model was done by using the videos of managing operators that were collected during the contextual enquiries and this gave a narrative of the work. The audio from the video collected was then transcribed using keyword analysis to categorize sections of text according to where they would fit on the abstraction hierarchy model.

2.4   The Work Domain Model

See Fig. 2.

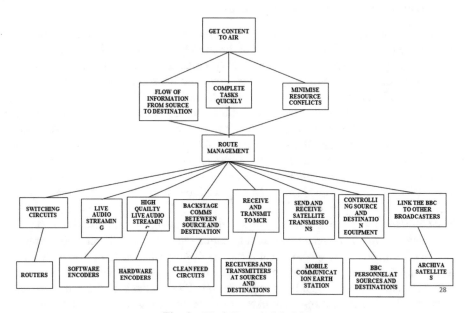

Fig. 2. Work Domain Model.

2.5   Top Levels of the WDM: Goals, Values and Priorities of the System

The goal of the system is to get content to air, the Values and priority measure level of the system relates in terms of the information that flows from one point to another. Information is generated at a source and transferred through a channel through to its destination. The mechanism that permits this transfer must ensure that there are available paths. In order to ensure the availability of paths that conform to hard technical constraints and soft constraints relating to preferences of the staff at a source or a destination the operators have set organizational priorities relating to routing tasks that allow for optimal allocation of routes for information to flow from source to destination.

2.6   Middle Level: Proactive Work Function of Route Management

The proactive work function of Route Management requires the operator to anticipate future route requests by using the schedule to look for future scheduled requests. The managing operator must also predict future Non-Scheduled requests by looking into the past. The reason for this is that an imminent future event which could conflict with his present decision to allocate, clear or retain a route and thus it's equipment that in the case of reallocation or retainment becomes locked for the duration of a broadcast. If the resource is cleared then this means additional time to reconnect all necessary routers which in can create a backlog of tasks that have not been completed on time.

2.7   Lower Levels: Technical Information

The data at this level represents the physical objects in a work system that afford the processes and functions at higher levels of the model. The physical objects can include an inventory of the physical equipment as is specified at the lowest level of the model. Equipment can involve various types of receiver and transmitter equipment located at BBC regional, national and foreign studios as well as BBC foreign, regional and national satellite trucks. Other technical equipment includes transmitters and receivers located internationally at BBC bureau. The equipment mentioned above helps to send and receive signals through the system thus enabling broadcasts. Receivers tend to be thought of as destinations while the Transmitters can be thought of as a Sources.

In addition, the system includes routers and circuits which permit the signal to travel from a source to a destination. Examples of routers and circuits are International Satellite Routers Regional Satellite Routers Regional Circuit Routers circuit types include high quality production channels and lower quality communication channels that allow a source to communicate with a destination and a destination with a source. The physical object level specifies that it can involve information that refers to how the objects are organized in relation to one another. In the case of this proactively controlled system the objects will be arranged temporally on a timeline in relation to each other in order to help determine the trends of usage and the technical specifications of the equipment in order to be able to predict what equipment will be requested in unscheduled requests over time.

2.8   Decision Ladder Analysis

The purpose of a decision ladder analysis is to identify the information processing stages and resultant knowledge states involved in the transformation of inputs into outputs transformation. We mapped the stages for the Work function of 'Route Management' identified in the WDA analysis onto a Decision Ladder. Decision ladders have been used in the redesign of an already existing system as is the case with [4]. The authors identified information requirements through conducting qualitative field work and categorized the data according to the Decision Ladder. There are two types of nodes, circular nodes involve information states, they are the resultant states of knowledge that originate from the data processing activities specified in the rectangular nodes.

We adopted the [4] method as they took a bottom up approach when using the decision ladder to model an existing system as is the case with our analysis of the BBC Broadcast system. Data gathering techniques used to identify the necessary professional terminology and the phrases in a work domain involved collecting data through interviews with the managing operator, observation of their work activities and Walkthroughs and Talkthroughs. The data collected during visits to the control rooms allowed us to form a set of activity elements that are prototypical and were defined either by the problem to solve or the situation in which to solve the problem. The different combinations of these collected during interviews with the managing operator served to characterize the activity within the work system. Interviewee responses were examined for information relating to all parts of the decision ladder. We conducted Keyword Analysis on the transcripts based on keywords were taken from [5] that express the various parts of the decision ladder. From the Decision Ladder we found that the system state is the most important element we are interested in due its relevance to the temporal organization of the various technical elements on the timeline. The System state node in the Decision Ladder Analysis is an understanding of the current situation or the state of the system and is dependent on what the overall goal of the work function is, which in this case is to get content to air on time while minimizing resource conflicts. In populating the system state node, we first considered what situation assessments are required to achieve the overall goal of a work function as stated by [3]. It is an awareness of a situation by the BBC operator, that awareness is based on the availability of information about the resources comprised of a combination of the technical specification for the types of route, this includes the intermediate equipment used to make the route as well as the equipment at the source and destination. The other side to this system state is the qualitatively different temporal state of the route which includes the future state of the route based on the usage profile of that route. The usage profile of a route over time includes the duration of the route, the time the route was made active, frequency that the route is made active and period over which the route is requested. The route usage profile and the technical identification information create a total of 44 system states (Table 1).

Table 1. System states types created by combining usage profiles and technical specifications.

System state	Route usage profile	Route technical spec
System state 1	New route that won't repeat	Technical type spec
System state 2	New route that will repeat	Technical type spec
System state 3	Historical repeating route	Technical type spec
System state 4	Historical final route	Technical type spec

2.9 Semantic Mapping of a Time Tunnel Visualization to a Novel BBC Broadcast Control Interface Demonstrator

In the Time Tunnel version shown below the horizontal line has been split into two separate lines. The fact the lines are not aligned demonstrates a sub-optimal allocation of resources which is linked to the flow of the data through the signal routing system.

The hinged design of this visualization (blue dots at either end of horizontal lines) is taken from [2] RAPTOR interface used for military command and control (Fig. 3).

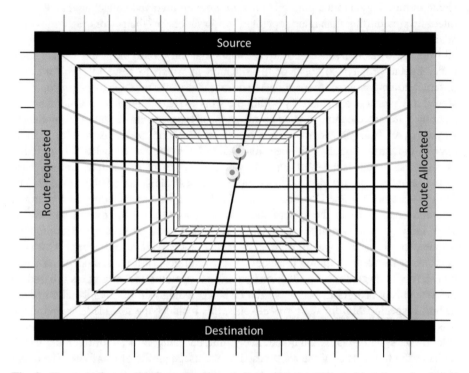

Fig. 3. Emergent feature interfaces time tunnel display format (temporal location of technical Equipment positioned on the left and the right of the time tunnel box.)

Semantic Mapping of high-level flow of information. The Work Domain Analysis identified the key concept of flow of information from a source to a destination and vice versa thus allowing the system goal of getting content to the air. The ethnographic study has identified that the operators do in fact use historical data to build up an overview of the situation in order to create routes through allocating resources to sources and destinations. The emergent feature of information flow coming into the system from a source and moving through the system to arrive at a destination is visualized using the emergent feature of linearity by connecting the source and destination using a vertical line. The linearity and orientation emergent features can be used in such a way to indicate an imbalance in the flow of information through the system and thus provide a representation of the performance metric used to define how well the system is performing.

Semantic Mapping of lower level physical resources: The physical resources are mapped to the left and right walls of the time tunnel, the Left side of the tunnel includes the technical resources that were requested (Bookings) e.g. a receiver and transmitter, a router or a circuit and the technical resources that were provided. (Packages).

The balance of the booking and package entities are shown by the horizontal lines connecting the right and left of the time tunnel, the alignment takes into consideration soft constraints, which are the organizational entities within the BBC who are involved with sending the data around the system. The horizontal lines express the package to booking suitability measure otherwise described as route suitability measure. The distance between the two lines denote the level of fit with greater distance indicating less fit. The horizontal line element of the visualization gives operators the overview of the emergent feature of suitability relating to allocation of routing equipment to a routing request by considering soft constraints that are usually considered through laborious investigation of numerous screens.

The information requirements we produced and semantically mapped onto the time tunnel format showed the time tunnel format is a suitable candidate for an interface that operators can use to give them an overview of the information flow through the system based on allocation of resources. Historical information can be viewed by the managing operator on the walls of the tunnel as historical usage profiles. It gives an indication of what technical equipment the crew out in the field has previously requested which is mapped to the left wall of the time tunnel. The equipment the managing operator provided is mapped to the right side of the tunnel. The suitability between the two is denoted by the horizontal emergent feature line that connects the usage profiles mapped to the left and the right walls. This presentation of variables on the time tunnel and the emergent relationships between the variables, aims to minimize misallocation decisions that create conflicts with repeating well suited future pairings and have a negative effect on the flow of information through the system. On the flip side operators can also predict poor future pairings based on the historical information that they can set a lower priority to the soft constraint of keeping the same booking – package pair, this is because the resources did not work well previously.

3   Discussion

This paper has provided a case study of a first of its kind use of the Work Domain Analysis and Decision Ladder Analysis in modelling a proactively controlled complex system domain. The Work Domain Model provided novel information requirements at multiple levels of abstraction that were then utilized in a time tunnel visualization of the domain. The modelling process was able to express this domain through the recon-ceptualization of the system physical objects along a temporal plane which is more relevant when analyzing proactively controlled systems that rely on trend-based decision making than just the spatial arrangement of objects.

This visualization can provide the managing operator with an integrated overview of the system state that moves through time and can be factored into the proactive decision-making process of route management. We propose that the time tunnel format can be used to indicate the goodness of fit in terms of how allocated resources (left and right of the time tunnel) affect the flow of information from a source to a destination (top to bottom) which is the success metric of the system. We feel the time tunnel is an untapped visualization that deserves more attention for the unique ways in which we can present data on it. Furthermore, we feel that the next step would involve a novel virtual reality component where users are able to be more involved with the visualization and capable of moving through the tunnel and experiencing the data in new ways.

References

1. Bennett, K.B., Payne, M., Walters, B.: An evaluation of a "time tunnel" display format for the presentation of temporal information. Hum. Factors **47**, 342–359 (2005)
2. Bennett, K.B., Posey, S.M., Shattuck, L.G.: Ecological interface design for military command and control. J. Cogn. Eng. Decis. Making **2**(4), 349–385 (2008)
3. Elix, B., Naikar, N.: Designing safe and effective future systems: a new approach for modelling decisions in future systems with cognitive work analysis. In: The 8th International Symposium of the Australian Aviation Psychology Association, Sydney, Australia with Cognitive Work Analysis. Proceedings of the 8th International Symposium of the Australian Aviation Psychology Association (2008)
4. Jenkins, D.P.: Cognitive Work Analysis: Coping with Complexity. Ashgate, Aldershot (2009)
5. Naikar, N., Moylan, A., Pearce, B.: Analysing activity in complex systems with cognitive work analysis: concepts, guidelines and case study for control task analysis. Theor. Issues Ergon. Sci. **7**(4), 371–394 (2006)
6. Naikar, N., Sanderson, P.M.: Evaluating design proposals for complex systems with work domain analysis. Hum. Factors **43**, 529–542 (2001)
7. Vicente, K.J.: Ecological interface design: Progress and challenges. Hum. Factors **44**, 62–78 (2002)

An Instrumented Software Framework for the Rapid Development of Experimental User Interfaces

Hesham Fouad$^{(\boxtimes)}$ and Derek Brock$^{(\boxtimes)}$

Naval Research Laboratory, 4555 Overlook Ave SW, Washington, DC, MD 20735, USA
{hesham.fouad,derek.brock}@nrl.navy.mil

Abstract. One of the more demanding aspects of formal evaluations in user-centered system design is the underlying requirement for an experimental platform that accurately reflects the functionality being tested and captures relevant user performance data. While we are unaware of any truly turn-key solutions for avoiding many of the inherent costs associated with the iterative cycle of user-centered system development and formal testing, in our mediated multitasking research, we have found that a multi-application host environment, instrumented for user tracking and a playback function for reviewing what individuals attended to, experienced, and did during the task-performance segments of interaction studies, could be implemented as a reusable infrastructure of services and control functions. In this short paper, we introduce the Testbed Framework System, a flexible and extensible software platform for use in the development and conduct of formal usability testing, and outline a number of its core functions and capabilities.

Keywords: User-centered design · Usability testing · Instrumented testbed · Spatial audio · User monitoring · Mediated attention management · Multi-application development framework · Multitasking

1 Introduction

End-user applications that support information gathering and retrieval, decision-making, and communications in many enterprise settings, including the military, industry, medicine, and commerce, are often considered mission critical [1]. It has long been recognized that human factors are pivotal in the design of effective user interfaces in these and other types of systems, as well as in the composition of their information processing functions. Optimal information applications structure how content is presented, accessed, and understood by individuals with intuitive, task-based constructs and affordances for different levels of expertise [2, 3]. Moreover, the caliber of a user-centered interaction design [4] can have measurable effects on an enterprise through reduced error rates and increases in the breadth of users' operational awareness [5, 6].

User-centered design is a conceptual outgrowth of applied human factors and has been widely advocated in system development for several decades. Most successful as an iterative process of refinement [7], one of its basic tenets, showing the efficacy of

T. Ahram (Ed.): AHFE 2019, AISC 965, pp. 389–398, 2020.
https://doi.org/10.1007/978-3-030-20454-9_40

candidate solutions, requires repeated and systematic evaluations with real users [8]. Designing empirical studies for this purpose requires formal methods and thorough planning. One of the more demanding parts of this undertaking is the underlying need for a specialized platform that implements or accurately simulates contrasting, proto-typical task requirements and captures user performance data in each experimental run of a study (or possibly several such platforms) [9]. The expense of realizing this requirement has, over time, given rise to a variety of so-called "discount" usability evaluation strategies [10]. Ranging from informal, heuristic-based inspection and piloting to expert reviews and automated testing, none of these methods are represented as, nor are they intended to be, sufficient or equivalent to the power of a formally structured comparative evaluation. Some have proven advantages for refining usability at various stages of user interface development [11], but can also have implementation requirements of their own and other costs that can undermine the merit of their utility [12]. Likewise, another response to the expense of formal testing has been the practice of "reusability" [13], here, the idea that assets developed for evaluation in a prior effort can be reused, repurposed, or leveraged in subsequent projects.

In our experience as researchers concerned with designs for mediated multitasking, there is no simple turn-key solution for avoiding many of the repeated costs associated with the iterative cycle of user-centered system development and formal evaluation. The range of tasks we work with, involving transactions with text, audio, and graphical information displays, have operational and media-specific interaction requirements that can only be evaluated with a suite of distinct testbeds. Except for efforts in which software reuse is clearly warranted, there are few time-and-labor-saving substitutes for the straightforward process of developing platforms tailored for specific application requirements for user testing. We have, however, found that a few of our more chal-lenging evaluation objectives continually recur. Among these is the need for a host environment in which separate interaction tasks involving different media can be prototyped, scripted, and synchronized. By itself, having an environment with these capacities would facilitate a number of informal evaluation methods over the course of a system development effort. Additionally, because of the mix of attentional moves people employ in multitasking, two more capabilities that we have needed to use are user tracking and a playback function for reviewing corresponding sequences of dis-play activity (including sound) and user interactions across concurrent interfaces. Ultimately, these repeating requirements in our work motivated the development of the Testbed Framework System (TFS) presented in this paper.

The Testbed Framework System is a flexible and extensible environment for use in application development and usability studies. Somewhat like a lightweight operating system, it provides a runtime service infrastructure that, together with a tailored set of control functions, is designed to allow researchers to quickly develop, and concurrently run, fully instrumented software applications for iterative testing, revision, and experimentation. It's key service functions are (1) an application interface for com-munication between concurrent applications and the TFS; (2) configurable time man-agement via a global clock for timed execution of system services and events; (3) passive detection of user application focus via head tracking; (4) a fully control-lable, full-featured spatial audio display environment; and (5) a linear, variable speed event review function that uses the TFS's event monitoring/logging services to playback recorded timelines of user, application, auditory, and head tracking events. In the following sections, we profile each of these system functionalities in greater depth, briefly outline some of our research experiences with the TFS, and close with a discussion of future goals.

2   The TFS Application User Interface (UI)

The TFS provides experimenters with an intuitive graphical user interface for creating, controlling, and monitoring user interaction experiments in real-time (Fig. 1). The system's UI consists of four primary areas: a toolbar, an editor for controlling the monitored applications that implement the user-centered task design(s) being studied, a three-dimensional canvas depicting the current state of the user and the experimental applications, and a log output window.

The key features of the toolbar are called out in Fig. 1 below:

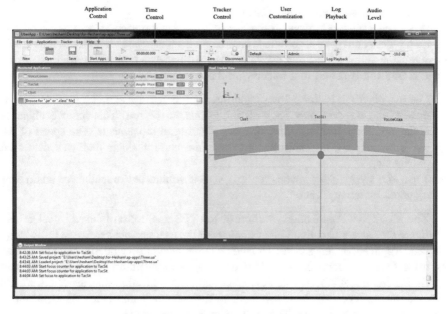

Fig. 1. The TFS application user interface.

- The Start Apps button (Application Control) launches all the monitored applications. This does not, however, start the global clock. Monitored applications are assumed to remain dormant until the global clock is started
- The Time Controls allow the experimenter to start the global clock, to view the current time, and to specify a time multiplier that accelerates and decelerates real-time based on that multiplier. The time multiplier applies during application execution as well as event log playback.

- The Tracker Controls consist of a Zero button that sets the head tracker's current azimuth and elevation measures as its origin; this effectively resets the head tracker. The Disconnect button disassociates the tracked head position from the audio spatialization function. This has the effect of making the position of directional audio cues invariant to user head motion.
- The User Customization section allows experimenters to select a Head Related Transfer Function (HRTF) that is optimized to the current user.

An HRTF consists of a set of digital filters that are used to simulate directional audio. Due to individual differences in the physiology of users' pinnae, each user has a unique HRTF. The human auditory system learns to spatialize sound based on a person's unique HRTF. Therefore, to achieve optimal results in the simulation of directional audio, individual HRTFs have to be measured for each user. This is not practical due to the specialized equipment and time involved in measuring a user's HRTFs. An alternative strategy, used in the TFS, is to use a structured HRTF selection process that selects the nearest fit HRTF out of a library of measured HRTFs [14]. This selection process is done using an application that is provided by the SoundScape3D spatial audio engine used by the TFS.

A second dropdown menu in the User Customization section selects the current user's profile. A unique user profile is created for each user and identifies that user by a unique name and specifies that user's gender. All log data collected during an experimental trial run is stored in a folder with that user's name.

- The log playback button allows experimenters to select a stored log of an experimental trial run and play back all the events in the run. This has the effect of replaying all the user's keyboard, mouse, and head movements. The speed of the log playback can be controlled by the time control slider that was discussed previously.
- The audio level slider provides control over the volume of the spatialized audio cues triggered by an application.

The Monitored Applications section of the TFS application is used to select and configure the Java application(s) that implement the user-centered software tasks being studied. Applications are added to a study's experimental user interface configuration by respectively selecting the Java JAR or class files containing each application and specifying the horizontal and vertical span and screen location that the application's user interface will occupy in the user's forward view (i.e., hemisphere). As applications are added, they appear in the TFS application's canvas window.

The function of the Head Tracker View is to render a depiction of the user's head orientation as well as the horizontal and vertical span occupied by each application within the user's frontal hemispheric view. The canvas also highlights the application being attended to by the user based on the current user's gaze vector (depicted in red).

3   System Architecture

The TFS system architecture is depicted in Fig. 2 below. The TFS system components were written in C++ so that low level operating system functions could be accessed. This is required for monitoring keyboard, mouse, and head motion events. C++ also provides runtime efficiency that makes the real-time logging of large amounts of data transparent to the operation of the system. In the following sections we discuss the principal features of each of the architectural components of the TFS system.

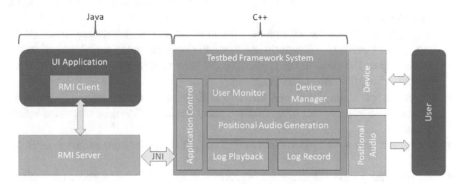

Fig. 2. The TFS system architecture.

3.1   Application Programming Interface

Communication between the TFS and applications comprising the user interface occur through a simple Java interface. The TFS uses Java's Remote Method Invocation (RMI) capability to facilitate communication between UI applications and the TFS system. The Java RMI implements a client/server model where multiple clients can connect to a single instance of a named RMI server to communicate. Once that connection is established, clients can invoke methods in the RMI server through an instance of an RMI client class. The TFS system creates an RMI server at start up and uses the Java Native Interface (JNI) to invoke methods within the server. The TFS system uses this capability to periodically update the global clock, update each application's state, retrieve applications' requests for service, and close applications.

The TFS application programming interface (API) also provides a full-featured auditory display facility that enables the playback of multiple spatialized audio streams. Control of the play state and position of audio sources is accomplished through the API. The TFS uses head tracking data to adjust the orientation of the virtual listener so that sound sources appear fixed in physical space. (Note that this functionality can be disabled via the Disconnect button described under "Tracker Controls" in Sect. 2.) The TFS also enables the selection of listener profiles so that spatial audio can be optimized for each subject (see "User Customization" in Sect. 2).

The RMI client class implemented by the TFS system provides applications with the ability to query information as well as control elements of the TFS system. Each method invocation results in the creation of a service request object that is periodically retrieved and processed by the TFS's Application Control. The methods provided follow:

- GetCurrentTime – Retrieves the current value of the global clock.
- GetIsRunning – Retrieves a Boolean value indicating whether or not the global clock is running. The TFS clock can be started and stopped either through the TFS application UI or by an RMI call from one of the applications.
- StartClock – Starts the TFS clock. The intent of giving an individual application the ability to control the global clock is that, in some instances, it may be desirable to have one application act as the clock master, starting and pausing the clock based on events in an experiment.
- Pause Clock – Stops the TFS clock without resetting the global time to zero.
- LogApplicationEvent – Inserts an entry into the application log file. Application log events contain the name of the application the generated the event as well as a string describing an application specific event.
- SetFocusApplication – Informs the TFS that the user should attend to a named application. The construct of application focus in the TFS will be discussed below in Sect. 3.2).
- GetApplications – Returns the name of all the active applications running under the TFS.
- GetApplicationState – Returns an object containing information about an application's current state. Application state is defined as follows:
  - Whether or not the application has user focus.
  - Whether or not the application was selected to have the user's focus.
  - The focus priority of the application (discussed in Sect. 3.2).
  - The last time that the application received the user's focus.
  - The last time that the application lost the user's focus.
- CreateAudioSource – Creates a named positional audio source. Audio sources can be audio files, audio inputs from an audio device, audio streams broadcast over a network, or audio channels generated by a ReWire enabled application [15].
- DeleteAudioSource – Removes a positional audio source.
- SetAudioSourcePosition – Set the position of a directional audio source. Position is specified in three dimensional space using x, y, and z coordinates.
- StartAudioSource – Starts the playback of a named audio source. A playback duration can be specified in seconds.
- StopAudioSource – Stops the playback of a named audio source and sets the playback position to zero.
- PauseAudioSource – Stops the playback of a named audio source but does not change the current playback position of the source.
- SetAudioSourceLevel – Sets the level of a named audio source. Source levels are expressed in decibels.
- SetAudioSourceDAF – Sets the distance attenuation factor of an audio source. The distance attenuation factor is a number n such that $0 < n <= 2$ and determines how an audio source's intensity behaves as the distance between the source and listener changes. A distance attenuation factor of 1 attenuates sound intensity as one would experience in real-world conditions. Values greater than 1 result in a faster drop off

of source intensity with distance. Values between 0 and 1 result in a more gradual drop off of source intensity with distance.

3.2   User Monitoring and Attention Management

One of the research areas where the TFS is being used is the exploration of user attention management for mitigating the negative effects of highly multitasked environments. In this context attention management is considered an assistive technology that reduces user cognitive load by monitoring events and information occurring in each of the applications comprising a system's user interface and cueing the user, either through visual or auditory channels, when new events or information occur in an application that is not being attended to by the user require attention. The TFS does not implement any specific attention management scheme; that is left to the experiment designer. It does, however, provide infrastructure that can be used to implement such a capability.

Any attention management scheme will generally require the following three elements:

1. A method to determine what the user is currently attending to.
2. A method to determine what the user should be attending to.
3. A method to cue the user that she may want to shift her attention elsewhere.

The TFS's User Monitoring component provides the first element above through user focus tracking. As mentioned earlier, the TFS positions the component applications making up an experimental user interface along a curved plane spanning the user's frontal hemisphere. An application's span is defined by an area of the frontal plane this is specified by minimum and maximum horizontal and vertical angles. The rational for specifying an application's area using angles instead of pixels is that the TFS does not have a priori knowledge of the size of the display area being used; it can consist of one or more application displays of varying sizes and resolutions. The TFS considers the collective size, in pixels, of all of these displays as a single continuous display spanning a wedge that is 90° high and 180° wide, where horizontal positions span −90 to 90° and vertical positions span −45 to 45°. The TFS places each component application's display within an area of the curved plane by mapping its position and dimensions, in pixels, to angles.

To detect which application a user is attending to, the TFS utilizes the user's head orientation to calculate a gaze vector. With the user looking straight ahead, the gaze vector lies along the Z axis. As the user's head moves, the gaze vector is transformed by the azimuth and elevation angles measured by the head tracker. The gaze vector is then intersected with the curved plane containing the application displays. If the intersection point falls within the area occupied by an application's window, the TFS flags that application as having user focus by modifying the application state objects to indicate which application currently has user focus.

The approach described above generally performs well, but some confounding factors must be taken into account. One example is the occurrence of momentary head movements that are not indicative of a change in the user's focus. A user may, for example, gaze at a clock on the wall momentarily. To handle this, the TFS allows the experimenter to specify a lag time that delays the switching of the focus application so that if user focus leaves and returns to the same application within the specified lag time, a change in focus application is not registered.

The second requirement for implementing an attention management scheme is a method for determining which application the user should be attending to. This problem is domain specific and a generalized solution is not practical. Instead, the TFS provides experimenters with ability to implement an attention scheduler using the SetFocusApplication, SetFocusPriority, and GetApplicationState client methods in the client interface.

An attention scheduler is a TFS application that does not provide a user interface. Instead, it monitors the state of each running application as well as the operational state of the system and determines the focus priority of each application. Focus priority can be one of four values:

- Low – indicates that the running application is in an idle state or doing background work.
- Normal – indicates that the application is a neutral state. Applications will be in this state most of the time.
- High – indicates that an event requiring the user's attention is upcoming.
- Critical – indicates that an event requiring the user's attention is pending.

The attention scheduler determines that a change in user focus is required when the application with the highest focus priority is not the current focus application as detected by the User Monitor. The scheduler would then call the SetFocusApplication method to inform that application that it is selected for attaining user focus. It is that application's responsibility to use either auditory or visual cues to indicate to the user that she should attend to that application. Once the application determines that it has received focus through its application state, it can terminate its user cues.

3.3   Event Logging and Playback

One of the most essential functions of an experimental testbed is event logging. The TFS Log Record component interacts with the Device Manager to provide the capability to record keyboard, mouse, and head motion events. Applications can also insert application specific events using the LogApplicationEvent method in the client interface. This provides experimenters with a rich data set for analysis.

One of the challenges in logging high volume data streams like mouse and head tracker events, is that the TFS system can become I/O bound causing event drop outs and lags in the responsiveness of the user interface. To address this issue, the TFS buffers all events, in their binary form, and then periodically writes them to disk during periods of low system activity. This ensures that all events are captured and the system remains responsive. A separate application is provided that reads the binary log files and transforms them into a human readable form.

Keeping the log files in binary form makes the log files more compact and also facilitates the log playback function. Log playback has proven to be one of most useful features of the TFS. The ability to efficiently scan all of a user's actions during an experimental run provides insight and intuition that is difficult to ascertain by looking at textual logs of those events.

4 Summary and Conclusion

The TFS has already been successfully used at our lab for exploratory user-centered design in mediated multitasking including studies that explored the use of spatial audio in the serialization and modulation of competing audio communications [16, 17], and mediated attention management in highly multi-tasked environments (unpublished work; see Fig. 3). Our intent is to make the TFS system available to other researchers in the hope that they might gain from the economy afforded by this system.

Fig. 3. The TFS being used to explore mediated management of user attention in a highly multi-tasked, simulated command-and-control information environment.

References

1. Mentler, T., Herczeg, M.: On the role of user experience in mission- or safety-critical systems. In: Weyers, B., Dittmar, A. (eds.) Mensch und Computer 2016 – Workshopband. Gesellschaft für Informatik e. V, Aachen (2016)
2. McGrenere, J., Ho, W.: Affordances: clarifying and evolving a concept. In: Proceedings of the Graphics Interface. Canadian Human-Computer Communications Society, pp. 179–186. Toronto (2000)
3. Hartson, H.R.: Cognitive, physical, sensory, and functional affordances in interaction design. Behav. Inform. Technol. 22(5), 315–338 (2003)
4. Norman, D.A., Draper, S.W. (eds.): User Centered System Design. Lawrence Erlbaum Associates, Hillsdale (1986)

5. Johnson, C.M., Johnson, T.R., Zhang, J.: A user-centered framework for redesigning health care interfaces. J. Biomed. Inform. **38**(01), 75–87 (2005)
6. Endsley, M.R., Bolte, B., Jones, D.G.: Designing for Situation Awareness: An Approach to Human-Centered Design. Taylor & Francis, London (2003)
7. Nielsen, J.: Iterative user-interface design. Computer **26**, 32–41 (1993)
8. Ritter, F.E., Baxter, G.D., Churchill, E.F.: Foundations for Designing User-Centered Systems: What System Designers Need to Know About People. Springer, London (2014)
9. Harte, R., Glynn, L., Rodriquez-Molinero, A., Baker, P.M., Scharf, T., Quinlan, L.R., ÓLaighin, G.: A human-centered design methodology to enhance the usability, human factors, and user experience of connected health systems: a three-phase methodology. JMIR Hum. Factors **4**(1), e8 (2017)
10. Nielsen, J.: Usability Engineering. Academic Press, San Diego (1993)
11. Edwards, K., Belotti, V., Newman, N.W.: Stuck in the middle. The challenges of user-centered design and evaluation of infrastructure. In: Proc. CHI 2003 Conf. on Human Factors in Computing Systems, ACM, New York (2003)
12. Pettit Jones, C.: Lessons learned from discount usability engineering for the federal government. Tech. Comm. **50**, 232–246 (2003)
13. Greenberg, S., Buxton, B.: Usability evaluation considered harmful (some of the time). In: Proc. SIGCHI Conference on Human Factors in Computing Systems (2008)
14. Seeber, B.U., Fastl, H.: Subjective selection of non-individual head-related transfer function. In: Proc. 2003 Int. Conf. on Auditory Display, pp. 259–262, Boston University, Boston, MA. (2003)
15. Developer - ReWire - Technical information. propellerheads.se. 28 Jan 2017
16. Brock, D., Peres, S.C., McClimens, B.: Evaluating listeners' attention to, and comprehension of, serially interleaved, rate-accelerated speech. In: Proc. 18th Int. Conf. on Auditory Display, Atlanta, GA (2012)
17. Brock, D., Wasylyshyn, C., McClimens, B.: Word spotting in a multichannel virtual auditory display at normal and accelerated rates of speech. In: Proc. 22th Int. Conf. on Auditory Display, Canberra, Australia (2016)

Using Virtual Reality and Motion Capture as Tools for Human Factors Engineering at NASA Marshall Space Flight Center

Tanya Andrews[1]([✉]), Brittani Searcy[2], and Brianna Wallace[3]

[1] NASA Marshall Space Flight Center, Huntsville, AL 35812, USA
Tanya.C.Andrews@NASA.gov
[2] Jacobs Space Exploration Group, 620 Discovery Dr. Suite 140,
Huntsville, AL 35806, USA
Brittani.R.Searcy@NASA.gov
[3] Wichita State University, 1845 Fairmount St, Wichita, KS 67260, USA
bewallace15@gmail.com

Abstract. NASA Marshall Space Flight Center (MSFC) Human Factors Engineering (HFE) Team is implementing virtual reality (VR) and motion capture (MoCap) into HFE analyses of various projects through its Virtual Environments Lab (VEL). VR allows for multiple analyses early in the design process and more opportunities to give design feedback. This tool can be used by engineers in most disciplines to compare design alternatives and is particularly valuable to HFE to give early input during these evaluations.

These techniques are being implemented for concept development of Deep Space Habitats (DSH), and work is being done to implement VR for design aspects of the Space Launch System (SLS). VR utilization in the VEL will push the design to be better formulated before mockups are constructed, saving budget and time. The MSFC VEL will continue forward leaning implementation with VR technologies in these and other projects for better models earlier in the design process.

Keywords: Human factors engineering · Human space flight · Virtual reality · Motion capture · Space launch system · Mockup · Simulation

1 Introduction

MSFC's HFE Team is responsible for all worksite analyses performed for the SLS pre-launch integration activities at Kennedy Space Center (KSC), as well as the analysis of DSH concepts. There is a wide variety of tasks, and it is important to verify early in the design process that the vehicle can be successfully integrated at KSC. If the ground support crew cannot complete the task, redesign efforts must be implemented. These worksite analyses are traditionally performed by inspection of drawings and construction of mockups to replicate the worksite. These mockups are most beneficial early in the design phase when changes to the design can be made to better meet the HFE requirements. However, the design is often so fluid that the mockup may not reflect the most recent changes.

This is a U.S. government work and not under copyright protection in the U.S.;
foreign copyright protection may apply 2020
T. Ahram (Ed.): AHFE 2019, AISC 965, pp. 399–408, 2020.
https://doi.org/10.1007/978-3-030-20454-9_41

Layout analysis of Deep Space Habitat (DSH) facilities is another field of investigation for the MSFC HFE team. These layouts are ever-changing, and VR is being used in the VEL (shown in Fig. 1) to examine possibilities and make design decisions. With a small team, the group has found a process to work with the Advanced Concepts Office and other departments to minimize efforts across the center for DSH work. Being able to bring together this innovative work in a way that can be rapidly re-planned has been most helpful in designing concepts that may be used for future deep space travel.

Fig. 1. Virtual Environments Lab (VEL) at NASA Marshall Space Flight Center. The VEL contains a 16 infrared Vicon motion capture camera system and VR capabilities with Oculus Rift and HTC Vive.

Being able to quickly review new designs without the expense of a mockup could also be very beneficial to SLS. Applying strategies to design development that include HFE from the beginning of the design process leads to better designs where functional tasks can be successfully accomplished. Doing this analysis initially with VR with input given back to designers will result in more accurate mockups when physical assessments are constructed. Integrating VR at design reviews will also expedite this process through immersive visualization of the object, rather than reading a print. Challenges with changing the way these reviews have been completed in the past are present, but work is being done to show the MSFC community how helpful VR can be.

2   Current Processes

MSFC HFE has developed a reputation of producing excellent quality physical mockups that are valuable to the engineering team. The fabrication shop has the ability to produce mockups of varying fidelity, dependent on budget and schedule. Assessment of these mockups have been used for decades to give HFE inputs and determine design direction. Engineers and managers outside of the HFE team have also used these mockups to demonstrate hardware issues and make design decisions. With a need for these mockups early in the design process, there are often major changes to the design after the mockups are constructed, resulting in the need for a rebuild. It is frequently a circular need in that the mockup is desired early, needs are identified, and design modifications result from the analysis. From this new design, the mockup has to be modified or be reconstructed altogether.

While the need for physical mockups is not going away, some work can be done earlier in the design process for designers and integrators to view the object in the appropriate scale through VR immersion. Being able to see the design as it is sized and better understand accessibility and workspace volumes at the beginning of the design process can lead to improved layouts from the beginning. If a better design is present well before the Critical Design Review (CDR), more accurate physical mockups will be built without as many iterations.

3   Physical Mockup vs VR Environment

VR provides the ability to view a scene on a 1-1 ratio without building a physical mockup. When looking at a space or object, engineers can better detect potential issues and produce solutions if the space or object is on a real-world scale. For this reason, it is beneficial to use a full-scale, three-dimensional, and intractable virtual model. It is also advantageous because an interactive VR environment offers a comparable experience to a physical mockup, while saving time and money. When using an interactive VR environment, physical materials are not required. This not only saves on materials, but also saves the time a person would use in looking at drafts, creating build procedures, and building the mockup. It can be more time-effective to create an interactive VR model compared to building a full-scale mockup. With adequate computing power, a CAD model can be translated inter an intractable VR environment in a days' time. And as the design evolves, models from the designers can be continuously incorporated into the VR environment; this saves the tremendous amount of time necessary to construct new physical mockups when designs become obsolete. A VR environment is often created in hours, compared to the weeks or months it can take to machine and assemble a high-fidelity mockup.

An advantage to this approach is the ability to manipulate and/or make changes to a model quickly in order to explore possible design changes. For example: a wall has a 10-inch diameter hole where a person needs to reach though and up to a box that is attached to the wall and 6 inches above the hole. In this scenario, a full-scale mockup can be built and an analysis can be performed using participants who are between the 5th percentile female size and the 95th percentile male range, as described in the Anthropometric

Survey of the US Army Personnel (ANSUR) [1]. These participants would each try to reach through the hole and up to the box while being asked questions about the task. With the physical mockup, the hole cannot be moved or changed without several hours of work. Contrasting, if this assessment were completed using the human factors analysis program, the wall and the hole dimensions could be changed relatively quickly.

A high-fidelity physical mockup is advantageous when factors such as heavy equipment and complex movements are needed. VR environments use visual and auditory senses only. If an assessment requires data from a participant's reactions to weight or touch, a VR environment alone will not be adequate.

4   VR for Deep Space Habitats

The SLS derived DSH is a concept mockup (seen in Fig. 2) that provides multiple analysis opportunities. It is a three-story habitat design with domed ends to provide for more living and storage space. The upper deck contains an exercise area and galley, the mid-deck provides living quarters, laboratory equipment, and storage space, and the lower deck provides a work area and operation components. The DSH serves as a test bed

Fig. 2. The physical SLS derived Deep Space Habitat (DSH) mockup. This mockup includes a galley and exercise area on the upper deck, crew quarters and laboratory equipment on the mid-deck, and a work area on the lower deck.

for various conceptual designs which might be used in future deep space habitats. It is more useful to think of the SLS derived DSH as a collection of parts rather than a whole. Some of its components will be utilized for NASA's eventual deep space habitat design, but likely not in the configuration that exists in the overall design of MSFC's DSH.

During the process of developing procedures to convert CAD files into VR environments useful for HFE analyses, the SLS derived DSH was converted into a VR format (shown in Fig. 3). The use of VR is particularly useful in the context of the SLS derived DSH for many reasons. Because the SLS derived DSH is used primarily as a conceptual test bed, there is less funding associated with it for the purpose of building mockups and conducting analyses. Since using VR is often more cost and time effective than building physical mockups, using VR in conjunction with the DSH allows for more work to be performed with the design.

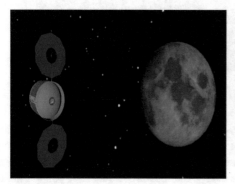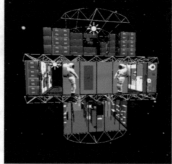

Fig. 3. Outside (left) and inside (right) views of the VR mockup of the DSH. Using VR allows for more design features to be included in the VR mockup (such as the end domes and solar panels) than are feasible in the physical mockup.

Another reason to use VR in conjunction with work on DSH models is VR's ability to allow for micro-gravity simulation. The DSH will operate in space, so its inhabitants will be living in a micro-gravity environment; simulating micro-gravity environments on earth is often difficult and requires expensive infrastructure, so the use of VR provides a relatively quick and inexpensive alternative. In the VEL, the HTC Vive (shown in Fig. 4) is used in conjunction with Unity 3D software. Unity 3D allows for features to be added through code in the languages C# and JavaScript; scripts written in these languages allow for micro-gravity to be utilized in VR environments.

The use of micro-gravity simulation in HFE analyses is very valuable when examining the way astronauts would use a system while in space. Many aspects of HFE analysis would be affected by the lack of gravity; being able to simulate microgravity allows for engineers to better anticipate design features which might need to be changed. Beyond the scope of HFE, VR micro-gravity simulations are also useful in the context of crew training and public education; VR technology today provides for an immersive experience which feels very realistic to the user.

Fig. 4. A participant using the HTC Vive VR console to view a VR model

There are some disadvantages to using micro-gravity simulations in VR. Creating an accurate micro-gravity environment through coding is difficult and time consuming during the development phase. Furthermore, some users find the simulation disorienting. Because VR feels so real to the user, it can be hard to feel as if you are floating in the virtual environment while still being affected by gravity in real life. It takes a bit of work by the user to adjust to the headset and avoid feeling motion sick; most users are able to get used to it rather quickly, but those prone to motion sickness may prefer to avoid these simulations.

The VR environments created for work on the DSH were built from scratch using Unity3D. A CAD model of the DSH existed in a different software, which was exported piece by piece and added to the environment in Unity 3D. Building the scene in this way allows the creator to tailor the environment to fit the specific needs of an assessment. As each piece of the model is added into the environment, the creator can add different characteristics to the piece and manipulate its behavior so that it closely mirrors the scenario being assessed. For example, an environment in low earth orbit would behave differently than an environment located on the moon, or on Mars; the VR environments must closely resemble the environment being assessed. The freedom associated with this method is very valuable when considering that NASA engineers often have a need to evaluate environments drastically different than what is experienced on earth. This method is time-consuming, as each part of the model must be given characteristics and added to the scene individually.

VR environments can also be created through the use of HFE software. Using HFE software allows VR environments to be created in very little time by simply translating the model into a different file format. This option has stricter limitations which depend on the parent HFE software, and is therefore much harder to customize. From the perspective of worksite analyses, using an HFE software should be preferred. Using HFE software allows engineers to examine a high-fidelity CAD model with many useful features and engineering tools that collect assessment data.

Creating a VR environment from scratch is more useful when examining a model in a conceptual way and while evaluating its general functions. It is also preferred when evaluating gravity environments other than that on earth. These models can be created to have very high fidelity, but the increased fidelity means more time spent on the project; in the fast-paced world of human space flight, by the time a VR environment is created in this way the design might be outdated. In the context of engineering analysis, HFE software provides an excellent option for generating VR mock-ups quickly and to the fidelity of thorough HFE assessments. It is this method which is used to evaluate the worksite analyses for NASA's upcoming SLS rocket.

5   VR HFE Analyses for SLS

In HFE at NASA MSFC, analyses are performed to assess if a task can be completed within the project requirements and performed without damage to the flight article. NASA considers personnel safety and flight article protection top priorities. It is important to understand that it is necessary to take measures to ensure the safety of the technicians and engineers involved in the product life cycle of SLS in order to safeguard mission success. Using VR in the design and verification processes allows the engineer to see and interact with an environment before it is built. This lets the engineer see potential issues early, allowing time to find a solution before a particular task needs to take place.

Because the SLS is such a large project, it is hard to know the anthropometrics of the person who will be performing a certain task at KSC. KSC is responsible for the ground processes of SLS including (but not limited to) stacking of the rocket, installation of payloads and hardware, and cable connections between hardware and elements. The SLS ground crew is comprised of a variety of people, therefore the requirements state that systems should be designed for assembly by 5th female to 95th male percentiles.

The HFE assessment begins with a breakdown of functions and their associated tasks to be performed, along with the type of analysis to be performed. Types of analyses can be an inspection where drawings or documents are evaluated, demonstrations such as physical mockups or VR experiences, and/or tests where a process is completed using mockups or intractable VR environments. The functions and tasks are then placed in a table where each task is lined up with SLS requirements to see if the requirement is applicable to the task. If nothing other than visual confirmation is needed, then a VR assessment will proceed. In this case, the CAD model is converted to an interactive VR environment for the engineer to experience.

The majority of HFE assessments at MSFC involve a reach analysis. For an SLS VR assessment, a SLS element CAD model will be converted and uploaded into the human factors program. Within this program, avatars can be inserted and manipulated to obtain an idea of task feasibility. These avatars are programmed to have the limitations of a human body. For example, a human arm cannot rotate 360 degrees around the shoulder socket. This limitation is applied to the avatars providing a more realistic scenario. After the avatars and necessary objects are added (such as platforms or tools), the virtual environment is converted into a VR environment. The user can then see the model on 1-1 scale with ability to explore and manipulate the environment and model.

In the VR environment, the participant can use the controllers to manipulate the environment, models, and avatars. The user may 'teleport' to any location in the environment then walk anywhere within the boundaries of the physical space where the user is operating the VR system, which changes the viewed environment to a new location within the model? The primary VR technology at MSFC is the HTC Vive headset using Steam gaming software and HP Z VR computers designed specifically for VR. The HTC Vive headset and controllers are tracked using two lighthouse sensors. The Vive controllers allow for several options. One of these options is the ability to separate parts of a CAD model. This is beneficial because in the HFE program the user can pick a specific part, pull it towards him/her, inspect all sides, and place it back on the model. This function saves time as there is no time required to build or assemble/disassemble a model in the VR environment. The program also allows the user to use tools such as a flashlight to illuminate shaded areas and a 'mark up' pen to save notes or design ideas.

The physical mockups and VEL technologies can also be combined for a more thorough assessment. By using physical mockups, the participant has the advantage of being able to feel a tangible object while performing the task or viewing the model. The physical feedback is often necessary when completing an HFE assessment, as it can be important to record realistic reactions from a participant. Some aspects of reality are lost while using the HFE program used for assessments: for example, the user is able to walk through hard boundaries, as the model doesn't obey the laws of physics. For example, when objects are released they do not fall to the floor of the model due to gravity nor do they stop when they interfere with a surface. Without this physical feedback, it can be difficult to record realistic reactions from participants. In the VR environment, the user can see interferences between the avatar and the model by the model changing color when it detects collisions, but there is no physical (haptic) feedback. Haptic feedback is important because it would let the participant know if he/she is reaching through a boundary (like a model wall or shell).

During a combined mockup/VEL assessment, recording the reactions and movements of a participant can be done using a MoCap system. The VEL contains 16 infrared Vicon cameras that track markers on a spandex suit specifically designed for use with MoCap (Fig. 5). The Vicon Blade program can track and record a participant's movements while simultaneously communicating with the HFE program. When mockups are used in the VEL they must built to allow optimum tracking, which is why mesh is often chosen to construct the mockups (example in Fig. 6). The mesh surface is necessary to permit the infrared light to continue tracking the markers. An example assessment is a participant opening a hatch door, setting it to the side, and stepping through the opening. The goal of this assessment is to determine if a participant can successfully remove the hatch door and step through the opening safely and easily. The participant would wear the MoCap suit and VR headset while performing the tasks. In the HFE program, the Blade program will track the movement of the participant (shown in Fig. 7), while he/she sees the hatch (as in Fig. 8). The mesh mockup real world environment and the VR environment boundaries are aligned so the participant touches the mockup as she/he sees in the VR environment. This allows the participant to feel as though they are in the environment where the task will actually take place at KSC. Both Blade and the HFE program can record scenes to be viewed later and used in design reviews, design feedback, and/or training.

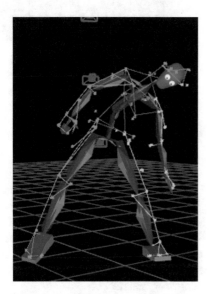

Fig. 5. The Blade program tracking the sensors on a MoCap suit

Fig. 6. HFE program avatar moving with the real-time MoCap tracking info sent from Blade

Fig. 7. A participant using the MoCap system wearing the Vicon suit and trackers.

Fig. 8. Mesh mockup of a hatch design used in a HFE assessment. Mesh mockups are compatible with the MoCap trackers.

6   Conclusions

The VR DSH work being performed at MSFC has allowed for fast-changing layouts to be analyzed by various departments without causing heavy impact to cost or schedule. These same techniques are being investigated for large programs, such as the SLS.

Implementing similar methods for SLS will allow for VR use in early design cycles, saving schedule and budget. HFE analysis of these early designs will allow designs to be more thoroughly evaluated before physical mockups are built and, ultimately, improve usability and provide for a superior design.

References

1. US Army NATICK Soldier RD&E Center: 2012 anthropometric survey of U.S. army personnel: methods and summary statistics: (NATICK/TR-15/007) (Updated December 2014)
2. Vicon Blade. https://www.vicon.com/products/software/blade
3. Wallace, B.: Virtual reality conversion manual technical report. NASA Marshall Space Flight Center (2018)

Towards a Coherent Assessment of Situational Awareness to Support System Design in the Maritime Context

Francesca de Rosa[✉] and Anne-Laure Jousselme

NATO STO Centre for Maritime Research and Experimentation,
Viale San Bartolomeo 400, 19126 La Spezia, SP, Italy
{Francesca.deRosa,
Anne-Laure.Jousselme}@cmre.nato.int

Abstract. Information systems in support to Situational Awareness (SAW), such as maritime surveillance systems, are an important family of tools that introduced automation with respect to the human cognitive activities. An integral part of the system design process is the testing and evaluation phase. The formal assessment of the mental state of SAW is a complex task. It appears that SAW assessment in testing and evaluation is often either overlooked by adopting a technology-focused approach or only partially addressed through the use of specific human factors methods. In this paper the authors will discuss how the testing and evaluation of maritime surveillance systems could account both for the system components enabling Situational Awareness and the human element. Furthermore, for such systems a simple and coherent list of key performance indicators and measures of performance is provided.

Keywords: Situational awareness · Systems engineering ·
Testing & evaluation · Key performance indicators · Maritime surveillance

1 Introduction

Nowadays we are assisting to a progressive automation of human tasks, both physical and cognitive. The newly introduced technologies can be characterised by different types and levels of automation [1].

An important family of tools that introduced a degree of automation with respect to the human cognitive activities is the one of information systems in support to Situational Awareness (SAW). Many definitions of SAW have been proposed, such as the well renowned one by Ensley [2]. This definition presents SAW as "the perception of the elements in the environment within a volume of time and space, comprehension of their meaning and the projection of their status in the near future". Therefore, it interprets SAW as an end-product of a cognitive process defined Situational Assessment (SA). SAW, together with decisions and actions form the building blocks of dynamic decision making, as depicted in the functional model in Fig. 1.

An interesting example of information systems are the military Command & Control (C2) systems, which aim at supporting the military commanders SA, enabling fast and

© Springer Nature Switzerland AG 2020
T. Ahram (Ed.): AHFE 2019, AISC 965, pp. 409–420, 2020.
https://doi.org/10.1007/978-3-030-20454-9_42

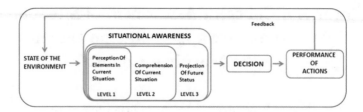

Fig. 1. Dynamic decision-making cycle (adapted from [3])

effective decision-making in complex environments. Maritime C2 information systems render information to the user through the Recognized Maritime Picture (RMP). Currently there is no formal definition of the RMP, however, following [4] the RMP can be interpreted as "a managed geographic presentation of processed all-source contact and information data, known at a given time, from all available assets, and compiled by an assigned RMP Manager. The RMP consists of all contacts in the maritime environment, both surface and subsurface, commercial, military and government platforms and vessels". The RMP, therefore, is concerned with reporting at the very least the location, kinematic information (i.e. tracks) and unique identities of the objects in an area of interest. This information could be further enriched, for instance with by the vessel identification labels (e.g. Friend, Assumed Friend, Neutral, Suspect, Hostile [5]).

Due to the growing volume, velocity and variety of information available and potentially relevant to the operator, automation through data and information fusion approaches is becoming instrumental to those systems [6].

A key step of the system design process is the testing and evaluation (T&E) phase, which consists in the testing of the new proposed design in the light of the requirements and specifications with the main objective of evaluating several criteria (e.g. performance and usability). T&E is an activity generally performed throughout the system design life cycle and should culminate in the operational testing and evaluation (OT&E), which corresponds to system employment under realistic operational conditions in real or simulated environments.

The formal assessment of SAW, as it is a mental state, is a complex task. It appears that this aspect in T&E is often either overlooked or only partially addressed. In fact, most of the time information system's T&E is technology oriented [7] or in other cases SAW is just measured through some subjective score (e.g. using the Situational Awareness Rating Technique [8]) without a benchmark or through the introduction of requirements loosely connected to SAW. Moreover, often OT&E is a performance-based testing. Although performances are closely related to SAW, as we will explain later, they should not be confused with an indicator of SAW. In fact, many *moderating factors* (e.g. training, personality, tactics or rules of engagement) intervene between the state of knowledge (i.e., SAW) and the action results that are measured through performances.

T&E should be based on the assumption that operational environments are composed by two main components: (i) a technological element, that enables SAW and (ii) a human element that detains SAW. Therefore, it is important to take a holistic approach and consider that SAW is very much related to the human-machine teaming aspect, also in the light of the *automation shift*. In fact, every time automation, physical or cognitive, is introduced we do not experience a mere substitution of tasks executor (from human to machine), but it intrinsically changes the operational environment and the way humans are integrated into it [9].

In this paper the authors propose a methodological approach to the T&E of systems in support to SAW, contextualising this approach in maritime surveillance applications, regardless of their civil or military nature. A simple yet coherent set of key performance indicators (KPIs) is provided, together with guidance on related measures of performance (MOPs) and methods of evaluation. The remainder of the paper is organised as follows: Sect. 2 reports on system assessment from a technology perspective, Sect. 3 provides the human factor perspective and in Sect. 4 a reconciled perspective is presented, together with relevant KPIs to be used in maritime surveillance systems T&E. Finally, the conclusions are summarised in Sect. 5.

2   The Technology Driven Perspective

As previously mentioned C2 information systems are increasingly integrating fusion techniques and researchers (e.g. [10–14]) are investigating the issue of defining standardised performance criteria and measures for systems adopting such techniques, with the specific aim of supporting T&E. A generic information fusion system consists of *low-level fusion* (LLF) and/or *high-level fusion* (HLF) components [10]. The former is concerned with objective and measurable quantities (e.g., location, speed, class) and the latter with more complex and abstract attributes, possibly subjectively estimated. Most of the work on metrics has taken place at LLF, while only more recently the attention has been shifted towards HLF [15], which is of paramount importance when performing T&E of the overall system [14], especially in the OT&E task. In [12] *accuracy, repeatability/consistency, robustness* and *computational complexity* are proposed as criteria to be adopted at the lower fusion level and *correctness in reasoning, quality of decision, intelligent behavior* and *adaptability in reasoning* at the higher levels. In [10, 16], instead, a performance-based approach to HLF assessment is proposed. This is centered around the constructs of *information gain, information* and *service quality* (i.e. *accuracy, timeliness, uncertainty, confidence, throughput, cost, credibility* and *reliability*) and *robustness* (i.e. over the number of use cases). It is interesting to notice how in [10] it is assumed that *quality* and *robustness* correspond to the *correctness* and to *intelligent support to decision making* proposed in [12]. Moreover, [10] views *decision making* as SAW level 2 (*understanding*). While the reader is referred to the relevant literature for specific details on the topic, in the next section we will discuss the importance of correctly selecting the construct to be analysed and the corresponding measurements to be performed in T&E of systems in support of SAW.

3   The Human Factors Perspective

In most cases the system design evaluation has been technology driven, however more research is nowadays devoted to the exploration of the human element of the operational environment, especially with reference to *human-automation systems*, composed by machines that include automation but require human interaction [17]. It is acknowledged that systems cannot be assessed independently from the human operators and, vice-versa, human performances cannot be assessed independently from automation [18]. An interesting approach to the assessment of human-automation systems is presented in [7]. The proposed framework highlights the different facets that should be measured for such assessment, starting from the inputs to the system up to the key outcomes constructs, which are linked through human processes and states. The inputs to the system are divided into: *human inputs* (i.e. cognitive competencies, interpersonal differences), *machine inputs* (i.e. level of automation, adaptiveness, reliability, transparency and usability) and *contextual inputs* (i.e. task variables, environment). The human processes and states are divided into: *attitudes* (i.e. trust), *behaviours* (i.e. reliance, monitoring and interaction with automation) and *cognition* (i.e. Situational Awareness and cognitive workload). Finally, the key outcomes are *performance* and *safety*. This work underlines how system designers often ignore the *attitude, behavior* and *cognition* factors (which include SAW). Although, Situational Awareness is presented as an important construct to be assessed in relation to the system, it is treated as a factor that bridges inputs to outputs. This example, like many others (e.g. [1]), highlights how system design assessment is generally *performance-based*.

When it comes to human-automation systems in support to SAW, *Situational Awareness* becomes obviously one of the key outcome constructs. Therefore, a consequent shift in the assessment should occur in order to avoid the pure *performance-based* approach. In fact, many researchers [19–22] have highlighted:

i. how SAW, although related, differs from other constructs such as attention, memory, cognitive workload;
ii. how SAW, although related, differs from other steps of the decision-making cycle (Fig. 1).

More specifically, [19] underlines how measurements of the *state of knowledge* are indicative of Situational Awareness, measurements about *behaviours* are indicative of decisions and measurements related to *performance* are indicative of actions. In fact, between each step of the decision-making cycle *moderating factors* might intervene (e.g. operator skills, knowledge, ability, tactics, training, system capabilities, etc.) [19], making *performance* and indirect and partial indicator of SAW.

Figure 2 shows an expanded dynamic decision-making cycle, together with the measurements corresponding to each step (rectangles with two rounded corners) and the possible issues (three lower rectangles) and moderating factors (the three higher rectangles) between the steps. The difference with Fig. 1 is that *state of the environment* is "exploded" in the *SAW information sources* as defined in [23]. More specifically, [23] defines the three *SAW information sources*, namely the *real world*, the *system knowledge* and the *interface knowledge*. The *real world*, provides information to the operator (through direct observation) or to the system. The knowledge acquired by the system is defined as *system knowledge*, which in the case of our maritime application would correspond to the RMP. Generally, the *system knowledge* does not correspond to the *real world* and similarly the *interphase knowledge* does not necessarily correspond to the *system knowledge*, because of loss of information between the transitions. Finally, the feedback loop is omitted for clarity of the figure.

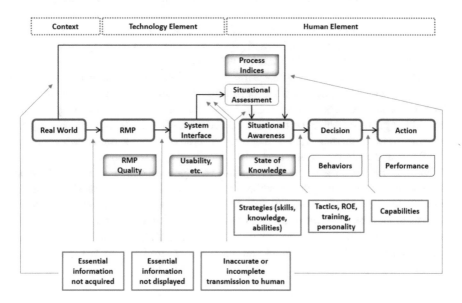

Fig. 2. Measurements for maritime surveillance systems T&E (inspired by [3, 19, 23])

3.1 Measures of the State of Knowledge

Several human factors methods have been proposed to assess and measure the *state of knowledge* of an operator, therefore to assess Situational Awareness [24]. Most techniques focus on the measurement of individual SAW by exploring physiological aspects, performance aspects, imbedded tasks, subjective ratings and questionnaires [1, 25]. The resulting measurements are often used as key performance indicators of the effectiveness of novel technologies.

In this paper we will focus on individual SAW, while the extension of the discussion to team and shared SAW is left to further work. Following [1] the individual SAW assessment techniques relevant to T&E are the: (i) freeze probe techniques; (ii) real-time probe techniques; (iii) self-rating techniques and (iv) observer-rating techniques.

Freeze probe techniques (*e.g.* SACRI [26], SAGAT [25] and SALSA [27]) require the participants to respond to SAW related queries administered during freezes of a task or scenario simulation in which the system is used. On the contrary, in real-time probe techniques (*e.g.* SASHA [28] and SPAM [29]), the queries are administered without freezing the simulation. Self-rating techniques (*e.g.* CARS [30], MARS [31], SARS

[32], SART [8] and C-SAS [33]) request the participants to self-rate elements related to SAW. Finally, in observer-rating techniques (*e.g.* SABARS [34]), an external expert observes and rates the participants SAW while they are performing a task. In most cases the SAW queries relate to elements of a specific task (e.g. aircraft altitude, speed and heading) and not to specific general SAW dimensions. Examples of exceptions are CARS, C-SARS and MARS in which the queries administrated correspond directly to the three SAW levels of the definition reported in Sect. 1. Therefore, those methods assume *perception, understanding* and *projection* as the underlining dimensions of SAW. Other methods (e.g. SART and SARS) attempt to define categories of elements to be rated, such as *familiarity of the situation, focus of attention, information quantity, information quality, concentration, reasoning abilities, plan execution, ability to effective use information, solicitation of information* and *communication of information*. Although those categories are all related to SAW, some of them are actually performance or behaviour related categories. Therefore, those should not be confused with SAW underpinning dimensions.

For most methods the single query ratings are aggregated in an overall SAW score. The selection of the most suitable method to the application at hand must consider the trade-offs related to their respective use. More specifically aspects such as complexity of the method, time to perform and analyse the data, cost associated to extensive use of experts are just some of the elements to consider. Moreover, some methods (e.g. CARS, MARS, SARS) give results that are possibly correlated to task performance, which as stated previously is a different construct than SAW.

4   Maritime Surveillance Systems T&E for Situational Awareness

4.1   Assessing Maritime Surveillance Systems for SAW

Reconciling the technology and human factors perspectives for T&E of systems in support of SAW, we suggest that the following elements should be approached in a holistic approach: (i) Operational picture quality evaluation, more specifically in the case of maritime surveillance systems the RMP quality evaluation; (ii) Interphase evaluation; (iii) SA evaluation; (iv) Workload evaluation and (v) SAW evaluation.

The corresponding measurements are reported in Fig. 2 as the shadowed rectangles with rounded corners. Although all those elements are correlated, they still need to be tested independently in order to evaluate if a system is enabling an adequate (or enhanced) SAW level. In fact, if SAW is an end-state, then its assessment alone is not sufficient to inform T&E. Therefore, it is important to explore all the building blocks that lead to that state. However, in this paper we will focus on the elements (i) and (v). The reason for this is that up to authors knowledge those two extremes have not been treated together, making it very hard for system designers to orient themselves when it comes to T&E, while it is quite common in literature to treat point (iv) and (v) as correlated elements. Similarly, interphase design and evaluation have been extensively treated in literature. Therefore, for the workload and interphase T&E the reader is referred to the relevant literature (e.g. [1]). Finally, the authors do not expand on (iii) as this element has received attention only very recently. In fact, researchers have started mentioning the importance to explore the SA process (e.g. [19]) and highlighted how different persons might reach the same level of SAW, but trough different reasoning paths. Exploring the process that leads to SAW could give important cues to understand which are the system elements that might have caused a low quality of SAW. Therefore, researchers have started looking at methods to explore SA (e.g. eye tracking, verbal protocols [35], scenario manipulation [36] and more recently analytical games [37, 38]), but it is still an open research topic.

4.2 SAW KPIs and Metrics for Maritime Surveillance System

In this section we propose a simple, but clear list of Maritime Situational Awareness (MSA) KPIs for maritime surveillance systems T&E in the light of SAW. On the basis of the elements exposed in the previous section those KPIs correspond to the *quality* dimensions of the RMP and the human element detaining SAW. The KPIs for the former category are: *accuracy, clarity, completeness, continuity, timeliness, consistency*. The KPI for the latter is the *overall level of SAW*. The RMP KPIs are based on [39, 40], which refer specifically to the construction of Single Integrated Air Pictures. However, the elements proposed can readily be used into the maritime domain.

Accuracy can be regarded as composed by *track accuracy* and *identification accuracy*. The first one is defined as the "measure of how accurately the [system] reports position and velocity values. The [...] picture is kinematically accurate when the position and velocity of each assigned track agree with the position and velocity of the associated object". The second one, instead, is defined as the "measure of the portion of tracked objects that are in the correct [identification] state. The [identification of the picture] is correct when all the tracked objects are in the correct [identification] state".

Ambiguous tracks are more than one track displayable to an operator, assigned to the same object and not correlated within a system. Moreover, *spurious tracks* are the ones that are not assigned to any object. Track identification is ambiguous if either its level of confidence is low or if it has several possible labels. T*rack clarity* can, therefore, be defined as the "measure of the portion of the [picture] that contains ambiguous tracks and/or spurious tracks. The [...] picture is clear when it does not include ambiguous or spurious tracks". *Identification clarity*, on the other hand, is defined as the "measure of the portion of the tracked objects that are unambiguously identified. The [identification] is clear if no tracked object is in the ambiguous [identification] state".

Similarly, *Completeness* is composed by *track completeness* and *identification completeness*. The former is defined as "portion of true [...] objects that are included in the [maritime picture]. The [maritime] picture is complete when all the objects are detected, tracked and reported", while the latter as "portion of tracked objects that are in an identified state. The [identification] is complete when all tracked objects are in an identified state".

Continuity is a "measure of how accurately the [picture] maintains track numbers over time. The [...] picture is continuous when the track number assigned to an object does not change". Continuity, like the previous dimensions, can be regarded as composed by two components: *track continuity* and *identification continuity*.

Timeliness points to the latency to obtain the data needed. We will refer to two components that build up the timeliness attribute of the RMP quality: the *track timeliness* (or currency) and *identification timeliness*. An additional element that links to the concept of timeliness is the latency of data dissemination, however we do not include this aspect here as it does not specifically refer to the construction of the RMP, which starts only once the data is received. Therefore, the basic assumption is that this aspect should be dealt with at system integration level.

Consistency is the quality dimension of the RMP defined as "the number of assigned tracks held by all participants at time t [such that] track number is the same for all participants" to "the number of assigned tracks held by at least one participant at time t". The participants can be individual units, platforms, the task group coordinator, or systems. From the above-mentioned definition, it is possible to notice that consistency can be regarded as composed by two components: *track consistency* and *identification consistency*.

Finally, the *overall level of SAW* is a measure indicative of the current operator mental state. This concept is proposed as a unified KPI, because as explained in Sect. 3.1 current research has not yet established a consolidated set of dimensions that build up SAW. The definition of such set would support the development of innovative and possibly more generic human factors methods for SAW assessment. Moreover, it would highly contribute to T&E, allowing to explore more in details the impact of the technology related dimensions (i.e. RMP quality dimensions) on the single SAW dimensions. For example, RMP *accuracy* impact on the different SAW dimensions could be investigated.

Table 1 reports MOPs for the proposed KPIs as defined in [39, 40]. Those are instantaneous MOPs, but could be extended to time average ratios if needed. Excepted consistency, MOPs are provided for single platforms (or single system), but can be extended to platform/system average. Finally, identification MOPs can be refined per identification type. Table 2, instead, summarises an example of SAW dimensions and the related method to rate them [8].

Table 1. RMP quality dimensions and measure of performance [39, 40]

Dimension	Measure of performance	
	Track	Identification
Accuracy	Mean track positional error (based on Euclidean error) over the number of detected tracks - It is important to notice that the ground truth need to be established, namely the real position and real velocity of the objects under analysis	Ratio of correctly identified tracks over the total number of detected tracks
Clarity	Time average fraction of ambiguous tracks over the total tracks	Time averaged fraction of the objects with a certain allegiance type in an ambiguous identification state over the number of tracked objects of a certain allegiance type
Completeness	Ratio of the number of objects with at least one assigned track and the total number of true objects	Ratio of the number of identified objects and the total number of tracked objects
Continuity	Rate of track number changes averaged over all objects	Percentage of time tracked objects depicted by only correctly identified tracks
Timeliness	Average time over the number of tracks between target detection and the report of a confirmed or deleted track	Average time over the number of tracks between initial target detection and its identification
Consistency	Ratio of tracks (common position) held in common by all participants compared to the total number of tracks	Ratio of common tracks with a common identification label compared to the number of common tracks

Table 2. Example of SAW dimensions and measure of performance from SART [8]

Dimension	Measure of performance
Familiarity of the situation	Likert scale, categorical scale or pairwise comparison
Focusing of attention	Likert scale, categorical scale or pairwise comparison
Information quantity	Likert scale, categorical scale or pairwise comparison
Instability of the situation	Likert scale, categorical scale or pairwise comparison
Concentration of attention	Likert scale, categorical scale or pairwise comparison
Complexity of the situation	Likert scale, categorical scale or pairwise comparison
Variability if the situation	Likert scale, categorical scale or pairwise comparison
Arousal	Likert scale, categorical scale or pairwise comparison
Information quality	Likert scale, categorical scale or pairwise comparison
Spare capacity	Likert scale, categorical scale or pairwise comparison

5   Conclusions

The proposed Maritime Situational Awareness (MSA) KPIs should be representatives of a state of knowledge necessary to make informed decisions and execute related missions such as conduction a maritime military operation or maritime surveillance. The authors explain how the MSA KPIs should account both for the system components enabling SAW and the human element. To address the first component the proposed set of MSA KPIs includes the Recognised Maritime Picture (RMP) quality dimensions (e.g. *accuracy, timeliness, clarity, completeness, continuity* and *consistency*). Those dimensions represent an objective and technology oriented assessment of the quality of information that is rendered to the operator. Although those are an important aspect to be assessed, they are not sufficient to have a complete evaluation regarding the enhancement (or degradation) of operators' SAW level. The proposed complementary MSA KPI is the *overall SAW level*, which is rather a subjective estimate which addresses the human side of the T&E portion related to SAW. The authors present guidance on possible MOPs, but highlight that although there is the attempt to define generic SAW dimension, future work should focus on the definition of a consolidated set, that could guide and harmonise future T&E.

The MSA KPIs discussed relate to RMP quality and SAW evaluation for T&E, however, further assessments should be performed in order to have a comprehensive T&E of the maritime surveillance system, such as interface evaluation, cognitive workload and Situational Assessment. Although researchers have acknowledged the usefulness of this last construct in providing guidance and valuable information for system design, it has only partially received attention. It is desirable that further research is conducted on methods to evaluate Situational Assessment, in order to further understand its potential with respect to systems T&E.

Acknowledgments. This research was supported by NATO Allied Command Trans-formation (NATO-ACT) through the DKOE programme of work.

References

1. Parasuraman, R., Sheridan, T.B., Wickens, C.D.: A model for types and levels of human interaction with automation. In: IEEE Transactions on Systems, Man, and Cybernetics – Part A: Systems and Humans, Vol. 30, pp. 286–297 (2000)
2. Endsley, R.M.: The application of human factors to the development of expert systems for advanced cockpits. In: Human Factors Society 31st Annual Meeting, pp. 1388–1392. Human Factor Society, Santa Monica, CA (1987)
3. Endsley, R.M.: Toward a theory of situation awareness in dynamic systems. Hum. Factors **37**(1), 32–64 (1995)
4. Nato Marcom, Marcom MSA Direction and Guidance (NATO UNCLASSIFIED). Technical Report Version 2.0, NATO MARCOM (2016)
5. NATO - STANAG 1241 Standard Identity Description Structure for Tactical Use (2005)
6. Mevassvik, O.M., Veum, K.: Distributed maritime situation picture production and data fusion. In: 4th International Conference on Information Fusion (2001)

7. Stowers, K., Oglesby, J., Sonesh, S., Leyva, K., Iwig, C., Salas, E.: A framework to guide the assessment of human–machine systems. Hum. Factors **59**(2), 172–188 (2017)

8. Taylor, R.M.: Situational awareness rating technique (SART): the development of a tool for aircrew systems design. In: Situational Awareness in Aerospace Operations (AGARD-CP-478), pp. 3/1–3/17, Neuilly Sur Seine (1990)

9. Sarter, N.B., Woods, D.D., Billings, C.E.: Automation surprises. Handb. Hum. Factors Ergon. **2**, 1926–1943 (1997)

10. Blasch, E.P., Valin, P., Bosse, E.: Measures of effectiveness for high-level fusion. In: 13th International Conference on Information Fusion (2010)

11. Waltz, E., Llinas, J.: System Modeling and Performance Evaluation. Multisensor Data Fusion Systems, Artech House, Boston (1990)

12. Llinas, J.: Assessing the performance of multisensor fusion processes. In: Hall, D., Llinas, J. (eds.) Handbook of Multisensor Data Fusion. CRC Press, Boca Raton (2001)

13. Theil, A., Kester, L.J.H.M., Bosse, E.: On measures of performance to assess sensor fusion effectiveness. In: 3rd International Conference on Information Fusion (2000)

14. Van Laere, J.: Challenges for IF performance evaluation in practice. In: 12th International Conference on Information Fusion (2009)

15. Costa, P., Laskey, K., Blasch, E., Jousselme, A.L.: Towards unbiased evaluation of uncertainty reasoning: The URREF ontology. In: 15th International Conference on Information Fusion (2012)

16. Blasch, E.P., Breton, R., Valin, P.: Information fusion measures of effectiveness (MOE) for decision support. In: SPIE 8050 (2011)

17. Sheridan, T.B.: Humans and Automation: System Design and Research Issues. Wiley, Santa Monica (2002)

18. Marquez, J.J., Gore, B.F.: Measuring safety and performance in human–automation systems: special issue commentary. Hum. Factors **59**(2), 169–171 (2017)

19. Endsley, M.R.: Situation awareness measurement in test and evaluation. In: O'Brien, T.G., Charlton, S.G. (eds.) Handbook of Human Factors Testing & Evaluation, pp. 159–180. Lawrence Erlbaum Associates, Mahwah (1996)

20. Endsley, M.: Situation awareness and workload: flip sides of the same coin. In: Jensen, R.S., Neumeister, D. (Eds.) Seventh International Symposium on Aviation Psychology, pp. 906–911 (1993)

21. Parasuraman, R., Sheridan, T.B., Wickens, C.D.: Situation awareness, mental workload, and trust in automation: viable, empirically supported cognitive engineering constructs. J. Cogn. Eng. Decis. Making **2**(2), 140–160 (2008)

22. Vidulich, M.A., Tsang, P.S.: The confluence of situation awareness and mental workload for adaptable human-machine systems. J. Cogn. Eng. Decis. Making **9**(1), 95–97 (2015)

23. Endsley, M.R.: Theoretical underpinnings of situation awareness: a critical review. In: Endsley, M.R., Garland, D.J. (eds.) Situation Awareness Analysis and Measurement. Lawrence Erlbaum Associates, Mahwah (2000)

24. Stanton, N.A., Salmon, P.M., Walker, G.H., Baber, C., Jenkins, D.P.: Human Factors Methods: A Practical Guide for Engineering and Design. Ashgate Publishing Company, Brookfield (2006)

25. Endsley, R.M.: Measurements of situation awareness in dynamic systems. Hum. Factors **37**(1), 65–84 (1995)

26. Hogg, D.N., Folleso, K., Strand-Volden, F., Torralba, B.: Development of a situation awareness measure to evaluate advanced alarm systems in nuclear power plant control room. Ergonomics **38**(11), 2394–2413 (1995)

27. Hauss, Y., Gauss, B., Eyferth, K.: SALSA - a new approach to measure situational awareness in air traffic control. Focusing Attention on Aviation Safety. In: 11th International Symposium on Aviation Psychology, Columbus (2001)

28. Jeannott, E., Kelly, C., Thompson, D.: The development of situation awareness measures in ATM systems. Technical Report, EATMP (2003)

29. Durso, F.T., Hackworth, C.A., Truitt, T., Crutchfield, J., Manning, C.A.: Situation awareness as a predictor of performance in en route air traffic controllers. Air Traffic Q. **6**, 1–20 (1998)

30. McGuinness, B., Foy, L.: A subjective measure of SA: the Crew Awareness Rating Scale (CARS). In: Human Performance, Situational Awareness and Automation Conference, Savannah (2000)

31. Matthews, M.D., Beal, S.A.: Assessing situation awareness in field training exercises. Research Report, U.S. Army Research Institute for the Behavioural and Social Sciences (2002)

32. Waag, W.L., Houck, M.R.: Tools for assessing situational awareness in an operational fighter environment. Aviat. Space Environ. Med. **65**(5), A13–A19 (1994)

33. Dennehy, K.: Cranfield - situation awareness scale user manual. Technical Report, College of Aeronautics, Cranfield University, Bedford (1997)

34. Matthews, M.D., Pleban, R.J., Endsley, M.R., Strater, L.D.: Measures of infantry situation awareness for a virtual MOUT environment. In: Human Performance, Situation Awareness and Automation Conference (HPSAA II), Daytona, LEA (2000)

35. Sullivan, C., Blackman, H.S.: Insights into pilot situation awareness using verbal protocol analysis. In: Human Factors Society 35th Annual Meeting 1, pp. 57–61. Santa Monica, CA: Human Factors Society (1991)

36. Sarter, N.B., Woods, D.D.: Situation awareness: a critical but ill-defined phenomenon. Int. J. Aviat. Psychol. **1**, 45–57 (1991)

37. de Rosa, F., Jousselme, A.-L., De Gloria, A.: A reliability game for source factors and situational awareness experimentation. Int. J. Serious Games **5**(2), 45–64 (2018)

38. Jousselme, A.-L., Pallotta, G., Locke, J.: Risk game: capturing impact of information quality on human belief assessment and decision making. Int. J. Serious Games **5**(4), 23–44 (2018)

39. SIAP System Engineering Task Force (SIAP SETF): Single Integrated Air Picture (SIAP) Attributes Version 2.0. Technical Report 2003-029 (2003)

40. Representative measures of a single integrated air picture. Technical Report, NAVSEA 05 (2000)

An IT Project Management Methodology Generator Based on an Agile Project Management Process Framework

Evangelos Markopoulos[✉]

HULT International Business School, 35 Commercial Road, Whitechapel,
E1 1LD London, UK
evangelos.markopoulos@faculty.hult.edu

Abstract. Information Technology Project Management and Software Project Management in particular depends heavily on the project's type and constraints. Quality, financial, technical, schedule, complexity and other constraints affect significantly the management process. Over the last two decades project management methodologies have been developed to support the project management effort. Many methodologies cover generic approaches emphasizing on the planning or estimation activities, others on tracking, others on quality and others on very specific management practices that could support the delivery of very specific projects. This paper introduces an adjustable (agile) project management framework for managing information technology projects of any type. The framework divides the management activities into systems engineering management and systems acquisitions management phases and operates as a methodology generator feed by the project constraints. The project management methodology that derives is a combination of management and engineering phases based on the needs and constraints of each project per case.

Keywords: Process · Project · Management · Agile · Engineering · Methodology · Acquisition · System · Software

1 Introduction

Managing information technology (IT) has always been a challenge, especially when managing software systems. The failure of many software projects in the 1960s and in the 1970s was the first indication of the upcoming process management difficulties in information technology and the software evolution. Software was delivered late, schedules were unpredictable, projects cost several times the original estimates and often experienced poor performance and quality characteristics [1]. Information technology projects fail at those times due to lack of engineering knowledge and expertise, but also due to the differences that existed between software and hardware engineering against other engineering disciplines [2]. IT and software projects, at those times in particular, were intangible, usually implemented without development standards and management processes, and were mostly considered 'one-off' projects with no repeatability.

© Springer Nature Switzerland AG 2020
T. Ahram (Ed.): AHFE 2019, AISC 965, pp. 421–431, 2020.
https://doi.org/10.1007/978-3-030-20454-9_43

This software crisis [3, 4] has been repeated in the 1980s and continued also in the 1990s and the 2000s as well [5, 6] in all types of organizations involved with software and information technology projects from either the developer / supplier (those who implement technology) perspective or from the customer (those who acquire technology) perspective.

2   The Process Adjustability Concern

The main cause of this software crisis that counts more than 40 years is primarily based on two major factors. First it is the need for adjustability on the management process processes to the project needs for any project, and second is the need for the requirements management process to be taken seriously and followed precisely [7].

Projects significantly vary on deterministic factors such as size, complexity, budget, time, etc., and nondeterministic factors such as development team maturity, acceptance criteria, process maturity, etc. The adjustability of the management process has a significant, and critical role not only on the technology providers, the ones developing the technology, but also on the customers, the ones acquiring the technology [8, 9]. The proper management process must have the characteristics that will support the efforts and goals of both parties involved in a project.

The management processes that are primarily focused on managing a project with emphasis on its technical challenges, or on its planning or tracking activities, create fuzziness in the interpretation of the implementation and management efforts that need to be placed.

The second factor in the software crisis is clearly focused on the requirements process, a concept that was [10] and will keep on being [11] closely related with the quality and success of a project [12, 13]. The quality of the requirements and the requirements management process is a barometer not only to the success of a project, but also to all planning and management techniques developed around the project. Engineering models, management methodologies, and operations environments are all heavily affected by the quality and the maturity of the project requirements which in turn define the project and impacts its implementation and management strategy.

Having analyzed the prime, and secondary factors of the software and technological crisis, an adjustable and unified project management framework is proposed that can be possibly used effectively and efficiently as a useful tool. However such a framework must have the capability to be scalable to the process and project requirements. This scalability, along with the identities of the framework [14] can be used as a methodology generator, producing adjustable process models, per case and when needed, for all type of projects.

The proposed approach redefines the term 'agility' and kind of renames it to 'adjustability' as the challenge is not only on being flexible, but mostly on being able to continuously adjust to the project needs in order to stay flexible, or agile.

3 VR the Adjustable Unified Project Management Framework (AUPMF)

The Adjustable Unified Project Management Framework (AUPMF) is a concept, which consolidates four project management dimensions into two framework dimensions giving this adjustable project management approach.

The first dimension of the framework is the Systems Engineering Management (SEM) dimension and derives directly from the IT and software engineering project management principles. The engineering framework dimension aims to manage a project from the engineering perspective. This perspective is based on the management of the technical quality, or qualitative management, of the project by managing the development method, the quality of the deliverables produced by the method and other technical documents, validations, verifications and milestones significant to software engineering under the software quality principles.

The second dimension of the framework is the Systems Acquisition Management (SAM) dimension and derives from the consolidation of the Planning, Tracking and Organizational project management approaches. This consolidation aims to manage a project from the pure managerial perspective under the total quality management principles. This is achieved by performing quantitative project management, which is exactly the opposite of the qualitative project management. The quantitative management approach is based on creating estimations and managing those estimations quantitatively by organizing management teams to track what has been planned against what has been done. The qualitative management approach on the other hand, is focused on the management that will achieve technical excellence, not necessarily within time and budget.

Figure 1 describes the formation of the two project management frameworks through the consolidation of the four project management dimensions.

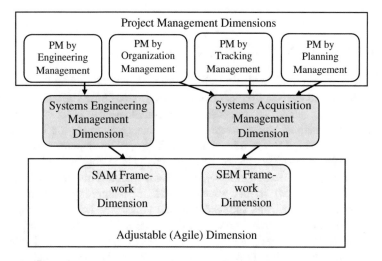

Fig. 1. Project management dimension consolidation in the AUPMF

The integration of the two project management frameworks (SEM and SAM) creates an Adjustable Unified Project Management Framework approach, which can cover the needs of almost all types of information technology projects, under all management goals and constraints.

The SEM (Systems Engineering Management) project management framework which is integrated in the AUPMF supports processes that can help the project management efforts and goals from the development point of view. Regardless the nature of the implementation process, managing the engineering process is very crucial to the success of the project or product that is being developed. Also regardless the way the implementation and the software development is executed, either it is in-house development, custom made projects, COTS (Components off the shelve), systems parameterization or systems implementation, the management scope is focused primarily on obtaining technical quality that meets operational expectations, manages constraints and stays within the deadlines. The processes included in the SEM framework are based on the development life-cycle of a system, something that is very critical to be followed and managed precisely in order to reach the expected quality which in this case is the prime goal of the SEM approach.

On the other hand, the SAM (Systems Acquisition Management) project management framework which is also integrated in the AUPMF, aims to meet the two other project management key expectations, which are based on the control of budget and the control of the time. Having in the SAM framework all the necessary processes that can provide accurate cost and schedule estimations, the management of a project relies then on the organizational structure, the management team and the tracking model that will verify the control of the estimations and will document the deviations.

The SAM framework approach is primarily used by not-technical project managers or by general project managers by profession, aiming to meet specific deadlines and constrains without much emphasis on the engineering dimension of the project. Meeting time and budget for SAM manages, is more important than meeting the quality of the project.

4   The AUPMF Dimensions Synergy

The benefit of the AUPMF and what characterizes it, can be outlined as the synergy among the SAM and the SEM project management dimensions. The framework allows and helps the project manager to select the proper combination of processes from the two management dimensions, and generate a project management approach based on the needs of each project. On the other hand, each project management dimension on AUPMF could also be used as a project management methodology as well, depending on the type of management desired per project implementation. This adjustability and, not agility, is what the AUPMF framework can offer to the mature information technology project managers who can think and lead, but not to ones who believe and follow.

This methodology generation process is described in Fig. 2. By defining the needs of the project to be managed, the phases and process that will be used towards the management of the specific project are selected from the management goals and expectations.

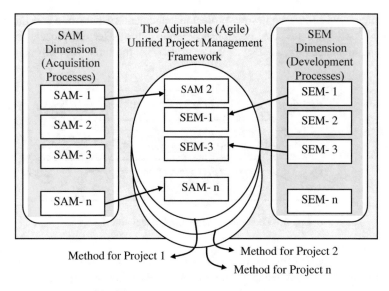

Fig. 2. Project dimension's process synergy

Taking for instance a hardware-oriented project with limited software applications. In this case the project management methodology will be significantly different from the one to be used in a custom-made software business application project. In the first case the management will be much more oriented in the planning and tracking of the project since most of the project components are well developed and tested. In the second case the management of the project will be quality oriented, based on detailed tracking on the engineering practices used to develop the software.

Other project parameters that affect the process structure of the management methodology are the results of an assessment that could take place before project initiation in order to identify the manager's needs and goals, in order for the framework to generate the most appropriate management method.

This methodology parameterization can be repeated for every project. A project manager has actually two sets of processes to work with (SEM and SAM) through which many project management methodologies can be generated if the project requirements and expectations are known, or even unknown in some cases.

5   The AUPMF Process Matrix

In order to simplify the complexity of process selection towards the creation and implementation of the desired project management methodology for each project specifically the Adjustable Unified Project Management Framework is based on a requirements interpretation matrix for process generation.

The matrix of AUPMF is a three-tier - two-dimensional matrix. The first tire indicates the available processes from the two process dimensions (SEM and SAM) of the framework. The second tire indicates the selected processes from each process dimension that will form the desired project management method for a specific project. The third tier indicates the relationship among the selected process and their implementation/execution strategy. Figure 3 describes the three AUPMF matrix tiers.

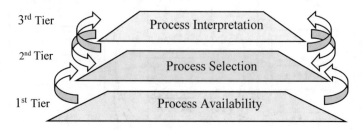

Fig. 3. Three-tier process implementation matrix

The interpretation of the two type of processes in the framework's dimensions form a two-dimensional matrix which is the AUPMF matrix layout as presented in Fig. 4. The processes of the engineering dimension in the matrix are vertically listed while the processes of the acquisition dimension are horizontally listed.

The way the matrix works is by initially selecting the processes that will be used in the desired project management methodology from each framework dimension (SEM or SAM). Not all SEM or SAM process are selected to be placed in the AUPMF matrix, but only the ones related with the project requirements and management goals.

The selection of the desired processes can be done initially in a conceptual, not precise, manner. By placing each desired process in the matrix, automatically the proposed project management method is generated.

If the project management goal, for example, is to manage the project using the engineering perspective, which is management for the development quality, then the

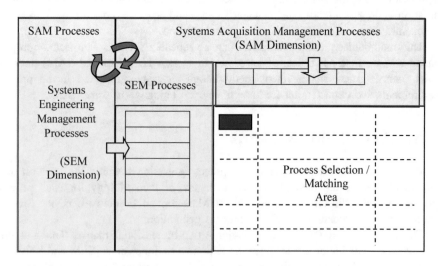

Fig. 4. The AUPMF matrix layout

completion of the matrix will start from the engineering processes filling the matrix left to right. Each matrix row indicates the implementation of the specific engineering process on the selected management processes.

If on the other hand the project management goal is to manage the project using the acquisition perspective, which is management by planning and tracking, then the completion of the matrix will start from the management processes, filling the matrix top to bottom.

6   Reading the AUPMF Matrix

An example of a semi-complete AUPMF Matrix is shown in Fig. 5. The interpretation of the project management methodology that derives from this matrix (with the limited indicative values) can be done from the SEM point of view or from the SAM point of view, depending what management approach is to be followed.

Application of the SAM Processes ....on		Systems Acquisition Management Processes						
Systems Engineering Mgmt Processes (SEM Dimension)	on the SEM Processes	All Mgmt Practices	Tracking	Change Mgmt	Quality Assurance	...	Contract Mgmt
	All Eng. Practices		0	0	1	0	0
	Req. Mgmt	0	1	0	0		0
	Syst. Analysis	0	1	1	0		1
	Coding	0	0	1	0		0
	...						
	Testing	0	1	0	0		1

Fig. 5. Example of AUPMF matrix in practice

The interpretation of the matrix from the SEM prospective can be done by reading the matrix rows, Left-to-Right, giving in this case the following project management methodology:

'All project processes in all phases will be subject to quality assurance inspection. In the requirements management process the customer will participate only with tracking activities in the requirements life cycle. In systems analysis, the customer will track the system, and any analysis changes done will be controlled by change management techniques that could affect contract management actions. The coding of the system will be under the programmer's control with no user involvement other that the responsibility to track code changes. The testing of the system will be under the system's tester control with the customer to track the test plan and proceed with contract changes when not satisfied.'

It must be noted that the above methodology derived by interpreting the '1's placed in the matrix. The '1's indicate that the specific cell will be activated, and the output will be the process that will derive by reading the SEM on the SAM process elements.

In a similar way, the interpretation of the matrix from the SAM perspective is based on reading the matrix columns, Top-to-Bottom, giving the following, in this case, project management methodology:

> 'Quality assurance activities will be applied to all engineering phases. The requirements will be tracked throughout their life cycle for completeness and correctness. Change management activities will be applied on the system analysis and system coding activities that will be delivered by the supplier in order to verify their completeness. Contract management activities will be applied on the systems analysis and systems testing activities in order to verify contractual agreements and possibly modify them if needed'.

The customer and supplier goals differentiate the way the SEM and SAM processes are interpreted. One customer, for example, might need to perform contract management towards tracking the deliverables of the project. if they are specified, in the contract, while another customer might need to perform contract management in order to keep on modifying or updating the contractual obligations of the supplier based on the quality and completeness of the project progress.

7 Benefits from Using the AUPMF

The AUPMF can contribute to organizational project management improvement plans in multiple ways. The entire framework is based on the concepts of adjustability and flexibility, in order to help the project's outcome to meet the information systems characteristics, and specifically the software quality characteristics [15].

Any framework, on the other hand, needs to be able to be used in all types of information technology projects, by all type of organizations under all types of technical, financial and management constraints. The structure of the AUPMF allows it to be easily and successfully used by all types of organizations and specifically by the SME's, or actually the SISMEs (Software Intensive Small and Medium Size Enterprises) [16]. SISMEs due to their size, time and budget constraints, face practical restrictions on either using specific project management methodologies or creating their own methodology to be used in each project per case.

Project differentiation creates process differentiation and therefore management differentiation. In general, any differentiation creates the need of adjustability, and adjustability requires changes, but changes are hard to be adopted and accepted. Resistance to change is a prime management consideration for applying management and organizational process control models towards process improvement and total quality management [17]. The AUPMF response to this challenge is its user-based adjustability per project and per case.

The framework specifically promotes user involvement since there are no mandating processes, basic models or minimum process requirements in it. The structure of the framework and the flexibility that provides are based totally on its operations model, which is user-driven and purely democratic. The proposed models that derive each time from the framework operations are accepted or enhanced / changed by the participants in the project, and not necessarily by the project managers.

Depending on the way each organization or project manager views the concept of success, the AUPMF has the flexibility to reach and support such success variations.

8   Risks from Using the UPMF

Every success factor can be turned into failure factor if not properly interpreted, understood and managed. What can be considered as a benefit can also be considered as a risky, if not properly approached. The freedom provided by the framework through its ability to be adjusted can be very harmful to the ones with no management and process engineering background, knowledge or experience.

Amateur and/or inexperienced project managers can create very complex management methods from using the framework in their attempt to make sure that their project will be well managed monitored and documented. However, the effort to manage such complex methods requires experience in both systems engineering and project management. On the other hand, project managers who might underestimate the complexity of a project can create a project management approach with processes that do not support major management activities and principles.

In order to bypass these risks, extensive training is needed on process improvement, project management and systems engineering only to matured and disciplined personnel.

Management commitment is another risk in the AUPMF. The management needs to support the methods deriving from the AUPMF even if they differ from project to project. After all, this is the major benefit of the framework. All the project participants need to be part of this process when the matrix is being completed; otherwise there are no guarantees that the derived management approach will work in practice. The senior management of the organization needs to support these activities regardless if they cost much or take productive time from the participant's busy work schedules.

Finally, the AUPMF is not a panacea. The benefits offered require process maturity and management commitment, to work out. Quality is free [18], but only if you do everything right.

9   Results

Information technology project management and specifically software project management is full of gray areas, unexpected situations, dependencies and ambiguous trade-offs [19]. Managing information technology projects is very difficult, but it is not impossible. Project management is more about understanding the management needs than the implementation processes, activities and milestones. If you do not know where you are going; no road will help you [20].

Successful project management is based on successful understanding of the project environment and requirements [21]. This diversity on the project management goals puts the project management concept in an endless loop seeking for the silver bullet in a continues evolving industry composed from new process, methods and best practices.

Undoubtedly all of the new contributions in the international project management community and discipline are working well, but only under specific conditions and limitations. On the other hand, most of them require significant expertise in order to be followed completely in order to be effective, and others require a bureaucratic mentality to get aligned with their standards [22].

The AUPMF presented in this paper can contribute towards managing software projects and information systems complexity. The framework matrix which is the key element in its operation and interpretation works actually as a methodology generator. A project manager with a defined set of requirements, can create through the framework, the proper process model that can be used towards successfully implementing this specific set of requirements.

The matrix on the other hand, and its process generation capabilities, allows the framework to be easily used for all type of projects regardless their size, volume and complexity. This capability comes to boost up the technocratic development visions of the SMEs and SISMEs, who silently today, support the larger part if the world's economy, but forbidden grow effectively by using proper project management methods and practices do to their size, budget, projects and even culture.

The AUPMF makes process engineering for process management affordable to anyone for anything, at any time.

References

1. Brooks, F.P.: The Mythical Man Month. Addison Wesley, Reading (1975)
2. Sommerville, I.: Software Engineering, 6th edn. Addison Wesley, Harlow (2001)
3. Mills, E.: Software metrics. SEI Curriculum Module SEI-CM-12-1.1 (1988)
4. Arthur, L.J.: Measuring Programmer Productivity and Software Quality. Wiley, New York (1985)
5. Glass, R.: Is there really a software crisis. IEEE Softw. 15(1), 104–105 (1998)
6. Gibbs, W.: Software's chronic crisis. Sci. Am. 271(3), 86–95 (1994)
7. Markopoulos, E.: Quality assurance best practices for the implementation of logistics information technology systems. In: Proceedings of the 17th International Logistics Congress, Vol. 1, pp. 532–557, Athens Greece, October 2001
8. Hoffmann, H., Geiger, J.: Quality management in action: a Swiss case study. Inf. Technol. People 15(1), 35–53 (1995)
9. Herbsleb, J.: Benefits of CMM-based software process improvement: initial results. Technical Report CMU/SEI-94-TR-13, SEI (1994)
10. Davis, G.: Strategies for information requirements determination. IBM Syst. J. 21(1), 4–30 (1982)
11. Wyder, T.: Capturing requirements with use-cases. Softw. Dev. 4(2), 37–40 (1996)
12. Leveson, N.: Software safety: why, what and how. ACM Comput. Surv. 18(2), 125–163 (1986)
13. Robertson, S., Robertson, J.: Mastering the Requirements Process. Addison Wesley, Upper Saddle River (1999)
14. ISO/IEC 12207:1995: Information technology-software life cycle processes – amendment 1 (2002)
15. Dunn, R.: Software Quality – Concepts and Plans. Prentice Hall, Upper Saddle River (1990)

16. Markopoulos, E.: An empirical adjustable software process assessment model for software intensive small and medium size enterprises. In: 7th European Conference on Software Quality, Conference Notes Tammerpaino, Tampere, Finland, 16–19 (2002)
17. Deming, W.E.: Out of Crisis. MIT Press, Cambridge (2000)
18. Crosby, P.: Quality Is Free: The Art of Making Quality Certain. McGraw-Hill, New York (1979)
19. Royce, W.: Software Project Management: A Unified Framework. Addison-Wesley Longman Publishing Co., Inc., Boston (1998)
20. Humphrey, W.S.: Managing for Innovation – Leading Technical People. Prentice Hall, Englewood Cliffs (1987)
21. Reel, J.S.: Critical success factors in software projects. IEEE Softw. 16(3), 106–113 (1999)
22. Emmerich, G.: Managing standards compliance. IEEE Trans. Software Eng. 25(6), 836–851 (1999)

Analysis on Visual Information Structure in Intelligent Control System Based on Order Degree and Information Characterization

Linlin Wang[1], Xiaoli Wu[1(✉)], Weiwei Zhang[1], Yiyao Zou[1],
and Xiaoshan Jiang[2]

[1] College of Mechanical and Electrical Engineering, Hohai University,
Changzhou 213022, China
wuxlhhu@163.com
[2] Trina Solar Energy Co. Ltd., Changzhou 213100, China

Abstract. Efficient intelligent control system helps enterprises to create greater economic value and social value, so the rationalization of information structure plays a decisive role in operators' cognition and operational efficiency. Based on the information characterization method, this paper presents a model of applying the order degree algorithm to the intelligent control system and applies it to the MES production line control system of an enterprise. It not only quantifies the advantages and disadvantages of the information structure by the order degree algorithm, but also provides the direction for the design of the information structure at the beginning of the design process from the microscopic perspective.

Keywords: Order degree algorithm · Information characterization ·
Information structure · Interface information layout

1 Introduction

As the only communication channel between the huge amount of information carried by the intelligent control system and the operator, the interactive interface affects the fluency of human-computer interaction. Different information structures lead to totally different search paths for operators, which greatly affect the efficiency and quality of human-computer interaction. Therefore, information structure, as the main influencing factor of operator's cognition and operation, needs to keep pace with the development of intelligent control system.

In recent years, scholars both at home and abroad have devoted time and effort to research on the interaction of intelligent control system. For instance, Yim et al. [1] found the way to present interface design of complex information in limited screen space by information hierarchy visualization. Paul [2] studied the overload of information complexity in complex digital interface and established the pattern decision-making model from the perspective of time pressure to analyze the execution time required by different amounts of information. Cheshire et al. [3] illustrated the significance of large data and showed how to describe the data flow in different time scales

© Springer Nature Switzerland AG 2020
T. Ahram (Ed.): AHFE 2019, AISC 965, pp. 432–444, 2020.
https://doi.org/10.1007/978-3-030-20454-9_44

based on London public transport system. Burns [4] and Carvalho [5] established evaluation model and design method for the human-computer interactive interface controlled by nuclear power plant. Domestic scholars Zhou et al. [6–8] constructed three different APP layouts of mobile phone interfaces and made a quantitative analysis of the order degree based on the micro information structure using the entropy theory. Li et al. [9–11] completed the research on the visualization structure mapping relationship of multi-dimensional attributes of information. Li [12] studied the design method combining high-dimensional data with visual structure. Wang et al. [13] used the four dimensions of point, line, plane and volume to strengthen the visualization structure of information entity and the sense of space in visualized structure. Li [14] explored the impact of information structure and content representation on user visual guidance. Ren [15] put forward the best method of presenting massive information in order from the perspective of information structure. Wang et al. [16] used eye-tracking technology to conduct quantitative experiments on the layout design of the display control interface in new generation of fighter jets. Zhang et al. [17] proposed the rank of operator's memory efficiency with different background colors. The above researches on information structure of human-computer interactive interface, information layout and information entropy theory provide scientific theoretical basis and technical means. There is less research on quantitative analysis of information structure while information entropy theory is only applicable in tree structure (such as interface layout in mobile phone). But intelligent control system interface carries a large quantity of information and its relevance of information is complex. The information structure of interface in intelligent control system is often multiple structures concomitant or disordered so it is difficult to extract the tree structure which is the basis for application of order degree algorithm. Therefore, there is a lack of the application method of the order degree algorithm especially in intelligent control system.

Based on the order degree algorithm, this paper establishes a model of applying the algorithm to intelligent control system and the model is applied to the MES production line control system in an enterprise. The model describes information characterization method through which complex information structure in intelligent control system can be characterized as a kind of tree structure. Besides, this model describes how to optimize the process of information structure design in intelligent control system through the order degree algorithm.

2 The Model of Order Degree Algorithm in Intelligent Control System

2.1 MES Production Line Control System of an Enterprise

The main research content of MES production line control system in this enterprise includes Module Process and Module Wip (Work In Process) Management. The function of Module Process is designing the production line independently so that the products can be processed automatically as the designed process. Module Process enters from the home page. Interface of hierarchy 1 includes three elements such as process setting production section and process category. Clicking element of process setting enters hierarchy 2. Interface of hierarchy 2 contains 4 signing states of order and 9 operations of specific process. 4 signing states and 9 operations form a tabular structure. After selecting a signing state and operation, the interface of hierarchy 3 will display the specific process information. The typical part of the original interface in the Module Process is shown in Fig. 1.

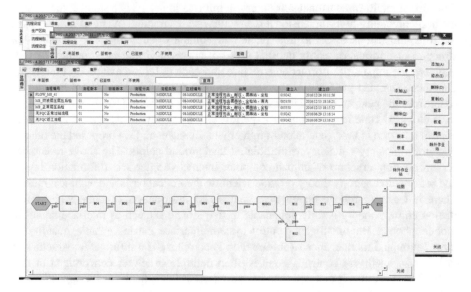

Fig. 1. Original interface of Module Process

The function of Module Wip Management is monitoring the production status of any wip on the production line timely. Module Wip Management enters from the home page as well. Interface of hierarchy 1 should be deleted because it is repeated with the superior interface. Interface of hierarchy 2 includes five elements such as raw code assignments, transfer laminated piece to work order, invalid component number, production code component and label printing and batch execution. Clicking element of batch execution enters hierarchy 3. Interface of hierarchy 3 contains many elements. The main contents are function of area number, function of section number and function of assignment number. The above functions require frequent entry and exit of hierarchy 3. After selecting the above three functions, the interface of hierarchy 4 will display specific production batch information. Each production batch corresponds to 4 specific stations which are the interface of hierarchy 5. The typical part of the original interface in the Module Wip Management is shown in Fig. 2.

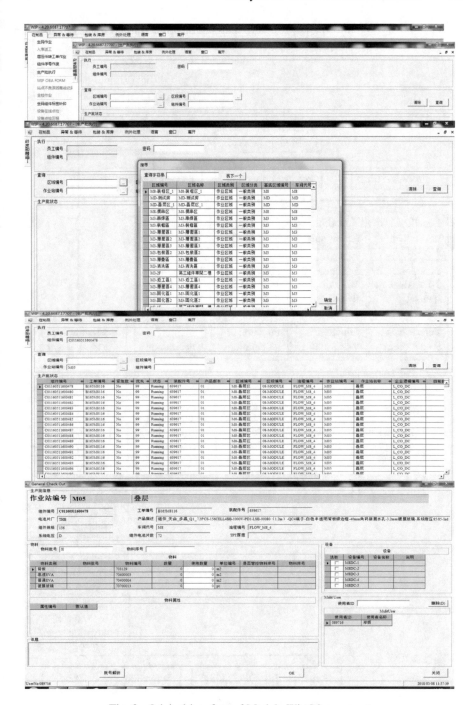

Fig. 2. Original interface of Module Wip Management

In summary, the information structure diagram of original interface is obtained, as shown in Fig. 3.

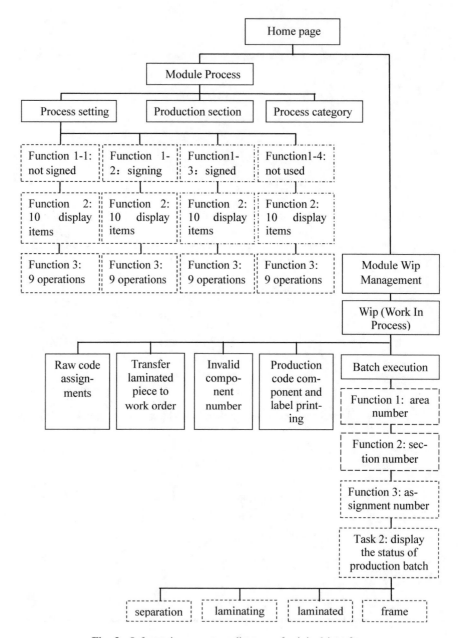

Fig. 3. Information structure diagram of original interface

2.2   Information Characterization Method

The order degree algorithm should calculate based on tree structure. In Sect. 2.1, the part shown by the dotted line in Fig. 3 has problems such as inter-hierarchy confusion, semantic ambiguity and unclear partition within one hierarchy, etc., so it is necessary to organize the information elements by information characterization method, so as to integrate and partition hierarchies.

The first step of characterization method is to extract all information elements to form a table. During the extraction, attention should be paid to recording information element features. In MES production line control system, the largest information feature is the function-task and presentation attribute of information elements.

The second step is to divide the extracted information elements into function areas and task areas. A function area is formed by information elements of related functions. For example, functions of elements "not signed, signing, signed and not used" are related to signed, so those 4 elements can form a group called a function area. In addition, relevant functional areas should be integrated into a new group called a task

Table 1. Elements characterization in Module Process

No.	Task area	Function area	Elements extracted	Presentation attribute
1	-	-	Process setting	Static
2	-	-	Production section	Static
3	-	-	Process category	Static
4	Task 1	Function 1:	not signed	Static
5		4 states	signing	Static
6			signed	Static
7			not used	Static
8		(Element)	Query	Dynamic
9		Function 2:	Process number	Static
10		10 display	Process version	Static
11		items	Current version	Static
12			Process classification	Static
13			Process difference	Static
14			Section number	Static
15			Instruction	Static
16			Creation people	Static
17			Creation date	Static
18			Graphic display area	Static
19	Task 2	Function 3:	Add	Dynamic
20		9 operations	Modify	Dynamic
21			Delete	Dynamic
22			Copy	Dynamic
23			Version	Dynamic
24			Audit	Dynamic
25			Attribute	Dynamic
26			Excluded station	Dynamic
27			Draw	Dynamic
28		(Element)	Close	Dynamic

area based on the operator's monitoring tasks. Takeing the Module Process as an example, the characterization results of the first two steps are shown in Table 1.

The third step is to integrate and partition hierarchies based on function areas and task areas. There are several design standards in the process of integration and partition: (1) According to the mathematical operation rules of model of the order degree algorithm, the hierarchy should be reduced reasonably; (2) Function areas and task areas should be considered as basic units. For instance, similar areas can be displayed in similar way while unimportant areas or single information element can be put together; (3) Displaying all hierarchies in one interface as far as possible to realize real-time monitoring in a giant screen. Some areas of the giant screen can be locally enlarged when operating.

Taking the Module Process as an example, all information elements are summarized as follows: There are 28 information elements to be integrated in the Module Process. Elements 1–3 belong to hierarchy 1 and are not displayed in hierarchy 2; Elements 4–28 is integrated into hierarchy 2 from the multiple hierarchies of original interface to reduce hierarchy. Elements 4–18 constitute task area 1 (including function area 1, function area 2 and 1 dynamic information element) while elements 19–28 constitutes task area 2 (including function area 3 and 1 dynamic information element). Among them, function area 1 and 2 form a tabular structure. According to the relevance of function-task area and the rule of visual flow direction, the layout of the hierarchy 2 in Module Process is shown in Fig. 4.

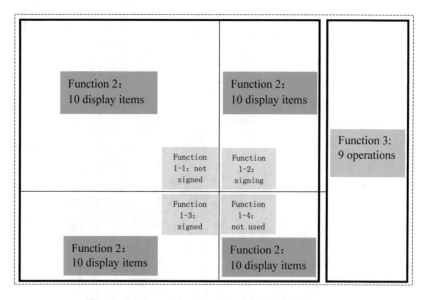

Fig. 4. Layout of the hierarchy 2 in Module Process

The specific analysis process in Module Wip Management is similar to Module Process and will not be illustrated here. The layout of the hierarchy 2 in Module Wip Management is shown in Fig. 5.

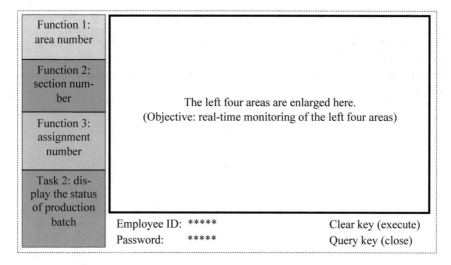

Fig. 5. Layout of the hierarchy 2 in Module Wip Management

Considering all the figures and tables in Sect. 2.2, the current interface information structure is shown in Fig. 6.

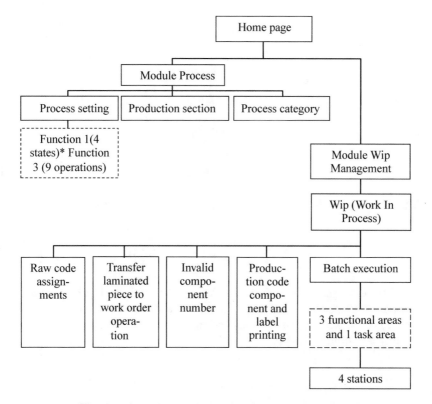

Fig. 6. Information structure diagram of current interface

2.3   The Model of Order Degree Algorithm

The timeliness of the information structure reflects the path length of operator to obtain the target information element while the timeliness entropy reflects the uncertainty of the timeliness of information flow. The formulas for calculating timeliness entropy are as follows.

$$H_T = \sum_i \sum_j R_{ij} \tag{1}$$

$$R_{ij} = -P_{ij} \log P_{ij} \tag{2}$$

$$P_{ij} = L_{ij}/L \tag{3}$$

$$L = \sum_i \sum_j L_{ij} \tag{4}$$

$$H_T^* = \log L \tag{5}$$

The quantity of information elements in a hierarchy affects the accuracy of search so the quality entropy reflects the quality of information transmission. The formulas for calculating quality entropy are as follows.

$$H_Q = \sum_i H_i \tag{6}$$

$$H_i = -F_i \log F_i \tag{7}$$

$$F_i = D_i/D \tag{8}$$

$$D = \sum_i D_i \tag{9}$$

$$H_Q^* = \log D \tag{10}$$

The order degree (R for order degree) can be expressed as follows.

$$R = \alpha \left(1 - \frac{H_T}{H_T^*}\right) + \beta \left(1 - \frac{H_Q}{H_Q^*}\right) \tag{11}$$

α and β are weight coefficients of timeliness effectiveness and quality effectiveness in information structure respectively. The larger the value of R is, the more effective and orderly the information structure is. The smaller the value of R is, the less effective and orderly the information structure is.

3   Calculation of Order Degree Algorithm Model in Intelligent Control System

The original interface structure diagram in Fig. 3 and the current interface structure diagram in Fig. 4 are converted into the abstract information structure diagrams for the convenience of the order degree algorithm calculation. The abstract structure diagrams of the original interface and the current interface are shown in Fig. 7.

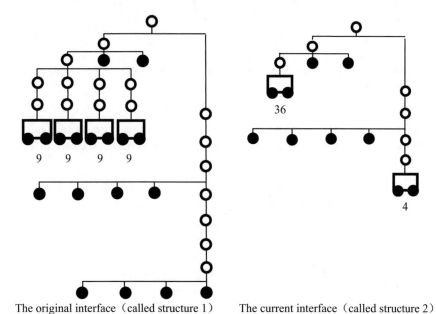

The original interface（called structure 1）   The current interface（called structure 2）

Fig. 7. The abstract structure diagrams of the original interface and the current interface

Information elements of the same number (that is, the number of black solid points of the two structures is the same) can be characterized by two different information structures so both the length of the information path and the number of information elements at each hierarchy in different structures are different. The characteristics of time effectiveness in two structures can be summarized, as shown in Table 2.

As can be seen from Table 2, there are a large quantity of intermediate hierarchies and the "keyhole effect" is serious in structure 1. The number of nodes in the information search process in structure 1 is greatly increased which requires frequent entry and exit of single hierarchy. Therefore, the timeliness of structure 1 is worse than that of structure 2.

Table 2. Time effectiveness calculation results of two information structures

Length of segment road of L_{ij}: r	Structure 1		Structure 2	
	S_r	P_r	S_r	P_r
1	63	1/866	52	1/330
2	61	2/866	50	2/330
3	57	3/866	46	3/330
4	48	4/866	5	4/330
5	43	5/866	4	5/330
6	6	6/866	0	0
7	5	7/866	0	0
8	4	8/866	0	0

Note: S_r is the number of r-segment road; P_r is the realization probability of hierarchy road.

According to the connectivity of information nodes in two structures, characteristics of quality effectiveness in two structures can be summarized, as shown in Table 3

Table 3. Quality effectiveness calculation results of two information structures

Number of D_i node: k	Structure 1		Structure 2	
	N_k	q_k	N_k	q_k
1	46	1/126	46	1/104
2	10	2/126	3	2/104
4	1	4/126	1	4/104
5	2	5/126	1	5/104
6	1	6/126	1	6/104
10	4	10/126	0	0
37	0	0	1	37/104

Note: N_k is the number of k-node; q_k is the realization probability of information elements.

As can be seen from Table 3, the quality effectiveness of structure 1 are similar to structure 2. However, the hierarchy is chaotic and the semantics are unclear in structure 1.

In this paper, based on the structural design principle of accurate and rapid, the weight coefficients of α and β were taken to be 0.5. The value of order degree can be obtained from Tables 1 and 2 and formulas (1)–(11), the order degree of the two information structures is 0.2061 and 0.2412 respectively. In general, the order degree of structure 2 is higher than that of structure 1 and structure 2 (the current interface structure) is more optimized than that of structure 1 (the original interface structure).

4 Conclusions

(1) This paper establishes a model of applying the order degree algorithm to the intelligent control system through the information characterization method. Besides, this paper applies the model to the MES production line control system of an enterprise.

(2) According to the mathematical calculation rules, in the process of calculating the ratio of time entropy (H_T) and maximum time entropy (H_T^*), under the premise of the same number of information elements, the fewer the hierarchies are, the smaller the denominator of P_r is, the larger the value of P_r is and the greater the time effectiveness of information structure is; in the process of calculating the ratio of quality entropy (H_Q) and maximum quality entropy $\left(H_Q^*\right)$, the larger the product of k, N_k and $\lg 1/q_k$ is, the greater the quality effectiveness of information structure is (let us not worry about k = 0 temporarily).

(3) All interfaces avoid graphic elements in order to avoid the influence of graphics on operators' cognition and focus only on the information structure itself. The mathematical calculation rule of order degree algorithm provides the direction for the design of information structure from the microcosmic perspective at the very beginning, which avoids the waste of design resources to the greatest extent.

Acknowledgement. This work was supported by Jiangsu Province Key Project of philosophy and the social sciences (2017ZDIXM023), Science and technology projects of Changzhou (CE20175032), Jiangsu Province Nature Science Foundation of China (BK20181159, BK2017-0304), the National Nature Science Foundation of China (Grant No. 71601068, 61603123, 61703140), Outstanding Young Scholars Joint Funds of Chinese Ergonomics Society- King Far (No. 2017-05), Overseas research project of Jiangsu Province (2017), and Fundamental Research Funds for the Central Universities (Grant No. 2015B22714).

References

1. Yim, H.B., Lee, S.M., Seong, P.H.: A development of a quantitative situation awareness measurement tool: Computational Representation of Situation Awareness with Graphical Expressions (CoRSAGE). Ann. Nucl. Energy **65**, 144–157 (2014)
2. Paul, S., Nazareth, D.: Input information complexity, perceived time pressure, and information processing in GSS-based work groups: an experimental investigation using a decision schema to alleviate information overload conditions. Decis. Support Syst. **49**(1), 31–40 (2010)
3. Cheshire, J., Batty, M.: Visualisation tools for understanding big data. Environ. Plann. B-Plann. Des. **39**(3), 413–415 (2012)
4. Burns, C.M., Skraaning Jr., G., Jamieson, G.A., et al.: Evaluation of ecological interface design for nuclear process control: situation awareness effects. Hum. Factors **50**(4), 663–679 (2008)
5. Carvalho, P.V.R., dos Santos, I.L., Gomes, J.O., et al.: Human factors approach for evaluation and redesign of human–system interfaces of a nuclear power plant simulator. Displays **29**, 273–284 (2008)

6. Zhou, L., Xue, C., Wang, H., et al.: Order degree analysis of digital interface microscopic information structure. J. SE Univ. (Nat. Sci. Ed.) **46**(6), 1209–1213 (2016)
7. Zhou, L., Xue, C., Tang, W., et al.: Aesthetic evaluation method of interface elements layout design. J. Comput. Aided Des. Comput. Graphics **25**(5), 758–766 (2013)
8. Zhou, L.: Research on digital interface layout design method based on visual pathway theory. Southeast University (2014)
9. Li, J., Xue, C., Wang, H., et al.: Encoding information of human-computer interface for equilibrium of time pressure. J. Comput. Aided Des. Comput. Graphics **25**(7), 1022–1028 (2013)
10. Li, J., Xue, C.: Color encoding research of digital display interface based on the visual perceptual Hierarchying. J. Mech. Eng. **52**(24), 201–208 (2016)
11. Li, J., Xue, C., Shi, M., et al.: Information visual structure based on multidimensional attributes of information. J. SE Univ. (Nat. Sci. Ed.) **42**(6), 1094–1099 (2012)
12. Li, A.: Research on data visualization structure model of big data and high-dimensional information interface. Southeast University (2016)
13. Wang, H., Xue, C., Shi, M.: Research on digital interface design of complex systems under the influence of automation. Packag. Eng. **4**, 36–39 (2014)
14. Li, J.: Study on visual guidance mechanism of interface information. Southeast University (2017)
15. Ren, S.: Research on design of natural interactive interface based on situational perception. Southeast University (2016)
16. Wang, H., Bian, T., Xue, C.: Experimental evaluation of fighter's interface layout based on eye tracking. Electromech. Eng. **27**(6), 50–53 (2011)
17. Zhang, H., Zhang, C., Zhang, Y., et al.: Study of information coding design of ATC system based on working memory. Sci. Technol. Eng. **17**(7), 46–51 (2017)

An Analysis of Mobile Questionnaire Layouts

Helge Nissen$^{(\boxtimes)}$, Yi Zhang, and Monique Janneck

Electrical Engineering and Computer Science,
Technische Hochschule Lübeck, University of Applied Sciences,
Mönkhofer Weg 239, 23562 Lübeck, Germany
{helge.nissen,monique.janneck}@th-luebeck.de,
zyellion@gmail.com

Abstract. This study examines the effects of mobile questionnaire layouts. The goal is to shed more light on mobile questionnaire usage. Furthermore, we aim to give guidance for researchers implementing online surveys. Researchers in a variety of fields use online questionnaires for their flexibility and efficiency. Poor usability on mobile devices may be associated with underrepresentation of certain target groups or lower data quality. In contrast, the use of well-designed smartphone-based surveys can open up new possibilities for researchers. We developed three different layout variants for comparison with an international sample of N = 204 smartphone users. The results show that grouping questions on separate pages works best with regard to missing values, dropouts, and completion time. However, results also suggest a possible distortion of answering patterns in this layout.

Keywords: Questionnaire · Smartphone · Layout · Design · Dropout · Completion time · Response behavior

1 Introduction

Online questionnaires are flexible and efficient data collection instruments and are therefore used in various research disciplines. In many cases, they have even replaced paper questionnaires. When setting up online questionnaires, researchers need to take into account that respondents will also use mobile devices such as smartphones. The usage of smartphones for answering questionnaires is steadily increasing. In 2011, 4% of respondents used smartphones, compared to 18% in 2014 [1]. Smartphones offer opportunities to researchers using online questionnaires. Mobile devices are frequently used in situations of waiting [2], so respondents might be willing and motivated to spend that time to fill out the questionnaire.

Furthermore, smartphones are especially common among younger people [3]. On the one hand, optimizing online questionnaires for smartphones might therefore enhance access to that target group. On the other hand, poor usability on mobile devices might be a reason for underrepresentation of younger respondents.

In this paper we aim to investigate whether and how questionnaire layouts influence the usage of questionnaires on smartphones. We developed and analyzed several mobile questionnaire layouts to derive design guidelines for optimizing online surveys for use on mobile devices.

© Springer Nature Switzerland AG 2020
T. Ahram (Ed.): AHFE 2019, AISC 965, pp. 445–455, 2020.
https://doi.org/10.1007/978-3-030-20454-9_45

This paper is structured as follows: In Sect. 2 we review related work on survey design for mobile devices. Section 3 presents the layouts developed and used in this study. Subsequently, research questions and methods are introduced. We present results in Sect. 5 and conclude this article with a general discussion.

2   Related Work

A general problem with online surveys is a high dropout rate, resulting in missing values and lesser data quality. This is especially true for smartphone use: Several previous studies identified higher dropout rates among respondents using smartphones [1, 4].

Processing time is a further challenge on smartphones. Various studies show that the time needed to fill out a questionnaire on smartphones is significantly higher [3–5]. Poorly designed layouts may be a reason for this. For example, questionnaire templates using tables are unsuitable for display on mobile devices [6]. Also, respondents using smartphones have been observed to do a lot of zooming for better interaction, also leading to longer completion times [7].

For optimization, attention should be paid to navigation paths within the questionnaire. This addresses aspects like the positioning of items on the pages as well as scrolling vs. navigation via buttons. According to Mavletova and Couper, scrolling-based designs, as opposed to page-by-page designs, can result in shorter completion times and lower dropout rates [8]. However, in their study they used a consistent layout across all devices instead of responsive designs optimized for mobile use. On the other hand, studies investigating optimized layouts for smartphones recommend one question per page, as scrolling is often burdensome and might rather lead to errors in extensive surveys [9, 10].

Users typically expect the possibility for backward navigation in addition to forward navigation [11]. Back buttons enable respondents to review and maybe correct their given answers [12].

The size of interaction areas is another important aspect on smartphones. Lai and colleagues found that input elements are often too small to be used conveniently [13]. Radio Buttons are very common input fields, but due to their size they seem to be less suitable for smartphone use. Larger input fields designed as buttons are more promising [9, 14].

3   Questionnaire Layouts

In order to develop design guidelines for mobile questionnaire layouts, we created three different layout variants for comparison in the empirical study described in Sect. 4. The widely used survey tool *Limesurvey* was chosen because of its openness for developing own templates [15].

3.1 Layout 1: Groups of Questions

The *Groups* layout displays several questions per page (Fig. 1). Participants, especially on smartphones, have to scroll down to see the next questions. Users navigate to the next page by hitting a button at the bottom of the page. Additionally, there is a back button to return to the previous page.

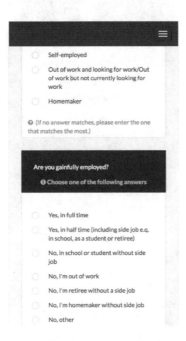

Fig. 1. Groups of questions

3.2 Layout 2: Manual Forwarding

The *Manual Forwarding* layout displays only a single question on each page. This presentation avoids scrolling (see Fig. 2), as previous studies have revealed negative effects of scrolling on smartphones [9]. A further advantage of the *Manual Forwarding* layout is that data is transferred after each question. Therefore, in case of dropout, all entries made up to that point are saved.

3.3 Layout 3: Automatic Forwarding

The *Automatic Forwarding* layout also displays one question per page, but introduces a new interaction concept: after answering an item, the user is automatically forwarded to the next page without having to press a button. This variant is promising to shorten completion time, which has been found to be substantially longer among smartphone users [3–5]. A back button enables users to navigate backwards (Fig. 3). This layout

Fig. 2. Manual Forwarding

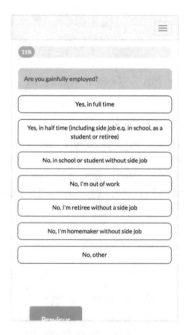

Fig. 3. Automatic Forwarding

also has the advantage that data is frequently transmitted to the server. In the *Automatic Forwarding* layout, there is no button for skipping an answer. This is one of the main differences to the other layouts. While *Limesurvey* supports implementing *Groups* and *Manual Forwarding, Automatic Forwarding* is not part of the software. Therefore, we created our own template, following the recommendations given in previous studies (see Sect. 2).

4 Research Questions and Methods

The aim of our work was to replicate – or possibly revise – previous findings regarding questionnaire layouts for smartphones, as previous studies are mostly several years old and technology as well as design patterns for mobile devices have progressed in the meantime. Furthermore, we wanted to compare respondents from different countries, as – to the best of our knowledge – there are no studies yet investigating possible intercultural differences regarding the use of questionnaires on mobile devices. To that end, we carried out an international online study using the three layouts described in Sect. 3.

4.1 Procedure

We used *Limesurvey* [15] to set up a comprehensive questionnaire containing 58 items, including – among others – a standardized questionnaire on personality traits [16] and several questions concerning international working experiences and computer usage. Likert scales (1–7, 1–5) were the main answer format. The questionnaire also included some open questions requiring text input, which are often avoided in other studies [9, 17]. We analyzed "small displays" with a screen width up to 768 px as current standards in media queries and breakpoints suggest this classification [18]. These display sizes indicate the use of a smartphone or tablet in portrait format (i.e. a mobile device). We created a typical situation for researchers, in which the device for participation was not specified, unlike in previous studies [3, 8].

4.2 Sample

Participants were recruited via Social Media and paid panels in Europe, China, and the USA. A total of 631 respondents participated in the survey. However, in this paper only smartphone users were considered, resulting in a sample of N = 204 relevant cases. We randomly deployed one out of three links to each participant that lead to one of the three layout variants (see Sect. 3). Despite this random allocation of participants, an approximation of group sizes cannot be expected given the relatively small total sample size. 29 people were assigned to the *Manual Forwarding* group, 59 persons to the *Groups* condition, and 116 persons to the *Automatic Forwarding* condition (see Tables 1 and 2).

70.1% of all participants were female (see Table 1 for gender distribution in the three groups). It is noticeable that there are many missing values (7.8%) for gender in *Automatic Forwarding* (Table 1).

Table 1. Gender distribution

Layout	Female	Male	Other	Missing	In total
Manual Forwarding	17 (58.6%)	10 (34.5%)	1 (3.4%)	1 (3.4%)	29
Groups	46 (78.0%)	11 (18.6%)	0	2 (3.4%)	59
Automatic Forwarding	80 (69.0%)	27 (23.3%)	0	9 (7.8%)	116
	143 (70.1%)	48 (23.5%)	1 (0.5%)	12 (5.9%)	204

Unfortunately, the distribution of participants across region of origin turned out to be extremely unbalanced in the smartphone sample. By far the most participants were located in the USA (75.5%), while only very few participants came from Europe (15.2%) and China (2.9%), respectively (see Table 2). Therefore, intercultural comparisons could not be calculated as intended.

Table 2. Regional distribution of participants

Layout	China	USA	Europe	Missing	In total
Manual Forwarding	5 (17.2%)	18 (62.1%)	5 (17.2%)	1 (3.4%)	29
Groups	0	54 (91.5%)	3 (5.1%)	2 (3.4%)	59
Automatic Forwarding	1 (0.9%)	82 (70.7%)	23 (19.8%)	10 (8.6%)	116
	6 (2.9%)	154 (75.5%)	31 (15.2%)	13 (6.4%)	204

4.3 Hypotheses

We put forward the following hypotheses based on findings of previous studies and the design discussions presented in Sect. 2:

H1: The choice of questionnaire layout will influence completion time on smartphones. Specifically, *Automatic Forwarding* and *Groups* are expected to yield lower completion times.

H2: The choice of questionnaire layout will influence dropout rates and missing values on smartphones. In particular, *Manual Forwarding* and *Automatic Forwarding* are expected to yield lower dropout rates, as they show only one question per page to make answering questionnaires on smartphones more comfortable.

H3: The choice of questionnaire layouts will influence response behavior on smartphones. In particular, *Manual Forwarding* and *Automatic Forwarding* are expected to show less distortion of response patterns, as they were specifically designed for a better overview of items on smartphones.

5   Results

In the following sections we present the results from the comparison of layout variants.

5.1   Missing Values

Table 3 shows average missing values for each layout. We calculated this variable by adding up all items that were not answered. The quota for missing values was then calculated as mean value of all cases in each group. In the *Groups* condition, there were about 19% missing values, compared to 29% in the *Manual Forwarding* and 33% in the *Automatic Forwarding* conditions.

Table 3. Differences in missing values between layouts

Display	N	Mean	Standard deviation	Standard error of mean
Manual Forwarding	29	17.03	21.946	4.075
Groups	59	10.98	19.674	2.561
Automatic Forwarding	116	19.28	23.284	2.162

Welch-ANOVA was calculated because variances were not homogenous, revealing a significant difference in the scores of missing values for the different layouts (Welch's $F(2, 74.852) = 3.071$, $p = .052$.).

The Games-Howell test was calculated additionally in order to verify this result, indicating a significant difference regarding missing values between the *Groups* and *Automatic Forwarding* layouts ((-8.293, $95\%-CI[-16.24, -.35]$), $p = .039$).

5.2   Dropouts

To analyze dropouts, value "1" was assigned when participants did not finish the questionnaire, while "0" was assigned if the questionnaire was completed. The dropout quota was then calculated as mean value of all cases in each group. Results show that dropout rates were lowest in the *Groups* condition. The *Automatic Forwarding* group, where participants had no choice of leaving questions unanswered, shows the highest dropout rate (see Table 4).

Table 4. Differences in dropouts between layouts

Display	N	Mean	Standard deviation	Standard error of mean
Manual Forwarding	29	.34	.484	.090
Groups	59	.22	.418	.054
Automatic Forwarding	116	.44	.498	.046

Again, Welch-ANOVA was calculated because variances were not homogenous. There is a significant difference in the dropout scores for the different layouts (Welch's $F_{(2, 74.186)} = 4.670$, $p = .012$). Games-Howell test was performed for post-hoc analysis, showing a significantly higher dropout rate for *Automatic Forwarding* compared to *Groups* $((.219, 95\%-CI[.05, .39]))$, $p = 007$).

Thus, Hypothesis H2 could not be confirmed as missing values and dropout rates were lowest in the *Groups* condition.

5.3   Response Behavior

For all numerical items, a general average response value was calculated, indicating the position of the chosen response (lower value - top position on the screen, higher value - bottom position of screen).

To compare the three groups, an ANOVA was calculated. The Tukey test was performed for post-hoc analysis because the variances were homogeneous.

The results reveal the highest mean in the *Automatic Forwarding* condition, followed by the *Manual Forwarding* and *Groups* layout. Namely, *Automatic Forwarding* and *Groups* differ significantly $((.373, 95\%-CI[.010, .736]))$, $p = .043$) (Table 5).

Table 5. Differences in response behavior between layouts

Display	N	Mean	Standard deviation	Standard error of mean
Manual Forwarding	19	3.408	.971	.223
Groups	46	3.249	.702	.104
Automatic Forwarding	65	3.621	.800	.099

Thus, Hypothesis H3 was confirmed as response behavior differs between the layout variants.

5.4   Completion Time

The time used to complete the questionnaire was recorded by *Limesurvey*. The dataset revealed some outliers, indicating that the questionnaire was possibly paused for a longer time. To exclude these outliers, only cases with a processing time up to 2200 s were considered, resulting in a sample of $N = 122$.

To compare the three groups, an ANOVA was calculated (Table 6).

Table 6. Differences in completion time between layouts

Display	N	Mean	Standard deviation	Standard error of mean
Manual Forwarding	17	1013.824	440.078	106.735
Groups	46	543.217	291.832	43.028
Automatic Forwarding	59	816.339	419.256	54.583

The Tukey test was performed for post-hoc analysis because equal variances could be assumed, indicating significant differences between the *Groups* and the *Manual Forwarding* layouts $((-470.606, 95\%-CI[.-726.167, -215.046])$ p $< .001))$ and between the *Groups* and *Automatic Forwarding* layouts, respectively $(-273.122, 95\% -CI[-450.221, -96.022])$ p $= .001)$.

Thus, Hypothesis H1 was partly confirmed as completion time is significantly shorter in the *Groups* condition. Contrary to the assumption, however, the processing time for *Automatic Forwarding* is not significantly lower.

6 Conclusion

In this paper we analyzed the effects of different questionnaire layouts on missing values, dropout rates, completion time, and response behavior in mobile online surveys. Unlike in previous studies [3, 8], participants were free to use the device of their own choice when answering the questionnaire. Among the three designs tested, the standard *Groups* of question layout performed best in terms of dropout rate, missing values and completion time. However, smartphone users in that condition also showed distorted answering patterns to a larger extent.

The *Manual Forwarding* layout – displaying only one item per page to avoid scrolling as recommended in prior studies [9, 10] – did not show satisfactory results in the categories analyzed here, especially regarding dropouts and completion time.

The newly developed *Automatic Forwarding* layout showed an especially high number of missing values. This might be due to the fact that in this layout participants were not able to skip individual questions if they preferred, for some reason, not to answer them. Instead, they were forced to quit the survey altogether. This is especially reflected in the first question of the questionnaire asking for participants' gender. In the *Automatic Forwarding* layout, almost 8% of participants refused to answer this question, resulting in early dropout. Another explanation might be that technical problems occurred (e.g. the website did not fully load). Nevertheless, the missing possibility to intentionally leave out questions constitutes a major drawback in this layout which needs to be investigated further. Furthermore, a design solution should be explored, e.g. adding a "prefer not to answer" alternative or an additional button allowing to skip a question. Overall, the design guideline to use large buttons on mobile devices, as recommended in previous studies [9, 14], does not seem to have a particular impact on dropouts and missing values.

Naturally, dropouts and missing values are related to each other, as especially early dropouts result in a larger number of missing values. Interestingly in this regard, the *Manual forwarding* layout, which allows participants to skip a question, yields fewer dropouts. However, there is no notable difference regarding missing values between the *Manual forwarding* and *Automatic Forwarding* layouts. Overall, however, the *Groups* layout was the most promising in terms of both dropouts and missing values although other studies recommend to display only one question per page on mobile devices [9, 10].

As longer surveys might discourage participants and increase dropout rates or missing values, completion times need to be taken into account. In the *Groups* layout, completion times were significantly lower compared to the other layouts. However, the analysis of response patterns in this group indicates that participants may have tried to finish quickly by simply selecting one of the first answer options. This would mean that completion times are lower because fewer possible answers are read by the participants, which would have a negative impact on data quality.

The current study has some limitations. As pointed out above, a majority of the respondents were paid for their participation, thus creating a special condition which cannot be transferred to other surveys. Presumably, especially dropout rates and missing values might be higher in surveys with purely voluntary participation. Second, we did not influence the choice of device or the assignment to a group. Therefore, the sample sizes of the three groups differ considerably. Third, the layout of *Automatic Forwarding*, unlike the others, provided no possibility to skip a question, which makes it difficult to compare dropout rates across the three conditions.

Furthermore, unfortunately, only few respondents from Europe and China participated via Smartphone, so intercultural comparisons could not be conducted as we had planned.

We will carry out further studies to investigate the issues raised in this article. Furthermore, we will improve the *Automatic Forwarding* layout by providing options to skip questions. Additionally, larger samples with roughly equal groups are required to confirm the results presented in this article. Moreover, we conduct further studies to confirm the results presented here with non-paid participants.

References

1. Sarraf, S., Brooks, J., Cole, J.: Taking surveys with smartphones: a look at usage among college students. In: AAPOR Annual Conference, Anaheim, CA (2014)
2. Thorsteinsson, G., Page, T.: User attachment to smartphones and design guidelines. Int. J. Mob. Learn. Organ. **8**(3), 201–215 (2014)
3. Lugtig, P., Toepoel, V., Amin, A.: Mobile-only web survey respondents. Surv. Pract. **9**(4), 1–8 (2016)
4. Mavletova, A.: Data quality in PC and mobile web surveys. Soc. Sci. Comput. Rev. **31**(6), 725–743 (2013)
5. Horwitz, R.: Usability of the ACS internet instrument on mobile devices 1. In: 2014 Proceedings of Statistics Canada Symposium (2014)
6. Couper, M.P., Peterson, G.J.: Why do web surveys take longer on smartphones? Soc. Sci. Comput. Rev. **35**(3), 357–377 (2017)
7. Olmsted-Hawala, E.L., Nichols, E.M., Holland, T., Gareau, M.: Results of usability testing of the 2014 American community survey on smartphones and tablets phase I: before optimization for mobile devices. Survey Methodology 3 (2016)
8. Mavletova, A., Couper, M.P.: Mobile web survey design: scrolling versus paging, SMS versus e-mail invitations. J. Surv. Stat. Methodol. **2**(4), 498–518 (2014)
9. Andreadis, I.: Web surveys optimized for smartphones: are there differences between computer and smartphone users? Methods, Data, Anal. **9**(2), 213–228 (2015)

10. De Bruijne, M., Wijnant, A.: Can mobile web surveys be taken on computers? A discussion on a multi-device survey design. Surv. Pract. **6**(4), 1–8 (2013)
11. Couper, M.P., Baker, R., Mechling, J.: Placement and design of navigation buttons in web surveys. Surv. Pract. **4**(1), 3054 (2011)
12. Hays, R.D., Bode, R., Rothrock, N., Riley, W., Cella, D., Gershon, R.: The impact of next and back buttons on time to complete and measurement reliability in computer-based surveys. Qual. Life Res. **19**(8), 1181–1184 (2010)
13. Lai, J., Vanno, L., Link, M., Pearson, J., Makowska, H., Benezra, K., Green, M.: Life360: usability of mobile devices for time use surveys. Surv. Pract. **3**(1), 3022 (2010)
14. De Bruijne, M., Wijnant, A.: Comparing survey results obtained via mobile devices and computers: an experiment with a mobile web survey on a heterogeneous group of mobile devices versus a computer-assisted web survey. Soc. Sci. Comput. Rev. **31**(4), 482–504 (2013)
15. Open Source Survey Application Limesurvey. https://www.limesurvey.org
16. Gosling, S.D., Rentfrow, P.J., Swann Jr., W.B.: A very brief measure of the Big-Five personality domains. J. Res. Pers. **37**(6), 504–528 (2003)
17. Buskirk, T.D., Andrus, C.: Smart surveys for smart phones: exploring various approaches for conducting online mobile surveys via smartphones. Surv. Pract. **5**(1), 3072 (2012)
18. Natda, K.V.: Responsive web design. Eduvantage **1**(1) (2013)

Human Interaction with the Output
of Information Extraction Systems

Erin Zaroukian[✉], Justine Caylor, Michelle Vanni, and Sue Kase

Army Research Laboratory, Computational and Information Sciences
Directorate, Adelphi, MD 20783, USA
erin.g.zaroukian.civ@mail.mil,
justine.p.caylor.ctr@mail.mil,
michelle.t.vanni.civ@mail.mil,
sue.e.kase.civ@mail.mil

Abstract. Information Extraction (IE) research has made remarkable progress in Natural Language Processing using intrinsic measures, but little attention has been paid to human analysts as downstream processors. In one experiment, when participants were presented text with or without markup from an IE pipeline, they showed better text comprehension without markup. In a second experiment, the markup was hand-generated to be as relevant and accurate as possible to find conditions under which markup improves performance. This experiment showed no significant difference between performance with and without markup, but a significant majority of participants preferred working with markup to without. Further, preference for markup showed a fairly strong correlation with participants' ratings of their own trust in automation. These results emphasize the importance of testing IE systems with actual users and the importance of trust in automation.

Keywords: Information extraction · Trust in automation ·
Reading comprehension · Deductive reasoning · Visual search · Workload ·
Usability

1 Introduction

With downstream processes for Information Extraction (IE), there is a tendency to consider only automated routines taking annotated text as input for computing co-reference, translating, populating a knowledge base, developing watch lists, or related tasks. Little attention has been paid to human analysts as downstream processors.

To evaluate progress in computer science for IE system-building research, the Natural Language Processing community has compared output to gold-standard data-sets curated by humans who have annotated the named entity items in text as being references to entities, in the form of token mentions of IE category types. IE research has made remarkable progress in this area using this intrinsic-measure framework.

Although researchers are always pushing the envelope, most systems for English, trained and tested on standard newswire, do very well, particularly in the area of Named Entity Recognition (NER) within IE systems. Intrinsic metrics are so high for

T. Ahram (Ed.): AHFE 2019, AISC 965, pp. 456–466, 2020.
https://doi.org/10.1007/978-3-030-20454-9_46

English NER that many consider IE a solved problem [1].[1] This work addresses the important issue of what needs to happen to have the technology serve situational awareness, decision making, and other cognitive requirements of human analysts, building a framework in which systems are compared against an extrinsic metric.

2 Experiment 1 - Testing an Existing IE Pipeline

In an experiment described in detail in [3], participants were presented sets of sentences describing a hypothetical adversarial attack, which they saw plain or with markup from an IE pipeline. The participant's task was to act as analyst and identify the perpetrator, target, time, and location of the attack, and their performance with and without markup was compared to determine whether the markup was helpful.

2.1 Participants

One hundred participants were recruited through Amazon Mechanical Turk to take part in this experiment. Each participant was compensated $2.00.

2.2 Materials and Equipment

The experiment was created using the Ibex tool for running behavioral psycholinguistic experiments (https://code.google.com/archive/p/webspr/) and run online through Amazon Mechanical Turk.

The text used in this experiment was drawn from the Experimental Laboratory for the Investigation of Collaboration, Information Sharing, and Trust (ELICIT) [4]. ELICIT is a simulated intelligence task containing a number of hypothetical adversary attack scenarios. Each scenario is a list of 68 simple sentences that together allow a reader to deduce the attacker, target, attack time, and attack location (*Who, What, When,* and *Where*) of an anticipated adversary attack.[2] These roles are identified in this experiment through seven dropdown menus (*When* is broken down into separate menus for month, date, time of day, and am/pm). See Fig. 1 for example sentences from an ELICIT scenario.

The markup presented in this experiment was generated using an IE pipeline developed at Rensselaer Polytechnic Institute [5, 6], which uses NER and event detection techniques. Recognized entities (e.g., person, vehicle, geo-political entity) and events (e.g., attack, enter) were shown via bracketing and subscripts, with mouseover revealing additional information (e.g., an event's arguments, the class an entity belongs to). See Fig. 1 for an example of ELICIT text marked up through this IE pipeline.

---

[1] See [2] though for outstanding issues in NER, such as "different definitions of NE, different types of text, different languages, and noisy data such as OCR and S2T."

[2] See also [7] for work with ELICIT and additional scenarios.

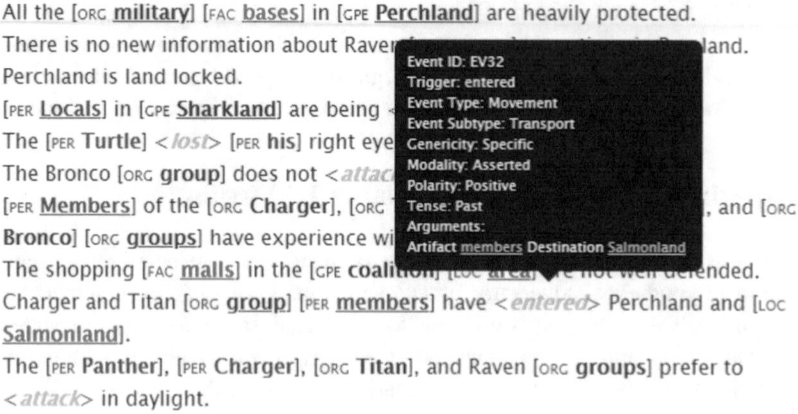

All the [ORG **military**] [FAC **bases**] in [GPE **Perchland**] are heavily protected.
There is no new information about Rave̶───────────────land.
Perchland is land locked.

> Event ID: EV32
> Trigger: entered
> Event Type: Movement
> Event Subtype: Transport
> Genericity: Specific
> Modality: Asserted
> Polarity: Positive
> Tense: Past
> Arguments:
> Artifact members Destination Salmonland

[PER **Locals**] in [GPE **Sharkland**] are being
The [PER **Turtle**] <*lost*> [PER **his**] right eye
The Bronco [ORG **group**] does not <*attac*
[PER **Members**] of the [ORG **Charger**], [ORG ─────────────, and [ORG
Bronco] [ORG **groups**] have experience wi
The shopping [FAC **malls**] in the [GPE **coali**───────────e not well defended.
Charger and Titan [ORG **group**] [PER **members**] have <*entered*> Perchland and [LOC
Salmonland].
The [PER **Panther**], [PER **Charger**], [ORG **Titan**], and Raven [ORG **groups**] prefer to
<*attack*> in daylight.

Fig. 1. Excerpt from an ELICIT scenario showing markup with mouse-over information for "entered".

This experiment also included a demographic questionnaire and a modified version of the NASA Task Load Index (NASA-TLX) [8]. The modified NASA-TLX asked participants to directly compare the two versions of the task (with and without markup) on a variety of workload measures as well as overall task-version preference. Participants responded to each question by choosing a point on a 21-point scale where the ends of the scale represent a strong preference for each of the versions.

2.3   Procedure

At the beginning of the experiment, participants completed a demographic questionnaire and read a page of instructions explaining the experiment. Before each test scenario, participants completed an abbreviated practice scenario to familiarize them with the task.

Each participant completed two test scenarios, one with markup (Markup condition) and one without (Plain condition), each preceded by an abbreviated practice scenario. Accuracy and response time were collected for each test scenario. At the end of the experiment, participants completed the workload and preference questionnaire.

2.4   Results

Accuracy and Response Time. Participants' accuracy and response times are shown for the plain and markup trials separately in Fig. 2. Overall, these results point to a surprising advantage for text without markup over text with markup.

Accuracy counts (the number of correctly identified attack roles for a trial, from 0 to 7) are shown on the y axis in Fig. 2. A Wilcoxon signed-rank test indicated that participants answered significantly more questions correctly in the plain condition (median = 5) than in the markup condition (median = 4.5, $p = 0.04$), with 46 out of 77 (60%) participants scoring higher in the plain condition (23 participants scored the same across conditions).

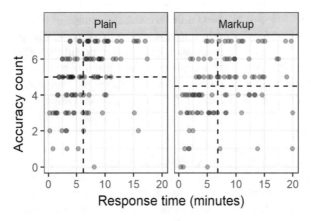

Fig. 2. Accuracy count (number of correctly identified attack roles) versus response time in minutes for each participant in each condition. Medians are shown as dotted lines.

Response time is shown on the x axis in Fig. 2. A Wilcoxon sign-rank test indicated that participants completed scenarios significantly faster in the plain condition (median = 6.19 min) than in the markup condition (median = 6.83 min, p = 0.02), with 58 out of 100 (58%) participants responding faster in the plain condition.

Workload and Preference. In the interest of space, workload scores will not be discussed here, but they are consistent with preference responses, see [3] for details. The question on preference, *"Overall, which version of the task do you prefer?"*, directly compared both versions of the task and so was binned as Plain preference versus Markup preference (scores of 11, indicating no preference, were excluded from analysis). A Pearson's Chi-squared test showed a significant preference for plain over markup ($\chi2(1, N = 96) = 13.5, p < 0.001$), with 66 participants preferring the Plain condition and 30 participants preferring the Markup condition.

While there is an overall preference for Plain trials, there is still a sizeable minority who prefer Markup trials, and, descriptively, participants prefer the version of the task that they performed better at, as summarized in Table 1.

Table 1. Accuracy and speed by preference

Preference	N	Condition	Median accuracy count	Median response time (min)
Plain	66	Plain	6	6.82
		Markup	4	7.15
Markup	30	Plain	4	4.10
		Markup	5	5.25

Furthermore, Mann-Whitney U tests shows that participants who prefer the Markup version completed the task significantly faster than participants who preferred the Plain version (p < 0.01), though their accuracy was not significantly worse (p = 0.33). While this test with its relatively small sample size has fairly low power, these results reveal some hope for the markup used here, at least with certain participants. Overall, however, participants appear to have found the Plain version easier to work with.

2.5   Discussion

While the IE pipeline tested here is intended to help the downstream human analyst, in this experiment, the pipeline's markup seems to hurt performance, both in accuracy and speed. Additionally, participants tend to find that markup leads to higher workload and is dispreferred in favor of plain, non-marked-up text. It is counterintuitive that markup would be categorically harmful to performance, so there may be forms of markup that are better suited to, and therefore more helpful in, this specific task.

Additionally, not all participants preferred and performed better without markup. This points toward the importance of providing options to participants, and it may be valuable to identify predictors for whether participants will work well with markup.

3   Experiment 2 - Testing an Ideal IE Pipeline

For this experiment, the aim is to design more relevant and accurate markup for ELICIT scenarios in an attempt to find conditions under which markup improves performance. Further, additional questions are included to provide predictive insight in determining which participants would prefer and perform better with or without markup.

3.1   Participants

This experiment treated Plain/Markup as a between-participants manipulation, so 200 participants were recruited through Amazon Mechanical Turk. Each participant was compensated $2.00.

3.2   Materials and Equipment

Like Experiment 1, this experiment was created using the Ibex tool for running behavioral psycholinguistic experiments (https://code.google.com/archive/p/webspr/) and run online through Amazon Mechanical Turk.

The text used in this experiment is the same text drawn from ELICIT used in Experiment 1.

The markup used in this experiment was generated by hand by the first author and checked by the other authors. It highlights phrases relevant to four types of responses (*Who*, *What*, *Where*, and *When*) that participants are required to provide. While there are many ways to judge relevance, the decision was made to highlight all and only potential responses (e.g., all and only country names were highlighted as possible *Wheres*). This strategy was chosen to make the markup more relevant than the markup in the first experiment without making it too computationally unrealistic or causing it to directly give away any answers. See Fig. 3 for an example of marked-up text. The markup here is expressed through background color instead of font color, as we felt this better allowed us to maintain four visually distinct categories (*Who, What, Where,* and *When*) without sacrificing the contrast between text and background color [9], and the bracketing and labeling used in the first experiment were dropped as participants often commented that they found this distracting.

Fig. 3. Excerpt from an ELICIT scenario showing hand-generated markup designed to be as accurate and relevant as possible.

Like Experiment 1, this experiment included a demographic questionnaire, but an additional question about participant occupation was included in hope of finding correlations between reported occupation and preference. Additionally, participants were required to enter a free-text response at the end of the experiment describing any strategies they used to solve the scenarios. This experiment also included an unmodified version of the NASA-TLX (because Plain/Markup was a between-participants manipulation, it would be difficult for participants to directly compare both conditions, so they only rated the version of the task that they completed). A preference question was again included, asking participants whether, were they to participate again, they would prefer the text to appear plain or with markup, indicating their preference by choosing a point on a 21-point scale where the ends of the scale represent a strong preference for each of the versions.

3.3 Procedure

As in Experiment 1, participants first completed a demographic questionnaire and read a page of instructions explaining the experiment. These instructions specified that any markup they see in the experiment was automatically generated (though in this experiment it was actually generated by hand). Participants then completed two practice scenarios, first in the Plain condition, then in the Markup condition. They then completed a Trust in Automation survey [10] asking for subjective ratings about systems like the one that generated the markup seen in the Markup practice scenario. Each participant completed two test scenarios, both in either the Markup or Plain condition. Accuracy and response time were collected for each test scenario. At the end of the experiment, participants again completed the Trust in Automation survey, provided their strategy descriptions, and completed the workload and preference questionnaire.

3.4   Results

Accuracy and Response Time. Participants' accuracy and response times are shown for plain and markup trials separately in Fig. 4. While Experiment 1 showed an advantage for text without markup over text with markup, the differences here are minimal.

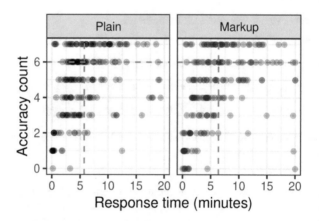

Fig. 4. Accuracy count (number of correctly answered questions) versus response time in minutes for all participants, separated by each condition. Medians (with the filtering criterion applied) are shown as dotted lines.

Concerns about speed and consequent quality of responses were raised in Experiment 1, so for the analyses below only participants with response times of 2 min or longer for each test scenario were included (150 out of 200 participants). One additional participant was removed due to a technical failure, leaving 80 participants in the Plain condition and 69 participants in the Markup condition.

A Wilcoxon rank sum test found no significant difference in the number of correctly answered questions between conditions (Plain median = Markup median = 6, W = 10976, p = 0.93, r = 0.005).

An additional Wilcoxon rank sum test found no significant difference in response time between conditions (Plain median = 5.73, Markup median = 6.38, W = 12005, p = 0.19, r = 0.08). While the effect size here is quite small, it suggests that with more power significantly faster response times may emerge in the Plain condition, as was seen in the Experiment 1.

Workload and Preference. Again, in the interest of space, workload scores will not be discussed here, be they were overall similar across conditions. The question of preference, "*If given the choice, which version of the task would you prefer to work with?*", was again binned as Plain preference versus Markup preference (with scores of 11 excluded from analysis). Responses were pooled across both conditions, and a Pearson's Chi-squared test showed a significant preference for Markup ($\chi2(1, N = 124) = 23.52$, $p < 0.001$), with 35 participants preferring the Plain condition and 89 participants preferring the Markup condition. This contrasts with the first experiment, where all advantages were in favor of Plain trials.

Demographics and Correlations. Participant responses to questions about whether their native language is English, their gender, their age, their level of education, and their occupation are summarized in Table 2.

Table 2. Summary of responses to demographic questions.

Question	Response category	N (All)	N (Filtered)
Language	No	27	16
	Yes	172	133
Gender	Female	86	66
	Male	112	82
	Other/prefer not to say	1	0
Age	18–29	78	45
	30–49	91	78
	50–64	25	22
	65+	5	4
Education	High school graduate or less	20	15
	Some college	70	55
	College degree or more	109	79
Occupation	Science and technology	67	56
	Arts, entertainment, and media	9	7
	Education	7	5
	Legal	13	13
	Sales	17	10
	Food preparation and serving	12	6
	Office administration and support	3	3
	Accounting and finance	20	17
	Healthcare and medical	7	2
	Industry and manufacturing	18	11
	Law enforcement	15	9
	Business management	11	10
	Other	0	0

Neither performance nor preference correlated well with any of the demographic information collected, with the exception of Language, where there is a medium positive correlation between accuracy and being a native speaker of English, as well as a medium negative correlative between response time and being a native speaker of English. Correlation coefficients are shown in Table 3, where one participant (Other/prefer not to say) was dropped from Gender correlations, and Occupation was pooled into Non-science and technology versus Science and technology.

Table 3. Correlation coefficients between collected demographic information and performance and preference for Plain and Markup trials. A coefficient is listed as 0 if it is less than 0.01 and greater than −0.01.

Covariates		Plain	Markup
Language	Accuracy	0.31	0
	Response time	−0.31	−0.31
	Preference	−0.08	−0.06
Gender	Accuracy	−0.10	−0.10
	Response time	0	−0.14
	Preference	0.05	0.01
Age	Accuracy	0.05	0.03
	Response time	0.03	0.14
	Preference	−0.22	−0.13
Education	Accuracy	−0.05	−0.20
	Response time	0.06	0.10
	Preference	−0.05	−0.18
Occupation	Accuracy	0.12	0
	Response time	0.21	0.06
	Preference	0.04	0.01

Preference for markup showed a fairly strong correlation with participants' ratings of their own trust in automation ($r = 0.39$). The correlation between trust in automation and objective performance measures, however, is very small (accuracy: $r = 0.06$, response time: $r = −0.05$).

3.5   Discussion

Experiment 1 asked participants to uncover hypothetical adversary attacks described in text documents with and without markup from an existing IE pipeline and found that, instead of helping, markup hurt performance and was dispreferred to plain text. While the markup used in Experiment 2 was hand-generated to be as helpful but realistic as possible, it still did not lead to better performance than plain text. This is an important warning to researchers trusting that actual automated markup will be helpful. This markup, however, was overall preferred to plain text, which is valuable for the overall user experience.

These results also emphasize that the trust in the automation that is used in an IE pipeline may be important for user experience and for encouraging users to opt to use these pipelines. However, the link between trust in automation and objective performance measures in the current study is very small, and experiments like this demonstrate that the automation need not improve performance. Much remains to be understood about the gap between IE technology and its human user for this technology to truly support human-computer interaction.

An additional consideration was highlighted by the unexpectedly high number of low-quality responses. These were responses that were too quick to represent true attempts to read the texts and identify the hypothetical adversary attack. The participants providing these responses were roughly twice as likely to report that English was not their native language (11/27 Non-native English speakers were filtered versus 39/172 native English speakers, shown in Table 3), and they often provided incoherent free-text strategy descriptions. This might indicate that workers on Mechanical Turk are generally not willing to put in the work necessary to do well at this task. However, regardless of their performance, this population of workers does not necessarily predict the performance of any other population, importantly, intelligence analysts. While workers on Mechanical Turk can be helpful due to their availability, it is important to include the specific intended end user in the testing cycle.

Acknowledgments. Many thanks to Stephen Tratz, Claire Bonial, Jeffrey Micher, Clare Voss, Jon Bakdash, Lucia Donatelli, and Jeff Hoye for their assistance in designing, deploying, and interpreting this work. This research was supported in part by an appointment to the Student Research Participation Program at the Army Research Laboratory administered by the Oak Ridge Institute for Science and Education through an interagency agreement between U.S. Department of Energy and ARL.

References

1. Cunningham, H.: Information extraction, automatic. In: Encyclopedia of Language and Linguistics, 2nd edn., pp. 665–677. Elsevier, New York (2005)
2. Marrero, M., Urbano, J., Sánchez-Cuadrado, S., Morato, J., Gómez-Berbís, J.M.: Named entity recognition: fallacies, challenges and opportunities. Comput. Stan. Interfaces **35**, 482–489 (2013)
3. Zaroukian, E.: Information extraction for optimized human understanding and decision making. In: Proceedings of ICCRTS (2018)
4. Ruddy, M.: ELICIT – the experimental laboratory for the investigation of collaboration, information sharing, and trust. In: Proceedings of the 12th ICCRTS (2007)
5. Li, Q., Ji, H.: Incremental joint extraction of entity mentions and relations. In: Proceedings of the 52nd Annual Meeting of the Association for Computational Linguistics, pp. 402–412. ACL, New York (2014)
6. Li, Q., Ji, H., Huang, L.: Joint event extraction via structured prediction with global features. In: Proceedings of the 51st Annual Meeting of the Association for Computational Linguistics, pp. 73–82. ACL, New York (2013)

7. Krausman, A.: Understanding audio communication delay in distributed team interaction: impact on trust, shared understanding, and workload. In: Proceedings of the IEEE CogSIMA Conference, pp. 1–3. IEEExplore (2017)
8. NASA: NASA Task Load Index (TLX), v. 1.0 Manual (1986)
9. Williams, T.R.: Guidelines for designing and evaluating the display of information on the Web. Tech. Commu. **47**(3), 383–396 (2000)
10. Jian, J.Y., Bisantz, A.M., Drury, C.G.: Foundations for an empirically determined scale of trust in automated systems. Int. J. Cogn. Ergon. **4**(1), 53–71 (2000)

Antenna Technology in Wireless Biometric Systems

Rafal Przesmycki[⊠], Marek Bugaj, and Marian Wnuk

Faculty of Electronics, Military University of Technology, Warsaw, Poland
{rafal.przesmycki,marek.bugaj,marian.wnuk}@wat.edu.pl

Abstract. The article presents basic medical research, as well as measuring devices and methods used in these studies. Human life parameters, which can be obtained after the examination, were also characterized.

The concept of a wireless biometric system consisting of a research module, a central unit and an antenna is presented. The first two elements were described theoretically while the antenna was developed in the CST Microwave Studio program. It is a microstrip antenna working in the frequency range from 2.3 GHz to 2.8 GHz. The energy gain of the designed antenna is from 3 dBi to 3.8 dBi. The physical model of the antenna meets the assumption of using it in a wireless biometric system.

Keywords: Microstrip antenna · EMC · Biometric system · CST · Wireless

1 Introduction

An active lifestyle is one of the most important features of every young person. Current times have forced the society to accept "being fit", even in some cases it can be said about "fitocracy". The fit culture is characterized by an appropriate choice of diet, active lifestyle, and the use of pro-health prophylaxis. Is it possible to minimize the time required to perform the necessary tests in health care? Can the only concept of "prophylaxis" be included as an open collection of measures and activities that can quickly and at an early stage detect dangerous diseases? These questions were the reason for the creation of this article. The main goal of the article is to prepare and present the concept of a biometric system that allows, in addition to measurements of select-ed human life parameters, to be sent wirelessly to specialists and medical facilities using unlicensed frequency bands [1].

2 Overview of the Measured Biomedical Quantities of a Human

The functioning of the human body is based on efficient information transfer. They must be transmitted not only inside the body, but also between the body and the environment. The main medium of information is in this case an electrical impulse, and its carrier ions distributed along cell membranes (mainly Na, K, Ca_2). The phenomenon of the flow of ions causing depolarization of the membrane can be easily compared to

© Springer Nature Switzerland AG 2020
T. Ahram (Ed.): AHFE 2019, AISC 965, pp. 467–479, 2020.
https://doi.org/10.1007/978-3-030-20454-9_47

the flow of electric current through a given medium, and thus all rights related to the flow of electric current apply.

2.1 Electrocardiography (ECG)

Electrocardiography, as a diagnostic procedure used to detect heart disease, is connected with the discovery at the beginning of the 19th century of a galvanometer, which was used to measure small amounts of electric current. In a modern form, the cardiographic study uses six unipolar precordial leads, allowing the measurement of electric field differences in the horizontal plane. The cardiologist, as a person specializing in ECG tests, chooses the method of conducting the examination. By default, it consists in laying the patient on the bed, placing the electrodes in certain places and starting the electrocardiograph. The measurement itself can take from several seconds to several minutes depending on the needs.

Another type of ECG is stress testing. The subject is placed on a specialized bike, called a cycloergometer or on a treadmill with an adjustable angle of inclination and speed. During the examination, the cardiologist gradually increases the load. The results obtained can be used to determine cardiac output, detect arrhythmia and the severity of coronary heart disease.

ECG examination, which is completely different from the previous two, is the Holter method. The process consists in a 24-h diagnosis of the patient during his daily life during exercise, work, during rest, as well as in stress and in sleep. In order to measure the heart rate, the subject receives a device recording the electrocardiogram (so-called Holter), together with the electrodes that are connected to his chest. The apparatus allows recording anomalies occurring during the test [2].

2.2 Electroencephalography (EEG)

The EEG test aims to record bioelectrical activities of the human brain. It is used for monitoring and diagnosis of epilepsy, sleep disorders, organic brain diseases, statements of poisoning with neurotoxic substances and in the adjudication of coma or death of the brain. It is a non-invasive method. It consists of placing nineteen electrodes (eight for each hemisphere and three on the midline of the skull), which record changes in the electrical potential resulting from brain activity. The voltage of currents with a frequency varying between 0.5 Hz and 250 Hz, ranges from a few to several hundred microvolts. Due to such a low value, in order to correctly read the results, an electroencephalograph is used, amplifying the signal about a million times and registering the potentials between the electrodes. The results are presented in the form of a graphic printout. The device averages the activity of neurons located in the area of a given electrode, instead of receiving an electrical impulse from one nerve cell [3, 4].

2.3 Electromyography (EMG)

Another type of research is electromyography (EMG) which consists in reading the electrical signal generated during muscle action. This signal is read during controlled and inert muscle contraction. The test is carried out to determine the muscles' ability to work and to identify diseases of the peripheral nervous system. However, it is not possible to obtain information on how much strength a muscle works with, how many meat fibers have been used at work, etc.

The EMG examination can be carried out by choosing one of two methods - clinically or kinesiologically. The division results mainly from the method of conducting the research. In the clinical trial, needle electrodes are used, which are directly inserted into the muscle belly (a large concentration of muscle fibers from which the skeleton muscles are created). This method is considered the most accurate because of the possibility of observation of the action of a single motor unit, observation of muscle located in deeper layers and analysis of the work of a single muscle, thanks to which it is possible to assess in depth the possible anomalous state. However, this involves the requirement of sterility of the place where the test is carried out, which can usually be obtained in a hospital setting. In addition, due to the interference in the abdomen of the examined muscle, the presence of a doctor is necessary [5, 6].

2.4   Electroneurography (ENG)

The next examination consists in the examination of nerve conduction, during which the activities of peripheral nerves, i.e. motor and sensory fibers, are evaluated in relation to the measurement of responses to controlled electrical stimuli. Appropriate surface electrodes on the skin of the patient are used, which transmits a current in the range from 0 to 100 mA and 2 Hz with a duration of 0.2 ms.

This signal is conducted along a specific nerve to a different, remote point. Then the measurement of the conduction velocity of the given electrical stimulus by the nerve is performed. Thanks to this research, it is possible to locate damaged nerves, determine the size of pathological changes, and also the type of their differentiation. During the examination, two electrodes are connected to the patient: a stimulating one, located above the stimulated and receiving muscles, which located on the stimulated muscle registers the moment of receiving the signal. The time of the signal flow between the electrodes is one of the parameters measured from the moment of applying the current to the moment of muscle contraction. An additional feature is also the difference in the distance between the two electrodes, the time of potential transfer in the nerve, the amplitude of the response to stimulation and latency [5, 6].

2.5   Temperature Measurement

The temperature of the human body is of great importance in the medical aspect, because it can indicate the condition of the subject. In addition, by taking regular measurements you can take an appropriate medical treatment plan. The source of the disease is the cause of the increase in body temperature, initiated by the immune system. Human pyrroles appear in human blood circulation, which affect the thermoregulatory center located in the hypothalamus, changing the biological temperature standard. Contrary to popular belief, there is no concept of "normal temperature" [7].

2.6   Blood Pressure Measurement

\For testing the pressure value a sphygmomanometer is used, generally called a blood pressure meter, while using the auscultatory method. The classic sphygmomanometers are made of a rubber band with an air chamber in the cuff, a pressure gauge and a pressure pump. In order to test the pressure, the pressure should be reached in the cuff to stop the blood flow in the artery, between the diastolic and systolic pressure. At the time of contraction, the heart pours blood into the arteries at high speed. At this moment, audible tones arise. Remembering the value at which the clatter appears (the so-called 1st phase of the Korot cycle), the systolic pressure value is obtained. The last value at which to listen to the clatter (so-called V-phase Korotkov) is diastolic pressure [7].

3   An Overview of the Frequencies and Antennas Used in Wireless Biometric Systems

The frequency selection for wireless biometric systems will be based on specific needs. It is worth remembering that if the used frequency will be higher, the relatively smaller will be the radio range. On the other hand, the higher the frequency, the greater its availability and the possibility of booking. When selecting the frequency, it is also worth paying attention to the available arrangement of channels. In the licensed bands, channels with the following width are available: 3.5 MHz, 7 MHz, 14 MHz and 28 MHz. In unlicensed bands, the channel width is usually 20 MHz. Depending on the modulation, the larger width of the channel usually means a higher bandwidth of the radio system. The channels are available as simplex and duplex. If we get one 28 MHz wide duplex channel, in reality we get 2 simplex channels with 28 MHz bandwidth, or 56 MHz spectrum. Simplex and duplex channels are associated with device operating modes. On the duplex channel with the FDD mode, the transmission is carried out in both directions simultaneously on different frequencies. The simplex channel is related to the TDD mode, in which the reverse-side transmission takes place within one frequency [8].

Wi-Fi networks and hardware use unlicensed frequencies. The most popular used ranges are 2.4 GHz and 5 GHz. The popular range, 2400.0–2483.5 MHz, is used by Wi-Fi devices operating in 802.11 b/g/n standards, and used mainly in rooms. The maximum radiated power can be 100 mW EIRP. The mentioned frequency range is divided into 11 channels, of which 3 do not overlap. Frequency can be disturbed by home electronic devices, including DECT wireless telephones and microwave ovens. Wi-Fi devices must therefore accept harmful interference from other systems operating at that frequency [8, 9].

Subsequent popular ranges that do not require a license are 5150–5350 MHz and 5470–5725 MHz In the first one it is allowed to work with a maximum power of 200 mW EIRP indoors. Devices should be equipped with DFS (Dynamic Frequency Selection) system to ensure compatibility with radar systems. The similar range is 5470–5725 MHz, which allows you to work with 1 W EIRP. Devices operating in this band should support the DFS mechanism, which allows dynamic selection of the channel (frequency) depending on the radio environment parameters. 5 GHz channels are available mainly for equipment operating in 802.11a /n standards. There is a significant number of non-overlapping channels here [8, 9].

In the case of a wireless biometric system, the proposed solution is to use the Wi-Fi band. 2.4 GHz technology ensures transmission of information over long distances (over 10 m) enabling data reading from objects moving at high speed (over 100 km/h), which is impossible in LF and HF technologies. It is mainly used for identification, registration of fast-moving objects - for example, management of the municipal transport fleet, registration of the passage of railway wagons. The waves used by the 2.4 GHz frequency are longer, so they are better suited for traversing larger distances or penetration through fixed objects such as walls. They are used by the Wi-Fi module for communication in the b, g, n standard. This frequency is also used by the Bluetooth module, which can be placed in the mobile device and is used to connect the mobile device - mobile phone and mobile phone - portable devices. Wireless networks rely primarily on the IEEE 802 standard. The 802.11 family includes three completely independent protocols focusing on coding (a, b, g). The first widely accepted standard was 802.11b, then 802.11a and 802.11g. The 802.11n standard is not officially approved yet, but more and more network devices are compatible with this technology.

The 802.11n standard was approved in September 2009. It can work on 2.4 GHz and 5 GHz frequencies. It allows reaching the maximum theoretical data transfer rate up to 600 Mb/s. Its operating range has been extended to 50 m in the room and over 100 m in the open space [9].

The quality of the Wi-Fi network is largely determined by the antenna-radio component, i.e. the physical side of communication. In the case of Wi-Fi networks in relation to the wired network, i.e. Ethernet, we can talk about a significant regression of parameters such as: reliability, performance and security - at least in the colloquial reception of these parameters by users, as well as the majority of people responsible for implementation and supervision over the network. The directionality of the antenna is related to its gain, the gain is achieved by shaping the signal. A typical non-directional antenna used in AP Wi-Fi has directivity at the level of 2–3 dBi [9].

Designing antennas in contact with the human body is extremely difficult. The first of the problems an engineer encounters is the fact that the human body absorbs energy from electromagnetic waves, transforming it into thermal energy (this phenomenon can be observed during a long conversation over a mobile phone - a warming ear). This results in the fact that when the antenna is placed near the human body, its performance drops significantly. For example, if you design a wearable antenna and measure its efficiency, which is −3 dB, when it is placed on the surface of the body, the efficiency can drop even to −13 dB (5%). Another obstacle is the user's convenience, and hence the minimization of size and the creation of flexible structures. At the moment there are several different antenna solutions available that allow achieving harmony between efficiency and the above-mentioned difficulties. It means the following antennas: wearable, microstrip, textile, EBG Woodpile, reconfigurable.

3.1 Wearable Antennas

Wearable antennas are not the only antennas whose purpose is related to the movement of the object. An example of this type of antennas can be those used in the so-called Smartwatches or GoPro cameras.

The first group is characterized by the fact that they have built-in Bluetooth antennas. The other one has built-in Wi-Fi, Bluetooth and GPS. It is worth noting that wearable antennas are increasingly used in electronics. Their design requires knowledge of electromagnetic properties, such as permeability and losses arising from contact with textile material. The antenna in question is made of conductive material such as Zelt, Flectron or polyester taffeta of pure copper. Such a material plays the role of a radiating element, while silk, felt and nonwoven are non-conductive substrates [10].

3.2 Microstrip Antennas

Microstrip antennas are characterized by a small mass, thin and planar structure, which enables the construction of complex antenna systems. In addition, their shape can be easily and accurately reproduced by means of a printed circuit. This results in low production costs and easy repeatability. They occur in two structures: single-layer and multi-layer. The microstrip antenna is made of metallic elements that have been milled or etched in a conductive layer. The radiator and mass were placed on opposite sides of the dielectric layer (Fig. 1) [10].

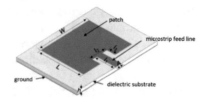

Fig. 1. A typical microstrip antenna design

3.3 Textile Antennas

Textile antennas (Fig. 2) are characterized by high flexibility and easy integration with the dress. When designing the analyzed antenna, in addition to the design itself, it is necessary to consider the impact of the material on which the antenna will be placed. The principle of their operation and construction is not significantly different from microstrip antennas.

3.4 Woodpile Antennas

Woodpile antennas are antennas with a structure resembling a pile of wood. They are based on the material EBG (Electromagnetic Band Gap), also known as photonic crystals. They give new possibilities to control and manipulate the flow of electromagnetic waves.

Fig. 2. An example of a textile antenna

They are made of dielectric structures that are periodic in both one and more dimensions. The complete EBG suppresses the propagation mode in all three dimensions and can only be obtained in the three-dimensional (3-D) phallic crystal. EBG materials can provide significant benefits for attenuation and control of radiation when using the antenna. Until now, most studies have focused on one (1-D) and two-dimensional (2-D) EBG materials. This was because of their simpler construction. However, due to the fact that this three-dimensional EBG materials give the possibility of greater control of antenna radiation properties, current research has been intensified and focused on them. Antennas based on this technology are characterized by low height, low losses and low side lobes [10, 11] (Fig. 3).

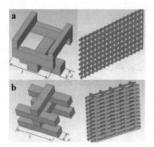

Fig. 3. 3-D structures Woodpile EBG: a. Elementary cell (left) and periodic structure of the wood stack, b. Elementary cell and structure of a cylindrical stack of wood

3.5   Reconfigurable Antennas

Reconfigurable antennas are antennas capable of dynamically changing frequency and radiation in a controlled and reversible manner. This type can perform frequency, polarization or radiation pattern reconfigurability, allowing radio systems to meet different communication standards and usage conditions. The term "reconfigurable antenna" refers to a system consisting of an antenna connected to a signal processor that attempts to adjust, e.g. radiation patterns, to select or isolate a usable signal. Reconfigurable antennas can be classified depending on the parameter that is dynamically corrected. Typically, this is the operating frequency, radiation pattern or polarization [10].

4 The Concept of a Biometric System for Measuring Human Life Parameters

The main goal of the biometric system is to create a measurement system that would allow the measurements of human life functions. Thanks to this, it would be possible to observe the vital parameters of the patient throughout the study period, outside the medical facility. For this purpose, you can use a classic pulse meter, which after measuring a given value, converts it from an analogue form to a digital form. Then, with use of a microstrip antenna, it would be possible to send a signal to the central unit, which would receive data, process it and save it. On the Fig. 4 the concept of the entire biometric system is presented.

Fig. 4. The concept of a biometric system

In the further part of the article, individual components of the biometric system with a detailed development of the design of the antenna designed for such a system were briefly discussed. The measuring equipment and the central unit were described only theoretically, while the antenna was developed in the CST Microwave Studio program and presented along with simulations and measurements of its electrical parameters and practical execution.

4.1 Measuring Equipment

In principle, the measuring device will use an automatic wrist blood pressure meter, because it is the most mobile and the simplest of all available methods. The analyzed module, due to its specificity, will be able to be used in two configurations:

- "on-body" in which the blood pressure monitor is always placed on the wrist of the patient
- "off-body" when the patient is forced to remove the device. In this case, before making the measurement, the portable unit will signal the sound along with the message on the screen, the need to attach the module on the wrist.

Switching between configurations can be done automatically via the cuff fastener sensor. When first turned on, communication with the mobile device will be established in order to pair and download the measurement configuration. The measurement configuration consists of:

- measurement frequency - how many minutes or hours to be measured
- measurement period - for how many minutes, hours or days to be measured

– notification - how many seconds, minutes before the planned measurement, signal the user the need to mount the device, volume and notification sound.

4.2 Central Device

The central unit will be located in the patient's home and requires permanent connection to the electricity grid and access to the Internet via Wi-Fi, ADSL or LAN. It is equipped with a charging station for a portable device and separate microUSB cables for charging each module. The device consists of a touch screen that allows access to data, updating the configuration of modules, as well as logging in to a special patient portal. Each central unit has an individual identification number assigned by the internal system that allows the medical doctor to monitor the patient's vital parameters on an ongoing basis and to introduce changes in the configuration of the modules.

5   Antenna Design

A microstrip antenna has been designed for the presented biometric system, which complements the measuring device and the central unit. This antenna sends the processed information packet to the central unit. In order to design it, CST STUDIO SUITE from Computer Simulation Technology was used [12].

As it was mentioned earlier, an antenna powered by a microstrip line was designed to create the system. The antenna powered by a microstrip line thanks to the easy-to-build structure settles the ability to control the fitting by changing the line connection location. In addition, it has a low level of unwanted radiation (about −20 dB), a narrow bandwidth and easier integration with electrical systems. The location of the radiator and the supply line on the same ground can cause reciprocal coupling and uncontrolled parameter changes. In addition, radiation from the antenna can interfere with electrical systems. However, this is also an advantage from the point of view of antenna production and the possibility of coupling with active systems. It is also impossible to reconcile optimal substrate parameters for the radiator and the microstrip line. Due to the efficiency of the antenna, it is desirable that the electrical permeability of the substrate is as small as possible [12].

The designed antenna has dimensions 158 mm and 155 mm. The length of its screen is 23 mm, width 29 mm. Due to the small size, it is possible to easily move and attach it in different places. The bandwidth of the designed antenna is provided on the frequency 2.45 GHz. It will be possible to use it mainly in home systems in which Wi-Fi connectivity operating on the 2.4–2.485 GHz band is used. This allows the entire system to work at home. In the designed antenna the reflection coefficient on the desired 2.45 GHz band obtained the value of about −57 dB (Fig. 5), which is a very satisfactory result and allows for successive work with the system.

Figure 6 shows a graph of the VSWR coefficient as a function of frequency. The orange color presents the results of a computer simulation, while the blue color presents the results of measurements of the physical model of the designed antenna. In the designed antenna the value of the reflection coefficient, on the required frequency 2.45 GHz, is close to 1, and the determined operating range from the measurements ranges from 2.2 GHz to 2.8 GHz.

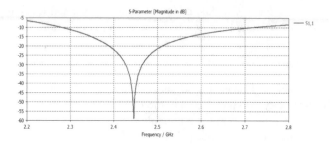

Fig. 5. Antenna reflective losses

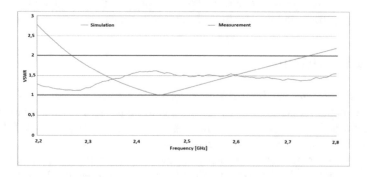

Fig. 6. Standing wave ratio in frequency function

The next parameter is the input impedance of the antenna, consisting of radiation resistance and loss resistance. The first component is associated with the power radiated by the antenna, and the second reflects thermal losses. As in the case of each resonant circuit in the resonant frequency, the impedance has only the real part. The antenna input reactance depends on its geometrical dimensions. Figure 7 shows the impedance results for the designed antenna. The orange color presents the results of a computer simulation, while the blue color presents the results of measurements of the physical model of the designed antenna. In the assumed antenna work band it is about 50 Ω for the real part and about 0 Ω for the imaginary part (Fig. 8).

The last determined parameter for the designed antenna is radiation pattern, from which it is possible to determine the directional gain of the antenna. For the designed antenna, the level of directional gain reaches the value of 3.86 dBi (Fig. 9). This is a satisfactory result that will ensure proper functioning of the biometric system at home.

With regard to the shape of the radiation pattern of the antenna, it can be observed that the designed antenna has an omni-directional radiation pattern (Fig. 10). Due to the operating mode, it is recommended that the antenna sends data to the central unit in any position.

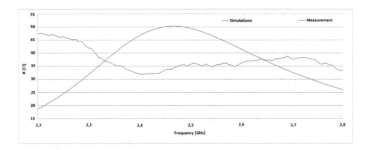

Fig. 7. The real part of the input impedance of antenna in the frequency function

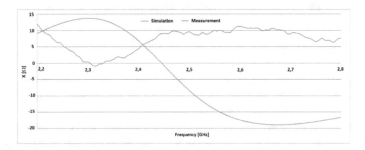

Fig. 8. The imaginary part of the input impedance of antenna in the frequency function

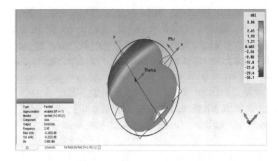

Fig. 9. Radiation pattern of antenna in three-dimensional space (3D)

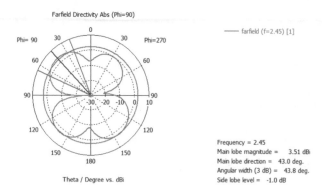

Fig. 10. Radiation pattern of antenna in polar coordinates

6   Conclusions

The main purpose of the article was to present the designed antenna and present its parameters, which could be used as an element of the system to measure human biometric quantities. The CST Microwave Studio program was used to prepare the antenna design. The antenna will work in the Wi-Fi wireless band, in order to cooperate with other devices connected to the home network. Based on available domestic and foreign literature, and using the CST software, the antenna was designed and modelled.

It is an antenna powered by a microstrip line, operating in the range from 2.2 GHz to 2.8 GHz. Its impedance is 50 Ω. Its main assumption was the ability to work on the Wi-Fi band, which is used in most households. This will allow the patient to be observed throughout the whole period without having to be in a medical ward. The antenna has an omnidirectional characteristic; it can be used as an integral part of the biometric system. Its task is to send a processed signal containing data about the subject to the central unit, without the need for continuous synchronization between the devices. Antenna gain is from 3 dBi to 3.8 dBi. The antenna made meets the assumption of using it in a wireless biometric system. In the Wi-Fi band, the antenna has satisfactory parameter values, small size and low weight.

The article analyses the research using the fact of the flow of electrical impulses in the human body. This allowed to explore the knowledge used to check and monitor the vital signs of a human being. In addition, it drew attention to problems that may arise during the implementation of the practical part of the work. The further part of the article presents the analysis of unlicensed frequency bands that can be used in biometric systems. A strong accent was imposed on limitations and disturbances that may occur during antenna operation in the home. In the following, the focus was on the proposed solution, i.e. the Wi-Fi band. His work band as well as advantages and disadvantages were described. Antennas that are used for this type of wireless communication and radiation characteristics and phenomena that accompany them are presented.

The article also presents the analysis of the type of antennas that are currently used in biometry. Due to the absorption of energy coming from electromagnetic waves through the human body, transformed into thermal energy, the design of antennas in contact with the human body is a complicated undertaking. One of the problems is also the comfort of the examined person, which forces the minimization of measuring instruments and their flexible construction.

References

1. Traczyk, W., Trzebski, A.: Fizjologia człowieka z elementami fizjologii stosowanej i klinicznej, pp. 56–60. Wydawnictwo Lekarskie PZWL (2004)
2. A (not so) brief history of electrocardiography, 11 April 2018. https://ecglibrary.com/ecghist. html
3. Stanley, F.: Origins of Neuroscience: A History of Explorations in Brain Function, pp. 41–42. Oxford University Press, New York (1994)
4. Quigg, M.: EEG w praktyce klinicznej; Wyd. Elsevier Urban & Partner, Wrocław (2008)
5. Bradley, W., Daroff, R.: Neurologia w praktyce klinicznej, pp. 592–627. Wyd. Czelej, Lublin (2006)
6. Andrzej Szczeklik (ed.): Choroby wewnętrzne. Przyczyny, rozpoznanie i leczenie, tom I, pp. 89–92. Wyd. Medycyna Praktyczna, Warszawa (2005)
7. McKenzie, I.S., Wilkinson, I.B., Cockroft J.R.: Nadciśnienie tętnicze; Wyd. Elsevier Urban & Partner, Warszawa (2006)
8. Ross, J.: Sieci bezprzewodowe. Przewodnik po sieciach Wi-Fi i szerokopasmowych sieciach bezprzewodowych, Wydanie II; Wyd. Helion, Gliwice (2009)
9. Leary, J., Roshan, P.: Bezprzewodowe sieci LAN 802.11. Podstawy; Mikom, Warszawa (2004)
10. Milligan, T.A.: Modern Antenna Design, 2nd edn. Wiley-IEEE Press, 11 July 2005
11. Brizzi, A., Pellegrini, A., Zhang, L. Hao, Y.: Woodpile EBG-based antennas for body area networks at 60 GHz. In: 4th High Speed Intelligent Communication Forum, Nanjing 2012
12. CST Studio Suite 2014, 11 April 2018. https://www.cst.com/2014

Parametric Study of Pin-Fins Microchannel Heat Exchanger

Mohammed Ziauddin, Fadi Alnaimat$^{(\boxtimes)}$, and Bobby Mathew

Department of Mechanical Engineering, The United Arab Emirate University,
P.O. Box 15551, Al Ain, United Arab Emirates
falnaimat@uaeu.ac.ae

Abstract. MEMS heat exchangers use microchannels to increase the surface area of contact between the moving fluids. MEMS heat exchangers are typically utilized for processor chip cooling applications. This study investigates the influence of geometric and operating parameters on performance of a pin-fins microchannel heat exchangers. ANSYS Workbench is used for modelling. The mathematical model consists of the continuity equation, Navier-Stokes equation, and energy equation. Water is taken to the fluid in this study. The boundary conditions of the model consist of the inlet temperature of the fluids, inlet flow rate of the fluids, fluid velocity on the walls (set v = 0), outlet pressures (set to Pout = zero), and temperature gradient at the outlet (set equal to zero). The influence of hydraulic diameter, pin fins diameters, structural material, and Reynolds number on the effectiveness of heat exchanger is studied. Studies are carried out for Reynolds number varying between 100 and 2000. The materials considered for this study are stainless steel, silicon, and copper. The effectiveness is determined using the inlet and outlet temperatures of the fluids. The thermal performance of MEMS heat exchanger is influenced by the hydraulic diameter and pin-fin diameter. The thermal performance of the MEMS heat exchanger increases with reduction in Reynolds number.

Keywords: Effectiveness · Microchannel heat exchanger · Counterflow · Computational fluid dynamics (CFD)

1 Introduction

Heat exchangers are required in the refrigeration and air conditioning systems for heat transfer applications. With the demand of reduction in weight and increasing cost of manufacturing, technology improvement with Microchannel Heat Exchanger (MCHX) is very demanding [1]. The heat exchangers of micro dimensions have more advantages such as savings material, space, and meeting the demand of heat transfer [2]. The design can affect the performance if the heat exchanger is in parallel or counterflow arrangements. The performance is also expected to be varying. The use of microchannel heat exchangers is very recent from past decade. Samsung electronics have implemented the use of microchannel heat exchangers since 2006 in cooling model of air conditioning systems [3]. A study conducted exploring the performance of counter flow microchannel heat changers with different channel cross-sections found that increasing number of channels improved the effectiveness and pressure drop [4].

© Springer Nature Switzerland AG 2020
T. Ahram (Ed.): AHFE 2019, AISC 965, pp. 480–487, 2020.
https://doi.org/10.1007/978-3-030-20454-9_48

In literature, very limited work is covered in this direction studying microchannel heat exchanger whether it is through simulations or experiments. In this study, counter flow microchannel heat exchanger is studied with square cross-sections. The temperature variations of running hot fluid and running cold fluid are observed in order to calculate the effectiveness of the heat exchanger. Simulation approach is adopted using ANSYS Fluent and performing Computational Fluid Dynamics (CFD) simulation. Simulation is done in order to reduce cost and make use of computational tool to get satisfactory results controlling various parameters.

This could contribute in modifying and finalizing the model in order to prepare, manufacture, and test the microchannel heat exchanger experimentally in the future.

2 Mathematical Modeling

Main parameters that are involved in determining the performance of heat exchangers are the heat transfer rate of hot fluid and cold fluid. This could also be understood from observing the temperatures at the inlets and outlets of both fluids. The heat exchanger is expected to cool the hot fluid running at temperature less than 100 °C. Hence, the specific heat capacity is assumed to be constant in theoretical calculations for the cold and hot fluids. The capacity ratio is considered as shown in Eq. (3) value equal to 1 [5].

$$C_c = \left(\dot{m}C_p\right)_c \tag{1}$$

$$C_h = \left(\dot{m}C_p\right)_h \tag{2}$$

$$C_r = \frac{C_{min}}{C_{max}} = \frac{\min(C_c, C_h)}{\max(C_c, C_h)} \tag{3}$$

Where, C_c and C_h are the heat capacity rate of cold and hot fluids respectively, C_p is the specific heat at constant pressure, and $\dot{m}$ is the mass flowrate. The rate of heat transfer can be determined by $\dot{Q}$ from the below Eqs. (4) and (5).

$$\dot{Q} = C_c\left(T_{c,out} - T_{c,in}\right) \tag{4}$$

$$\dot{Q} = C_h\left(T_{h,in} - T_{h,out}\right) \tag{5}$$

$$\dot{Q}_{max} = C_{min}\left(T_{h,in} - T_{c,in}\right) \tag{6}$$

Where, $T_{c,in}$, $T_{c,out}$, $T_{h,in}$ and $T_{h,out}$ are the temperatures of cold and hot fluids at the inlets and outlets respectively. The log-mean temperature difference (LMTD) can be calculated using the following Eqs. (7), (8) and (9). The heat transfer effectiveness can be determined using NTU method and using the following Eqs. (10) and (6). The effectiveness is the ratio of the actual heat transfer rate to the maximum heat transfer rate possible [5].

$$\Delta T_1 = T_{h,in} - T_{c,out} \tag{7}$$

$$\Delta T_2 = T_{h,out} - T_{c,in} \tag{8}$$

$$\Delta T_{lm} = \frac{\Delta T_1 - \Delta T_2}{\ln\left(\frac{\Delta T_1}{\Delta T_2}\right)} \tag{9}$$

$$\varepsilon = \frac{\dot{Q}}{\dot{Q}_{max}} \tag{10}$$

ANSYS Fluent simulation software is used to design and analyze microchannel heat exchanger with square microchannels and rectangular fins inside the channel. The modeled section of the heat exchanger is shown in Fig. 1. Overall dimension of the analyzed unit heat exchanger is 0.5 mm in width, 0.5 mm in height and 2 cm in length. Two square microchannels were modelled with 100 μm by 100 μm cross-sectional area. One carrying hot fluid and other carrying cold fluid in a counter flow arrangement. Each channel was attached with 20 fins of 20 μm by 20 μm square cross section that wee equidistant through located through the length of the channel. The working fluid used is water. The Reynolds number was varied from 0.0005 to 50 and the effectiveness was calculated for each case.

2.1   Counterflow Microchannels

In Fig. 1, the CAD model of the heat exchanger is shown that is consisting of the hot fluid domain on the top channel and cold fluid domain in the bottom channel.

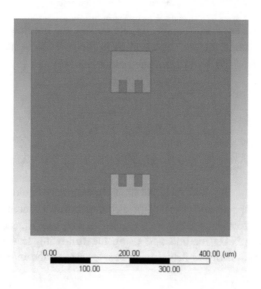

Fig. 1. CAD Model of microchannel heat exchanger.

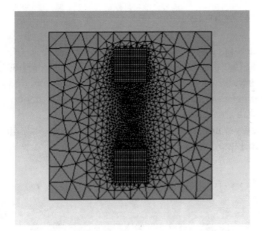

Fig. 2. Meshed model of microchannel heat exchanger.

In Fig. 2, the meshed model is shown, and it was done applying mesh sizing tool through sizing the edges. Mesh elements were made smaller and coarse near the microchannel edge compared to other regions ensuring more elements located at the hot and cold fluids domains.

3 Setup

The model was analyzed and solved as steady state conditions using energy equations and viscous model (realizable k-epsilon model considering standard wall functions). For the fluid domain, water-liquid was considered as the working fluid. For the shell domain, aluminum material was considered in the model. The temperature of hot inlet was kept at 70 °C,whereas, the temperature of the cold inlet was set at 15 °C. The thermal conditions were kept as convection at the shell wall with heat transfer coefficient as 20 W/m^2.K.

4 Data Collection

The results were mainly collected to obtain the centerline temperatures of the hot fluid and cold fluid domains. As the Reynolds number was varied from 5 to 300 for the hot fluids and respective velocity obtained from capacity ratio was inputted for cold fluid velocity in the counterflow. The temperature distributions were recorded, and effectiveness was calculated for each case based on the outlet temperature readings. For viewing the contours of temperature distribution in both fluid and shell domain, a plane was created in the mid-section of the heat exchange where it can show the temperature contours of both domains. However, only one legend could be viewed, so the hot fluid temperatures were viewed as those were more significant.

5 Results and Discussion

The results are plotted against the hot fluid Reynolds number. Observing through the contours, it was clear that the amount of temperature loss in the hot fluid was similar to the temperature gained by the cold fluid. For Reynolds number equal to 5, the results are illustrated in the below Fig. 3, the legend shows the temperatures of hot fluid with maximum of 343 K in the inlet and 321.06 K in the outlet.

The centerline temperatures of hot and cold fluid domains are plotted for each of the Reynolds number in Figs. 4 and 5 respectively.

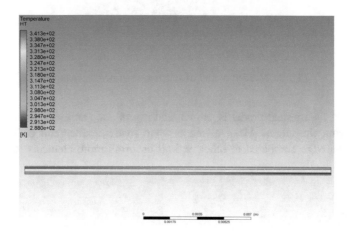

Fig. 3. Temperature contours (section view)

These results also showed that the temperature variations are closely opposite to each other indicating the temperature drop in the hot fluid side is close value of the temperature increased in the cold fluid side. Figures 4 and 5 provide an evidence that the heat capacity of both hot and cold fluids remains similar which is leading to assumption of capacity ratio equal to 1. In order to calculate the effectiveness, the value of $\dot{Q}$ obtained from either Eqs. (4) or (5) will give precise results.

According to Figs. 4 and 5, the Reynolds number of 100 showed less temperature difference with respect to inlet and outlet. Heat transfer is effective in the small thermal entrance length as compared to heat transfer observed for Reynolds numbers 5 to 25. The change in temperature is greater from Reynolds number 5 to 25. For all the cases of Reynolds numbers, the temperature profile is observed as linear from inlet to outlet. It can also be depicted that by further increase in Reynolds number beyond 50 would not result in effective temperature as temperature difference from inlet to outlet gets lower. As the effectiveness is the ratio of actual to maximum heat transfer rate, the decrease in temperature difference from inlet to outlet can decrease the heat transfer effectiveness.

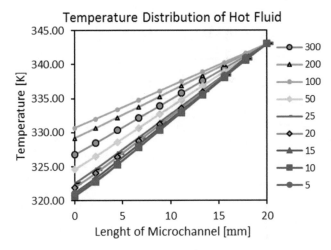

Fig. 4. Temperature distribution of hot fluid.

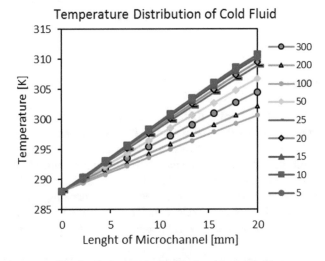

Fig. 5. Temperature distribution of cold fluid.

The Fig. 6 shows the plot of heat transfer effectiveness with respect to Reynolds number for the analyzed unit of microchannel of heat exchanger. From observation, the effectiveness is increasing from 5 to 10 and then decreasing with higher Reynolds number. For Reynolds number 5, 10, and 15 the effectiveness remains above 0.65. The highest value of 0.706 heat transfer effectiveness was found for the Reynolds number equal to 10. That is corresponding to 4.13 m/s for hot and 4.048 for cold fluid. This range of the Reynolds number can yield the most effective performance of the microchannel heat exchanger.

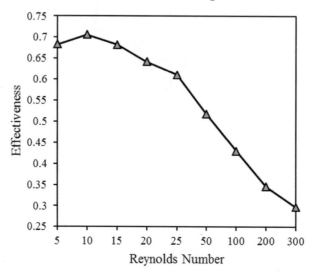

Fig. 6. Heat transfer effectiveness.

6 Conclusions

In this study, the simulation using ANSYS Fluent was achieved. Because there is still less work upcoming in this field of research, it is always better to start with most feasible tools. The results obtained in this work show that heat capacity of both hot and cold fluids remains similar which is leading to assumption of capacity ratio equal to 1 for the counterflow arrangement. The effectiveness is high in certain range of the Reynolds numbers and these can be considered as best operating conditions for good performance of microchannel heat exchanger. This simulation study provided a basis for analyzing the microchannel heat exchanger and indicating best operating conditions for analyzed unit.

In upcoming follow up work in this direction, simulation of different shapes of microchannels, and studying variations in the other parameters including the dimensions of the microchannel itself, and the impact of using different material is expected to be explored as well. Also, experimental analysis is expected to be covered later after subsequent CFD analysis for validating. Moreover, the effect of velocity and effectiveness of the heat exchanger can be observed. All these will surely add to credibility in this field of research adding to literature.

Acknowledgments. The authors acknowledge financial support received from the UAE University for Grant no. 31R153, and the ADEK Award for Research Excellence (AARE) for Grant no. 21N220-AARE18-089.

References

1. Hana, Y., Liu, Y., Li, M., Huang, J.: A review of development of micro-channel heat exchanger applied in air-conditioning system. In: 2nd International Conference on Advances in Energy Engineering (ICAEE2011) (2012)
2. Khan, M.G., Fartaj, A.: A review on microchannel heat exchangers and potential applications. Int. J. Energy Res. **35**(7), 553–582 (2010)
3. Hayase, G.: Development of Micro Channel Heat Exchanger for Residential Air-Conditioners. In: International Refrigeration and Air Conditioning Conference, Purdue (2016)
4. Hasana, M.I., Rageb, A.A., Yaghoubi, M., Homayoni, H.: Influence of channel geometry on the performance of a counter flow microchannel heat exchanger. Int. J. Therm. Sci. **48**(8), 1607–1618 (2009)
5. Fakheri, A.: Heat exchanger efficiency. J. Heat Transf. **129**, 1268–1276 (2007)

Evaluation of the Saudi Construction Industry for Adoption of Building Information Modelling

Obaid Aljobaly[1] and Abdulaziz Banawi[2(✉)]

[1] Department of Architecture, King Abdulaziz University Jeddah,
Jeddah, Saudi Arabia
[2] Department of Architectural Engineering, King Abdulaziz University Rabigh,
Jeddah, Saudi Arabia
abanawi@kau.edu.sa

Abstract. Building Information Modelling (BIM) has shown great benefits in the building industry on a different level. It helps enhance collaboration, increase efficiency, and reduce waste. However, to achieve good results, BIM requires a customized work plan, a set of regulations, and adequate infrastructure to succeed. The aim of this study is to evaluate the capabilities of the Saudi Construction Industry to adopt BIM. The authors apply results of a previous study completed recently to identify and examine the success factors of BIM implementation in Saudi Arabia. Three main categories are chosen to be examined including infrastructure, process, and existing building policies and regulations. Each category will be further divided into subcategories and evaluated individually. For that purpose, a questionnaire is designed and administered to 150 individuals who are closely involved with the construction market of Saudi Arabia. A preliminary result revealed that many of the participants agreed that the lack of a written procedure is the major reason BIM is not adopted by many. The study further showed that stakeholders are not embracing BIM because it requires many changes to function properly. Lack of knowledge and awareness is another reason BIM is not adopted by many in the field. The output of this study will be used in a future study to help develop a BIM framework that is customized to the Saudi Construction industry.

Keywords: Building information modelling · Critical success factor · Implementation

1 Introduction

The construction industry has been evolving globally at a slow rate, even within developed countries. Saudi Arabia is considered a developing country, where the construction sector is a significant contributor to the country's gross domestic product (GDP). The Ninth Development Plan by the Saudi Ministry of Economy and Planning showed that the construction sector contribution to the GDP in 2008 was 6.9% (compared with 6.7% in 2004) with an annual growth rate of 4.9% [1]. Developments within the sector have appeared in every stage of a building project, with enhancements

© Springer Nature Switzerland AG 2020
T. Ahram (Ed.): AHFE 2019, AISC 965, pp. 488–498, 2020.
https://doi.org/10.1007/978-3-030-20454-9_49

effecting changes in tools, processes, and policies based on the individual phases of a building [2]. Furthermore, the architecture, engineering and construction (AEC) community in Saudi Arabia has suffered many problems in both the public and private sectors – such as project delays and overrunning budget – many of which exist because of the individuality of every stage of the project [3]. The combination of new and disparate developments in different areas, project difficulties faced by industry professionals, and the significance of the sector highlight the need to adopt new technologies.

Building Information Modelling (BIM) represents an opportunity to solve the issues of coordinating elements from different phases of a building's construction, mitigating possible problems, and strengthening a sector that is an important part of Saudi Arabia's economy. BIM has a proven capability to empower projects through its collaborative features, by linking the designers with the contractor in real time and avoiding miscommunication in documents [4], and achieving improved results with regard to the simulation of cost and time of the project in early stages. Furthermore, BIM has shown great potential to enhance the building industry with regard to high integration among all phases of a project [5]. Transforming from the old methods of building design and construction (such as using computer-aided-design (CAD) and hand documentation) to BIM requires a change in the existing regulations, policies and tools used in order to develop the industry. Defining the factors needed to adopt the BIM methodology successfully will enhance the Saudi building sector and identify the areas most needed to be developed.

Many previous works have investigated the barriers to BIM implementation within the Saudi sector. The authors believe it is essential to define the critical factors for successful BIM adoption in order to accelerate the adoption process. In this paper, the applied methods to achieve such a transformation shall be evaluated and measured as Critical Success Factors (CSFs) to implement BIM as a path to assure effective adoption of the process within the industry.

Focusing on the CSFs will provide practical results contributing to the adoption of BIM within the Saudi building sector. The CSFs will define the critical factors that need to be changed in order to facilitate the creation of a framework to adopt BIM in both the private and public sectors. The authors reviewed literature on the adoption processes and the barriers to implementation, developing a list of questions that were then used to structure a survey to evaluate the Saudi construction sector and define the CSFs for effective BIM adoption.

The challenges that the AEC community are facing to develop the industry in Saudi Arabia are varied, and there are barriers to the implementation of BIM. Measuring the market through a survey including the determined CSFs in the Saudi construction sector provides a practical result to associate with the adoption of BIM. This research aims to:

- evaluate the capability of the Saudi construction industry to adopt BIM;
- evaluate the factors that will affect the implementation of BIM; and
- facilitate the creation of a BIM framework and an implementation strategy.

This research provides decisive factors that can assist in the adoption of BIM in Saudi Arabia, overcoming barriers to implementation, in order to facilitate creation of a

BIM framework. This will contribute to solving the problems of the building industry in Saudi Arabia.

1.1 BIM Adoption

Many countries now mandate the use of BIM in their respective construction sectors, including the United States of America (USA), the United Kingdom (UK), Finland, and Singapore [6]. More than 48% of the architectural offices within the USA have adopted BIM [7].

During the adoption process, many countries developed an implementation procedure to solve any obstacles and/or issues that appeared, including: structuring a BIM framework; choosing BIM champions to lead the adoption of this new system; creating databases to store newly generated information; and developing processes to facilitate the adoption of BIM. Many organizations tried to standardize the frameworks, for example the International Organization of Standardization (ISO) issued a framework for providing specifications for BIM (SO/TS 12911:2012).

The Saudi Arabian construction industry is considered a growing sector, with many new projects built in recent years. Saudi Arabia has been one of the large contributors to the oil industry; recently, however, the government decided to move forward with Vision 2030 to minimize reliance on oil revenues [8]. As a result efficacy is required, whereas the building sector has a low performance and huge amount of wastage in time and cost [9]. Therefore, the adoption of new processes and developments in the building sector is significant, but this transformation requires changing policies and minimizing or eliminating expected barriers facing the transformation.

1.2 Barriers to Implementing BIM in Saudi Arabia

Globally many issues have appeared in the AEC industry with the implementation of BIM. Industry professionals have encountered varying levels of resistance and have faced a range of problems, some cultural in nature and others technical. Each country began to create and develop their own frameworks and regulations to address the problems of adopting BIM specific to their circumstances. For example, the United Arab Emirates (UAE) faced several issues: technological, the interoperability between applications; organizational, the coordination between industry professionals, vendors and training; and attitude, including interest in learning BIM, BIM awareness, willingness to use BIM, and the perceived cost of BIM technology and associated platforms [10].

Many companies in Saudi Arabia have implemented BIM within their businesses at some level and there have been individual initiatives to implement BIM in small and medium-sized firms. Based on the research of Saud [11], barriers to implementing BIM in Saudi Arabia can be categorized as legal, business, human, and technical. According to Banawi [12] there are additional barriers, including client knowledge of BIM benefits, market readiness, and legal issues.

Further, a lack of market readiness, meaning the market is not prepared to adopt the technology, causes owner concerns in the building contracts; insufficient human resources leads to a lack of professionals capable of running the software to create an integrated model; and many projects require substantial owner involvement. These likewise appear as major problems in the market. A case study of BIM implementation in Saudi Arabia conducted by Banawi found that there is a lack of the fundamental resources to successfully complete a BIM project and meet the design objectives [12].

In conclusion as with the adoption of any new technology, difficulties will always arise. In the case of BIM, most of the problems are related to the technology: the applications, tools, and Information and Communications Technologies (ICT) used; cultural barriers also contribute: AEC community constraints, and a lack of skills and training among industry professionals with regard to the processes and tools of implementation.

1.3   CSFs of BIM Implementation

The CSFs of BIM implementation are a set of critical areas measuring outcomes that drive all major practitioners to transform from the traditional project delivery (using object-oriented computer-aided design (CAD)) to successfully implementing BIM collaboratively [13]. The purpose of using CSFs is to highlight the most important factors that should be changed during the implementation process. Accordingly, the CSFs are validated more for implementation reasons, offering deeper insights into the problems associated with BIM adoption. According to Liao and Teo [14] the construction industry in Singapore witnessed duplication in modeling projects between the designers and contractors; an investigation of the CSF to enhance the implementation of BIM among the involved parties resulted in the creation of a conceptual framework for enhancing the adoption process [13]. According to Amuda-Yusuf [15] there are 28 CSFs for IT adoption of BIM in the construction industry of emerging economies, related to business processes, standardization, education, software and stakeholder awareness.

Yaakob, Ali, and Radzuan [16] categorized CSFs into four areas – technology, organization, process, and legal. The *technological factors* cover the software, complicated by the fact that working with BIM requires many more than one software program [16]. Additionally, there is resistance to changing to new software by AEC industry professionals; and there are security problems concerning the protection of model information and data.

Organizational factors are related to education about the protocols and standards; training for the use of new applications and software; and leadership of the adaptation process in many countries. BIM champions are needed to lead the adoption process and coordinate in modifying rules and regulations while directing institutes towards BIM by associating with rules and regulations modification.

The *process factors* of implementation are focused on regulation. Standardization prevents low performance of the design and build processes, and changes polices to assure the proper management of standards implementation. The *legal factors* are considered in terms of managing access to the model that is in use by the client or the BIM manager; securing the model and the information within it from a legal perspective; and critically ensuring the awareness and commitment of the owner, considering his role and responsibilities in running the model during the project lifecycle.

2   Methodology

This paper focuses on evaluating the Saudi construction sector through defining CSFs of BIM implementation, in order to associate future related studies to BIM adoption in Saudi Arabia. A literature review was conducted to understand the adoption process and the value of BIM implementation, as well as an assessment of the current status of the market. The study also covered the related literature on the barriers to completion of BIM adoption in Saudi Arabia, which added another layer to highlight the CSFs for adopting BIM and created a deeper insight into the factors that will enhance the Saudi construction sector overall.

The literature review, in addition to the authors' experience and other informational sources, was considered in the creation of an online survey. The target of the questionnaire was to generate data that would inform issues related to the adoption, status and barriers to BIM adoption, along with the determination of the CSFs. The survey results were analyzed by the response ratio, with each factor targeted in an individual question requiring a scaled response. The scale of the agreement to each factor was considered in defining each CSF.

3   Results and Analysis

The survey was published for five days and distributed to over 150 individuals, with responses from 56 participants from a range of backgrounds and experience, their characteristics shown in Figs. 1 and 2 below.

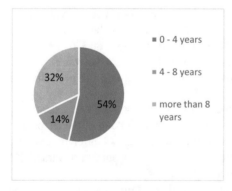
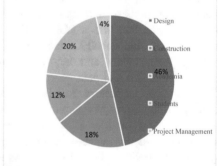

Fig. 1. Background of respondents   **Fig. 2.** Experience of respondents, in years

The majority of the respondents were from the design field (46%) and had fewer years of experience (53.6% at 0–4 years). Regarding knowledge of BIM, results showed that the considerable majority of respondents (73.2%) claimed to have a good knowledge of BIM. This suggests that the market is ready for BIM adoption, which was confirmed by respondents, 73.8% of whom felt that the sector is either "ready" or "somewhat ready" (39.3% and 33.9%, respectively).

A significant number of the respondents (46.4%) have already worked with BIM in Saudi Arabia, in major cities including Jeddah and Riyadh. Other responses reported BIM experience in the UAE, USA, and Europe.

The CSFs for BIM implementation were arranged into three factors: Infrastructure Factor (IF), Process Factor (RF) and Policies and Regulation Factor (RF). To develop an appropriate survey, a series of fifteen specific potential issues or barriers to BIM implementation were determined that were derived from the literature review and aligned with specific factors and subfactors, shown in Table 1.

The survey questions were designed to follow these criteria, which each question measuring a CSF in order to develop a practical reading for each factor.

3.1 Process Factors

The flow of the BIM processes among AEC professionals is considered essential [17]. Managing information and defining roles and responsibilities allows high collaboration, and according to Azhar [18] in order to accelerate the use of BIM, the collaboration features must increase. The Process Factors target the facilitation of this workflow and include leadership, roles and responsibilities, and cultural issues.

The first two questions of the survey addressed leadership subfactors, targeting the designer's role in raising awareness and educating stakeholders about the benefits of BIM, and half of the respondents (50%) agreed that designers were not taking leadership roles in promoting BIM effectively. This suggests that competition within the Saudi construction sector could play a major role for the adoption of BIM. The competition among companies in support of the government role was addressed in the survey and it showed that again a strong majority (68% strongly agree or agree) believe that there currently is no competition. Concerning the ability of the stakeholders to commit to BIM management, and 46% of the respondents agreed while 37.6% were neutral.

The difficulty in explaining BIM to stakeholders is high, as a clear majority (71.5%) felt that BIM and its processes are complicated and difficult to explain to customers. The overall acceptance of transformation in technology usage, as a cultural phenomenon, was nearly balanced between agree and neutral, at 37.5% and 33.9% respectively.

Table 1. Potential issues or barriers to BIM implementation in Saudi Arabia.

Potential issues or barriers	Factor	Subfactor
BIM /Construction processes are complicated for people to understand	PF	Cultural
Designers do not convey the benefits of implementing BIM empirically, which would facilitate decision-making	PF	Leadership
There is no competition in the construction sector between companies, which would improve the Government's agenda	PF	Leadership
Customers are capable and willing to commit to managing BIM models	PF	Commitment
The building sector is resistant to implementing BIM	PF	Cultural
There is a shortage of BIM skilled workers in the construction market	IF	Qualification
The market needs to change contracts and procurements systems, to fulfil BIM requirements	IF	Qualification
There is a shortage in the appropriate technology & equipment for the contractors/designers	IF	Applications
The data management platform is not ready to deal with BIM technology	IF	Systems
BIM applications are more expensive than the conventional ones in terms of initial cost, payback period & repairing and maintenance	IF	Applications
Few potential customers know the benefits of BIM in Saudi Arabia	RF	Awareness
Government incentives are not sufficient to encourage BIM adoption	RF	Mandate
Financial Risks for building using BIM are greater compared to conventional buildings	RF	Awareness
There is an insufficient number of investors for construction with BIM equipment	RF	Mandate
The BIM requirements/policies required are challenging	RF	Standards

3.2 Policies and Regulations Factors

Policies and Regulations factors centered on awareness, standards and mandates as subfactors. Those factors are to be standardized around the world as the industry establishes foundation classes (IFC) and ISO frameworks. Many countries mandate the use of BIM to accelerate its adoption [19]. The subfactors focused on measuring the applied standards and policies.

The first question measured stakeholder awareness, with the survey showing that the vast majority of respondents, 80.4%, strongly agreed or agreed that few potential customers know the benefits of BIM. However, when looking at the risk factors to the adoption of BIM building compared to conventional building, results were quite close with 25% agreeing and 26.8% disagreeing with the high risks associated with BIM usage. The current state of BIM polices in the Saudi construction sector is challenging as 37.5% of respondents answered. This increases the difficulty of achieving the government's mission, and the leadership role becomes a significant factor.

Therefore, the following question focused on governmental support for the adoption of BIM, with most respondents (71.4%) strongly agreeing or agreeing that there is no governmental support for such movement. Of those who have worked with BIM, 51% have not used official guidelines and/or protocols in the designing process.

3.3 Infrastructure Factor

The infrastructure factor focused on three subfactors, including qualification, systems and application issues. Qualification encompasses technical and legal aspects, with compatibility among systems and applications essential for BIM performance. The first topic was related to the alignment of contracts and a procurement system to match BIM requirements, where 80.3% strongly agreed or agreed to the necessity of modifying contracts. The survey as regards infrastructure factors, specifically the level of qualifications, showed 82% of the respondents strongly agreeing or agreeing that there is a shortage in BIM skilled workers in the construction market.

Turning to ICT platforms and the BIM information exchange, considering existing equipment and technical capability in the sector, 71.4% of respondents strongly agreed or agreed that there is a lack of equipment and technology to manage the adoption of BIM. Regarding used systems the results were closer, with 30% agreeing to readiness and 32.1% neutral. Application costs vs revenues are also considered a valid concern for stakeholders, and there was parity (30.4% both agreed and disagreed) regarding whether the cost of investment in BIM technology was balanced out by the project benefits. The final topic, approximately half of all respondents (51.8%) agreed that there is an insufficient number of investors for a viable movement to BIM technologies.

3.4 Critical Success Factors

After completion of the survey and a review of the results, sixteen CSFs were determined from the initial issues or potential barriers to the implementation of BIM in Saudi Arabia. These are organized by category in Table 2 below.

Table 2. CSFs for the Saudi construction sector

CSF number	CSF
PF 1	BIM understanding by Saudi stakeholders
PF 2	Designers' role in explaining BIM benefits
PF 3	Corporate competition to improve Government agenda
PF 4	Stakeholder commitment to manage BIM
PF 5	Technical improvement of transformation acceptance
IF 1	BIM training for workers in the construction industry
IF 2	Compatibility of construction sector contracts with BIM
IF 3	Use of technology and equipment within the sector
IF 4	ICT platforms preparedness for BIM processes
IF 5	BIM applications – initial costs, payback and maintenance
RF 1	Stakeholder awareness
RF 2	Government roles to encourage BIM adoption
RF 3	Financial factors associated with BIM
RF 4	Investors' platform preparedness for BIM transformation
RF 5	Flexibility in changing policies
RF 6	Standards and guidelines for BIM

4   Conclusion

This study investigated the CSFs of implementing BIM in the Saudi construction sector, to enhance the Saudi construction industry and solve problems related to project time and cost, and resolve issues revolving around the individuality of project phases. The literature review informed the BIM adoption process and possible barriers to implementation, while the analysis identified the categories for the CSFs, to effect practical indicators for the adoption. A distributed survey to the local AEC community resulted in 56 respondents who were tasked with evaluating the sector in terms of the potential CSFs within the structured survey.

The findings covered the critical areas for development: processes, infrastructure, and polices and regulations; and highlighted the most needed factors for adopting BIM. As a result, 16 CSFs were determined to accelerate adoption of BIM in the Saudi construction sector.

5   Discussion

The Saudi construction sector is facing major changes in the past few years, the used systems in the industry are not connecting to the majority of the involved parties, this allows difficulty to enhance the performance of the sector. The studies in the field of construction are not providing practical solutions, as a result of the lack in defining a vision to develop the industry. So, there was a need for finding processes and technologies work collaboratively, to associate a full transformation for the sector.

Regard to that the authors believed the possibility to integrate the systems should start from a comprehensive method as BIM. Defining the CSF with from three categories will be considered to be a step for dividing the fields of the researcher in the area to specify solutions for each subfactor to provide a series of clarifications for the role of BIM to improve the industry. Also, this study facilitates future study for establishing a framework for BIM adoption in Saudi Arabia, with consideration of the CSFs as an indicator for the adoption process.

References

1. Ministry of Economy and Planning (2018). https://www.mep.gov.sa/en/ AdditionalDocuments/PlansEN/9th/Ninth%20Development%20Plan%20-%20Chapter% 2013%20-%20Building%20And%20Construction.pdf
2. Ozorhon, B., Karahan, U.: Critical success factors of building information modeling implementation. J. Manag. Eng. **33**(3), 04016054 (2016)
3. Otaibi, S., Osmani, M., Price, A.D.: A framework for improving project performance of standard design models in Saudi Arabia. J. Eng. Proj. Prod. Manag. **3**(2), 85 (2013)
4. Doumbouya, L., Gao, G., Guan, C.: Adoption of the building information modeling (BIM) for construction project effectiveness: the review of BIM benefits. Am. J. Civil Eng. Archit. **4**(3), 74–79.0 (2016)
5. Azhar, S., Khalfan, M., Maqsood, T.: Building information modelling (BIM): now and beyond. Constr. Econ. Build. **12**(4), 15–28 (2015)
6. McAuley, B.: BICP Global BIM Study - Lessons for Ireland's BIM Programme. Dublin Institute of Technology, Dublin (2017)
7. Schlueter, A., Thesseling, F.: Building information model based energy/exergy performance assessment in early design stages. Autom. Constr. **18**(2), 153–163 (2009)
8. Fattouh, B., Sen, A.: Saudi Arabia's Vision 2030, Oil Policy and the Evolution of the Energy Sector. Oxford Institute for Energy Studies, Oxford Energy Comment, July 2016. https:// www.oxfordenergy.org/wpcms/wp-content/uploads/2016/07/Saudi-Arabias-Vision-2030-Oil-Policy-and-the-Evolutionof-the-Energy-Sector.pdf
9. Sarhan, J.G., et al.: Lean construction implementation in the Saudi Arabian construction industry. Constr. Econ. Build. **17**(1), 46 (2017)
10. Mehran, D.: Exploring the Adoption of BIM in the UAE Construction Industry for AEC Firms. Elsevier Ltd. (2016)
11. Alhumayn, S., Chinyio, E., Ndekugri, I.: The Barriers and Strategies of Implementing BIM in Saudi Arabia (2017)
12. Banawi, A.: Barriers to implement building information modeling (BIM) in public projects in Saudi Arabia. In: International Conference on Applied Human Factors and Ergonomics. Springer (2017)
13. Antwi-Afari, M., et al.: Critical success factors for implementing building information modelling (BIM): a longitudinal review. Autom. Constr. **91**, 100–110 (2018)
14. Liao, L., Teo, E.A.L.: Critical success factors for enhancing the building information modelling implementation in building projects in Singapore. J. Civil Eng. Manag. **23**(8), 1029–1044 (2017)
15. Amuda-Yusuf, G.: Critical success factors for building information modelling implementation. Constr. Econ. Build. **18**(3), 55 (2018)

16. Yaakob, M., Ali, W.N.A.W., Radzuan, K.: Critical success factors to implementing building information modeling in Malaysia construction industry. Int. Rev. Manag. Market. 6(8S), 252–256 (2016)
17. Sebastian, R.: Changing roles of the clients, architects and contractors through BIM. Eng. Constr. Architect. Manag. 18(2), 176–187 (2011)
18. Azhar, S.: Building information modeling (BIM): trends, benefits, risks, and challenges for the AEC industry. Leadersh. Manag. Eng. 11(3), 241–252 (2011)
19. Teo, X.Q.: A study of building information modeling (BIM) in Malaysia construction industry. UTAR (2012)

An Experimental Trial of a Novel Ticketing System Using Biometrics

Jun Iio[1(✉)] and Tadashi Okada[2]

[1] Faculty of Global Informatics, Chuo University, 1-18 Ichigaya Tamachi,
Shinjuku-Ku, Tokyo 162-8478, Japan
iiojun@tamacc.chuo-u.ac.jp
[2] Principles of Informatics Research Division, National Institute of Informatics,
2-1-2 Hitotsubashi, Chiyoda-Ku, Tokyo 101-8430, Japan
okada_t@nii.ac.jp, tadashiman@mrj.biglobe.ne.jp

Abstract. To prevent ticket scalping, we implemented fast identity online (FIDO) technology to authenticate and ensure the correct ticket holder identities at a basketball game held in the city of Tsukuba, Ibaraki, Japan. This paper describes an overview of our system and describes the advantages of our method compared with other ticketing systems. The system was implemented using an open-source electronic commerce platform. The addition of the FIDO module to the platform enables the biometric authentication function. Moreover, we carried out a simple evaluation of the service by asking participants to respond to a questionnaire. We also present the results of the evaluation analysis.

Keywords: Electronic ticketing system ·
The fast identity online (FIDO) technology ·
Personal verification system using biometrics ·
Electronic commerce web portal system

1 Introduction

Ticket scalping is a social problem that is expected to be solved by information communication technology (ICT). If biometric identity authentication is available, an electronic ticket (e-ticket) sold to an individual can be strictly ascribed to the person by the verification system. The ticket is only valid when used in conjunction with the biometric information of the authorized person. Therefore, the system fundamentally prevents resale.

The challenge arises in easily configuring the authentication system to use biometrics. Currently, personal verification methods using biometrics such as fingerprint readers, facial recognition software, and iris scanners are widely used. However, they are primarily utilized for personal devices or select systems that require high-level security, and they are not commonly used in web-based applications.

The fast identity online (FIDO) technology [1] provides biometrics-based personal authentication. Currently, most web applications requiring sign-in require a password to authenticate individuals. Hence, the users of such systems must remember numerous pairs of personal identifications (IDs) and passwords. FIDO provides freedom from this

© Springer Nature Switzerland AG 2020
T. Ahram (Ed.): AHFE 2019, AISC 965, pp. 499–507, 2020.
https://doi.org/10.1007/978-3-030-20454-9_50

troublesome situation by using biometric technology instead of the ID-and-password pair. If a web application adapts the FIDO architecture, users are not required to enter the magic phrases.

In this study, we have utilized the biometric authentication technology "FIDO" to control seat management for a sports event. At the three-by-three basketball game held in the city of Tsukuba, in the Ibaraki prefecture on November 11, 2018, the event organizer adopted the ticketless entrance management system, which was based on individual biometric recognition.

The base of the system was an open-source software package, "EC-CUBE," which provides an electronic commerce (EC) web portal.[1] A cooperative company developed an EC-CUBE plug-in module that works with the particular FIDO application offered by the company, and subsequently implemented it into the ticket-vending website.

Spectators wanting to buy a ticket for the game are required to install the FIDO application onto their smartphones in advance. When they arrive at the entrance to the game venue, their tickets are authenticated by the FIDO application using their biometric information. First, the record of their confirmed ticket purchase is sent from the server to the FIDO application. Subsequently, the verifying information is matched with their biometric data using the authentication devices with which their smartphones are already equipped.

In this paper, an overview of the architecture of our system is presented, which includes the FIDO module implemented into the EC-CUBE web portal. Further, we conducted a simple evaluation of the operation. Questionnaire sheets were distributed to the audience, and 29 answers were collected. Accordingly, we also present the results of the evaluation analysis.

2   System Overview

This section describes an overview of the system. Fig. 1 explains the initial steps required to use the system, buy a ticket, and prepare the ticket and user's information on the smartphone application.

Initially, a user must provide registration information, such as account details, name, address, and payment information.[2] If the registration is successful, the user can purchase event tickets and other products sold at the virtual store available through the EC web portal. Following several necessary sequences to buy a ticket, the system sends the receipt to the user via e-mail.

To enable biometric authentication, the user must install a smartphone application provided by us to the user. Before activating the user-purchased ticket information, the application requires registration of the fingerprint. The application utilizes the fingerprint to authenticate the user and confirms whether they are the correct ticket holder.

---

[1] EC-CUBE can be downloaded from its website: https://www.ec-cube.net/.

[2] The system requires a credit card number, the card-holder's name, the card's expiration date, and its cvc-code as the payment information.

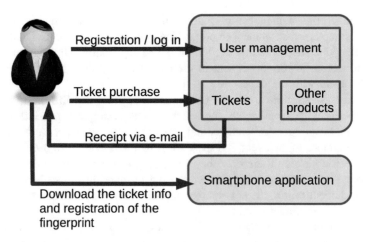

Fig. 1. The first sequences necessary to purchase a ticket for an event. Initially, a user (*top-left*) must register various information. Next, the user logs in to the system and purchases a ticket. After the ticket purchase transaction, the receipt will be sent to the user. The user then downloads the ticket information to the smartphone application that also stores the user's fingerprint.

Figure 2 shows the system architecture. The server-side application, that is, the EC-cube based web portal, has a database in which customers' data and their purchase data are stored.

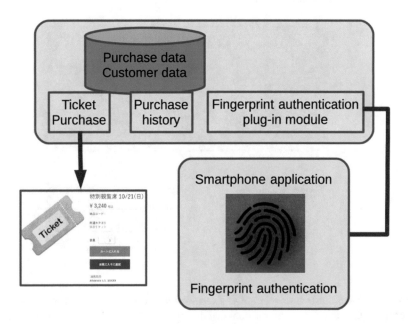

Fig. 2. The system architecture. The web portal application has a database that stores customers' data and their purchase data. The system also provides the fingerprint authentication function in cooperation with the smartphone application.

The fingerprint authentication plug-in module developed by our cooperative company is implemented in the web portal. The system provides the user authentication function in cooperation with the user's smartphone application, which confirms that the ticket information is correct, as sent to the customer, and that they have the right to access the event with the ticket.

A typical-use scenario for ticket confirmation is illustrated in Fig. 3. The confirmation process at the event venue was designed to be as simple as possible because the person in charge at the event entrance must manage many spectators in a short period.

At the entrance to the event, the spectator is required to confirm their right to enter with a specific ticket. Therefore, they attempt fingerprint authentication with the smartphone application and the plug-in module installed in the web portal. After a successful authentication, the smartphone application displays its one-time password, presented as a two-dimensional code.

Finally, the ticket-checking staff read the one-time password for confirmation, using their own devices. If they confirm that the customer possesses a proper ticket without any issues, they permit the spectator to enter the event venue.

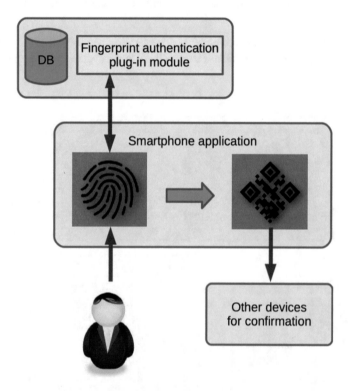

Fig. 3. This figure illustrates a typical-use scenario at the event venue. At the entrance the user can confirm that the ticket information that he or she presents is valid using fingerprint authentication. After the confirmation, the smartphone application shows a one-time passcode in QR-code format to enable validation of the user by the ticket collection authority.

3   Advantages and Disadvantages of the Proposed System

Currently, several ticketing systems similar to ours have been proposed around the world, although our system uses novel technologies including the biometric authentication, web portal, and application software in consumers' mobile devices. However, we cannot expect the users (spectators at sporting events) to be familiar with such new technologies. Hence, we must design the applications and services such that the users can conceive the benefits provided by the new services.

Table 1 summarizes the advantages and disadvantages of three authentication methods: physical identification, password authentication, and biometric verification. Physical identifications such as paper tickets or tokens are very simple to authenticate, and constitute conventional methods. Such methods are widely used all over the world, and both users and event organizers are accustomed to the authentication process operation. However, the system presents a serious problem in the ease of ticket scalping.

Table 1. Advantages and disadvantages for each ticketing system with differing methods of user authentication.

Authentication method	Advantage	Disadvantage
Physical identification, e.g., paper ticket	Easy to handle; a conventional method; both users and event organizers are accustomed to the operation	It cannot prevent ticket scalping
Password/passphrase	Using passwords is a simple and conventional way to authenticate in the case that the ticketing system is implemented online	It cannot prevent ticket scalping
Biometrics	These types of authentication can prevent ticket scalping. No need to keep any passwords nor key information	The system tends to be slightly complicated. Users and event organizers are not accustomed to the operation

The second option is password or passphrase authentication. Using passwords is a simple and conventional method for online systems. If the ticketing system is implemented as an online system, it is quite natural to choose this type of user authentication. Although password authentication is considered to have many disadvantages, it is prevalent among many services presently provided via the internet. Unfortunately, this method also cannot prevent ticket scalping by way of passing the id and password information to a third party.

The last method is the biometric validation technique adopted by our system. The main advantage in using biometrics is to prevent ticket scalping. The ticket holder's information is uniquely connected to the user's characteristics such as a fingerprint or iris patterns. If the users allow the use of biometric authentication, they have the additional advantages of not needing to preserve any physical items (such as tickets), or having to remember their id and password pair. One of the disadvantages of this method is that the system tends to be complicated. Furthermore, the users and the event organizers may not be accustomed to the operation, and they may have limited knowledge about the system processes due to lack of experience with such systems.

4 Evaluation

The system was implemented and utilized for the sports game event held in the city of Tsukuba, on November 11, 2018. At the event, we conducted a simple evaluation by asking the spectators to complete a questionnaire.

Table 2 presents the questions we asked the spectators.

Table 2. The questions asked to the audience to assess the usability of the ticketing system proposed in this study.

Questions	Options
Gender	male, female
Generation	$10 \sim 19$, $20 \sim 29$, $30 \sim 39$, $40 \sim 49$, $50 \sim 59$, $60 \sim 69$, over 70
Information-technology literacy	very high, high, medium, low, very low
How did you find out about the event?	website, school, friends, other
Satisfaction level	very high, high, low, very low
The reason for satisfaction	(Free answer)
Experience as an audience participant at the games of the team	yes, no (this is the first time)
Frequency of participation	more than three times a year, one or two times a year, fairly infrequently, never
Difficulty of purchasing e-tickets	very hard, hard, medium, easy, very easy
The reason for difficulty	(Free answer)
Difficulty of installing the smartphone application	very hard, hard, medium, easy, very easy
The reason for difficulty	(Free answer)
Convenience level of this ticketing system using biometrics	very high, high, medium, low, very low
The reason for convenience	(Free answer)

We collected 29 answers from 15 male and 12 female participants. Two participants did not submit their gender information. The ages of spectators ranged from teenagers to people in their fifties. No replies were gathered from the sixties and over seventy groups.

Regarding information technology (IT) literacy, the most frequent answers were medium and low. Ten audience members selected each of those two options. While one person answered that there was a problem that made them dissatisfied, the remainder of the participants replied that they were either very entertained or satisfied.

This paper focuses on the difficulties in the proposed system's ticketing process and the convenience in utilizing the biometric authentication. Thus, we pay careful attention to the latter three questions: the difficulty in purchasing e-tickets, the difficulty in installing the smartphone application, and the convenience level of this biometrics-based ticketing system.

Three figures (Figs. 4, 5, and 6) depict the results of these three questions. Unfortunately, more than half of the respondents answered that the ticket purchasing process was difficult (Fig. 4). Further, two respondents disclosed that they could not receive their e-ticket due to the complexity of the ticket issuing process.

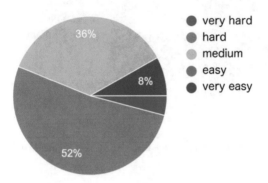

Fig. 4. The results for the question: "Is the e-ticket purchasing process difficult or easy?"

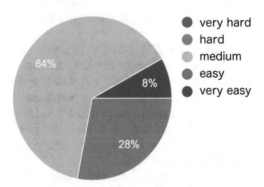

Fig. 5. The results for the question: "Is the installation of the smartphone application difficult or easy?"

Conversely, the installation of the smartphone application did not appear to be difficult (Fig. 5). In response to the question asking about the difficulty of the installation, more than two-thirds of respondents answered that it was of medium difficulty or easier.

Regarding the opinions about the convenience of using the proposed system, approximately two-thirds of respondents replied that it was convenient (Fig. 6). Eight respondents stated that the convenience level was medium, and only one person answered that it was inconvenient. This result implies that the authentication using biometrics is convenient, and there is potential for its acceptance if awareness of the technology is increased.

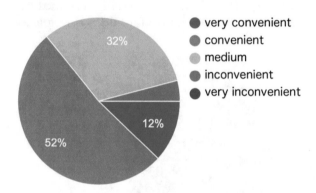

Fig. 6. The results for the question: "Is the ticketing system biometric authentication convenient or inconvenient?"

5   Related Work

As previously described in Sect. 3, several ticketing systems similar to ours have been proposed. Most kinds of those e-Ticket systems adopt the near field communication (NFC) technology.

Stopka *et al.* [2] discussed the NFC-based e-Ticketing system in public transportation. They argued the characteristics of the NFC technology, the NFC enabled devices, and the necessity of standardization and development. In addition, they evaluated its user acceptance of the new technologies, which indicated the requirements of a high level of usability and ease of use. Wu and Lee [3] proposed e-Ticketing system implementing with off-line authentication. They also used the NFC technology to authenticate the ticket holders. Other similar systems using the NFC were proposed by Guasch *et al.* [4] and Gudymenko [5].

An interesting trial on the integration of FIDO and OAuth for the internet of things (IoT) are introduced by Tschofenig [6]. IoT devices have some limitations in the processing power, the network bandwidth, and the smaller user-interface. With such restrictions, it discussed how the security issues are kept by these technologies.

6   Conclusions and Future Work

In this study, a novel ticketing system was proposed, which uses FIDO technology as the biometric authentication method. The primary objective of the system was preventing ticket scalping. The access right of the ticket holder in this system is uniquely connected to the holder's biometric information. Therefore, it is difficult to transfer the right to a third person without any permission from the ticket provider. The system was practically implemented at a basketball game event held in Tsukuba, to confirm the usability of the system's services.

This paper explained the overview of the system, and the advantages and disadvantages of the authentication method in comparison with other ticketing systems. In addition, we conducted a simple evaluation through an audience questionnaire. The results discussed in this paper especially focused on the usability of our ticketing and authentication process. According to the results, there areas for improvement in the process. The most significant area for development is in mitigating the difficulty in the e-ticket purchasing process. Making the system more user-friendly, even for users with low IT literacy, is a necessary task for future work.

References

1. Lindemann, R.: The evolution of authentication. In: Reimer, H., Pohlmann, N., Schneider, W. (eds.) ISSE 2013 Securing Electronic Business Processes, pp. 11–19. Springer Vieweg, Wiesbaden (2013)
2. Stopka, U., Schäfer, G., Kreisel, A.: NFC-enabled eTicketing in public transport—aims, approaches and first results of the OPTIMOS project. In: Kurosu M. (eds.) Human-Computer Interaction. Interaction Contexts. HCI 2017. LNCS, vol. 10272. Springer, Cham (2017)
3. Wu W.J., Lee W.H.: An NFC E-ticket system with offline authentication. In: The 9th International Conference on Information, Communications & Signal Processing, IEEE. Taiwan (2013). https://doi.org/10.1109/icics.2013.6782911
4. Guasch, A.V., Payeras-Capellà, M.M., Puigserver, M.M., Castellà-Roca, J., Ferrer-Gomila, J. L.: A secure E-ticketing scheme for mobile devices with near field communication (NFC) that includes exculpability and reusability. IEICE Trans. Inf. Syst. **95-D**(1), 78–93 (2012)
5. Gudymenko, I.: On protection of the user's privacy in ubiquitous E-ticketing systems based on RFID and NFC technologies. In: Proceedings of the 3rd International Conference on Pervasive Embedded Computing and Communication Systems. SciTePress (2013)
6. Tschofenig. H.: Fixing user authentication for the Internet of Things (IoT): integrating FIDO and OAuth into IoT. Datenschutz und Datensicherheit **40**(4), 222–224 (2016)

The Role of Change Readiness in Determining Existing Relationship Between TQM Practices and Employee Performance

Timothy Laseinde[1], Ifetayo Oluwafemi[2,3(✉)], Jan-Harm Pretorius[2],
Ajayi Makinde O[3], and Jesusetemi Oluwafemi[4]

[1] Mechanical and Industrial Engineering Tech Department,
University of Johannesburg, Johannesburg 2006, GT, Republic of South Africa
otlaseinde@uj.ac.za
[2] Postgraduate School of Engineering Management, University of Johannesburg,
Johannesburg, Republic of South Africa
{ijoluwafemi,jhcpretorius}@uj.ac.za
[3] College of SMS, Afe Babalola University, Ado Ekiti, Ekiti State, Nigeria
maksonjay04@yahoo.com
[4] Department of Quality and Operations Management,
University of Johannesburg, Johannesburg, Republic of South Africa
jesusetemiayeni@yahoo.com

Abstract. The significance of Total Quality Management (TQM) in fast changing industries such as the Information Technology (IT) sector cannot be overemphasized. The study was designed to determine the existing relationship between employee performance and Total Quality Management practices within a technological driven sector. Coupled with this, employee willingness to change and Individual Change Readiness (ICR) were explored while considering employee performance. Quantitative study of Software firms in Nigeria was carried out using descriptive statistical method, and the results were subjected to correlation analysis and structural equation modeling, for the data analysis. From the results, it is seen that TQM practices, as well as ICR, serve as a considerable benchmark for measuring employee performance. The findings observed favors the facilitative function of ICR. These observations are critical so long the degree of employee's readiness to change contributes immensely to organizational transformation. This study contributes to the body of knowledge by revealing the significance of ICR from a TQM perspective as it influences employee performance.

Keywords: Productivity · Work environment · Job satisfaction · Satisfaction · Quality management

1 Introduction

For every organization, change is constant and is a continuous process. Therefore, to ensure effective implementation of TQM practices, an environment that accepts change is needed [1–3]. TQM can be regarded as a quality improvement model aimed at

© Springer Nature Switzerland AG 2020
T. Ahram (Ed.): AHFE 2019, AISC 965, pp. 508–522, 2020.
https://doi.org/10.1007/978-3-030-20454-9_51

improving the quality of products as well as enhancing employee performance [4–6]. The focus of any organization aiming to implement change to their processes/products that are already developed, should be on individual change readiness (ICR) [7]. Similarly, having the ability, enthusiasm and the readiness to embrace change is highly imperative [8]. It is required that the management of organizations, particularly software firms work hand in hand with their employee on how to adapt, acclimatize to change at all time. This will bring about the willingness and readiness to change in such organisation, also, the employee needs to be determined to successfully implement TQM practices in such organization. The efficiency and productivities of any organizations is a function of its workers Individual Change Readiness (ICR) [9–11], which can be improved by interfacing it with TQM practices. Following the incessant change in technological advancements, competition in software firms has increased, consequently requiring the workers to put in more effort as compared to their predecessors. Product quality can also be used to assess the performance of individuals employee and the organization as a whole [12, 13]. In Nigeria software firms, it has been observed that they remained unresponsive to these practices, generally showing disinterest in implementing TQM practices and promoting readiness for change amidst its employees. Several scholars have elucidated TQM practices as well as employee performance in numerous context [14–16]. This research contributes to the body of knowledge by measuring the facilitative role of an individual readiness for change amidst TQM practices as well as employee performance in an emerging country. Likewise, it puts in place strategic pathways to ensure these concepts are successfully implemented in Nigeria software firms.

Owing to the nature of the study, well-structured research questions were developed to ascertain the role of ICR in determining the relationship that exists between TQM practices as well as employee performance, this is illustrated below:

  i. What impact do TQM practices have on employee performance?
 ii. How do TQM practices correlate with ICR?
iii. What impact does ICR have on employee performance?
 iv. Does ICR facilitate the connection between TQM practices as well as employee performance?

2 Review of the Literature

2.1 The Connection Amid TQM Practices As Well As Employee Performance

TQM can be regarded as a systemic method aimed at enhancing quality improvements in both products, service and enhances quality improvements and, also assures customer satisfaction [17]. Several TQM factors which have the potential of maximum quality of products and services assurances has been identified [18]. These dimensions, which was divided into several practices, are best tailored for manufacturing as well as the service sectors. A study carried out by Mahanti and Anthony [19], which was an insight on the Indian IT sector, his study examined the software industry and concluded that the software industry depends on TQM as a critical success factor to improve their quality and their employee performance. He further emphasised how imperative it is for their IT sector to adopt the TQM program appropriately to sustain their quality reputation.

Employee performance refers to the actions pertaining to the assignments anticipated from a member of staff and how it is being carried out [20]. Some factors have been identified as the major cause of employee's poor performance, such factors include; sluggishness, lack of motivation and poor commitment to work. Employee satisfaction shows the pleasure and certain level of gratification an employee derives from his or her work [21]. In any organization, employees are represented by their responsibilities based on their involvement and liability [22, 23]. It was furthermore discovered that the efficiency of an organization highly depends on the harmony of staff members who are the key elements of the organization [24]. TQM practices are the major factors that influence employee performance in an organization [25]. Implementation of TQM practices in the service sector is highly significant on employee performance. The goal of TQM is to improve the efficiency of an organization [26–28]. TQM practices have a significant contribution towards achieving an enhanced employee performance as well as staff members deriving satisfaction in their job. It has been observed that to improve quality and efficiency of an organisation, the aptness of employees is an important factor. TQM helps in implementing laws in an organization, which enhances efficiency, quality improvements in the organization, on the part of the employee, improves their productivity [29]. An average employee performance will not be enough when there is high competition among contemporaries' organizations. In a bid to increase competition, every element of employee performance should be observed closely [30]. Furthermore, quality improvement teams are put in place to ensure employees have respect for their superiors and their work. Several studies have shown that TQM practices and employee performance are positively related. From literature considered, it is observed that in Nigeria software firms, there is a gap linking TQM practices as well as employee performance. This research, therefore, intends to unfold the connection that exists between TQM practices as well as employee performance, especially in Nigeria software firms. Several hypotheses were founded based on this objective and its illustrated below:

H1. There is an affirmative connection between TQM practices as well as employee performance in Nigeria software firms.

2.2   Correlation Amid Individual Change Readiness as Well as TQM Practices

An Individual willing to embrace Change at all level, must be ready to continually commence and embrace change in diverse avenues, with the motive of achieving quality and performance enhancement, and to also, reduce risk [31]. Readiness can be defined as the ability to sense as well as plan change, once its advantages and disadvantages have been critically evaluated. Mantos and Paula [32], considers resistance as well as readiness to change to be incisive. How a personality responds to change, or his/her readiness is referred to as the internal attitude of the individual, resulting in etiquette for molding gradually to change or being resistant towards change [33].

Organizations need change to survive in this time where competition is very high. If there is a case of a change in an organization, its proper implementation and practice are dependent on the members' readiness for change [34]. A measurement scale for ICR which has multiple factors such as self-effectiveness, personal valence, inconsistency and organizational valence. To avoid failed application of change, the imperative role of every employee is required to be known and the junior staff members must realize the advantages of change. In a bid to pursue creative ideas, ICR can efficiently result in organization change and a positive result can be attained through these sequel decisions. According to Haffar [7], he posits that ICR is directly linked to TQM, argues that they augment each other. In his subsequent studies, He emphasized that TQM practices will be successful in an organization where change readiness prevails in their organizational culture. In self-motivated environments, an organizational culture that promotes change in such an environment can serve as a framework for implementing and ensuring Quality management in such an environment. It becomes highly imperative for all individual change agents to come together to ensure quality intervention in such an organizational culture. Besides, the relationship that exists amid TQM practices as well as organizational change can never be set aside as its implementation can result in increased competitiveness of an organization. This study attempts to gratify the gap in the active software industry; a gap which is seen in several developing countries. The hypothesis is, therefore:

H2. TQM practices are said to have an affirmative connection with ICR in Nigerian software firms.

2.3 Correlation Between Employee Performance as Well as Individual Change Readiness

On organizational change, two key elements have been identified from the literature which are; individual change readiness as well as employee performance. The positive or negative connection that exists amid individual change readiness as well as employee performance, is dependent on the extent of employee readiness. Employee's stake on change readiness has a vital influence since the level of understanding among employees varies.

Robbins [35], stated that the threat of employees' jobs caused by abrupt changes is not good for an organization. Organizations must always be ready to respond to change and this should be achieved using performance evaluation management systems [36]. Based on literature, there exists a gap on the connection amid individual change readiness as well as employee performance in Nigerian business organizations such as software firms. Thus, this present research aims to examine how employee performance, as well as individual change readiness, are related to Nigerian software firms. The hypothesis therefore is:

H3. Employee performance has an affirmative connection with ICR in most Nigeria software firms.

If the employees of organization are properly informed on the content of change, change process, as well as change readiness will be adequately facilitated [37]. Change readiness facilitates the connection amid transformational leadership as well as commitment to change. Past literatures reviewed above, justifies the use of these variables for this research which is being assessed in Nigeria software firms. The present study focuses on the research gap in how ICR facilitates TQM practices as well as employee performance. Therefore, except employees are prepared to admit and apply the proposed change using TQM practices, there will only be slightly enhancement in employee performance. The hypothesis here is:

H4. ICR has been a major element of change that positively facilitates TQM practices in Nigerian software firms, and, also, enhances employee performance.

This research, thus, seeks to address the connection that exists amid TQM practices as well as employee performance in Nigerian software firms with the consideration of ICR.

3   Methodology

The positivism method which includes verification of different experiences is used in this research rather than an intuitive method. Also, this research was qualitative in nature. There are over 500 software firms in Lagos and Abuja having about 2300 on-site and off-site employees. It is approximated that 15 to 30 employees work in an individual software house. The targeted population of this research are the staffs of registered software firms in Lagos and Abuja city. The Nigeria Software Export database website supplied the list of software firms to researchers. The software firms are closely located in the major cities of Lagos province. Data were received from employees of these software firms by the researchers. The only employees who were considered are those that had a minimum of a graduate degree, having at least a one-year experience of the software development process. The site workers are the ones directly involved in product quality maintenance and they were the ones considered. The purpose of these criteria was to ensure that only employees who had adequate information and are directly relevant to organizational change readiness and TQM practices are involved. Out of the 2300 employees, the only ones that could the meet requirements were 750 and questionnaires were administered to them. The paper-based questioner was used in ensuring the retrieval of data. The researcher revisited the organization for follow-ups to collect the questionnaire that had been filled. The response rate after the follow-ups was 60%, with only 450 questionnaires returned. From the 450 retrieved, only 340 useable responses were utilized while others were discarded due to incompletion errors noticed in the filling.

4   Results and Findings

SPSS and Smartpls software were used to apply the test statistics for verifying the research model and hypothesis. For dimension reduction, exploratory factor analysis (EFA) was applied and this was confirmed by confirmatory factor analysis denoted as

CFA. After checking multicollinearity amid the variable, by acceptance as well as VIF index, results were found to be in a satisfactory range. A check for data validity was done by skewness as well as kurtosis gauge, the entire variable has their values within the range of ± 2; which means the values are suitable for statistical equation modeling.

Table 1. Demographics

		n	%
Sex	Male	242	71.17
	Female	98	28.82
Age	Below 25	84	24.7
	25–30	90	26.47
	31 yrs or above	166	48.82
Academic qualifications	First Degree	174	51.17
	Second degree (M.Sc)	116	34.11
	Other Degree's	50	14.7
Work experience (in yrs)	1–3	40	11.8
	4–7	141	41.5
	5 years Above	159	46.8
State surveyed	Abuja	92	27.1
	Lagos	139	40.9
	Port	61	17.9
	Kano	48	14.1
Salary on monthly basis N = 340	Less than 20,000	19	5.6
	20,000–50,000	198	58.2
	More than 50,000	123	36.2
	Total	340	100

4.1 Descriptive Analysis

In Table 1 above, the involvement of male as well as female in this study is shown. Female involvement is 31.80% and this means in Nigerian software firms, females are the least involved in technical office jobs. Also, most participants in this survey are within the age group of 31–35, a 46.80% of them have more than 5 years working experience, meaning that they take keen interest in the industry and wish to build a career in it. The educational qualification of the participants in this study varies, with 50% of the workers being bachelor's degree holders and about 34% are master's degree holders.

4.2   Factor Analysis

Exploratory factor analysis (EFA) and confirmatory factor analysis (CFA) were done to eliminate irrelevant dimensions that do not contribute to the parent constructs. Eigenvalues > 1 were secured, to be used in the completion of the factor extraction process. KMO index was used to confirm the appropriateness of the acquired data for variable extraction. KMO assesses sampling suitability, which is 0.922 and must be well over 0.6 (p < 0.05), also, it was observed that there was an interconnection amongst the list of issues. The data is suitable for analysis if the KMO index value is more than 0.6. with the use of Smartpls, a measurement model was done for confirmatory factor analysis as shown in Fig. 1 below. The measurement model is a validity as well as a reliability check. The factor loading aggregate of every item should be well over 0.5. Any item whose factor loading aggregate is below 0.5 would be eliminated. A few of the item had to be eliminated due to their factor loading aggregate being lower than 0.5.

Table 2 shows the outcomes of factor piling aggregate in which its average variance was haul out. The fitness of the model was check by several fitness indices. Table 3 illustrates the appropriateness of the fit indices for the dimension model. The model is fit as the value of all directories are greater than the threshold value.

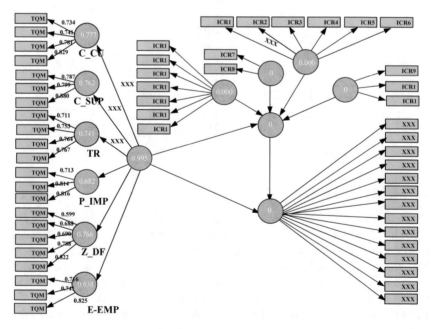

Fig. 1. Dimension model

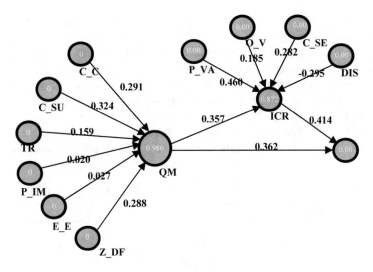

Fig. 2. Illustration of the structural model.

4.3   Reliability and Validity

The measures of reliability are Cronbach's α and composite reliability, though composite reliability is seen as the uppermost culmination of Cronbach's α score. The factor loading aggregate is the template for the calculation of the composite reliability aggregate. As seen in table illustration below (2), composite reliability aggregates of the variables are obviously well above the threshold of 0.7, with all being within 0.806–0.915. The Convergent, as well as the discriminant validity, were utilized for the verification of the strength of the variables adopted in this research. when ascertaining the discriminant validity, it is important that the square root of the Average Variance Extracted (AVE) be more than the correspondence of the variables adopted in this model. According to Bagozzi et al. [38], AVE value must be greater than 0.5 when examining the convergent validity. Most of the variables shown in Table 2 have their AVE value within 0.521 and 0.678, thereby confirming the convergent validity satisfactory.

4.4   Result of the Inferential Analysis

The inferential analysis shows the outcomes of the *structural equation modeling* denoted as *SEM* by Smartpls software. Hypotheses was tested by the SEM results. Figure 2 shows the structural model. SEM technique is used to evaluate the correlation between TQM practices in relations to employee performance as well as Individual Change Readiness. The structural model was tested after the confirmatory factor analysis (CFA). The values of fit indices for the structural model are in the appropriate range, which are RMSEA = 0.05, PNFI = 0.76, CFI = 0.96, and PGFI = 0.74 respectively. This means that the model is perfectly fit. The result indicated an affirmative relationship exists between ICR as well as employee performance.

The relationship amid TQM practices, employee performance as well as Individual Change Readiness is elaborated by the path model.

A positive and significant connection amid TQM practices as well as employee performance is shown by regression loads of structural model. Also, a positive and significant connection amid TQM practices as well as individual change readiness was seen in the results. In addition, results still show that the ICR as well as employee performance are positively related. Sobel's test was carried out by the study to calculate the worth of the facilitating variable which is the ICR, and the outcomes are shown in Table 4. Also, Table 5, shows the illustration of the summary of the hypotheses testing.

Table 2. Illustration of the factor loading (λ), Cronbach's α, AVE, and composite reliability

Variables	Items	λ	α	AVE	CR
TQM Practices					
Closer to Customers	TQM1	0.734	0.777	0.6	0.857
	TQM2	0.741			
	TQM3	0.829			
	TQM4	0.781			
Closer to Suppliers	TQM5	0.88	0.762	0.678	0.863
	TQM6	0.787			
	TQM7	0.799			
Training	TQM8	0.753	0.741	0.561	0.836
	TQM9	0.764			
	TQM10	0.711			
	TQM11	0.767			
Employee Empowerment	TQM12	0.713	0.682	0.613	0.825
	TQM13	0.814			
	TQM14	0.816			
Process improvement	TQM15	0.69	0.766	0.521	0.843
	TQM16	0.688			
	TQM17	0.788			
	TQM18	0.599			
	TQM19	0.822			
Zero-defects Mentality	TQM20	0.741	0.638	0.581	0.806
	TQM21	0.825			
	TQM22	0.716			
Individual Change Readiness					
Change self-efficacy	ICR1	0.683	0.743	0.54	0.824
	ICR2	0.727			
	ICR3	0.693			
	ICR4	0.623			
	ICR5	0.661			
	ICR6	0.58			

(continued)

Table 2. (*continued*)

Variables	Items	λ	α	AVE	CR
Discrepancy	ICR7	0.814	0.524	0.677	0.807
	ICR8	0.832			
Personal	ICR9	0.787	0.802	0.525	0.853
Valence	ICR10	0.581			
	ICR11	0.647			
	ICR12	0.639			
	ICR14	0.465			
	ICR13	0.609			
	ICR14	0.579			
	ICR15	0.54			
Organizational Valence	ICR16	0.797	0.645	0.584	0.807
	ICR17	0.792			
	ICR18	0.699			
Employee Performance EP					
	EP1	0.704	0.9	0.537	0.915
	EP2	0.518			
	EP3	0.682			
	EP4	0.684			
	EP5	0.734			
	EP6	0.761			
	EP13	0.731			
	EP7	0.768			
	EP8	0.768			
	EP9	0.762			
	EP10	0.735			
	EP11	0.744			
	EP12	0.707			
	EP13	0.764			
	EP14	0.725			

Table 3. Results of the goodness of fit indices for the measurement model.

Goodness of fit statistics	Measurement model	Empirical standard
x^2/df	428/137 = 3.05	<3.0
RMSEA	0.06	<0.08
PGFI	0.61	>0.5
PNFI	0.73	>0.5
CFI	0.81	>0.9

Table 4. Mediation analysis through Sobel's t

Path tested	A	S.Ea	B	S.Eb	Sobel's t value
TQM → ICR	0.357**	0.070			
ICR → EP			0.414**	0.092	
TQM → ICR → EP					4**

**P < 0.001.

Table 5. Summary of hypotheses testing.

Path description	Estimates	S.E	C.R	Hypothesis	Results
TQM → EP	0.362	0.078	4.593	H1	Accepted
TQM → ICR	0.357	0.070	5.094	H2	Accepted
ICR → EP	0.414	0.092	4.498	H3	Accepted
TQM → ICR → EP	0.148	0.043	3.374	H4	Accepted

**P < 0.001.

5   Discussion

TQM practices, from the results shown, are positively and significantly related to employee performance in Nigerian software firms. Therefore, by the SEM technique, the first hypothesis, H1, is recognized based on regression significances. Over 36% regression is shown on employee performance because of TQM practices. It has been confirmed under different conditions that TQM offers rules, as well as regulations in any organization, were its use has been adopted. And employee performance in such an organization is being promoted by TQM practices [39]. An average performance will not be enough when there is increased competition. It has also been confirmed in this study that an affirmative connection exists amid TQM practices as well as employee performance, in the context of Nigerian software firms. Though only factors such as process improvements, employee training, customer focus as well as zero defect policy, which are highly relevant in software industries, have been measured. All the TQM practices which have been mentioned are very important for employee performance and its implementation can increase performance during tasks.

Nigerian software firms have equally shown positive as well as an affirmative connection amid TQM practices as well as individual change readiness (ICR). Therefore, by the SEM technique, the second hypothesis, H2, is admitted based on regression weights. Over 36% regression is shown on individual change readiness because of TQM practices. The study shows that the positive connection amid TQM practices as well as employee performance in Nigerian software firms is dependent on the employee's attitude to change. This result correlates with [7, 9, 40]. ICR supports the relationship between TQM practices and employee performance. The efficiency of an organization is employees' enthusiasm to change and this is the only way to promote the connection amid TQM practices as well as employee performance.

Nigerian software firms have also shown positive and significant connection amid individual change readiness (ICR) as well as employee performance. Therefore, by the SEM technique, the third hypothesis, H3, is admitted o based on the regression significations. Over 41% regression is shown on employee performance because of ICR. Readiness means that the workers in the organization will not reduce their performance and that they are willing to adopt the organizations change policy. If the employees are aware of any change occurring in the organization, it will positively affect their performance. This study's results correlate with [35, 41] The positive connection amid ICR as well as employee performance is confirmed in this research in the context of Nigerian software firms.

From the perspective of Nigerian software firms, the mediating impact of ICR in improving employee performance is confirmed by this study. Therefore, by the virtue of the SEM technique, the fourth hypothesis, H4, is admitted based on the result of regression significances. which means change readiness partly facilitates and improves the effect of TQM practices on employee performance by almost 15% in Nigerian software firms. Therefore, TQM practices can be appropriately implemented when the culture of the organization has incorporated change readiness.

6 Conclusion

The literature on TQM and ICR has been extended by describing the facilitative function of ICR amid TQM practices as well as employee performance. Findings has shown that TQM practices are helpful to the entire service sector. Though only factors such as process improvements, employee training, customer focus as well as zero defect policy, which are highly relevant in software industries, have been measured. All the TQM practices which have been mentioned are very important for employee performance and its implementation can increase performance during tasks. High product quality can only lead to a higher competitive edge to the service industry, especially the software firms, and the only way to achieve this is by implementing TQM practices. Therefore, this can be considered as a competitive tool. It is possible for Nigeria software firms to gain the same benefits as stated in this study, which is a substantial improvement in employee performance. Likewise, the facilitative function of ICR in respect to the connection amid TQM practices as well as employee performance is confirmed by this study. In Nigeria software firms, ICR can increase the connection amid TQM practices as well as employee performance. According to Walker et al. [42], ICR provides conditions for performance improvement. The outcomes of this research correlate with preceding studies on ICR, TQM practices, as well as knowledge sharing. Hopefully, this research will create awareness for further studies to be carried out in area of TQM, as added empirical evidence is needed in this area. The research affirms and expands the scope of research in respect to affirmative connection that exists amid change readiness as well TQM and employee performance.

The gaps found in literature on the facilitative role of change readiness on employee performance improvement in software firms has been clearly identified empirically by this study. Furthermore, Nigeria software firm's organizational change will be enhanced by employee change readiness. Nigeria Software Export Board (PSEB), software firms and their managers will have a better understanding of the performance of their employees as a result of this study. This study can help facilitate subsequent research in the subject matter of individual change readiness and TQM practices in different sectors. The paper and its findings are also relevant to practitioners—encouraging the adoption of TQM and change readiness training in software and other tech-related operational settings in developing economies.

The limitation to the study is that only software firms within a region were considered, which may not reflect the conditions of all software firms across the entire country. Few dimensions of TQM were considered due to the scope of the research. Also, employee bias cannot be totally ruled out while answering questions about their performance and TQM practices. However, effort was made to expunge outliers from the data set.

Based on the research framework adopted, it is recommended that for future works, a larger sample size from multiple regions may be considered in order to bring generalizability of results. Equally, a triangulated approach should be adopted for minced authenticities which were not captured in this research because the research scope was limited to quantitative data mining.

References

1. Clair, J., Milliman, J.: Best environmental HRM practices in the US. In: Greening People, pp. 49–73. Routledge, New York (2017)
2. Ruiz-Moreno, A., Haro-Domínguez, C., Tamayo-Torres, I., Ortega-Egea, T.: Quality management and administrative innovation as firms' capacity to adapt to their environment. Total Qual. Manag. Bus. Excellence 27(1–2), 48–63 (2016)
3. Zibarras, L.D., Coan, P.: HRM practices used to promote pro-environmental behavior: a UK survey. Int. J. Hum. Resour. Manag. 26(16), 2121–2142 (2015)
4. Honarpour, A., Jusoh, A., Md Nor, K.: Total quality management, knowledge management, and innovation: an empirical study in R&D units. Total Qual. Manag. Bus. Excellence 29(7–8), 798–816 (2018)
5. Putri, N.T., Yusof, S.M., Hasan, A., Darma, H.S.: A structural equation model for evaluating the relationship between total quality management and employees' productivity. Int. J. Qual. Reliab. Manag. 34(8), 1138–1151 (2017)
6. Chang, J.F.: Business Process Management Systems: Strategy and Implementation. Auerbach Publications, Boca Raton (2016)
7. Haffar, M., Al-Karaghouli, W., Ghoneim, A.: An empirical investigation of the influence of organizational culture on individual readiness for change in Syrian manufacturing organizations. J. Organ. Change Manag. 27(1), 5–22 (2014)
8. Holt, D.T., Armenakis, A.A., Harris, S.G., Feild, H.S.: Toward a comprehensive definition of readiness for change: a review of research and instrumentation. In: Research in Organizational Change and Development, pp. 289–336. Emerald Group Publishing Limited, Bingly, UK (2007)
9. Haffar, M., Al-Karaghouli, W., Irani, Z., Djebarni, R., Gbadamosi, G.: The influence of individual readiness for change dimensions on quality management implementation in Algerian manufacturing organizations. Int. J. Prod. Econ. 207, 247–260 (2019)

10. Rusly, F.H., Sun, P.Y., Corner, J.L.: Change readiness: creating understanding and capability for the knowledge acquisition process. J. Knowl. Manag. **19**(6), 1204–1223 (2015)
11. Seggewiss, B.J., Straatmann, T., Hattrup, K., Mueller, K.: Testing interactive effects of commitment and perceived change advocacy on change readiness: investigating the social dynamics of organizational change. J. Change Manag. **19**(2), 122–144 (2019)
12. Gutierrez-Gutierrez, L.J., Barrales-Molina, V., Kaynak, H.: The role of human resource-related quality management practices in new product development: a dynamic capability perspective. Int. J. Oper. Prod. Manage. **38**(1), 43–66 (2018)
13. Hannington, T.: How to Measure and Manage Your Corporate Reputation. Routledge, London (2016)
14. Anil, A.P., Satish, K.: Investigating the relationship between TQM practices and firm's performance: a conceptual framework for Indian organizations. Procedia Technol. **24**, 554–561 (2016)
15. Jaca, C., Psomas, E.: Total quality management practices and performance outcomes in Spanish service companies. Total Qual. Manage. Bus. Excellence **26**(9–10), 958–970 (2015)
16. Wang, M., Shieh, C.: The impact of TQM implementation on employee's performance in China–an example of Shanghai Fu-Shing company. Qual. Control Appl. Stat. **52**(6), 641–642 (2007)
17. Beckford, J.: Quality: a critical introduction. Routledge, London (2016)
18. Ross, J.E.: Total Quality Management: Text, Cases, and Readings. Routledge, New York (2017)
19. Mahanti, R., Antony, J.: Six Sigma in the Indian software industry: some observations and results from a pilot survey. TQM J. **21**(6), 549–564 (2009)
20. Mone, E.M., London, M.: Employee Engagement Through Effective Performance Management: A Practical Guide for Managers. Routledge, New York (2018)
21. Huang, Q., Gamble, J.: Social expectations, gender and job satisfaction: front-line employees in China's retail sector. Hum. Resour. Manage. J. **25**(3), 331–347 (2015)
22. Durkheim, E.: The division of labor in society. In: Inequality, pp. 55–64. Routledge (2018)
23. Shamir, R.: Between self-regulation and the alien tort claims act: on the contested concept of corporate social responsibility. In: Crime and Regulation, pp. 155–183. Routledge (2017)
24. Ishihara, A.: Relational contracting and endogenous formation of teamwork. Rand J. Econ. **48**(2), 335–357 (2017)
25. Valmohammadi, C., Roshanzamir, S.: The guidelines of improvement: relations among organizational culture, TQM and performance. Int. J. Prod. Econ. **164**, 167–178 (2015)
26. Al-Dhaafri, H.S., Al-Swidi, A.K., Yusoff, R.Z.B.: The mediating role of TQM and organizational excellence, and the moderating effect of entrepreneurial organizational culture on the relationship between ERP and organizational performance. TQM J. **28**(6), 991–1011 (2016)
27. Raja Sreedharan, V., Raju, R., Srivatsa Srinivas, S.: A review of the quality evolution in various organisations. Total Qual. Manage. Bus. Excellence **28**(3–4), 351–365 (2017)
28. Calvo-Mora, A., Picón-Berjoyo, A., Ruiz-Moreno, C., Cauzo-Bottala, L.: Contextual and mediation analysis between TQM critical factors and organizational results in the EFQM Excellence Model framework. Int. J. Prod. Res. **53**(7), 2186–2201 (2015)
29. Dahlgaard-Park, S.M., Reyes, L., Chen, C.: The evolution and convergence of total quality management and management theories. Total Qual. Manage. Bus. Excellence **29**(9–10), 1108–1128 (2018)
30. Fallah Ebrahimi, Z., Wei Chong, C., Hosseini Rad, R.: TQM practices and employees' role stressors. Int. J. Qual. Reliab. Manage. **31**(2), 166–183 (2014)

31. Amis, J.M., Aïssaoui, R.: Readiness for change: an institutional perspective. J. Change Manage. **13**(1), 69–95 (2013)
32. Matos Marques Simoes, P., Esposito, M.: Improving change management: how communication nature influences resistance to change. J. Manage. Develop. **33**(4), 324–341 (2014)
33. Smollan, R.K.: The multi-dimensional nature of resistance to change. J. Manage. Organ. **17**(6), 828–849 (2011)
34. Rusly, F.H., Corner, J.L., Sun, P.: Positioning change readiness in knowledge management research. J. Knowl. Manage. **16**(2), 329–355 (2012)
35. Robbins, S.P.: Organisational Behaviour: Global and Southern African Perspectives. Pearson South Africa (2001)
36. Mustain, J.M., Lowry, L.W., Wilhoit, K.W.: Change readiness assessment for conversion to electronic medical records. J. Nurs. Adm. **38**(9), 379–385 (2008)
37. Young, W., Davis, M., McNeill, I.M., Malhotra, B., Russell, S., Unsworth, K., Clegg, C.W.: Changing behavior: successful environmental programmes in the workplace. Bus. Strategy Environ. **24**(8), 689–703 (2015)
38. Bagozzi, R.P., Yi, Y.: On the evaluation of structural equation models. J. Acad. Mark. Sci. **16**(1), 74–94 (1988)
39. Soltani, E., Wilkinson, A.: TQM and performance appraisal: complementary or incompatible? Eur. Manage. Rev. (2018)
40. Lameei, A.: Assessment of organization readiness for TQM implementation. Iran. J. Public Health **34**(2), 58–63 (2005)
41. Holt, D.T., Bartczak, S.E., Clark, S.W., Trent, M.R.: The development of an instrument to measure readiness for knowledge management. Knowl. Manage. Res. Pract. **5**(2), 75–92 (2007)

Modeling Decision-Making with Intelligent Agents to Aid Rural Commuters in Developing Nations

Patricio Julián Gerpe[1](✉) and Evangelos Markopoulos[2]

[1] Enpov, Silk Lafto, Kebele 02, New, 5600, Addis Ababa, Ethiopia
patricio@enpov.com
[2] HULT International Business School, 35 Commercial Road, Whitechapel,
E1 1LD London, UK
evangelos.markopoulos@faculty.hult.edu

Abstract. More than a billion rural merchants in the developing world depend on hiring on-demand transportation services to commute people or goods to markets. Selecting the optimal fare involves decision-making characterized by multiple alternatives and competing criteria. Decision support systems are used to solve this. However, those systems are based on object-based approaches which lack the high-level abstractions needed to effectively model and scale human-machine communication. This paper introduces AopifyJS, a novel agent-based decision-support tool. We developed a two-agent simulation. One agent makes a request, then another takes a dataset of a stratified sample of 104 Ethiopian commuter criteria preferences and a dataset of fare alternatives. The second agent computes HPA and TOPSIS algorithms to weight, score, rank those alternatives. Once we run the simulation, it returns an interpretable prescription to the first agent, storing all interactions in an architecture that allows developers to program further customization as interactions scale.

Keywords: Agent-based modeling · Agent-oriented programming ·
Multi-criteria decision-making · TOPSIS · Social innovation ·
Interpretable artificial intelligence

1 Introduction

Rural transportation has been frequently regarded as a critical infrastructure for economic development. [1, 2]. According to the latest Rural Access Index of the World Bank, over a billion people has not access to all-weather roads in a range of fewer than two kilometres [3]. In Ethiopia, rural commuting has its own challenges. One critical job for rural merchants is selecting the most convenient fare alternative to get them to the market. Several factors influence the merchant's decision making (DM), such as price, time and the trust they have in the driver. This type of problem can be classified as a multi-criteria decision making (MCDC) problem. MCDC is a subject that is a common concern for researchers in the field of decision science [4, 5].

This concept refers to a type of problem where an agent is required to make a choice based on different conflicting criteria. Different techniques have been used to

© Springer Nature Switzerland AG 2020
T. Ahram (Ed.): AHFE 2019, AISC 965, pp. 523–532, 2020.
https://doi.org/10.1007/978-3-030-20454-9_52

address these problems such as AHP, TOPSIS, ELECTRE, and PROMETREE [6–9]. Those techniques are indeed useful to build the so-called 'decision-support systems'. We claim that traditional object-oriented paradigms, due to the lack of high-level abstractions, are quite limited and unscalable when it comes to model human communication. That is why we propose to introduce a novel approach to tackle DM problems combining agent-oriented software engineering (AOSE) with MCDC methods.

Our approach, which we named AopifyJS, is the first AOP open-source framework written in Javascript language to fully integrate methods widely used in MCDM. For the sake of clarity, this paper is structured as follows: we will first detail the problem that rural merchants face in Ethiopia; then we will provide a literature review of MCDC and AOSE scientific literature; after that, we will detail our agent-based tool; and finally we will run our two-agent simulations based on the exemplary case given in this paper.

2 Field Study: Inefficiencies in Commuting in Rural Areas

Ethiopia has a total population of around 104 million people [10], where 70% of the population relies on agriculture. 36.7% of GDP comes from agriculture [11]. Yet, the country deals with critical challenges in transportation infrastructure.

Smallholder farmers commonly sell their produce to the internal market. To do that, rural merchants are required to wait on rural roads for small motorized vehicles called 'Bajajs' or sometimes simply horse-drawn carts, which regularly pass downrural roads seeking customers. We first conducted preliminary field observations with the purpose of acquiring a first-hand understanding of the commuting process of farmers in Ethiopia.

During 2017 and 2018, we spent two days in Wacho Belisso, Ethiopia, and seven days in the Duken Region where we interviewed a sample of 28 rural commuters asking about their commuting routine.

The key insights from those observations were that (A) on average, transportationseems to cost 30% of the farmers' income and, (B) commuters usually go to main roads, waiting near KM markers for vehicles to pass by.

Apart from the field interviews, we visited different institutions in Ethiopia to contrast information from the ATA (Agricultural Transformation Agency), the Transportation Bureau, and the national agricultural ministry in the city of Addis Ababa. In summary, rural commuters are often immersed in complex decision-making environments constrained by scarce resources and multiple alternative courses of actions wherein information is often incomplete and outcomes are difficult to predict.

Figure 1 illustrates the whole process. As already outlined, in this paper we will be focusing on solving the second step of this flow chart.

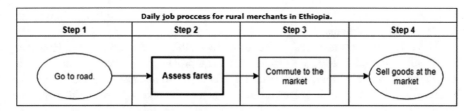

Fig. 1. Commuting process flowchart.

3 Literature Review

3.1 Overview of Agent-Oriented Software Engineering Research

The term 'agent-oriented programming' was formerly coined by computer scientist Soham in 1993 [12]. This concept refers to the developing paradigm in which the center of applications are agent entities. Those agent entities provide to the developer high-level abstractions which include but are not limited to: beliefs, motives, skills, and policies.

Table 1 describes Shoham AOP definition, as an extension of Object-Oriented programming OOP [13].

Table 1. Table of comparison. OOP and AOP.

Framework	OOP	AOP
Basic unit:	Object	Agent
Parameters defining state of basic unit:	Unconstrained	Beliefs, commitments, capabilities, choices
Process of computation:	Messages passing and response methods	Message passing and response methods
Types of message:	Unconstrained	Inform, request, offer, promise, decline
Constraints on methods:	None	Honesty, consistency

Agents have the following properties: autonomy, social, reactivity, and proactivity. This means that an agent's memory is self-contained; they have a language in which they can share information with other agents; they are proactive in that they pursue goals and act to achieve them; and they are reactive in that they produce new actions in response to stimuli in a given environment.

In Javascript, some *general-purpose* attempts to create frameworks have already been made. For instance, SieborgJS [14], is a proposed framework for multi-agent simulations in Javascript. Nonetheless, when reviewing the available open code repositories, the code does not appear to be open-source. AgentSimJS [15] is a comprehensive framework for 3D spatial agent simulations and EveJS [16] is a framework to build multi-agent systems in web ecosystems, integrated with websockets, https, http and JSON RPC protocols.

These last two frameworks do have an accessible open-source code. However, based on its commit history, we deduce that those projects were merely academic and are no longer actively maintained or tested. Unlike those attempts, we seek to build a scalable toolkit that not only solves problems such as the one outlined in this paper but can also turn into a standard in the IT industry, shifting object and function-oriented programming into an approach specifically designed to consider human factors when aiding decision problems. Apart from that, our proposed framework is the only one in this language to provide integration with MCDC algorithms.

3.2   Overview of Multi-criteria Decision Making Literature

Multi-criteria decision making (MCDC) is an area that has caught the attention of scholars from fields such as operation research, ergonomics, management, and decision science. Stanley Zionts popularized the acronym MCDC back in 1979 [17].

Throughout recent decades, several methods have been proposed. A number of researchers have compared these methods [18]. The most common methods include but are not limited to: AHP, ANP, TOPSIS, ELECTRE, PROMETHEE, and VIKOR. In recent years, with the increase in computing power and the availability of data, machine learning techniques have been widely incorporated into the creation of decision support systems, especially optimization algorithms such as Markov chains, naive Bayes and decision trees that had been used in the so-called recommender systems [19]. To tackle our problem we will use AHP and TOPSIS algorithms.

The Analytic Hierarchy Process (AHP) was proposed by Saaty in 1979. Essentially, the algorithm takes a matrix of comparative judgment between competing criteria, it normalizes that matrix and then computes an eigenvector considering each column data. That eigenvector represents the weighting of preference that each criterion has in the decision-maker.

The Technique for Order of Preference by Similarity to Ideal Solution (TOPSIS) was proposed by Hwang and Yoon in 1981. The algorithm is based on the premise that the most desirable solution out of a set of alternatives is the one with the closest Euclidean distance to the hypothetical ideal solution and the farthest Euclidean distance to the hypothetical anti-ideal solution.

4   AopifyJS: Our Agent-Based Tool

Unlike previous attempts, our tool integrates MCDC methods for prescriptive analytics with agent-based modeling (ABM). Due to its high-level abstractions, we argue that this ABM design allows the early incorporation of ergonomics and human factors when it comes to deploying decision-support systems.

We decided to build our system in Javascript because (A) it has extensive and on-growing support from the developer community, (B) it allows developers with cross-platform developments and (C) its built-in asynchronous methods can be reutilized for our event-based communication engine.

To tackle our fare-selection problem, our datasets include: (A) one matrix containing fictional data on six available drivers and its core attributes (time, price and trust – where trust is an integer number from 1 to 5 representing a rating system such as those commonly used in on-demand reputation systems) and (B) a dataset of 104 matrixes that contains an assessment of those three criteria based on a comparative subjective judgment from real rural merchants we interviewed in Ethiopia. In other words, each farmer we interviewed compared each criterion against the others. To collect that data, we used a simplified version of the Saaty Scale [20].

Table 2. Fare alternative matrix.

Alternative	Time	Price	Trust
Driver 1	2	5	5
Driver 2	60	26	4
Driver 3	20	20	4
Driver 4	500	2	4
Driver 5	50	23	3
Driver 6	25	10	1

Table 2 illustrates our alternative matrix. To build our dataset of criteria assessment we asked 104 rural merchants in the Dukem region, through direct field interviews, the three questions below:

Is price more important for you than time?
Is price more important for you than trust?
Is time more important for you than trust?

Considering cognitive and educative aspects of our respondents, we used a simplified version of a Saaty questionnaire to ensure greater enthusiasm and comprehension by them.

For each interview we built a matrix and we inserted that matrix into a new row of a csv file.

Therefore, our csv dataset file looks as Fig. 2 illustrates. The full dataset is available in our open-source repository in Github. If the value is one, that means the criteria was perceived to have equal preference against the other.

A number higher than one means higher preference and, a lower one means lower preference. Afterwards, we created an agent entity that incorporated both AHP and TOPSIS algorithms in its methods in order to aid fare requests from rural merchants. Our agent entities and methods are mainly inspired in EveJS, yet, tailored to facilitate decision-making like the one dealt by merchants.

OBVERVATIONS

#	Time vs Time	Time vs Price	Time vs Trust	Price vs Time	Price vs Price	Price vs Trust	Trust vs Time	Trust vs Price	Trust vs Trust
1	1,00	3,00	3,00	0,33	1,00	3,00	0,33	0,33	1,00
2	1,00	3,00	0,33	0,33	1,00	0,20	3,00	5,00	1,00
3	1,00	3,00	3,00	0,33	1,00	0,20	0,33	5,00	1,00
4	1,00	3,00	0,33	0,33	1,00	5,00	3,00	0,20	1,00
5	1,00	3,00	0,33	0,33	1,00	1,00	3,00	1,00	1,00
6	1,00	3,00	1,00	0,33	1,00	5,00	1,00	0,20	1,00
7	1,00	0,33	3,00	3,00	1,00	5,00	0,33	0,20	1,00
8	1,00	3,00	3,00	0,33	1,00	5,00	0,33	0,20	1,00
9	1,00	1,00	3,00	1,00	1,00	0,20	0,33	5,00	1,00
10	1,00	3,00	3,00	0,33	1,00	0,20	0,33	5,00	1,00

Fig. 2. Dataset: structure of the criteria assessment dataset.

```
class Agent extends EventEmitter {
    constructor(name) {
        super();
        this.id = uuid();
        this.name = name;
        this.isAlive = false;
        this.interactions = new Set();
    }
}
```

Our agents communicate with each other via an event-based architecture. That is why we are utilizing the built-in module of NodeJS: 'EventEmitter'. Each agent has a name (string) and an id (uuid). The 'isAlive' property indicates whether the agent has been initialized. If the agent is not initialized then it cannot receive/send any message. Each time our agent interacts with another agent, we will store that information in a Javascript Set. This architecture, unlike traditional non-agent approaches it is self-contained in each agent, allowing developers to provide further user customization and user experience optimization as we scale.

The Agent class has two elemental methods to initialize or deinitialize the agent (start and kill) and four methods that allow it to communicate and capture data from its environment: (tell, on, store and decide) where 'on' is already inherent in the EventEmitter class. Table 3 illustrates all these methods:

All these methods are written in the most declarative way possible, allowing human interpretability not just from the user-side but also from the developer side. Start() is the method that initializes our agent so it starts listening to the events to which it is subscribed. Kill() is the method that deletes our agent from its environment. On() is the method of subscribing to one particular event. Tell() is the method to send a particular message to another agent. Store() is used to tell the agent to store information from a message it received, and Decide() is the method that imports a prescriptive analytics function to solve a problem of MCDC.

Our agent system is prepared to compute AHP algorithm for criteria weighting and TOPSIS for alternative selection. Both came from a self-coded repository (Ahp-lite and

Table 3. Agent class methods.

Method	Purpose	Use case
decide()	To enable agents to resolve complex decision-making problems	Agent aiding another agent to make a decision
tell()	To enable communication between agents	Agent sending its insights to another agent
on()	To enable agents to capture stimuli from environment	Agent waiting for another agent's information
store()	To enable agents to store data about their interactions	Agent storing the information it received

Topsis named respectively) available in NPM package management system of the NodeJS server-side javascript framework.

Consequently, we can proceed to set-up our simulation. We initialize two agent entities. One represents a rural merchant in Ethiopia that needs to go to market and the other one represents a computer agent that handles his or her request.

```
const ai = new Agent('ai');
const human = new Agent('human');
```

Prior to initializing the communication between these two agents, we will make the assistant agent pre-process our criteria assessment data collected through field interviews using the AHP algorithm.

Each assessment has the form illustrated in Eq. 1. Where M stands for criteria matrix and c represents each criterion. Therefore, c1 is time, c2 is price and c3 is trust. And, as previously outlined, each merchant was asked to compare all criteria against each other.

$$M = \begin{bmatrix} c_1/c_1 & c_1/c_2 & c_1/c_3 \\ c_2/c_1 & c_2/c_2 & c_2/c_3 \\ c_3/c_1 & c_3/c_2 & c_3/c_3 \end{bmatrix}. \tag{1}$$

Moving onward, for each assessment the AHP algorithm returns us an eigenvector (ev) with the weighting of each criterion for a particular rural merchant that we interviewed.

The eigenvector formula is indicated in Eq. 2. Where w_1, w_2 and w_3 represents the weighting of each criterion. Saaty recommends that we check the consistency of the judgment by computing the consistency ratio [21]. To maintain the quality of the data we will only include in our analysis the judgements that were consistent. The lamda max operator is used to that end.

$$ev = \begin{bmatrix} w_1 & w_2 & w_3 \end{bmatrix}. \tag{2}$$

Once we process all judgements, we will proceed in synthesizing all judgements into one eigenvector. As Saaty also recommends, we will use the geometric mean to that end.

Equation 3 indicates the results. This means that for an average merchant in the Dukem region of Ethiopia, the time criterion is estimated to have a weighting of 37% importance, price a 29% weighting and trust a 25% weighting.

$$ev = \begin{bmatrix} 0.37 & 0.29 & 0.25 \end{bmatrix}. \tag{3}$$

That being processed, we are now ready to program our communication engine.

We will program our assistant agent ('ai') to subscribe to our human agent 'request' events.

Every time the agent receives a request from its human partner it will take as arguments the alternative matrix of available drivers and the eigenvector resulting from the processing the dataset of jugements, to compute the TOPSIS algorithm and return a prescription:

```
ai.on('request', (msg) => {
  const res = ai.decide('topsis', data);
  ai.store(msg, 'request', human.id, ai.id);
  msg = `AGENT: The best fare for you is this one. The
rating is ${res[2]} stars. You will reach location in
around ${res[1]} minutes and the cost is ${res[0]}
birrs.`;
  ai.tell({ name: 'response', msg }, human);
});
msg = 'HUMAN: I need to find a ride to market!';
human.tell({ name: 'request', msg }, ai);
```

5   Results

Once we run our programme and we inspect our agent instance in our console we will see this Javascript class (Fig. 3):

Where the full '*msg*' text is: 'AGENT: The best fare for you is this one. The rating is 5 stars. You will reach location in around 5 min and the cost is 2 birrs.' Therefore, that means, our most desirable solution was Driver 1.

As can be seen, almost instantly we computed our response from the agent and all interactions remained self-contained within the agent entity. Driver 1 evidently provides the optimal solution as it has a high trust score (benefit criteria to maximize), extremely low price (cost criterion to minimize) and it is extremely quick (time was a cost criterion to minimize).

The self-contained storage architecture of our machine-human communication engine allows developers to collect valuable data in real time to further customize interactions between agents.

```
Agent {
  _events: [Object: null prototype] { request: [Function] },
  _eventsCount: 1,
  _maxListeners: undefined,
  id: 'dc29ebb3-d602-4976-96c8-7aa1116cc16b',
  name: 'ai',
  isAlive: true,
  interactions:
   Set {
     Interaction {
       from: 'b52347b0-2de1-4ab8-bfe4-c97bb188cd96',
       to: 'dc29ebb3-d602-4976-96c8-7aa1116cc16b',
       evnt: 'request',
       msg: 'HUMAN: I need to find a ride to market!',
       time: '2019-02-09T10:21:05+00:00' },
     Interaction {
       from: 'dc29ebb3-d602-4976-96c8-7aa1116cc16b',
       to: 'b52347b0-2de1-4ab8-bfe4-c97bb188cd96',
       evnt: 'response',
       msg:
        'AGENT: The best fare for you is this one. The rating is 5 stars. You will reach location in around 5 minutes
   d the cost is 2 birrs.',
       time: '2019-02-09T10:21:05+00:00' } },
  birthday: '2019-02-09T10:21:05+00:00' }
```

Fig. 3. Console results after running our simulation.

To conclude, the major contributions of this paper can be summarized as follows:

- A literature review of MCDC and agent systems in Javascript.
- An open-source tool that sets the basis for interpretable intelligent agents from both the user and developer side that can be used for research simulations or industrial applications.
- A novel integration of high-level abstractions from AOSE and MCDC methods in Javascript that enables developers to program incremental interactions between computer agents and both computer and human agents.
- Proof of concept with a concrete use case to improve a social outcome.

6 Discussion and Further Research

This research was designed with a view to providing easy-to-deploy standardized methods using agents system architectures for general purpose MCDC problems in a scalable and ergonomic manner. We believe that this integration may have a wide arrange of potential use cases in management science and business analytics.

A matter for further study is the implementation of more human-like abstractions for machine argumentation and communication between human and intelligent agents, especially assessing this architecture in simulations with a much larger number of agents and events. Further improvements will include integrations with machine learning algorithms, reward and utility functions, three-valued logic inference systems and other general-purpose AI capabilities. All those features are still in progress. So the results will be published soon.

References

1. Gollin, D., Rogerson, R.: Agriculture, roads, and economic development in Uganda (No. w15863). National Bureau of Economic Research (2010)
2. Wondemu, K.A., Weiss, J.: Rural roads and development: evidence from Ethiopia. Eur. J. Transp. Infrastruct. Res. **12**(4), 417–439 (2012)
3. Roberts, P., Kc, S., Rastogi, C.: Rural Access Index: A Key Development Indicator. World Bank, Washington, DC (2006)
4. Belton, V.: A comparison of the analytic hierarchy process and a simple multi-attribute value function. Eur. J. Oper. Res. **26**(1), 7–21 (1986)
5. Pawlak, Z., Sowinski, R.: Rough set approach to multi-attribute decision analysis. Eur. J. Oper. Res. **72**(3), 443–459 (1994)
6. Saaty, T.L., Erdener, E.R.E.N.: A new approach to performance measurement the analytic hierarchy process. Des. Methods Theor. **13**(2), 62–68 (1979)
7. Hwang, C.L., Yoon, K.: Methods for multiple attribute decision making. In: Multiple Attribute Decision Making (pp. 58–191). Springer, Berlin, Heidelberg (1981)
8. Roy, B. Classement et choix en présence de points de vue multiples (la méthode ELECTRE). *La Revue d'Informatique et de Recherche Opérationelle (RIRO)* (8): 57–75 (1968)
9. Brans, J.P., Vincke, P.: A preference ranking organisation method: the PROMETHEE method for MCDM. Manag. Sci. **31**, 647–656 (1985)
10. Ethiopia Population, total. World Bank Group (2017). https://data.worldbank.org/country/ethiopia
11. Annual Report 2016/17. Agricultural Transformation Agency (2017)
12. Shoham, Y.: Agent-oriented programming. Artif. Intell. **60**(1), 51–92 (1993)
13. Shoham, Y.: Agent oriented programming: an overview of the framework and summary of recent research. In: International Conference on Logic at Work, pp. 123–129. Springer, Berlin, Heidelberg (1992)
14. Lukic, A., Luburi, N., Vidakovi, M., Holbl, M.: Development of multi-agent framework in JavaScript (2017)
15. Calenda, T., De Benedetti, M., Messina, F., Pappalardo, G., Santoro, C.: AgentSimJS: a web-based multi-agent simulator with 3d capabilities. In: WOA, pp. 117–123 (2016)
16. De Jong, J., Stellingwerff, L., Pazienza, G.E.: Eve: a novel open-source web-based agent platform. In: Systems, Man, and Cybernetics (SMC), 2013 IEEE International Conference on (1537–1541). IEEE (2013)
17. Zionts, S.: A survey of multiple criteria integer programming methods. Annals of Discrete Mathematics **5**, 389–398 (1979)
18. Zavadskas, E.K., Turskis, Z., Kildienė, S.: State of art surveys of overviews on MCDM/MADM methods. Technol. Econ. Dev. Econ. **20**(1), 165–179 (2014)
19. Portugal, I., Alencar, P., Cowan, D.: The use of machine learning algorithms in recommender systems: a systematic review. Expert Syst. Appl. **97**, 205–227 (2018)
20. Saaty, T.L.: Exploring the interface between hierarchies, multiple objectives and fuzzy sets. Fuzzy Sets Syst. **1**(1), 57–68 (1978). https://doi.org/10.1016/0165-0114(78)90032-5
21. Aczél, J., Saaty, T.L.: Procedures for synthesizing ratio judgements. J. Math. Psychol. **27**(1), 97 (1983). https://doi.org/10.1016/0022-2496(83)90028-7

Self-similarity Based Multi-layer DEM Image Up-Sampling

Xin Zheng, Ziyi Chen, Qinzhe Han, Xiaodan Deng,
Xiaoxuan Sun, and Qian Yin[✉]

Image Processing and Pattern Recognition Laboratory,
College of Information Science and Technology,
Beijing Normal University, Beijing, China
{zhengxin, chenziyi, yinqian}@bnu.edu.cn,
hanqinzhe@mail.bnu.edu.cn, dengxiaodancyc@163.com,
sunxiaoxuan922@163.com

Abstract. As one of the basic data of GIS, DEM data which expresses the surface elevation data is widely used in many fields. How to obtain a wide range of high-precision elevation data is a big challenge, the simple interpolation algorithm currently used is less accurate. Due to the fractal data characteristics of terrain data, DEM data shows strong self-similarity. Based on this feature, this paper proposes a multi-layer Dem image up-sampling method. Image up-sampling is performed multiple times in layers on the low-resolution DEM image, therefore, high-precision DEM information with less error is obtained. In this paper, elevation data of 30 m is expanded to elevation data of 10 m by gradually using this method. Experimental results show that the algorithm can achieve good results and has a small deviation from the real elevation data of 10 m.

Keywords: Up-sampling · DEM · Self-similarity

1 Introduction

The digital elevation model is an entity ground model that expresses the ground elevation in the form of an ordered array of numerical values. In short, it is a digital representation of the ground form. In economic construction and national defense construction, it has important utilization value and has been widely used in many fields such as surveying and mapping engineering, civil engineering, geology, fine agriculture, urban and rural planning.

The DEM also plays a very important role in the pre-alarming and prevention of hill-flood disasters. Hill-flood disasters often cause heavy losses to the national economy and people's property by floods caused by heavy rainfall and mudslides and landslides caused by flash floods in sub-basins of hilly areas. Owing to its suddenness and great destructive power, it often leads to the destruction of houses, roads and bridges, and the dams and mountain ponds, and even cause casualties. Therefore, on the basis of available rainfall data, if accurate information on terrain altitude and slope can be obtained, disaster losses can be effectively prevented.

© Springer Nature Switzerland AG 2020
T. Ahram (Ed.): AHFE 2019, AISC 965, pp. 533–545, 2020.
https://doi.org/10.1007/978-3-030-20454-9_53

With the development of DEM technology, the high-precision grid point gradient and altitude data combined with rainfall data will be able to estimate the rainfall threshold for mountain flood disasters, which become an important indicator for monitoring mountain flood disasters. Limited by the accuracy of the acquisition equipment, the high-precision elevation data are often difficult to obtain. If the up-sampling precision of the low-precision DEM image is expanded, higher-precision elevation data information can be obtained.

The simplest method of improving image resolution is to use the analytical interpolation formula to predict new pixels, such as Bilinear interpolation or Bicubic interpolation, but the simulated data obtained by these methods is excessively smooth and different from complex real data. In order to increase randomness, a method for realizing high-precision image expansion using an instance-based Markov random model is proposed. This model uses a set of common module to predict the pixel information that the image needs to be expanded. Due to the lack of matching module and the error of search correlation, it will produce a lot of noise and cause irregular data images. Ebrahimi and Vrscay [1] propose to use the input image itself as a source of matching module information acquisition. Although this method can only get a limited number of reference matching modules, it has more correlation with the image itself.

In 1975, Mandelbrot proposed fractal geometry [2], which used fractal dimensions to describe broken, chaotic objects and express complex forms in nature. Contrary to Euclidean geometry, fractal geometry describes the measurement of disordered, irregular objects, such as the length of the natural coastline [3], reservoir capillary channels [4], bottom contour lines, curved river channels, etc. [5].

The application of fractal in terrain and geomorphology evaluation mainly includes two methods: self-similar fractal and multi-fractal analysis. The application of self-similar fractal in terrain analysis means that the part is similar to the whole in some form [6]. Mathematically it is a fractal under the function of a uniform linear transformation group. Within a certain range, self-similarity described by a fractal dimension is an important feature of fractal shape. If the natural surface has a good self-similarity, its self-similarity curve should maintain a good linearity in the whole metric space [7]. Topographic surface is the product of long-term natural evolution and has remarkable self-similar fractal characteristics.

In this paper, a new DEM image up-sampling method based on surface self-similar fractal features is proposed. According to the terrain self-similarity, inspired by Ebrahimi and Vrscay, the terrain self-similarity rule is obtained and the high-precision upsampled DEM image is matched by intercepting the example sliding window of the input image, and finally the high-precision DEM data image is obtained. At the same time, in order to avoid the problem of insufficient up-sampling data, this method proposes the idea of multi-level segmentation up-sampling, which decomposes a high-multiplication up-sampling coefficient into the product of multiple low-multiplication up-sampling coefficients. High-power image up-sampling is realized through multiple low-power up-sampling processes. The experimental results show that multiple decompositions will better preserve the characteristics of the image itself and eliminate the problem of less input image data. In this paper, by inputting the DEM elevation data of 30 m accuracy and using this up-sampling method just mentioned to expand step by step and realize the accuracy simulation of 10 m elevation data. Experiments show that the algorithm achieves better results and has smaller deviation compared with the real 10 m data.

2 Related Work

Image up-sampling is widely used in computer vision, computer graphics and other fields. Different up-sampling methods have differences in the input image content and method model.

The simplest interpolation method in up-sampling is nearest neighbor interpolation [8]. In this algorithm, the value of each interpolated sample point is the value of the sample point closest to it in the input image. This interpolation method has a very small amount of computation and simple calculation. For precise image, the block effect can be clearly seen in the image. At the same time, the value of the nearest neighbor interpolation method is not good. From its Fourier spectrum, it can be seen that it has a large difference from the ideal low-pass filter. In many cases, the results are also acceptable. However, when the image contains fine structures with varying gray levels between pixels, that is, high-frequency components, the amplification process performed by this method produces a significant serrated effect in the image.

The output pixel of the bilinear interpolation method [9] is the average value of its 2 × 2 field sampling points in the image, which is interpolated in both horizontal and vertical directions according to the gray values of four pixels around a pixel. Compared to the nearest neighbor difference method, bilinear interpolation algorithm produces a more satisfactory effect, but the program is a little more complicated. However, this method only takes the influence of the gray values of the four direct neighbors around the sample to be tested into consideration, and ignores the influence of the rate of change of the gray value between adjacent points, so it's equivalent to a low pass filter. which results in the high-frequency components of the image after the scaling are lost, as well as the edges of the image become blurred to some extent. The output image scaled by this method still has a problem of impaired image quality and low calculation accuracy due to poor design consideration of the interpolation function. This method is also often cited by commercial software. The basic geographic software can realize the precision simulation expansion of DEM elevation data, but it can only achieve the effect of spatial smooth transition without considering any terrain properties.

There is a further method to adjust the interpolation weight based on the adjacent content, that is, the inverse distance weight interpolation method [10]. It estimates the interpolation result by averaging the sample data points in the neighborhood of each sample to be measured. The closer a point is to the sample point, the greater its impact or weight in the averaging process. Intuitively, the weight of each sample point on the interpolation results decreases as the distance increases. The inverse distance weight interpolation method mainly relies on the power value of the inverse distance. The power parameter can control the effect of a known point on the interpolation result based on the position from the output point. The higher the power, the greater the influence of the nearest point. As a result, Adjacent data will be most affected, and the surface will become more detailed and uneven. As the power value increases, the output will gradually approach the value of the adjacent point. Specifying a smaller power value will have a greater impact on the points that are farther away, resulting in a smoother surface. Since the inverse distance weight interpolation formula is not related to the actual physical process, it is impossible to determine whether the specific power value is too large. If the power value is not chosen correctly, an incorrect result may be generated.

For fitting the power value, the Kriging interpolation method [11] is further proposed. In addition to the distance between the sample and the predicted point, the distance feature between the sample points should also be considered, and the corresponding weights are obtained by fitting the variogram model. The method can obtain the interpolation result between the sample and the sample, and between the sample and the predicted position. The Kriging interpolation method often needs random sampling to fit in a specific range, creating variograms and covariance functions to estimate the correlation of different fitting models, so as to predict the weight value. Due to the huge amount of computation, slow speed and large amount of CPU rates, only a part of random sample points are often used for fitting. So the self-similarity of the whole terrain is not considered.

The image up-sampling method proposed by Freeman [12] et al. based on the self-similarity principle to find features in the original image to complement the missing details. The method first uses an interpolation method to interpolate the input image to a higher resolution. Finding the optimal match in the original image according to the low frequency component of the example window, and then superimposing the corresponding high frequency information to generate texture details in interpolated image. However, due to the correlation problem of the matching window, image noise may be generated and certain irregularity may occur.

Based on the correlation of the image itself, Ebrahimi and Vrscay [13] decomposed the input image into multiple small-scale windows as the source of the matching window. Although this method limits the number of matching windows, the results show that there is a more significant correlation between these matching windows and the input images.

3   Methodology

Based on the idea of classification and the significant self-similarity of topographic data, the algorithm in this paper improves the existing module matching algorithm to make it more suitable for the synthesis of super-resolution images of DEM data. First, different from the above method, in which the source of the sample window directly decomposes the input image into multiple small-scale windows, we first down-sampled the original image window to get a lower resolution image, and then subdivided the image into small-scale sample windows. Second, we have different rules for correlation matching. Finally, our method of expansion is different from the existing method. After matching the execution window with the sample window, we extend the precision of the execution window in accordance with the rules of the sample window.

In this method, we sample the original as the source of the sample window. Then we finds the sample window that best matches the window to be expanded. Due to the self-similarity of terrain, there is a strong correlation between the low resolution sample window and the window to be expanded. Therefore, the self-similarity law of the sample window before down-sampling can be applied to the execution window. At this time, the execution window can obtain the data with higher resolution, that is, the image expansion.

Because of the self-similarity of the image, we believe that the image still keeps the self-similarity rule after the original image's resolution is reduced. Under topographic conditions, the self-similarity is more likely to occur. Therefore, we propose a method that the input image can be scaled to a certain scale, and the rule of the smaller scale image itself can be used as the reference rule of the window to be expanded (Fig. 1).

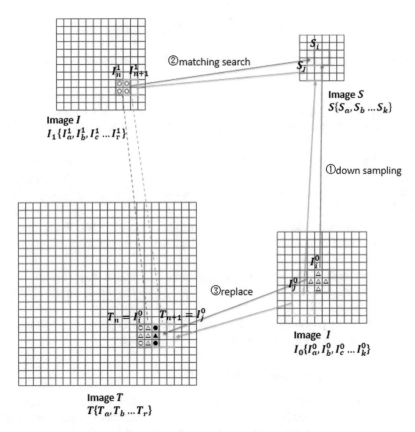

Fig. 1. Flow chart of our method. This shows the overall flow chart, which is mainly divided into three steps: sampling, window matching and window replacement. The line box represents the window, the $\bigcirc$ represents the known vertex data, $\triangle$ is expanded data after window replacement. There is overlap in window sliding, and the solid black part is the overlapping part of the window. At the $\blacktriangle$, the data obtained by two different Windows need to be averaged.

In the experiment of this paper, we first sampled to obtain the window set. According to the above sampling target multiple of λ, DEM image I is compressed into an image S with lower resolution. Then, low-resolution Dem image S is segmented into fixed-size Windows. In this step, the image is down-sampled. Then the window originally containing data information of size n × n is compressed into a window of size m × m (n > m).

We slide the window on low-resolution DEM image S, and divide it into windows with fixed size m × m. Then put them into data source S as the example windows. After sampling, there is a corresponding position relation between the window of image S and the window of original image I. Suppose mapping relationship is f (), in which window S_i is mapped to window I_i^0, the mapping is as follows:

$$f(S_i) = I_i^0 . \tag{1}$$

$$f(S\{S_a, S_b \ldots S_k\}) = I_0\{I_a^0, I_b^0, I_c^0 \ldots I_k^0\} . \tag{2}$$

where I_i^0 is the window of original image I, $I_0\{I_a^0, I_b^0, I_c^0 \ldots I_k^0\}$ is the all windows after mapping.

At the same time, we construct a window set of images to be expanded. We slide the window on the original DEM image I with step size 1. The window has the same size as window to be expanded. Then we obtain the set of windows to be expanded $I_1\{I_a^1, I_b^1, I_c^1 \ldots I_r^1\}$.

Next, We search for the best match. After constructing window set, we extract the height difference rule for each window in the collection and normalize them. After normalization, data source set, original set and the set of image to be expanded are respectively $S'\{S_a', S_b' \ldots S_k'\}, I_0'\{I_a^{0'}, I_b^{0'}, I_c^{0'} \ldots I_k^{0'}\}, I_1'\{I_a^{1'}, I_b^{1'}, I_c^{1'} \ldots I_r^{1'}\}$. After normalization, self-similarity matching will be performed between windows. We search the set S' to find the normalization sample window, which has the highest correlation matching with window to be expand $I_n^{1'}$.

Finally, we replace the window. According to the mutual mapping relation, the position of the window in the collection I_1 is replaced by the window in I_0. Finally, we get the extended set $T\{T_a, T_b \ldots T_r\}$.

In the following paragraphs, we focus on the process of optimal window matching in the second step and window replacement in the third step.

3.1   Optimization of Window Matching

Normalization. For every example window with size n × n in data source S, we extract and normalize their height difference rules, and map them to the interval of [0, 1]. Then the normalized two windows are matched by self-similarity.

We slide to traverse the image to get the n × n data window, and take one of the windows to be represented by the n-dimensional matrix I. In the process of normalizing I, we search for the maximum and minimum values in each window firstly. The maximum value is denoted as I_{Max}, the position is denoted as $P_I =)$, the minimum value is denoted as I_{Min}, and the position is denoted as $Q_I =)$.

For the maximum value I_{Max} in the matrix I, when $I_{Max} = 0$, the matrix is not normalized. When I_{Max} does not equal 0, subtract the minimum value I_{Min} from each element in the matrix I and divide by the maximum value I_{Max}. We obtain matrix I' as follows:

$$I' = (I - I_{Min})/I_{Max}. \tag{3}$$

After normalizing, the maximum of matrix I' is 1, and the minimum value is 0. The positions of the maximum and minimum values remain unchanged and have the following properties:

$$P'_I = P_I = \begin{cases} R_{Max} = R'_{Max} \\ C_{Max} = C'_{Max} \end{cases}. \tag{4}$$

$$Q'_I = Q_I = \begin{cases} R_{Min} = R'_{Min} \\ C_{Min} = C'_{Min} \end{cases}. \tag{5}$$

After the normalization operation, the original height data in the window matrix is transformed into data reflecting the height difference rule. The original set $I_0\{I_a^0, I_b^0, I_c^0 \ldots I_k^0\}$ is normalized to set $I'_0\{I_a^{0\prime}, I_b^{0\prime}, I_c^{0\prime} \ldots I_k^{0\prime}\}$. The data source collection $S\{S_a, S_b \ldots S_k\}$ is normalized to set $S'\{S'_a, S'_b \ldots S'_k\}$. The set to be Extended $I_1\{I_a^1, I_b^1, I_c^1 \ldots I_r^1\}$ is normalized to set $I'_1\{I_a^{1\prime}, I_b^{1\prime}, I_c^{1\prime} \ldots I_r^{1\prime}\}$. Papers not complying with the LNCS style will not be considered or delayed. This can lead to an increase in the overall number of pages. We would therefore urge you not to squash your paper.

Correlation Judgment. Two normalized matrices can be used for correlation judgment. Set the two matrices after normalization are respectively I'_0, I'_1,

$$I_0 = \begin{pmatrix} a_{11} & a_{12} & \cdots & a_{1n-1} & a_{1n} \\ a_{21} & a_{22} & & a_{2n-1} & a_{2n} \\ \vdots & & \ddots & & \vdots \\ a_{n-11} & a_{n-12} & \cdots & a_{n-1n-1} & a_{n-1n} \\ a_{n1} & a_{n2} & & a_{nn-1} & a_{nn} \end{pmatrix}. \tag{6}$$

$$I_1 = \begin{pmatrix} b_{11} & b_{12} & \cdots & b_{1n-1} & b_{1n} \\ b_{21} & b_{22} & & b_{2n-1} & b_{2n} \\ \vdots & & \ddots & & \vdots \\ b_{n-11} & b_{n-12} & \cdots & b_{n-1n-1} & b_{n-1n} \\ b_{n1} & b_{n2} & & b_{nn-1} & b_{nn} \end{pmatrix}. \tag{7}$$

We define that when calculating the correlation of two matrices, the following conditions shall be met first:

$$\left(P'_{I^0} = P'_{I^1}\right) \cap \left(Q'_{I^0} = Q'_{I^1}\right). \tag{8}$$

The correlation judgment value of the two matrices was set as C, and the calculation method of I was set as the sum of squared errors of the corresponding positions,

$$e = \left(a'_{11} - b'_{11}\right)^2 + \left(a'_{12} - b'_{12}\right)^2 + \cdots + \left(a'_{1n} - b'_{1n}\right)^2 + \left(a'_{21} - b'_{21}\right)^2 + \cdots + \left(a'_{nn} - b'_{nn}\right)^2. \quad (9)$$

$$C_{I'_0 I'_1} = \left\{ \begin{array}{l} \frac{1}{e} (e \neq 0) \\ +\infty (e = 0) \end{array} \right. . \quad (10)$$

As we know, the greater the correlation value C of the two Windows is, the greater the matching degree will be (Fig. 2).

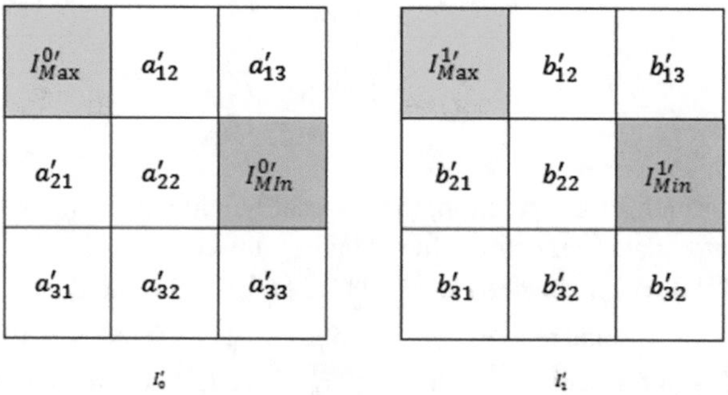

$I^{0'}_{Max}$	a'_{12}	a'_{13}
a'_{21}	a'_{22}	$I^{0'}_{Min}$
a'_{31}	a'_{32}	a'_{33}

$I^{1'}_{Max}$	b'_{12}	b'_{13}
b'_{21}	b'_{22}	$I^{1'}_{Min}$
b'_{31}	b'_{32}	b'_{32}

I'_0 I'_1

Fig. 2. a' is the element in the matrix I'_0 , b' is the element in the matrix I'_1, I_{Max}, I_{Min} is the maximum value and minimum value in the matrix. This figure shows that when judging the correlation of Windows the first need to meet is that the maximum and minimum positions of matrices I'_0 and I'_1 are the same. Then the correlation value C is calculated.

Window Matching. For any window $I^{1'}_n$ in set to be expanded I'_1, we do correlation judgment with each sample window in set S'. The process of correlation judgment is as follows:

$$g = max\left(C_{S'_i I^{1'}_a}, C_{S'_i I^{1'}_b} \ldots, C_{S'_i I^{1'}_c}, C_{S'_i I^{1'}_r}\right). \quad (11)$$

By comparing, we obtain sample window S'_i which is the most correlative window with $I^{1'}_n$. We define the correlation function as $h()$,

$$h\left(I^{1'}_n\right) = S'_i. \quad (12)$$

The corresponding relation $h()$ of the existing data source is expressed as follows:

$$h\left(S'\{S'_a, S'_b \ldots S'_k\}\right) = I'_1\{I^{1'}_a, I^{1'}_b, I^{1'}_c \ldots I^{1'}_r\}. \quad (13)$$

Given that there is a mapping relationship between the data source and the original image, the set to be expanded can be correlated with the original image set, the mapping is as follows:

$$I_1' = h\big(f^{-1}(I_0')\big). \tag{14}$$

3.2 Replacement of Window

Because there are mapping relationship between set I_0' and I_1', the window could be replaced. After replacing each window in set $I_1'\{I_a^{1\prime}, I_b^{1\prime}, I_c^{1\prime} \dots I_r^{1\prime}\}$, we obtain the expanded collection $T'\{T_a', T_b' \dots T_r'\}$. The replaced image window will get more normalized terrain data rule information, namely precision expansion.

$$T'\{T_a', T_b' \dots T_r'\} = T'\{I_u^{0\prime}, I_v^{0\prime} \dots I_w^{0\prime}\}. \tag{15}$$

Since the replaced normalization window only represents the data rule information, it needs to be converted into the real data height. At this time, the normalized data needs to be restored.

The maximum value in matrix I is I_{Max} and the minimum value is I_{Min}. The process of recovering normalized matrix I' is as follows:

$$I = I' \times I_{Max} + I_{Min}. \tag{16}$$

The maximum value and minimum value of $I_1\{I_a^1, I_b^1, I_c^1 \dots I_r^1\}$ have been recorded. We obtain the expanded windows set $T\{T_a, T_b \dots T_r\}$ after recovering $T'\{T_a', T_b' \dots T_r'\}$. After the normalization operation is restored, the height difference rule information in the window is transformed into height data.

The windows containing data information in the expanded image window set are restored to the expanded DEM data images according to the original sliding order of the windows on the image to be expanded. It should be noted that since the original sliding step of the window on the image to be expanded is 1, there will be overlapping data parts in the replacement window after the window corresponds to the image (Fig. 3).

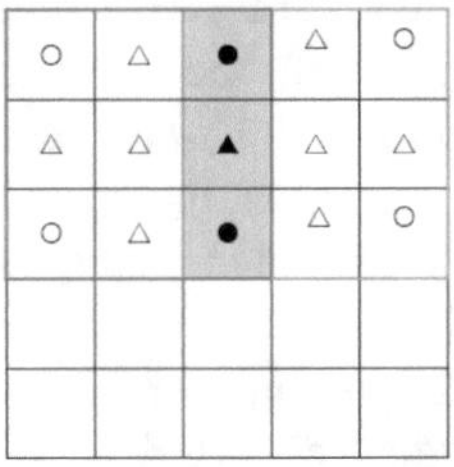

Fig. 3. The line box represents the window, the ○ represents the known vertex data, △ is expanded data after window replacement, the shaded part indicates that the window sliders overlap, the solid black part is the data of the overlapped part of the window. We take the average value for the overlapped data to get the final result.

Finally, the extended image points are re-assigned. Position mapping points exist in the expanded image and the image to be expanded. Since the entire window replacement will change all the extended images into predicted values, the data points on the image to be expanded need to be re-assigned to the expanded image according to the mapping relationship.

3.3   Small Scale Expansion Factor

Because the sampling multiple may be large and the sample data is too small, the multi-level piecewise sampling structure can be adopted to reduce the errors that may be caused by the prediction data. When calculating the enlarged image, the large scaling factor is way to reduce the input image. However, in this process, the original image features will become closer, and the isolated singularities in the unique geometric shapes may disappear.

In our experiment, we scaled the image by decomposing sampling multiples. GILAD et al. [14] explored the degree of maintaining local self-similarity under various scale factors. Large scale factor produces stronger smoothness. In the process of magnification, the image feature property is weakened globally and obvious artifacts will be generated. Therefore, we prefer to perform multiple multiplication steps of small factors to enlarge the image, so as to achieve the final desired magnification. We will also explore the impact of different scale factors on image magnification.

4   Result

In this paper, MATLAB software was used, 30 m elevation data value was adopted, and the image accuracy was gradually expanded to 10 m with this method. In this expansion, we use a gradually increasing ratio to perform the magnification.

In order to explore the effect of different scaling factors on image magnification, we try two methods to achieve precision expansion.

In the first method, we first sample the original image with the scaling factor of 1/2, and slide the window with size of 2 × 2 on low-resolution DEM image S into the data source as the example window. Similarly, the original DEM image I is slid to obtain the window with size of 2 × 2 in the set to be expanded. Each example window corresponds to the original picture window of 3 × 3. The enlarged image is enlarged by twice. Then we sample the original with the zoom factor 3/4, and get low-resolution DEM image S. Window with size of 3 × 3 is slid as an example window into the data source. The image we expand twice is as image to be expanded I. Since the points known on the original 10 m DEM precision map are still used as the window vertices of the extended set, we obtain the expanded set of windows with size of 3 × 3 using sliding step 2. According to the proportion coefficient of down-sampling, each example window corresponds to the window of size 4 × 4 in original image. The enlarged image will be enlarged by 3/2 times. After two steps, the extended image is three times more accurate than the original image.

In the second method, we still sample the original image with the scaling factor of 1/2. The size of the sliding example window is 2 × 2, so each example window corresponds to the window with size of 3 × 3 in original. It is known that this mapping method will double expand the image, and after we repeat this operation twice, the accuracy will be four times that of the original image. After that, we conduct down-sampling operation to obtain DEM elevation data images with an accuracy of 10 m.

After testing, we find that the second approach performs better. In the first method, during the down-sampling process of the original image, due to the scaling factor 3/4, the integer and linear average operations during the scaling will lead to severe image distortion after down-sampling. Taking the down-sampling image with severe distortion as the sample data source will lead to a large error, which is larger than that of the large expansion factor.

The extended 10 m DEM precision elevation data value was compared with the real elevation data of 10 m. Due to the need to calculate the monitoring early warning value when mountain flood occurs through the altitude data value, we mainly need to calculate the percentage of the deviation within 2 m in the image, as well as the standard deviation with the real data. The results are as follows:

Table 1. Comparison of three different methods

Interpolation method	Method 1	Method 2	Bilinear interpolation
Percentage	0.3543	0.5240	0.3751
Standard deviation	50.0057	42.2673	46.1299

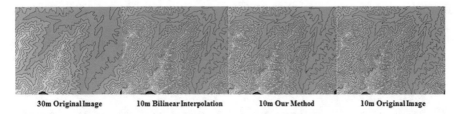

| 30m Original Image | 10m Bilinear Interpolation | 10m Our Method | 10m Original Image |

Fig. 4. This shows DEM images of three different methods. Elevation data of 30 m is expanded to elevation data of 10 m by gradually using different methods.

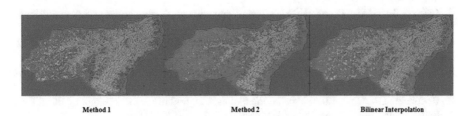

| Method 1 | Method 2 | Bilinear Interpolation |

Fig. 5. This shows special point identification DEM images of three methods. For 10 m DEM images using different methods, use the * symbol to identify points that differ from the actual original image by more than 10 m.

By comparing the data results, we found that our method is significantly better than the bilinear interpolation method. Most of the predictions within the height of 2 m deviation can be achieved, and small data deviation can be obtained (Table 1, Figs. 4 and 5).

5 Conclusion

We propose a new method of image enlargement based on sample DEM data, which is based on the self-similarity of land surface and expands the original image by taking the image itself as the sample sliding window. At the same time, this method adopts a multi-level piecewise up-sampling structure. This method decomposes the higher up-sampling coefficient into the product of multiple lower up-sampling coefficients, and realizes the process of multiple low-power up-sampling. The experimental results show that the method is effective and less bias is achieved.

In the future research, we consider to adopt more complex fractal rules to extract data information, so as to achieve better image accuracy expansion.

Acknowledgments. The research work described in this paper was fully supported by the National Key R&D program of China (2017YFC1502505) and the grant from the National Natural Science Foundation of China (61472043). Qian Yin is the author to whom all correspondence should be addressed.

References

1. Ebrahimi, M., Vrscay, E.R.: Solving the inverse problem of image zooming using "self-examples". In: Kamel, M., Campilho, A. (eds.) ICIAR 2007. LNCS, vol. 4633, pp. 117–130. Springer, Heidelberg (2007)
2. Mandelbrot, B.B.: The Fractal Geometry of Nature, vol. 51, p. 286 (1983). ISBN: 0-7167-1186-9
3. Mandelbrot, B.B.: Stochastic models for the Earth's relief, the shape and the fractal dimension of the coastlines, and the number-area rule for Islands. Proc. Natl. Acad. Sci. U. S. A. **72**, 3825–3828 (1975)
4. Mandelbrot, B.: How long is the coast of Britain? Statistical self-similarity and fractional dimension. Sci. **156**, 636 (1967)
5. Lathrop Jr., R.G., Peterson, D.L.: Identifying structural self-similarity in mountainous landscapes. J. Landsc. Ecol. **6**, 233–238 (1992)
6. Boming, Y.U., Jianhua, L.I.: Fractal dimensions for unsaturated porous media. Fractals **12**, 17–22 (2004)
7. Jin, Y., Li, X., Zhao, M., Liu, X., Li, H.: A mathematical model of fluid flow in tight porous media based on fractal assumptions. Int. J. Heat Mass Transf. **108**, 1078–1088 (2017)
8. Rouphael, T.J., Cruz, J.R.: A spatial interpolation algorithm for the upsampling of uniform linear arrays. IEEE Trans. Signal Processing **47**, 1765–1769 (1999)
9. Kirkland, E.J.: Bilinear interpolation. In: Advanced Computing in Electron Microscopy, vol. 4, pp. 103–115. Springer, Boston (2010)
10. Yu, L., Lin, W., Guangheng, N., Zhentao, C.: Spatial distribution characteristics of irrigation water requirement for main crops in China. Trans. CSAE. **25**, 6–12 (2009)

11. Woodard, R.: Interpolation of spatial data: some theory for Kriging. Technometrics **42**, 2 (2000)
12. Freeman, W.T., Jones, T.R., Pasztor, E.C.: Example-based super-resolution. In: IEEE Computer Graphics and Applications, vol. 22, no. 2, pp. 56–65. March–April (2002)
13. Xin, Z., Chenlei, L., Qian, Y., Ping, G.: A patch analysis based repairing method for two dimensional fiber spectrum image. Opt. Int. J. Light. Electron Opt. **157**, 1186–1193 (2018)
14. Jin, Y., Zhu, Y.B., Li, X.: Scaling invariant effects on the permeability of fractal porous media. J. Transp. in Porous Media. **109**, 433–453 (2015)

Automated Clash Free Rebar Design in Precast Concrete Exterior Wall via Generative Adversarial Network and Multi-agent Reinforcement Learning

Pengkun Liu[1], Jiepeng Liu[1(✉)], Liang Feng[2], Wenbo Wu[2], and Hao Lan[3]

[1] Department of Civil Engineering, Chongqing University, Chongqing, China
{pengkunliu, liujp}@cqu.edu.cn
[2] Department of Computer Science, Chongqing University, Chongqing, China
{liangf, wuwenbo}@cqu.edu.cn
[3] Chongqing University Industrial Technology Research Institute, Chongqing University, Chongqing, China

Abstract. The adoption of precast concrete elements (PCEs) are becoming popular in civil infrastructures. Since quality of connections determines the structure property, design of rebar in PCEs is a mandatory stage in constructions. Due to large number of rebar, complicated shapes of PCEs and complicated rules for arrangement, it is labor-intensive and error-prone for designers to avoid all clashes even using computer software. With the aid of BIM, it is desirable to have an automated and clash-free rebar design. Taking this cue, we introduce a framework with generative adversarial network (GAN) and multi-agent reinforcement learning (MARL) for generating design and automatically avoiding clash of rebar in PCES. We use GAN to generate 2D rebar designs. Then, 2D rebar designs are transformed into digital environments for MARL. In addition, layout of rebar is modelled as path planning of agents in MARL. An illustrative example is presented to test the proposed framework.

Keywords: Building information modeling · Generative adversarial network · Reinforcement learning · Multi-agent · Rebar design · Clash free · Precast concrete elements · Precast concrete exterior wall

1 Introduction

The adoption of precast concrete elements (PCEs) are becoming popular in civil infrastructures. Since the quality of connections between adjacent elements determines the structure property of precast structures, the design of rebar in PCES is a mandatory stage in construction projects [1]. Due to large number of rebar, complicated shapes of PCEs and complicated rules for arrangement of rebar, it is labor-intensive and error-prone for designers to avoid all collisions (hard clash) or congestions (soft clash) even using computer software [2]. In addition, current clash detection software like Autodesk Navisworks Manage and Solibri Model Checker, have realized detection and

© Springer Nature Switzerland AG 2020
T. Ahram (Ed.): AHFE 2019, AISC 965, pp. 546–558, 2020.
https://doi.org/10.1007/978-3-030-20454-9_54

visualization of the clash members [3]. However, the current software mainly focuses on the clash identifications of construction members after the design stage. It cannot automatically avoid the clash of the rebar or offer implementation resolution for solving clashes, which thus are lack of automatic arrangement in rebar design.

Recently, building information modeling (BIM) for structural engineering design makes process in the current Architecture, Engineering and Construction (AEC) industry. BIM technology allows us to represent the detailing of rebar digitally and transfer the detailing information to structural analysis software [4]. However, automated resolution of rebar clashes is lacking in the existing BIM software packages. Therefore, developing a framework for solving the problem of clash detection and resolution for automated design of rebar connects with the exiting BIM technology will be significant value in the AEC industry.

With the aid of BIM, several studies in the past have tried to solve the clash detection and resolution problem for rebar design, in the literature, Park [5] proposed a BIM-based simulator to automatically determine the sequence of rebar placement, and the clashes of rebar were identified by a developed application programming interface. Nevertheless, it focuses on the simulation of the placement sequence and the spatial clash has to be solved manually. Next, Navon et al. [6] proposed a system to detect rebar congestion and congestion. In fact, it provided a manual resolution for resolving limited clashes for rebar. Further, Radke et al. [7] proposed a demonstration system for mechanical, electrical and plumbing (MEP) systems. The system offers methods such as moving one of the entities to solve spatial conflicts and prompting users to choose a viable candidate. In fact, limited types of clashes are manually resolved in the system. Next, Wang et al. [3] carried out a knowledge representation for spatial conflict coordination of (MEP) systems. The clash knowledge representation included description, context, evaluation and management details. However, the developed presentation pattern only provided a documentation to store clash information without any clash resolution strategy for identified clashes. Mangal and Cheng [4] proposed a framework based on BIM and genetic algorithm (GA) to realize rebar design and avoid rebar clashes at RC beam-column joints. However, the proposed framework only offered clash resolution strategy for moving components by using GA and can only applied for regular shaped RC structures. In particular, the optimized path of rebar cannot bend to avoid the obstacles, which thus limits its practicability in real-world complex PCEs. The main drawbacks of existing approaches can be summarized as follows: (1) Due to the complex design codes of rebar, most of the above studies cannot meet design constraints after avoiding clash by moving one object. (2) Most of the above studies lack automatic and intelligent identification and resolution of rebar clash for real-world complex PCEs.

In machine learning, the Generative Adversarial Network (GAN) proposed by Goodfellow et al. [8] is a model framework in machine learning. It's specially designed to learn and generate output data with similar or identical characteristics. Pix2pix built by Isola et al. [9] is a modified version of the Generative Adversarial Network (GAN) that learns image data in pairs and generates new images based on the input. Since GAN is a powerful tool in dealing with image data, its application in architecture, especially in recognizing and generating architectural drawings is put forward [10]. Therefore, we applied pix2pix in recognizing and generating rebar design in construction detailing drawings for PCEs, marking structural components with different colors and then generating rebar design through two convolutional neural networks.

However, the generating rebar designs cannot avoid clash. In light of its strength, RL algorithms have achieved many important achievements in the field of complex adaptive systems such as mobile robot path planning. What's more, a multi-agent reinforcement learning (MARL) system [11] can be used as an efficient and effective tool for path planning problem-solving. Furthermore, the clash detection and resolution problem for the rebar design can be treated as a path planning of multi-agent in order to achieve automatic arrangements and bending of rebar to avoid obstacles. The similarity between the path planning of multi-agents and the arrangement of rebar, enlightens our work in this paper. Therefore, we propose a framework with GAN and MARL and BIM for automatically and intelligently provide clash resolution of rebar design in PCEs. To the best of our knowledge, this is the first modeling clash detection and resolution problem for the rebar design as a path-planning problem of multi-agent in the literature.

In particular, Pix2pix is used to generate the construction detailing drawings for precast concrete exterior walls (PCEWs) and the origins and targets of multi-agents are determined by generated results. Meanwhile, a MARL system is proposed for modifying rebar designs and avoiding rebar clash, and generate the three-dimensional design of rebar as per design code in PCEWs. In addition, the building information stored in BIM model about geometry and constraint conditions of PCEWs can be extracted from BIM models into the proposed framework. Furthermore, the design reward, punishment and some specific strategies in MARL are presented for build-ability constraints. Next, the generated design result of the layout of rebar in PCEWs are automatically visualized in BIM model in order to generate construction detailing drawings. To evaluate the efficiency and effectiveness of the proposed MARL, comprehensive simulations about PCEWs have been conducted. Lastly, the obtained results including the success rates confirm that the proposed system is effective and efficient.

The contribution of the present study can then be summarized as follows: (1) To the best of our knowledge, this is the first modeling clash detection and resolution problem for rebar design as a path-planning problem of multi-agent in the literature. (2) To achieve automatic rebar design in PCEWs, by employing Q-learning as the reinforcement learning engine, we design the particular form of state, action, and rewards for the reinforcement MARL. (3) To the best of our knowledge, this is the first using GAN to generate the construction detailing drawings of rebar design. (4) Comprehensive experiments on PCEWs are performed to verify the effectiveness of the proposed framework.

2 Preliminary

Section 2.1 introduces the basic module of reinforcement learning (RL). Section 2.2 presents the basic module of generative adversarial network (GAN). In Sect. 2.3, we describe the formulation of multi-agent path planning for rebar clash problem at PCEWs.

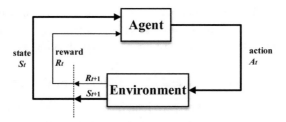

Fig. 1. Basic module of reinforcement learning

2.1 Introduction to Reinforcement Learning

A multi-agent reinforcement learning (MARL) system [11] can be developed as effective tools for path planning problem-solving. Reinforcement learning (RL) is a natural learning paradigm to both single-agent and multi-agents as presented in Fig. 1. It creates an autonomous agent that learns and then adjusts its behavior through the action feedback (punishment and reward) from the environment, instead of explicit teaching. Following the framework of a Markov decision process (MDP), a RL agent performs learning through the cycle of sense, action and learning [11]. In each cycle, the agent obtains sensory input from its environment representing the current state (S), performs the most appropriate action (A) and then receives feedback in terms of rewards (R) from the environment. It is important to note that how to turn a real-world environment into digital environment with clear reward signals is a key point to carry out RL.

2.2 Introduction to GAN (Generative Adversarial Network)

In the past years, Generative Adversarial Networks (GAN), as one type of machine learning algorithm, has achieved a lot of progress for generative tasks. Goodfellow *et al.* [8] were known as the first team to propose the GAN in machine learning. By providing training data in pairs, the program finds the most suitable parameters in the network so that the discriminator (D) has the least potential to distinguish the generated data (G) from the original data (Fig. 2).

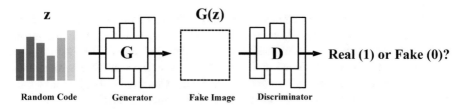

Fig. 2. Basic module of generative adversarial network

After the creation of GAN, Isola *et al.* [9] continued to work on pix2pix by generating a real photo from a partly-damaged photo, a colorful map from a black-and-white map, and an image with texture and shadow from a linear sketch. In pix2pix, the input D is a pair of images rather than a single image, and the task of D becomes the evaluation of whether those two images are the same or not [10]. After training, we can input an image and tell the program to generate the most possible corresponding output image.

2.3   Formulating Rebar Design as Path Planning of Multi-agent System

In particular, by treating each rebar as an intelligence reinforcement learning agent, we propose to model the rebar design problem as a path-planning problem of multi-agent system. It can be further modeled with a team of agents tasked to navigate towards defined targets safely across a PCEW that is gradually filled with obstacles which are rebars generated in the previous steps. In the task, agents can choose one of the five possible actions, namely, up, down, forward move, left, and right at each discrete time step. The task or objective of the agent is to navigate successful through the joint towards assigned targets within the stipulated time, without hitting any obstacle. With the proposed MARL, the three-dimensional coordinates of the clash free rebar design are then obtained by collecting the traces of the agents.

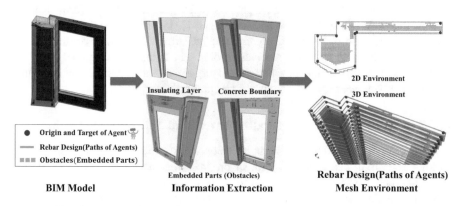

Fig. 3. Problem formulation and mesh environment processing for PCEW

2.4   Mesh Environment Processing

In addition, PCEWs have to be transformed into a digital environment that is suitable for MARL system. Furthermore, we transform the BIM model of PCEW into tessellated mesh environments approximating the geometry of the PCEW with known boundary conditions, as illustrated in Fig. 3. Then in tessellated mesh environment, a team of agents tasked to navigate towards defined targets safely. The origins and targets of agents in each mission are decided by generating results of GAN.

Each tessellated mesh dimension Di of environment is the dimension of a single square mesh, which can be calculated as:

$$Di = \min(d_c \text{ and } d_t). \tag{1}$$

Where d_c denotes the diameter of longitudinal compressive rebar, and d_t denote the diameter of longitudinal tensile rebar.

Therefore, the size of tessellated mesh environment Sz depends on Di and the dimension of PCEW,

$$Sz = floor(D/Di). \tag{2}$$

Where D denotes the dimension of PCEW, which can be length, width and height of RC members, and $floor()$ denotes the integer rounding down function in order to limit the range of Sz.

3 Proposed GAN with BIM for Generating Rebar Design in PCEWs

Since GAN is a powerful tool in dealing with image data, its application in construction, especially in generating construction detailing drawings has good potential for development. A process of training and generating between construction detailing drawings and its corresponding labeled maps was carried out by the author in Python and TensorFlow. In addition, to simplify the study, only a dataset of colorful PCEWs collected from the practical engineering was tested in order to remove the influence of varying scales and styles of the drawings.

In pix2pix, by providing training data in pairs, the program finds the most suitable parameters in the network so that the discriminator (D) has the least potential to distinguish the generated data (G) from the original data After training, we can input an image and tell the program to generate the most possible corresponding output image (Fig. 4).

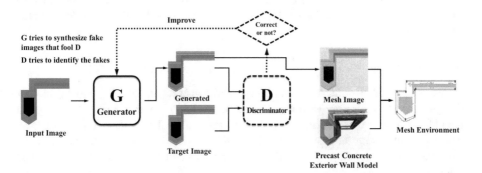

Fig. 4. Workflow of Pix2pix (modified GAN)

3.1   Labeling Principles

First of all, a labeling rule was created which uses different colors to represent areas with different functions (Fig. 5). Colors with RGB values of only 0 or 255 were commonly used in the labeling map in order to differentiate the labels as far as possible, so all together 4 combinations of RGB values can be achieved, which are used to two kinds of insulating layer, concrete and rebar. Because there are two kinds of insulating layer, so R:0 G:0 B:0 is used for A insulating layer and R:255 G:128 B:255 is used for B insulating layer (Fig. 5).

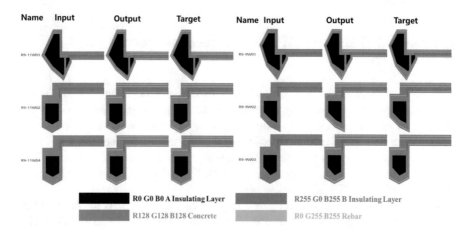

Fig. 5. Generating results by pix2pix

75 image pairs such as Fig. 5 were selected, sized to a fixed plotting scale, and carefully marked. Based on this dataset, map-to-design training (inputting color labeled maps and generating rebar design drawings), were tested and will be introduced in the following page.

3.2   Generating

By using color labeled maps as input images and construction detailing drawings (rebar design) as output images to train another network. When evaluating, the program should generate a rebar design according to the input labeled map. Figure 5 shows selected results from the testing set. All six selected images show clear generation of the rebar design including accurate positions and correct direction of rebar (Fig. 5). The high quality of these results is not surprising since there is not much uncertainty in the positioning of rebar in the training set.

In conclusion, the network has the potential to learn the rules of design effectively. Both the very certain rules that a design needs to follow and the uncertain situations that provide flexibility can be reflected by the network.

4 Proposed MARL with BIM for Solving the Path-Planning Problem of Clash Free Rebar Design in PCEWs

4.1 Architecture of Each Agent

The architecture of each agent takes the form of Q-learning [11], which has four modules: (1) **State**, (2) **Action** (3) **Reward** and (4) a **Q-Table** (Fig. 6). **State** module is for saving and representing current agent states, a module named **Action** is for representing the available actions and a feedback module **Reward** is representing the feedback values received from the environment and the internal states of an agent. **Q-Table** is a lookup table where agents calculate the maximum expected future rewards for action at each state.

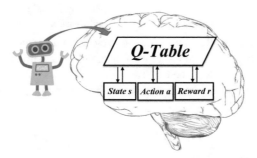

Fig. 6. Architecture of each agent

State Module. MARL system involves numbers of agent equipped with a set of sonar sensors that has a 180° forward view. Meanwhile, input attributes of sensory (state) vector consist of obstacle (path of other agent) detection, other agent position detection and the bearing of the target from the current position. Therefore, without a priori knowledge of the three-dimensional coordinate information of the obstacle and targets, each agent is equipped with a localized view of its environment (Fig. 7).

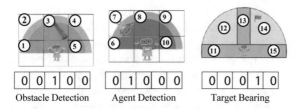

Fig. 7. Illustration of states in 2D

Action Module. In MARL system, the agent can choose one of the five possible actions (left, forward move, right, up and down at each discrete time step) (Fig. 8).

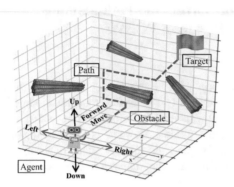

Fig. 8. Illustration of five possible actions in 3D

Fig. 9. Illustration of multi-agent path planning including reward, punishment and mission endings in 2D

Reward Module. In MARL, the design of reward, punishment and some specific strategies are presented for build-ability constraints. In particular, the reward and punishment strategies are described in Table 1:

A reward of +1 is given when the agent reaches the target without hitting obstacles

Table 1. Reward and punishment strategies for agents.

Reward and punishment strategies	
Reach targets without hitting obstacles	+1.0
The distance between agents and targets decreases	+0.4
Hit obstacles (paths of other agents)	−1.0
Hit other agents	−1.0
Within the range of other agents' paths	−1.0
Run out of time	−1.0
Take actions (left, right, up and down)	−0.5
Take action (forward move)	0

and running out of time. A reward of +0.4 is given when the agent takes action that can get close to the target to encourage agents search for defined targets. A punishment of −1 is given when the agent hits an obstacle (paths of other agents), collides with another agent or runs out of maximum time in order to avoid clash of rebar. A punishment of −1 is given when the agent moves into the specified range (1.5 × diameter of rebar) of paths or positions of other agents, therefore the spacing demand of rebar is satisfied. A punishment of −0.5 is given when the agent takes actions including left, right, up and down in order to make sure agent to move as straight as possible, therefore the layout of rebar is most likely to be a straight line unless obstacles are encountered. A reward of 0 is also assigned when the agent moves forward and does not find the target in the maximum allowable time.

Q-Table. Q-Table ("Q" for action-utility function) is a lookup table where agents calculate the maximum expected future rewards for action at each state. Specifically, this table will guide agents to the best action at each state. In terms of computation, this environment can be transformed into a table.

In the Q-Table, the columns will be the available actions. The rows will be the states. The value of each cell will be the maximum expected future reward for that given state and action. Each Q-table score will be the maximum expected future reward that the agent will get if it takes that action at that state. In order to learn and improve each value of the Q-table at each iteration during the iterative process, Q-Learning algorithm [11] is used.

4.2 Q-Learning Algorithm in MARL

The thought of Q-learning is that agents evolve from learning by a sequence of trials and adjust its behavior. The Q-learning algorithm in the proposed MARL is summarized in Algorithm 1. Given the current state S and a set of available actions A, the Q-Table is used to predict the value of performing each available action. The value functions are then processed by an action selection strategy to select an action. Upon receiving a feedback (if any) from the environment after performing the action, Q-learning is used for each agent learns the association from the current state and the chosen action to the estimated reward value.

Algorithm 1 The Q learning algorithm
Initialize an agent and the value function table $Q(s,a)$
Get current state s, for each action a_i in action set A:
$Q(s, a_i) = Predict(s, a_i)$
Select an action a_i in action set A:
$a_i = Scheme(Q(s,a))$
Perform a_{next}:
$(s', r) = Environment(Perform(a_{next}))$
where s' is the observed next state, r is the reward from environment when agent performs a_{next}
Estimate the value function $Q(s,a)$ by:
$Q(s,a)^{(new)} = Q(s,a) + \Delta Q$

In Step 3, we adopt an ε-greedy strategy in MARL to balance agent exploration and exploitation, which selects an action of the highest $Q(s,a)$ value with probability of $1 - \varepsilon(\varepsilon \in [0, 1])$, or takes a random action otherwise. In addition, it is beneficial to have a higher value of ε in the initial stage to encourage agents exploring new possibilities and a lower value of ε in the later stage to make agents performance converge gradually. Therefore, the value of ε is usually set gradually reduced over time.

The key to estimate the value function $Q(s, a)$ in Step 5 is using a temporal difference equation (bounded Q-learning):

$$\Delta Q(s, a_{next}) = \alpha TD_{err}. \tag{3}$$

where $a \in [0, 1]$ is the learning rate parameter, and TD_{err} is a value function of the current Q value predicted:

And using the TD formula in Q-learning, TD_{err} is computed by:

$$TD_{err} = r + \gamma max_{a'} Q(s', a') - Q(s, a_{next}). \tag{4}$$

Where r is the immediate-reward value, $\gamma \in [0, 1]$ is the discount parameter, and $max_{a'} Q(s', a')$ denotes the maximum estimated value of the next state s'.

4.3 Proposed MARL System

Algorithm 2 Pseudo Code of MARL System
Initialization: Generate the initial m agents
While (a mission ending conditions are not satisfied)
For each agent
If (agent dose not fail or not arrive the target)
Perform Q-learning algorithm
Else
Stop training
End If
End For
End While

The basic steps of the proposed MARL system are outlined in Algorithm 2. In the first step, a population of m agents is initialized. An agent fails when hitting obstacles, exceeding 30 sense-act-learn cycles (running out of time). A mission ends when all agents fail or arrive at the target successfully. A mission will also be deemed to have failed if an agent collides with another, as depicted in Fig. 9.

5   Illustrative Example

In this section, the empirical study is established to study the effectiveness of the proposed framework. One illustrative example about PCEW with 187 rebars as shown in Fig. 10(a) will be used to test the proposed framework. Furthermore, the tessellated mesh environment transformed from BIM model and the simulation results about paths of agents are shown in Fig. 10(b).

The automated 3D BIM outputs of the rebar in PCEW are given in Fig. 10(c). Design of rebars of the PCEW are based on the result of the proposed framework. It can be observed that there is no rebar clash in PCEW by clash detection of the 3D BIM output.

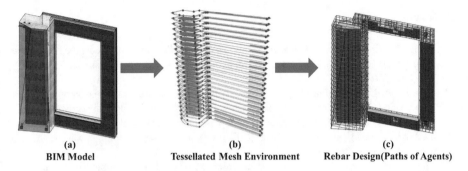

(a)   (b)   (c)

BIM Model   **Tessellated Mesh Environment**   **Rebar Design(Paths of Agents)**

Fig. 10. The simulation result of considered PCEW

Fig. 11. 3D output of rebar design in the considered PCEW

6 Conclusions and Future Research

In this paper, we model the clash detection and resolution problem for the rebar design in PCEs as a path planning problem of agents in order to achieve automatic arrangements and bending of rebar according to obstacles. Therefore, a framework via GAN and MARL with BIM has been proposed to identify and avoid rebar spatial clash in PCEWs. GAN is applied for generating rebar design in construction detailing drawings in PCEWs and the generating results are transformed into an environment for MARL. Next, according to MARL, the agent selects the suitable action and reaches the defined targets without hitting obstacles or running out of time. Subsequently, agents converge gradually to the global optimum along with the experimental training. Finally, the paths of agents are extracted to BIM model generating the rebar design. The simulation study in terms of the result of generating by Pix2pix and the paths of MARL have shown the effectivity and efficiency of the proposed system on the layout of rebar in PCEWs. However, the proposed framework via GAN and MARL with BIM still has a few limitations, such as it only applied for regular precast concrete exterior walls. Therefore, extending the system for more kinds of PCEs will be considered in the future work.

References

1. Wang, Q., Cheng, J.C.P., Sohn, H.: Automated estimation of reinforced precast concrete rebar positions using colored laser scan data. Comput. Aided Civ. Infrastruct. Eng. **32**(9), 787–802 (2017)
2. Tabesh, A.R., Staub-French, S.: Case study of constructability reasoning in MEP coordination. In: Construction Research Congress 2005: Broadening Perspectives (2005)
3. Wang, L., Leite, F.: Formalized knowledge representation for spatial conflict coordination of mechanical, electrical and plumbing (MEP) systems in new building projects. Autom. Constr. **64**(1), 20–26 (2016)
4. Mangal, M., Cheng, J.C.: Automated optimization of steel reinforcement in RC building frames using building information modeling and hybrid genetic algorithm. Autom. Constr. **90**, 39–57 (2018)
5. Park, U.: BIM-based simulator for rebar placement. J. Korea Inst. Build. Constr. **12**(1), 98–107 (2012)
6. Navon, R., Shapira, A., Shechori, Y.: Methodology for rebar constructibility improvement. In: Construction Congress VI, pp. 827–836 (2000)
7. Radke, A., Wallmark, T., Tseng, M.M.: An automated approach for identification and resolution of spatial clashes in building design. In: IEEM 2009-IEEE International Conference on Industrial Engineering and Engineering Management, pp. 2084–2088, Hong Kong (2009)
8. Goodfellow, I.J., Pouget-Abadie, J., Mirza, M., Xu, B., Warde-Farley, D., Ozair, S., et al.: Generative adversarial networks. Adv. Neural Inf. Proc. Syst. **3**, 2672–2680 (2014)
9. Isola, P., Zhu, J.Y., Zhou, T., Efros, A.A.: Image-to-image translation with conditional adversarial networks (2016)
10. Huang, W., Hao, Z.: Architectural drawings recognition and generation through machine learning. In: Proceedings of the 38th Annual Conference of the Association for Computer Aided Design in Architecture, Mexico City, Mexico (2018)
11. Feng, L., Ong, Y.S., Tan, A.H., Chen, X.S.: Towards human-like social multi-agents with memetic automaton. In 2011 IEEE Congress of Evolutionary Computation (CEC), pp. 1092–1099, June 2011

Cognitive Ergonomic Evaluation Metrics and Methodology for Interactive Information System

Yu Zhang[1], Jianhua Sun[1], Ting Jiang[2(⊠)], and Zengyao Yang[1]

[1] Department of Industrial Design, School of Mechanical Engineering,
Xi'an Jiaotong University, Xi'an 710049, China
{zhang.yu,yangzengyao}@xjtu.edu.cn,
sunjianhua0810@foxmail.com
[2] National Key Laboratory of Human Factors Engineering,
China Astronauts Research and Training Center, Beijing 100094, China
jting@aliyun.com

Abstract. In the face of complex information interactive system, it is essential to evaluate products achieve system performance within users cognitive capacity. Most of the research about ergonomic evaluation mainly focus on the macro ergonomic method, which not focus on concrete design problem at the micro level. This paper focuses on how to identify and predict cognitive ergonomic problems based user action and cognitive model and establishes the mapping relationship between cognitive ergonomic problems and real-time continuous measured data in order to let the evaluation results play a direct role in the design. The methodology was applied to evaluate the ergonomic quality of IETM used by astronauts in the space station, which including make flight plans, do experiments, in-orbit maintenance, and so on. A series of standardized evaluation procedures were designed to explore the possibility of remote ergonomic measurement for long-term orbiting operation.

Keywords: Cognitive ergonomics · Evaluation · Quantitative analysis · Eye-tracking · Interactive system

1 Introduction

Ergonomics has increasing become a critical factor in work-related safety and efficiency. Ergonomics consists of physical, cognitive, and organizational ergonomics, thus applying to all aspects of human activity [1]. Most of the early researches focus on physical activities, but now we are turning more attention to the cognitive element with information technology development and increasing complexity of human-machine interactive systems [2]. Cognitive ergonomics refers to how mental processes take place and is associated with memory, sensory-motor response, and perception [3]. Ergonomic evaluation is used to evaluate whether it can achieve a user's desired goals by a user using products and whether it can achieve system performance by human-machine synergy. We should think about definitely human's needs, limits, and characteristics, to assure complete tasks productively.

© Springer Nature Switzerland AG 2020
T. Ahram (Ed.): AHFE 2019, AISC 965, pp. 559–570, 2020.
https://doi.org/10.1007/978-3-030-20454-9_55

Generally, a software system can be evaluated concerning different aspects, for example, functionality, reliability, usability, efficiency, maintainability, portability [4]. To assess a user's performance with the software system, we concentrate on the aspect of usability from an ergonomic point of view. This aspect has gained particular importance during the last two decades with the increasing use of interactive software. It aims to assess a system's quality for specific users, for specific tasks, and in a specific environment [5]. Functional, task-oriented and user-oriented issues are to be covered by an evaluation, where the user-oriented perspective is the focus of the consideration.

In this paper, we focus on the interactive information system used by astronauts working and living in the space station, which including electronic manuals, log management, material management, and so on. Especially the Interactive Electronic Technical Manual (IETM) has become an indispensable supporting system for daily tasks. The electronic manual can be regarded as the astronauts' operational guide to complete flight plans, individual experiments, in-orbit maintenance and a series of functions such as complementary interactive tools. It plays a crucial role in improving the work efficiency of astronauts and ensuring the completion of space missions. However, the user demands and interface usability of handheld IETM systems have not received enough attention. For example, when locating fault information, how to display information can help astronauts quickly understand and make decisions; when searching information, how to set up a navigation catalog can help astronauts promptly found the information they need; when scanning data, how to make it easy for users to understand the relationship among different data; and so on. These usability issues fall under the category of cognitive ergonomic evaluation.

The study aims to establish a set of ergonomic evaluation metrics and methods to figure out design problems and to improve the usability of the interactive information system in complex task environments. Firstly, it is essential to analyze ergonomic requirements in the space station. We analyzed cognitive ergonomic requirements in terms of user's action process and cognitive characteristics. Secondly, we present an experimental study to structure a framework of cognitive ergonomic evaluation in terms of multi-source data measurement. The multi-source data came from the user's behavior recording, software logging, and physiological parameters. Thirdly, we design ergonomic evaluation experiments for typical information interactive systems. Our research focuses on how to identify and predict human errors with continuous real-time data and quantitative analysis. Respective metrics could serve as likely markers for usability problems. As well as a set of real-time recording and analyzing tools have been developed based on multi-dimensional measurements such as physiological signals, logging, and behavioral parameters. We established the mapping relationship between objective measurement data and usability issues. Finally, we discuss the implications of our findings for the ergonomic evaluation methodology.

2   Related Research

Over the past decades, studies on ergonomic evaluation of human-system information interaction focus on the establishment of evaluation criteria, the improvement of evaluation methods and techniques used in the evaluation process to improve and assure easy-to-use user interfaces and systems.

In ISO/IEC9126-1, Usability as one of the software quality characteristics is defined "A set of attributes that bear on the effort needed for use, and on the individual assessment of such use, by a stated or implied set of users", which including understandability, learnability, operability, and attractiveness [4]. Currently, the usability evaluation criteria of software are mainly based on the ISO 9241 - efficiency, effectiveness, and satisfaction as the evaluation criteria of usability [5], and usability frameworks that be established by some ergonomic experts, such as Nielson defined "Usefulness" as learnability, efficiency, memorability, errors, and satisfaction [6], Shackel defined Usability as effectiveness, learnability, flexibility and attitude (acceptable and satisfaction) [7]. Usability engineering has developed more than 20 kinds of methods, including performance measurement, cognitive walkthroughs [8], heuristic evaluation [6], thinking aloud protocol, feature inspection, lab observation, questionnaire, etc. There are many challenges of measuring usability, such as to distinguish and empirically compare subjective and objective measures of usability, to use both micro and macro tasks and corresponding measures of usability, and so on [9].

Mental workload (MWL) is another construct greatly invoked in Human Factors. Mental workload is defined as"the total of all assessable influences impinging upon a human being from external sources and affecting it mentally" [10]. Mental workload is the necessary complement that applied to evaluate and predict human performance for designing the interaction of the human with technological devices, interfaces, and systems [11–13], as well as another way to measure the usability and acceptability of human-machine system [14]. Overload and underload may have short-term consequences and long-term consequences, such as on the one hand, when MWL is at a low level, the human may feel boring and monotonous, on the other hand, when MWL is at a high level, the human may feel extremely frustrated, and pressured [15]. Many measurement methods and evaluation criteria have been established around the mental workload. They include subjective evaluation scale [16], such as NASA-TLX [17], SWAT [18], MRQ [19], etc.; behavioral index, which assess the performance of the operator's behavior in the task, such as task completion time, error rate and reaction time [20]; and physiological index including brain function, eye function, such as heart rate, pupil dilation and movement and so on [21]. O'Donnell proposed the related criteria for workload measurement methods: sensitivity, diagnosticity, interference, equipment requirements and operator acceptance [22]. Mental workload and usability are two none overlapping constructs, and they can be jointly employed to evaluate and predict human performance [9].

Scapin and Bastien presented a set of ergonomic criteria for interactive systems consists of eight main criteria as followed: guidance, workload, explicit control, adaptability, error management, consistency, the significance of codes, and compatibility [23].

There are some methodological variations for ergonomic evaluation in the area of interactive systems. They cover more ground than the definitions of usability and focus on different aspects of human performance. The methods may be broadly classified as quantitative or qualitative approaches. The quantitative methods predict the speed of performance (e.g., critical path analysis), errors (e.g., systematic human error reduction and prediction approach and task analysis for error identification), and speed and errors (e.g., observations). The qualitative methods predict user satisfaction (e.g., questionnaires and repertory grids), device optimization (e.g., checklists, link analysis, and layout analysis) or user and device interaction (e.g., heuristics, hierarchical task analysis, and interviews) [24].

However, there are still limitations of criteria and methods for evaluating interactive systems in the complex situation as follows,

(I)   Ergonomic issues cannot be reduced to sets of dimensions and recommendations [23], most of these criteria are 'expert-oriented'facing to evaluating, which not accord with user's mental process, especially overlook the cognitive and emotional state of the users at the time of operating the product [25]. It is difficult to establish 'user-oriented' evaluation process in terms of these criteria.

(II) Many criteria lack a unified structure so that it is difficult to consistency between different measurement methods [26]. We believe that a more comprehensive framework of evaluation factors needs to be established based on user action and cognitive models.

(III) As well as most of the traditional ergonomic testing in a lab is not adequate for the real working environment, which emphasizes the influence of the situational factors on users. In recent years, researchers tend to adopt an integrated method with combining advanced measurement techniques to collect user data noninvasively and continuously and quantitative analysis in different phases of the user-centered design and development process. Eye tracking [27], short-term or long-term logs study and remote testing have been widely used. However, there are still many problems to be resolved. For example, eye tracker or logs data mainly evaluated the performance of user operational or physiological criteria. It is impossible to identify cognitive ergonomic problems accurately from the individual data, and recording user data for long-term and mining data in practical scenarios has always been a technical difficulty.

3   Methodology

3.1   Cognitive Ergonomic Criteria

Taking IETM as an example, the astronauts need IETM as an assistant tool to check flight procedures, review operational guides, and synchronizing data, etc. So common cognitive ergonomic requirements of users depend on each cognitive stage of the tasks as shown in Table 1. We need to take into account both action factors and cognitive factors during the ergonomic evaluation. In terms of the common cognitive ergonomic requirements, the cognitive ergonomic criteria can be described as two categories:

design-oriented and user-oriented. To ensure the consistency and reliability between design and evaluation, we present the design-oriented criteria involved the main design principles of human-system interaction, as well as the user-oriented criteria, included the main errors occurred in the user's operation processing. The categories of the user errors as shown in Table 1 cover the perception and cognitive characteristics with the action mainline.

Table 1. Cognitive ergonomic criteria.

Cognitive ergonomic requirements	Evaluation criteria	
	Design-oriented	User-oriented (Cognitive ergonomic problems)
Intention	Functional directivity	Intention transformation error
Planning-searching	Interface layout logicality	Function entry not be found
Planning-identification	Controls saliency	Location of the command not be found
Planning-understanding & memorizing	Meaning consistency	Icon/function meaning not be understood
Implementation-searching & identification	Controls guidance quality	Next step not be found
Implementation-understanding & memorizing	Understandability	Operational command not be understood
Feedback	Navigability	Insufficient feedback information

3.2   Statistical Variables

To match the relationship between cognitive ergonomic problems and continuous data, we defined a set of representative variables in terms of related user-oriented evaluation criteria. The study selected eye-tracking data and behavioral log data as measure variables. Eye tracking measures are an efficient way to measure monitoring performance as previous research has found associations between cognition and eye tracking behavior [28]. There are seven statistical variables were selected to help identify the ergonomic problems according to the eye-tracking data and user interactive data collected in the experiment, as showed in Table 2.

(I)   Two pre-defined data by tasks: the number of standard steps, and the number of function entry steps;

(II)   Three user behavior data: the number of user's steps reached, the number of user operational chains, and operational chains where the function completed;

(III)   Two eye-tracking data: the number of fixation points in interest areas, and the average fixation duration in AOI.

Table 2. Statistical variables.

Statistical variables	Sign	Definition	Approaches	types
Standard steps	Cstd	The number of standard operations required from start to complete an operational command	Pre-defined by tasks	discrete variable
Function entry steps	Estd	The number of standard operations required from start to the function entry	Pre-defined by tasks	discrete variable
User's steps reached	Cact	The number of user actual operations from start to complete task.	Count user clicks	discrete variable
Number of user operational chains	N	The user from start to end or by pressing the "Return" button is referred to as an operation chain.	Count user clicks	discrete variable
Operational chains where the function completed	n	The chain number at which user operates a complete operation command	Count user clicks	discrete variable
Number of fixation points in AOI	Ef	It indicates that the user is aware of the target command	Eye-tracking	discrete variable
Average fixation duration in AOI	Et	The average duration of each gaze point in AOI	Eye-tracking	continuous variable

3.3 Discriminant

A mapping relationship between the seven types of ergonomic problems and statistical variables. Figure 1 shows these steps involved in applying a set of discriminants (as Table 3 shown) to classify the cognitive ergonomic problems in a task.

3.4 Experiment Design

Participants. Three participants aged 30–50 years old participated in the experiment. The participants carried out 30-minute training for the purpose and operation of IETM and one-hour training for the use of eye tracker and the click data acquisition program, an assessment was also made on the use of experimental equipment, every participant reach to the standard.

Experiment materials and devices. The experiment was conducted on three Pads with 1600×2560 pixels. An eye-tracker (Tobii Glasses 2) will be used for recording the participants' gaze data at a sampling rate of 50 Hz. Click data Acquisition Program will be used to collect the real-time clicking data. EV screennap will be used to record the task operating video.

Experiment tasks. This experiment has designed ten tasks according to the IETM daily tasks as Table 4 shown.

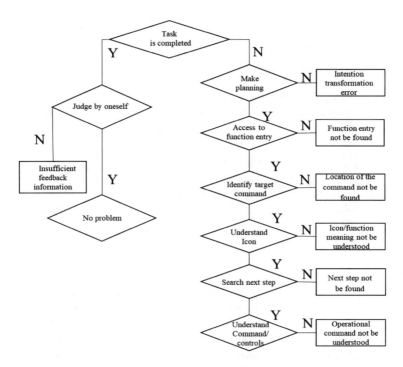

Fig. 1. Cognitive ergonomic problems discrimination flow chat.

Table 3. Statistical variables.

Cognitive ergonomic problems	Discriminant
Intention transformation error	Cact-Cstd < 0 and n-N = 0
Function entry not be found	Cact-Estd < 0
Location of the command not be found	Cact-Estd = 0 and(Ef = 0 or Et < 100)
Icon/function meaning not be understood	Cact-Estd = 0 and Ef > 0 and Et > 100
Next step not be found	Cact-Estd > 0 and(Ef = 0 or Et < 100)
Operational command not be understood	Cact-Estd > 0 and Ef > 0 and Et > 100
Insufficient feedback information	Cact-Cstd = 0 and n-N < 0

3.5   Data Collecting and Processing

For the eye-tracking data collected by the Tobii Pro Glasses Controller (x86),we use the FixationAnalysis 2.0 software to draw the AOI for the function control and check the number of fixation points in the AOI to determine whether the user percepted icons or words of the control or not.

The IETM is based on the Android platform. A collecting program coded by Python, call the "adb shell getevent -tt | findstr 00[237a][234569ef] >" to gather the real-time click hexadecimal coordinates based on the screen.

Table 4. Experiment tasks

Task number	Task description
1	Open the IETM and log in with your own account
2	Check your daily plan 5 days later
3	View the to-do tasks of the simulation cabin running for the 46th day
4	Next to the task 3, see if the third task in the to-do task list has an operation guidelin, and if so, open it
5	Check out today's group master plan
6	Check your schedule for this week
7	Check your schedule for this month
8	View the fault countermeasure of the alarm monitoring experiments
9	Check the Catalogue of Experiment Catalogues
10	Log out

The video, recorded by EV screencap is the mp4 format, FFmpeg is the leading multimedia framework, able to record, convert audio and video. OpenCV provides the resize function to change the size of the image. We called the FFmpeg commend line "ffmpeg -ss positon -i filename -frames:v 1 -avoid_negative_ts 1 [out].png" circularly to output a series of specified timestamp frame as JPG. Then adjust the image resolution to 1366 × 768 through the "res=cv2.resize(img,(1366,768), interpolation=cv2.INTER_CUBIC))". Finally the red circle which present the clicking coordinates on the JPGs was drawn through the code "res=cv2.resize(img,(1366,768), interpolation=cv2.INTER_CUBIC))".

As shown in the Fig. 2, a series of specified frame as JPG format with clicking coordinates, the name of each image consist of the click type and the click time were output, then the task operating path can be output.

ietm-2-01-2018
-09-10_09_42_3
2.mp4

297.314-home-button.jpg   299.422-ST.jpg   304.763-ST.jpg

Fig. 2. Visualization of task operating path.

4   Results and Discussion

Comparing user's operation path with standard operation path, classifying the discovered ergonomic problems according to the discriminant, the user error classification is as Fig. 3 shown.

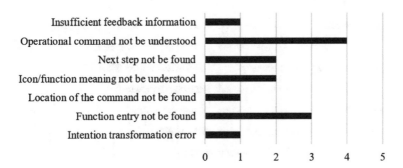

Fig. 3. Cognitive ergonomic problems classification.

The accordance ratio of discriminant by the aboved mentioned methods and experience judgment by ergnomical experts is 92.85%. There is a difference in the classification of one type of problem:

For task 5, we found that inconsistent classification was found in the operation of participant 02. The result of discriminant is "Icon/function meaning not be understood", but the result of experience judgment is "Location of the command not be found". Through the eye-movement data analysis of operation for taske 4, we found that the eye-tracking data of participant 02 have fixation on the AOI, but there is no further operation, so this error can not be classified as "Can't find the control" but should be classified as "Icon/function meaning not be understood". In view of whether users can see this control, the reliability of checking whether there have fixation point in the AOI is higher than the experience judgment.

Because of the difference between tasks difficulty, operation, and user's operation characteristics, it is necessary to establish a unified index for each user's operation of each task. We can use the unified index to compare and evaluate the operation of different users' tasks, rating the severity of errors. Tasks with a high degree of severity should be improved preferentially.

Smith proposed the concept of "Lostness" in the usability study of websites [29]. This paper makes some modifications to the " Lostness." The modified formula is as follows, covering the extent to which the task is approaching success, N represents the number of unique controls being clicked, S represent the number of controls Clicks.The modified formula is as follows, covering the extent to which the task is approaching success, N represents the number of unique controls being clicked, S represent the number of controls Clicks.

$$Lm = sqrt[\ (N/S - 1)^2 + (Cstd/Cact - 1)^2\ ]. \tag{1}$$

According to the formula, L is a value between 0 and $\sqrt{2}$. The lower the Lm value, the smaller the user's lostness. Ideally, the L value is 0, representing the user to complete the task successfully without extra steps (Fig. 4).

From the above figure, we can see that the lostness reflects the difficulty of users to complete the operation task. The higher the lostness, the more impossible it is for users to complete the task.

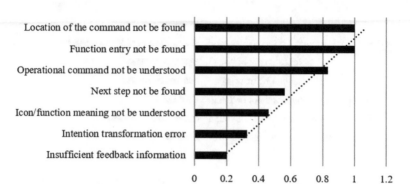

Fig. 4. The lostness (average) of the cognitive ergonomic problems.

The lostness reflects the difficulty of the user to complete the task. The higher lostness represents that the user can not complete the task, the worse the user experience.

Combined with the error rate and the average lostness of the task, we can grade the severity of the task, and the task with higher error rate and higher lostness can be selected to improve and iterate, from Fig. 5, we know task 4, task8, task9 needs to be modified firstly.

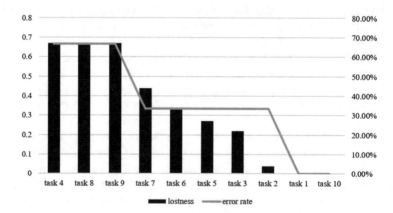

Fig. 5. The lostness and the error rate.

5   Conclusion and Future Research

By task testing, this paper introduces user eye movement data and click data, proposes a user experience measurement method based on user eye movement and interactive manipulation, classify the user errors, verify the effectiveness of the discriminant classification through the experienced judgment.

The coincidence ratio between discriminant classification and the experienced judgment is 92.85%, indicating that the discriminant method can largely reflect the cognitive characteristics and action characteristics of user's operation. At the same time, through the analysis of the task existed discriminating differences, it is found that the method can eliminate the impact of individual cognitive differences.

This study verified the rationality of data collection and data analysis process and preliminarily accumulated experience in data collection and analysis of on-orbit software's long-term tracking test in a closed environment. In the following, the process and method of on-track implementation of this project should be further studied and designed, and the synchronous collection and analysis method of recorded screen data and click data should be discussed to improve the efficiency of data processing.

Acknowledgment. This research has been supported by the Open Funding Project of National Key Laboratory of Human Factors Engineering, Grant NO. SYFD170051809 K.

References

1. What is ergonomics? International Ergon Associates (IEA) (2012). http://iea.cc/01_what/What%20is%20Ergonomics.html
2. Hollnagel, E.: Cognitive ergonomics: it's all in the mind. Ergonomics **40**(10), 1170–1182 (1997)
3. Berlin, C., Adams, C.: Production Ergonomics: Designing Work Systems to Support Optimal Human Performance, pp. 83–106. Ubiquity Press, London (2017)
4. ISO/IEC 9126. Information technology - software product evaluation – quality characteristics and guidance for their use. ISO/IEC (1991)
5. ISO 9241. Ergonomic requirements for office work with visual display terminals, Part 8. Requirements for displayed colors. ISO (1994)
6. Nielsen, J., Mack, R.L.: Heuristic evaluation. In: Usability Inspection Methods. Wiley, New York (2010)
7. Shackel, B.: Usability- context, framework, definition, design and evaluation. Interact. Comput. **21**(5–6), 339–346 (2009)
8. Lewis, C., Polson, P., Wharton, C., et al.: Testing a walkthrough methodology for theory-based design of walk –up-and-use interfaces. In: CHI'90 Proceedings of the SIGCHI Conference on Human Factors in Computing Systems, pp. 235–242. ACM, New York (1990)
9. Longo, L.: Experienced mental workload, perception of usability, their interaction and impact on task performance. PLoS ONE **13**(8), e0199661 (2018). https://doi.org/10.1371/journal.pone.0199661
10. ISO 10075-1:2017. Ergonomic principles related to mental workload – Part 1: general issues and concepts, terms and definitions. ISO (2017)
11. Young, M., Brookhuis, K., Wickens, C., Hancock, P.: State of science: mental workload in ergonomics. Ergonomics **58**(1), 1–17 (2015)
12. Hancock, P.: Whither workload? Mapping a path for its future development. In: International Symposium on Human Mental Workload: Models and Applications, pp. 3–17. Springer (2017)
13. Wickens, C.: Mental workload: assessment, prediction and consequences. In: International Symposium on Human Mental Workload: Models and Applications, pp. 18–29. Springer (2017)

14. Ariza, F., Kalra, D., Potts, H.W.: How do clinical information systems affect the cognitive demands of general practitioners? Usability study with a focus on cognitive workload. J. Innov. Health Inform. **22**(4), 379–390 (2015)
15. Cain, B.: A review of the mental workload literature. In: Defence Research & Development Canada, Human System Integration (2007)
16. Rubio, S., Diaz, E., Martin, J., Puente, J.M.: Evaluation of subjective mental workload: a comparison of SWAT, NASA-TLX, and workload profile methods. Appl. Psychol. **53**(1), 61–86 (2004)
17. Hart, S.G.: NASA-task load index (NASA-TLX); 20 years later. Proc. Hum. Factors Ergon. Soc. Annu. Meet. **50**(9), 904–908 (2006)
18. Reid, G.B., Nygren, T.E.: The subjective workload assessment technique: a scaling procedure for measuring mental workload. Adv. Psychol. **52**, 185–218 (1988)
19. Boles, D.B., Bursk, J.H., Phillips, J.B., Perdelwitz, J.R.: Predicting dual-task performance with The Multiple Resources Questionnaire (MRQ). Hum. Factors **49**, 32–45 (2007)
20. Eggemeier, T., O'Donnell, R.: A conceptual framework for development of a workload assessment methodology. In: Defense Technical Information Center OAI-PMH Repository (United States) (1998)
21. Charles, R.L., Nixon, J.: Measuring mental workload using physiological measures: a systematic review. Appl. Ergonomics **74**, 221–232 (2019)
22. O'Donnell, C.R.D., Eggemeier, F.T.: Workload assessment methodology. In: Measurement Technique, Ch42, pp. 42-5 (1986)
23. Dominique, L.S., Christian Bastien, J.M.: Ergonomic criteria for evaluating the ergonomic quality of interactive systems. Behav. Inf. Technol. **16**(4–5), 220–231 (1997)
24. Neville, A.S., Mark, S. Y., Catherine, H.: Guide to Methodology in Ergonomics Designing for Human Use, pp. 9–76. Taylor & Francis, London (2014)
25. Romaric, M., Andre, W.K., Marie-Catherine, B.Z., Elizabeth, M.B.: Insights and limits of usability evaluation methods along the health information technology lifecycle. Stud. Health Technol. Inform. **210**, 115–119. EEMI (2015)
26. Hornbaek, K.: Current practice in measuring usability: challenges to usability studies and research. Int. J. Hum. Comput. Stud. **64**(2), 79–102 (2006)
27. Alper, A., Duygun, E.B., et al.: Evaluation of a surgical interface for robotic cryoablation task using an eye-tracking system. Int. J. Hum. Comput. Stud. **95**, 39–53 (2016)
28. Imbert, J.p., Hodgetts, H.M., Parise, R., Vachon, F., Dehais, F., Tremblay, S.: Attentional costs and failures in air traffic control notifications. Ergonomics **57**, 1817–1832. (2014)
29. Smith, P.A.: Towards a practical measure of hypertext usability. Interact. Comput. **8**, 365–381 (1996)

A System for User Centered Classification and Ranking of Points of Interest Using Data Mining in Geographical Data Sets

Maximilian Barta[✉], Dena Farooghi, and Dietmar Tutsch

Bergische Universität Wuppertal, Rainer-Gruenter-Str. 21,
42119 Wuppertal, Germany
{barta,dfarooghi,tutsch}@uni-wuppertal.de

Abstract. In this paper we propose a system to automatically extract and categorize points of interest (POIs) out of any given geographical data set. The system is then modified for a user centered approach to work in a web environment. This is done by customizing the order, amount and contents of the previously created categories on a per user basis in real time. The aim of this system is to provide users with a more flexible and less error prone approach to POI data that can then be used in geographical routing and navigation applications, replacing the conventional existing solutions that need to be manually administrated. The generated results are validated using preexisting, manually created, point of interests and their corresponding categories.

Keywords: Data mining · Geographical data sets · Navigation ·
Points of interest · Systems engineering · User centered · Web technologies

1 Introduction

As mobility is an ever increasing topic in society, not only the means of transportation need to be evolved, but also the way of how humans and vehicles are navigated in the always changing urban and rural environments. One such component of navigation can be the so called points of interest (POIs) - a POI is a location that may be of certain interest, like a gas station, a charging point for electrical vehicles, a supermarket, etc. This paper will deal with the identification and classification of these locations using methods from the interdisciplinary field of data science, and will describe a system that is able to automatically generate new categories based on the POI usage in a navigation application. Conventional POIs have to be categorized manually, meaning that given data sets had to be enriched by humans who defined what location may or may not be a point of interest; also the categorization or classification of such points has to be performed by hand. Obviously this can lead to errors and in some cases may also unintentionally incorporate the subjective opinion of the editor. The purpose of this study is to minimize errors made by the human factor, while at the same time provide additional benefits to the users and navigation solutions such as automatic updates on map data information, and identifying new, previously unconsidered, POI categories. In this paper we will propose a method on how to extract POIs and their categories

© Springer Nature Switzerland AG 2020
T. Ahram (Ed.): AHFE 2019, AISC 965, pp. 571–579, 2020.
https://doi.org/10.1007/978-3-030-20454-9_56

using a data mining, or more precisely a text mining, approach based on the geographical map data set of the OpenStreetMap (OSM) project. An exemplary implementation for the region around the administrative district of Düsseldorf (Germany) is conducted and results for the implementation will be presented. The generated results, as well as the methods used for data preprocessing and the data mining methods are discussed. The implementation is integrated into a web based navigation system that allows uniquely identifying users. The focus thereby lies at creating a per user experience: Not only the number of routes, but also the start points, optional intermediate targets and end points a user sets for each individual route, will have an impact on how these points (which for example lie on or near a user selected point) are prioritized, categorized and presented to the user. To achieve this, each such user interaction will be logged and a model is created that can aggregate the collected data to create personalized POI categories.

2   Methodology

In order to extract POIs and their categories out of the given geographical data set, several data mining methods, discussed in Sect. 2.2, are applied. However before data mining algorithms can be applied to the data set, the data needs to be transformed into a compatible format – these methods are explained in Sect. 2.1.

2.1   Data Preprocessing

As data sets based on the OSM project are used, the data format worked upon is based on the human readable XML-format (see Fig. 1). In this study we focus on the *tag*-elements of each existing *node*-element respectively. OSM data is user generated, this means that error can and will occur. By choosing robust techniques, errors (such as typos, language mismatches or wrongly formatted data) in the data set can be, at least in part, identified and corrected.

```
<node id="616996551" visible="true" version="4" changeset="…"
    timestamp="2014-02-22T15:28:53Z" user="…" uid="…"  lat="51.2395813"
lon="7.1631659">
    <tag k="highway" v="bus_stop"/>
    <tag k="name" v="Rainer-Gruenter-Straße"/>
    <tag k="operator" v="Wuppertaler Stadtwerke (WSW)"/>
    <tag k="shelter" v="no"/>
</node>
```

Fig. 1. Example of an element (node) in the XML-based OSM-format. Personal data has been replaced with "…".

To transform the data into a format that can be used by the data mining algorithms a slightly modified procedure as proposed by Han et al. [1] is used, comprising of the following four steps: Selection, Cleaning, Integration and Transformation.

Selection. The contents of interest are determined in this step, this means that the entire data set is filtered for *node*-elements, afterwards each *node* is isolated (i.e. viewed as an entity) and the relevant attributes such as the id, geo location and tags are extracted.

Cleaning. Stop words [2] (i.e. *a, but, is, the, to,* etc.) as well as unnecessary characters, numbers and punctuation are removed. Capitalization is also normalized. Also stemming – in this case the Porter stemming algorithm [3] is used – is applied to the resulting data.

Integration. The data is combined for a number of regions relevant for further analysis and split into subsamples.

Transformation. The resulting documents are tokenized and presented as a matrix of words, a bag-of-words model as well as a term frequency – inverse document frequency matrix [4, 5].

Regarding the size of the data set and the runtime of the aforementioned procedures only the data set for the district of Düsseldorf (Germany) with a set size of 2,888 MiB is processed, as compared to 53,297 MiB for the data set of the entire federal state of Germany.

Overall the resulting matrix of words contains 79,978 proto-POIs (all *nodes* that remained after the preprocessing steps) containing 1,416 unique words.

The bag-of-words model therefore spans a matrix of 79,978 * 1,416 entries – see Table 1 for the top ten words in the bag-of-words model.

Table 1. Top ten words in the data set for the district of Düsseldorf (Germany) after preprocessing has been applied (note that words have been transformed due to the applied stemming algorithm).

1. highwai	23,174	6. bu	7,270
2. amen	18,144	7. turn	6,826
3. barrier	12,283	8. circl	6,786
4. stop	12,060	9. traffic	6,608
5. cross	9,793	10. signal	6,606

The term frequency – inverse document frequency (tf-idf) matrix has the same dimensions as the matrix of the bag-of-words model, but instead of the values representing the number of occurrences of a word in a document, the tf-idf is calculated using the formula [6] displayed in (1).

$$tf.idf_{t,d} = n_t \sum_k {}_k n_k * \log(N/df_t). \tag{1}$$

With N being the number of documents, df_t being the number of documents containing a certain term t as well as nt being the frequency of a certain term t and $\sum_k n_k$ being the length of the document.

2.2   Data Mining

After preprocessing has been performed, the data is then applied to several data mining methods. Before the different applications are performed however a reference classification is created. This reference classification is used as a benchmark and the baseline for the generated classes. As the OSM project already created such a classification, albeit limited to the *tag*-element *amenity*, we will use the projects' recommended categories as described in Table 2.

Table 2. Categories for POIs based on the recommended classification by the OSM project.

Sustenance	Education
Transportation	Financial
Healthcare	Other amenities
Entertainment, Arts & Culture	

As stated before, this manual classification scheme is limited to only one specific *tag*-element. In order to circumvent this, the approach is extended via set theoretic methods to span over every existing *tag*-element. Hereby every *node*-element is fitted to one of the existing categories by looking for similarities of all of the present *tag*-elements. Afterwards new categories are generated by grouping up words with at least a threshold of a frequency of 1% (meaning that words need to have a frequency of 1:100 when compared with the entire data set). Using this method, 15 new classes are created. This procedure allows us to generally classify every existing *node*-element – precise classification will be done in the following steps. The set theoretic approach can be seen as a more general and comprehensive extension to the manual approach (Table 3).

Table 3. Keywords with a frequency of at least 1% over the entire set of the district of Düsseldorf (Germany).

highway	23,166	tourism	2,471
barrier	10,882	entrance	2,416
public_transport	6,416	denotation	1,564
power	5,803	genus	1,078
railway	5,204	historic	1,061
crossing	4,332	emergency	841
shop	3,104	leisure	657
natural	2,877		

Vector Space Model using Latent Semantic Analysis. In the Vector Space Model [7] similarities between documents are presented as angles between two vectors. Each document is therefore transformed into a set of vectors wherein each word represents a vector. The Vector Space Model (VSM) uses the aforementioned tf-idf matrix as its basis. Because of its very high dimension count (79,978 * 1,416) the Latent Semantic Analysis (LSA) [8] is used to reduce the data sets' dimensions. The algorithms are fed with the domain recommended parameters [9], resulting in two matrices, one vocabulary matrix and one documents score matrix having dimensions of 1,416 * 100 and 79,978 * 10 respectively.

Clustering. As with the VSM, Clustering also relies on the use of LSA. The optimal number of clusters is determined using the Calinski-Harabasz criterion [10] – a cluster is hereby identical to a category. This paper uses the algorithm of Lloyd [11], a centroid based k-means method, as the implementation of the clustering algorithm. based If the model is fitted to use n = 100 components, as set in the VSM, the optimal number of clusters is determined to be 10 (for n = 50, 20 clusters is the optimum, n = 200 results into 17 clusters being the optimum). Applying Clustering allows for an exhaustive classification. Overall 8 out of the 10 generated classes also show up in the set theoretic approach.

Latent Dirilichet Allocation – Probabilistic Model. The Latent Dirilichet Allocation (LDA) [12] is used to identify topics in the collection (data set) by calculating word-likelihoods. To find out the optimal number of topics, the perplexity for different amounts of topics is calculated beforehand, this is done using the Log-Likelihood [13]. As seen in Fig. 2, 20 topics are suggested as an optimal number of topics.

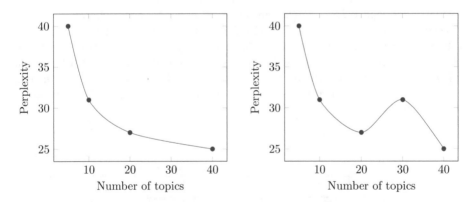

Fig. 2. Perplexity and number of topics for the LDA model – left: normalized data, right: non-normalized data.

It is noteworthy that the results of the LDA are not robust, therefore 100 runs were performed and the average of those runs was computed, resulting in the following 20 categories (Table 4).

2.3 Dynamic Model – User Centered Optimizations

All the aforementioned models generated similar category outputs that can be used in navigation software to display points of interest and to use them in routing algorithms,

Table 4. Topics generated with LDA based on normalized data.

1 vend machin cigarett	2 park bicycl stand hunt	3 restaur inform cafe
4 turn cicl	5 uncontrol cross level	6 bench switch
7 tree urban	8 signal traffic cross	9 stop platform bu
10 stop posit bu	11 barrier cycl	12 memori artwork station water sculptur
13 post box	14 stadtwerk street lamp gaslight straight	15 stop bu platform
16 bollard	17 recycl contain	18 tower main
19 gate	20 turn cicl	

but have one disadvantage: Categorization is static – meaning that user preferences and tastes are not taken into account.

To come by this limitation, we created a mechanism that leverages the discussed data mined results and uses them as the data source for a new dynamic approach (as well as the initial setup for a user's POI category list). At first the set theoretic model is used to match the user queries against (multiple user input words are treated like a Boolean AND). Every word of the query is compared to every word in the collection. Words that are contained in the document are counted as a match, words that are not contained in a document are counted as a miss. Additionally the Vector Space Model, including the applied LSA, is incorporated as an additional document into the entire collection. Here the cosine similarity is used to compare query words to the actual data set. For the cosine similarity a value of 0.5 was chosen [14] as the threshold for a hit condition. Afterwards a list of nodes is generated consisting of entries where both criteria generated a match, as well as a secondary list with suggestions, where only one of the two conditions matched – see Fig. 3, *same word* refers to the set theoretic model, *same context* to the VSM.

		Same context	
		yes	no
Same word	yes	Match	Common words, different context
	no	Same context, no common words	No match

Fig. 3. Criteria to determine the relevance of words of a specific query.

Should a user, for example, search for the phrase *bicycle* the user will be presented with matched results that are mapped to the word *bicycle* in the set theoretic model, but will also get results as suggestions if they share the same context (as determined by the VSM. Another case would be where no matches are generated via the set theoretic model. For example if the query would be *wlan* no match would be generated via the set theoretic approach, but the VSM will return results containing phrases like *internet* or *Wi-Fi*.

Implemented into the web based application users can also be uniquely identified. This enables us to also log the search history for POIs and use the gathered data to generate user centered categories which have not been present in the beginning. Should a user search for specific terms numerous times the term will be tailored to its own category for the user. Also when a user creates a route POIs will automatically be detected that are near the start, end, and intermediate points of the route and will receive a higher priority when searched upon in the future.

3 Results

Overall the set theoretic approach is best suited for a defined number of categories with heterogeneous content and few semantic similarities. Intriguingly dimension reduction using LSA did not yield significantly better results. By applying the Vector Space Model combined with LSA 99.07% of the resulting data set fits the set theoretic approach (Table 5).

Table 5. Results for the applied methods on the entire data set – reference is the set theoretic approach.

Approach	Classes	Coverage	Global accuracy
Set theoretic approach	18	95.06%	100% (initial)
Clustering (LSA – 100 components)	10	100%	86.81%
Clustering (LSA – 200 components)	18	100%	74.64%
Clustering (LSA – 300 components)	18	100%	80.00%
LDA	20	100%	61.33%
Vector Space Model (LSA – 100 components)	7	22.47%	99.07%

For the dynamic approach a combination of the set theoretic model and the Vector Space Model with LSA is used to distinguish between word similarities and contextual similarities. For data containing very homogeneous classes and content, Clustering (k-means++) delivers suitable results. Normalizing the data set by applying the Porter stemming algorithm results in small benefits in global accuracy.

The initially existing, OSM provided, categories are not symmetrical regarding the number of entries in each category which leads to some classes being neglected when applying automatic classification processes. As an example, when applying automatic classification methods, the combination of the classes *transportation*, *highway*, *railway* and *crossing* as well as the combination of *natural* and *waterway* (among others) are recommended.

As the whole system is applied to the administrative district of Düsseldorf (Germany) a regional bias cannot entirely be ruled out – results for different regions (e.g. the entire Federal Republic of Germany, the European Union or the entire world) might yield different results regarding the generated classification (Fig. 4).

Fig. 4. Map extract with activated POIs for food and drink, bus stops and parking sites (green markers).

4 Outlook

Despite receiving positive results, some limitations apply. The OSM data set, in theory, being uniform and culture neutral, sometimes has errors (like for example nodes that have no references whatsoever) and/or some local and regional specific oddities (meaning that some meta data does not exist in culture neutral form). Such oddities and errors can be identified by the applied preprocessing steps, but were not further inspected. In future iterations the applied methods could be used to correct errors in the OSM data set.

5 Conclusion

In this paper we described a system that effectively combines multiple data mining approaches and methods to the data set of the OSM project. The aim of the research was to extract and generate a classification scheme for points of interest. By extracting the existing *tag* information such POIs are created and multiple criteria for the dynamic user-centered approach are generated. The applied approaches consist of criteria based approaches (set theoretic approach and VSM) as well as data driven approaches (Clustering and LDA). We conclude that with the presented system, classification for POIs can be generated in a more flexible and faster way than by using conventional methods which are based on manual processing. The resulting data set can be used in navigation applications to enrich the customizability regarding POIs.

References

1. Han, J., Pei, J., Kamber, M.: Data Mining: Concepts and Techniques. THE Morgan Kaufmann Series in Data Management Systems. Elsevier Science (2011). ISBN: 9780123814807
2. Luhn, H.P.: Keyword-in-Context Index for Technical Literature (KWIC Index). Yorktown Heights, NY: International Business Machines Corp. (1959). https://doi.org/10.1002/asi.5090110403
3. Porter, M.F.: An algorithm for suffix stripping. Program **14**(3), 130–137 (1980)
4. Salton, G., Buckley, Ch.: Term weighting approaches in automatic text retrieval. Technical report TR87–881, Department of Computer Science (1987). Information Processing and Management 32, 431–443
5. Manning, C.D., Raghavan, P., Schütze, H.: Introduction to Information Retrieval. Cambridge University Press (2008). ISBN: 9781139472104
6. Miner, G., et al.: Practical Text Mining and Statistical Analysis for Non-structured Text Data Applications. Elsevier Science (2012). ISBN: 9780123869791
7. Singhal, Amit, et al.: Modern information retrieval: a brief overview. IEEE Data Eng. Bull. **24**(4), 35–43 (2001)
8. Deerwester, S., et al.: Indexing by latent semantic analysis. J. Am. Soc. Inf. Sci. **41**(6), 391–407 (1990)
9. Landauer, T.K., Foltz, P.W., Laham, D.: An introduction to latent semantic analysis. Discourse Processes **25**(2–3), 259–284 (1998)
10. Caliński, Tadeusz, Harabasz, Jerzy: A dendrite method for cluster analysis. Commun. Stat. Theory Methods **3**(1), 1–27 (1974)
11. Lloyd, Stuart: Least squares quantization in PCM. IEEE Trans. Inf. Theory **28**(2), 129–137 (1982)
12. Blei, D.M., Ng, A.Y., Jordan, M.I.: Latent Dirichlet allocation. J. Mach. Learn. Res. **3**(1), 993–1022 (2003)
13. Wallach, H.M., Mimno, D.M., McCallum, A.: Rethinking LDA: why priors matter. In: Bengio, Y., et al. (eds.) Advances in Neural Information Processing Systems 22, pp. 1973–1981. Curran Associates, Inc. (2009)
14. Tata, S., Jignesh Patel, M.: Estimating the selectivity of tf-idf based cosine similarity predicates. SIGMOD Rec. **36**(2), 7–12 (2007). ISSN: 0163-5808. https://doi.org/10.1145/1328854.1328855

NASA Marshall Space Flight Center Human Factors Engineering Analysis of Various Hatch Sizes

Tanya Andrews, Rebecca Stewart$^{(\boxtimes)}$, and Walter Deitzler$^{(\boxtimes)}$

NASA Marshall Space Flight Center, Huntsville, AL 35812, USA
{Tanya.C.Andrews,Walter.Deitzler}@NASA.gov,
Rsl840@MsState.edu

Abstract. The NASA Docking System (NDS) is a 31.4961-inch (800 mm) diameter circular hatch for astronauts to pass through when docked to other pressurized elements in space or for entrance or egress on surface environments. The NDS is utilized on the Orion Spacecraft and has been implemented as the International Docking System Standard (IDSS). The EV74 Human Factors Engineering (HFE) Team at NASA's Marshall Space Flight Center (MSFC) conducted human factors analyses with various hatch shapes and sizes to accommodate for all astronaut anthropometries and daily task comfort. It is believed that the hatch, approximately 32 inches, is too small, and a bigger hatch size would better accommodate most astronauts. To conduct human factors analyses, four participants were gathered based on anthropometry percentiles: 1st female, 5th female, 95th male, and 99th male.

Keywords: Human factors engineering · Human factors analyses · NASA ·
Common berthing mechanism · Jacobs space exploration group ·
Self-Contained atmospheric protective ensemble · US space and rocket center ·
Anthropometries · Underwater astronaut training · Microgravity

1 Introduction

Human Factors Engineering (HFE) has a key role in any system that contains human interaction with hardware. The purpose of this project was to conduct HFE analyses on various hatch shapes and sizes. Futuristic deep space missions need a standard size and shape of a hatch or common berthing mechanism (CBM) to connect modules or serve as an entryway or exit. CBMs are pressurized hatch connections between pressurized elements (PE). The five hatch sizes that were analyzed were 32", 42", 50 × 50", 50 × 50" 45°, and 62 × 50" (Fig. 1). The 32" hatch is in place on the Orion Spacecraft and the 50 × 50" CBM is currently being used on the International Space Station (ISS). The Advanced Concepts Office tasked the EV74 HFE team with conducting analyses to collect data that contributes to changing the standard from the 32" hatch to a larger, more accommodating hatch for future missions.

Each analysis was conducted in both a gravity and a microgravity environment. Surface analyses, performed in a gravity environment, were conducted at Marshall Space Flight Center's (MSFC) Building 4649. The tank analyses, performed in a

© Springer Nature Switzerland AG 2020
T. Ahram (Ed.): AHFE 2019, AISC 965, pp. 580–591, 2020.
https://doi.org/10.1007/978-3-030-20454-9_57

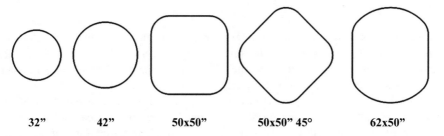

Fig. 1. The 32" and 42" shapes are circular hatches and the 50 × 50", 50 × 50" 45°, and 62 × 50" hatches are CBMs.

microgravity environment, were conducted at the Underwater Astronaut Training Facility (UAT) at the US Space and Rocket Center (USSRC). All hatches were analyzed in both docked and undocked configurations. The docked configuration contains two hatches parallel to each other to simulate when two PE's are connected for astronauts to pass from one module to the other. An undocked configuration contains only one hatch. This simulates when there is only one PE used for astronaut entry or egress onto a surface. Because of this, participants wore Self-Contained Atmospheric Protective Ensemble (SCAPE suits) to represent suits that astronauts would wear in space. All analyses were observed for task difficulty, adequate volume, reach difficulty, visual access, and overall comfort.

1.1 Participants

Analyses were performed with four participants of different anthropometric dimensions. To accommodate all astronauts using these passageways, the 1st and 5th percentile female and 95th and 99th percentile male height was used.

The Orion Spacecraft expanded the anthropometric dimensions to range from the 1st to 99th percentile, compared to the previous range of 5th to 95th percentile dimensions. The 1st and 99th percentile participant heights are very close to the accepted value. The 5th and 95th percentile participant heights are one or two inches different, but within the accepted value (Table 1).

Table 1. Participant Anthropometries.

Participant	Accepted value	Participant height
1st percentile female	4'10.5"	4'10.5"
5th percentile female	5'2"	5'3"
95th percentile male	6'2.8"	6'1"
99th percentile male	6'4.6"	6'4.5"

1.2   Safety

All participants were asked to thoroughly read and sign a consent form before the project began (Appendix A). For all surface analyses, participants were spotted while performing step-throughs with each hatch. The environment was prepared beforehand to ensure a clean and safe working space.

For all tank analyses, participants read and signed a waiver from the USSRC. The USSRC Dive Team then discussed diving basics, communication hand signs, safety hand signs, and questions that new divers had. After getting into the water and preparing dive equipment, participants went through training at the surface of the tank. They were shown diving basics and special skills to successfully dive in the UAT. While the test administrator, participants, and project assistants were in the tank, there were always enough USSRC divers to provide supervision inside and outside of the tank.

2   Surface Analyses

All surface analyses were conducted in MSFC's Building 4649. All hatches were analyzed in both docked and undocked configurations by all participants. Participants were asked to step through the hatch both frontwards and sideways, stepping through to the other side and back to the original position for each pass-through.

An intern and full time employee were responsible for the procurement, designs, and construction for the high-fidelity wooden mockups used.

2.1   Designs

The 32" and 42" hatches were already assembled from previous analyses. The 50 × 50" and 62 × 50" hatch designs were obtained from other departments within MSFC.

All hatches were designed with a specific tunnel length and depth. The hatch depths were found in the obtained designs. The tunnel lengths were either collected from designs or estimated by the test administrator and builder. The CBM tunnel lengths, all 15", were calculated using the 99th percentile shoe size, also considering clothing and boots worn. A wooden platform was used for the 50 × 50" docked configuration to help participants step through the hatch safely. Table 2 shows each hatch's tunnel length, hatch depth, and docked distance. The platform was 15" in depth and 8" in height.

Table 2. High-Fidelity Mockup Dimensions.

Hatch size	Tunnel length (in.)	Hatch depth (in.)	Docked distance
32"	10"	6 ¼"	16 ¼"
42"	10"	6 ¼"	16 ¼"
50 × 50"	15"	½"	15 ½"
50 × 50" 45°	15"	½"	15 ½"
62 × 50"	15"	4 ¼"	19 ¼"

2.2 Construction

The 32" and 42" high fidelity hatches were constructed prior to this project; however, there were only one of each. For the docked configurations, two of each hatch were needed, so low fidelity PVC structures served as the vehicle side hatch. The 50 × 50", 50 × 50" 45°, and 62 × 50" hatches were constructed using a CNC machine. All pieces were built using ¼" plywood sheets, painted, and attached to a Cygnus mockup in Building 4649. Reconfigurations between hatches took approximately 15 min.

3 Tank Analyses

All tank analyses were conducted in the USSRC's UAT which is 24 feet deep. All five hatches were analyzed by all four participants. Participants were asked to propel themselves through each hatch by pushing off the tank wall. Participants then turned around and pushed themselves off the center structure in the tank to go back through the hatch. The test administrator and supporting NASA high school interns were responsible for the procurement, designs, and construction for the PVC structures used.

3.1 Designs

A universal base design was created, allowing for simple reconfiguration for each hatch design. 1 ½" PVC was used for the universal base and for the 50 × 50" 45° hatch. The other hatches were built using ¾" CPVC. Fittings and adaptors were incorporated into designs for construction of each hatch.

3.2 Construction

PVC structures were constructed by hand. Both small and large pipe cutters were used to cut the PVC and CPVC pipe. The circular/ovular hatches were bent by hand, sometimes mounted while volunteers used force to form the correct shape and angle. Heavy duty primer and glue were used on the piping to secure into place and withstand strong chemicals in the UAT.

4 Results

4.1 Methodology

All analyses were observed and analyzed by the test administrator and surveys were given to participants after each analysis (Appendix B). As stated previously, the survey covered five topics: task difficulty, volume, reach difficulty, visual access, and overall comfort while performing the task of crossing through each hatch. The survey had five possible answers ranging from Strongly Disagree to Strongly Agree, with a scoring system ranging from one to five respectively. The questions were intentionally written so that higher scores would represent higher satisfaction with the task of passing through the hatch participants analyzed.

4.2   Data

All question scores were totaled for each participant. Each question counted for five points, making the maximum score per participant a 25. Each participants score was then totaled for all five hatches, making the maximum overall score a 125. Scores were taken as a percentage out of 125. Percentages for each participant were analyzed for each hatch configuration – surface docked, surface undocked, and microgravity analyses (Appendix C). Microgravity analyses were done in only one configuration because participants were floating through the hatches. This data was used in two different ways to show the results. First, a bar graph was made for each hatch configuration showing the overall scores per participant for all hatches (Fig. 2). Both the surface undocked and docked configurations mimicked a bell curve. The first percentile always scored the configurations the lowest and the 5th and 95th percentile scores were always greater than the 1st and 99th percentile scores. The 99th percentile score for the microgravity analyses was unexpected and therefore does not follow the same pattern as the surface analyses.

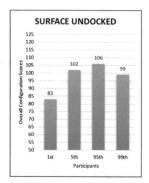 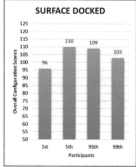 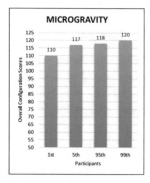

Fig. 2. Configuration scores based on each anthropometry.

The data was also used to create a line graph for each percentile that contains all five hatch scores for all three hatch configurations (Fig. 3). Scores increased as the hatch size grew larger from the 32" to the 50 × 50" hatch; however, results became constant as the hatch increased from 50 × 50" to the 50 × 50" 45° hatch.

Data was also analyzed by compiling total participant scores per hatch for each configuration (Appendix D). Each question had a maximum score of 20 and each hatch had a maximum score of 100. The percentage was calculated for each hatch in each configuration. Results were used to compile three bar graphs to show the increasing scores as hatch size grew larger (Fig. 4). As hatch size increased, total participant scores increased for surface analyses. Total participant scores increased from the 32" to the 42" hatch for the microgravity analyses; however, the results grew constant from the 50 × 50" to the 50 × 50" 45° hatch.

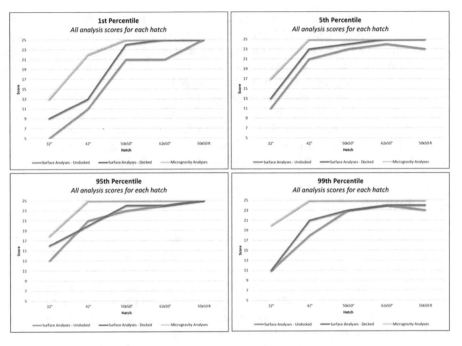

Fig. 3. Hatch scores based on individual anthropometries.

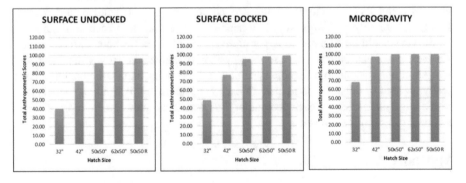

Fig. 4. Configuration scores based on all anthropometries.

5 Conclusion

As hatch size increases, total participant scores increase as well (Fig. 3). This shows a direct correlation between hatch size and comfort for all anthropometries (Fig. 2). Participant satisfaction increases as hatch size increases from the 32" to the 50 × 50" hatch; however, the 50 × 50", 62 × 50", and 50 × 50" 45° have very similar scores, resulting in the graphs flat lining.

Although the scores increase as the hatch size increases for the surface analyses, the same pattern does not occur with the microgravity analyses (Fig. 4). The only significant difference of scores for the microgravity analyses occurs between the 32" and 42" hatch. As the hatch grows larger from the 42" hatch, the score barely increases and remains approximately the same for the three larger hatches.

As hatch size increases, all anthropometries will be better accommodated; however, for future deep space missions, the largest hatch size (62 × 50") presented very similar data to the 50 × 50" hatch. For NASA's purposes, smaller hatches are more efficient overall. The results show that scores are constant once the size reaches the 50 × 50" hatch. A 50 × 50" or greater size hatch will better accommodate all anthropometries.

6   Future Work

This project was completed in approximately 10 weeks. If this project is extended and continued in the future, several factors should be considered and implemented.

Considering the hardware configurations for both surface and tank analyses, handles could be implemented to better simulate realistic hatch pass-throughs. For surface analyses, future participants could use the specifically placed handles for stability and handholds while stepping through the hatches. For tank analyses, future participants could use the handles to propel themselves through. This would better simulate microgravity environments, as opposed to pushing off the tank wall and center structure. High fidelity mockups would also be necessary for all hatch configurations. Lack of time and machine resources for this project contributed to some hatches for surface analyses using PVC structures for the docked configuration. If studied further in the future, high fidelity mockups would be needed for each hatch in each configuration.

Surveys could be adjusted to target more specific factors for both surface and tank analyses. Also, instead of using a scoring system to analyze the survey data, statistical analysis could be done to find more specific trends, outliers, and deviations in the data. Lastly, instead of only one participant of each anthropometry, it would be beneficial to have a large sample size in each percentile. This would cause the data to represent more accurate and significant results.

Appendix A

Participant Consent Form
Informed Consent for CBM Human Factors Assessments

Test Administrator: Becky Stewart   (rebecca.a.stewart@nasa.gov)
Department/Organization: EV74 Human Factors Engineering
Location: Marshall Space Flight Center, Building 4649 & US Space and Rocket Center

Mentors: Eric Staton   (eric.j.staton@nasa.gov)
         Tanya Andrews   (tanya.c.andrews@nasa.gov)

Part I: Information Sheet

Introduction: As a Human Factors Engineering Intern for the summer of 2018, I have been assigned with the task of performing human factors assessments on various common berthing mechanisms and hatches of various shapes and sizes. Each hatch will be tested in both docked and undocked configurations and in both gravity and microgravity environments. Both wooden and PVC structures have been built to represent the dimensions of all hatches. Assessments will be done in Building 4649 and in the Underwater Astronaut Training environment at the US Space and Rocket Center.

Purpose: These analyses are being conducted to determine which hatch shape and size will be the most objectively and subjectively accommodating to all people for future deep space modules. Participants of different anthropometries will be used in order to account for all heights.

Research: The participants will be informed and trained in a meeting prior to any analyses. The test administrator will inform the participants about the project in more depth and will instruct them what to do for each analyses. During each analyses, the test administrator will be observing how each participant steps (or floats) through each hatch. The volume, reach envelope, height, visual access, and comfort of each hatch will be observed for each participant. After each assessment, all participants will be asked to provide feedback. This will be done by a survey given by the administrator. The participants will be asked factual questions about the task as well as subjective questions like comfort, ease, and overall satisfaction.

Participant Selection: Participants of 4 anthropometries and one videographer were selected for the analyses. Participant height and experience was used to find volunteers, and specific heights and weights were used to select individuals. The four participants needed are listed below. Participants with the most similar heights to the standards were chosen.

1st percentile female	4'10.5"
5th percentile female	5'2.0"
95th percentile male	6'2.8"
99th percentile male	6'4.6"

Height will be recorded for each participant.

Voluntary Participation: Participation for this assessment is voluntary. Participants have complete authority to stop the assessment at any given time for any reason. Even after signing this form, participants can still choose not to participate in this assessment.

Risks:

Gravity Analyses: There are no major risks associated with the analyses held in 4649. Participants will simply step through various hatches. This may cause participants to bend over, crouch, or duck their heads. Closed toe shoes are required.

Microgravity Analyses: The analyses at the US Space and Rocket Center are somewhat dangerous. Those who have asthma should not participate. Proper equipment will be provided and each participant will be subject to a training course from the USSRC Aquatics Manager. The Aquatics Manager will be in the tank at all times, and two lifeguards and divers will be at the tank at all times. Diving has the potential to cause participants to be nervous and/or minor claustrophobia. All divers should pay close attention during training and remain calm and focused while performing analyses.

Benefits: The data and results gathered from these analyses will be used by NASA and the Advanced Concept Office in determining futuristic hatch designs and decisions. The participants will get to contribute to these important findings and perform analyses in the astronaut training facility at the USSRC.

Privacy/Confidentiality: Information collected from the participants will not be shared. All the information the EV74 Human Factors team collects will be kept confidential. If the data is published or presented, names will not be included. Participant information may be stored for future projects relating to the Common Berthing Mechanism, but will only be used as a resource for interns and the Human Factors team.

Multimedia Release: Photographs, video and/or audio recordings will be taken during the assessments. These photographs and videos will not be published unless given written approval in the statement below by the participants. Participants cannot participate in the assessment if multimedia release is refused. Participant names will not be stored with any photos, videos, or audio. Please initial next to your decision below:

_____ I agree to have video/audio recorded and photographs taken during my participation.
_____ I DO NOT agree to have video/audio recorded and photographs taken during my participation.

Right to Refuse or Withdraw: You do not have to take part in this research if you do not wish to do so. You may also stop participating in the research at any time you choose without any negative effects. It is your choice and all of your rights will be respected.

Who to contact: You may ask Becky Stewart any questions related to your participation before you sign this form. This procedure has been approved by Tanya Andrews. Please contact her with any additional concerns related to this research study.

Part II: Certificate of Consent

I have read and understood the information on this form. I've had the opportunity to ask questions, and any questions I have asked have been answered to my satisfaction. I consent voluntarily to be a participant in this assessment.

Print Name of Participant: _____
Signature of Participant: _____ Date: _____
Printed Name of Administrator: _____
Signature of Administrator: _____ Date: _____

Appendix B

Participant Survey Form

Please elaborate on any responses marked (Neutral), (Disagree), or (Strongly Disagree)

Please answer quickly; extensive thought should not be required, as these are first impressions.
Task: pass through specified hatch
1.   I was able to perform the task without difficulty.
(Strongly Disagree)   (Disagree)   (Neutral)   (Agree)   (Strongly Agree)

2.   I felt I could complete the task in the allocated volume.
(Strongly Disagree)   (Disagree)   (Neutral)   (Agree)   (Strongly Agree)

3.   I did not encounter any reach difficulties when completing the task.
(Strongly Disagree)   (Disagree)   (Neutral)   (Agree)   (Strongly Agree)

4.   I had adequate visual access necessary to perform the task.
(Strongly Disagree)   (Disagree)   (Neutral)   (Agree)   (Strongly Agree)

5.   I felt comfortable inside the hatch.
(Strongly Disagree)   (Disagree)   (Neutral)   (Agree)   (Strongly Agree)

Appendix C

Data: Configuration scores based on each anthropometry

SURFACE ANALYSES UNDOCKED	Participants			
Hatch	**1st**	**5th**	**95th**	**99th**
32"	5	11	13	11
42"	11	21	21	18
50 × 50"	21	23	23	23
62 × 50"	21	24	24	24
50 × 50 R	25	23	25	23
Sum:	83	102	106	99
Percentage:	66.4%	81.6%	84.8%	79.2%
SURFACE ANALYSES DOCKED	Participants			
Hatch	**1st**	**5th**	**95th**	**99th**
32"	9	13	16	11
42"	13	23	20	21
50 × 50"	24	24	24	23
62 × 50"	25	25	24	24
50 × 50 R	25	25	25	24
Sum:	96	110	109	103
Percentage:	76.8%	88.0%	87.2%	82.4%
MICROGRAVITY ANALYSES	Participants			
Hatch	**1st**	**5th**	**95th**	**99th**
32"	13	17	18	**20**
42"	22	25	25	25
50 × 50"	25	25	25	25
62 × 50"	25	25	25	25
50 × 50 R	25	25	25	25
Sum:	110	117	118	120
Percentage:	88.0%	93.6%	94.4%	96.0%

Each score is the individual participant score given for each hatch. Each table is a different configuration or environment. The maximum score for each participant for each hatch was 25. Scores in bold are unexpected results.

Appendix D

Data: Configuration scores based on all anthropometries

UNDOCKED							
Hatch	**UD Q1**	**UD Q2**	**UD Q3**	**UD Q4**	**UD Q5**	**TOTAL**	**PERCENTAGE**
32"	5.00	10.00	8.00	12.00	5.00	40.00	40%
42"	12.00	18.00	15.00	14.00	12.00	71.00	71%
50 × 50"	17.00	20.00	15.00	20.00	19.00	91.00	91%
62 × 50"	19.00	19.00	18.00	18.00	19.00	93.00	93%

(*continued*)

(continued)

UNDOCKED							
50 × 50 R	18.00	20.00	19.00	20.00	19.00	96.00	96%

DOCKED							
Hatch	**D Q1**	**D Q2**	**D Q3**	**D Q4**	**D Q5**	**TOTAL**	**PERCENTAGE**
32"	6.00	12.00	11.00	14.00	6.00	49.00	49%
42"	14.00	18.00	14.00	18.00	13.00	77.00	77%
50 × 50"	19.00	20.00	17.00	20.00	19.00	95.00	95%
62 × 50"	20.00	20.00	19.00	20.00	19.00	98.00	98%
50 × 50 R	20.00	20.00	20.00	20.00	19.00	99.00	99%

MICROGRAVITY							
Hatch	**D Q1**	**D Q2**	**D Q3**	**D Q4**	**D Q5**	**TOTAL**	**PERCENTAGE**
32"	10.00	14.00	16.00	15.00	13.00	68.00	68%
42"	19.00	19.00	20.00	20.00	19.00	97.00	97%
50 × 50"	20.00	20.00	20.00	20.00	20.00	100.00	100%
62 × 50"	20.00	20.00	20.00	20.00	20.00	100.00	100%
50 × 50 R	20.00	20.00	20.00	20.00	20.00	100.00	100%

The sum of participant scores is shown for each question for each hatch. The maximum total score possible is 20.

Ontological Description of the Basic Skills of Surgery

Kazuhiko Shinohara[(✉)]

School of Health Sciences, Tokyo University of Technology,
Nishikamata 5-23-22, Ohta, Tokyo 1448535, Japan
kazushin@stf.teu.ac.jp

Abstract. Artificial intelligence (AI) has recently been receiving increasing attention in the medical field. For AI to be successfully applied to surgical training and certification, ontological and semantic description of clinical training procedures will be necessary. This study proposes a method for the ontological analysis of basic endoscopic surgery skills. Surgical procedures learned during basic training using a training box for exercises such as pegboard, pattern cutting, and suturing, were analyzed and described as ontological procedures. Surgical maneuvers for endoscopic surgery performed in the training box were successfully classified and described by referencing ontological concepts. Such ontological descriptions can be applied in many ways, for example, computerized evaluation of medical training and certification, automated robotic surgical systems, and medical safety check and alert systems. The need for simple and practical methods for the ontological description of clinical medicine was also revealed in this study.

Keywords: Endoscopic surgery · Basic skill training · Ontology

1 Introduction

For artificial intelligence (AI) to be successfully applied to surgical training and certification, ontological and semantic description of clinical training procedures will be required. However, to date, few studies have adopted an ontological approach to clinical surgery. Therefore, this study aimed to describe, in an ontological manner, the basic skills required for the use of training instruments for endoscopic surgery.

2 Materials and Methods

Surgical procedures learned during basic training using a training box for exercises such as pegboard, pattern cutting, and suturing, were analyzed and described as ontological procedures. The pegboard exercise is used to train surgeons in the use of forceps by transferring a rubber ring under the guidance of a two-dimensional endoscopic image. The pattern cutting and suturing exercises are used to train surgeons to quickly and precisely cut and suture under the conditions of limited space and minimal tactile sensation that are inherent to endoscopic surgery. These exercises have been

© Springer Nature Switzerland AG 2020
T. Ahram (Ed.): AHFE 2019, AISC 965, pp. 592–597, 2020.
https://doi.org/10.1007/978-3-030-20454-9_58

widely used in surgical training and certification since the dawn of endoscopic surgery in the 1990s (Figs. 1 and 2).

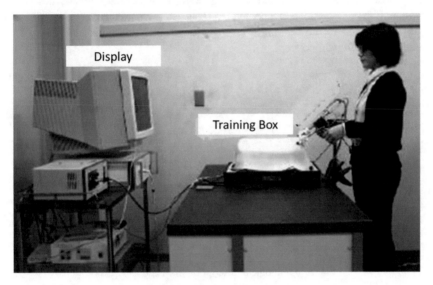

Fig. 1. Training of basic skills for endoscopic surgery by training box

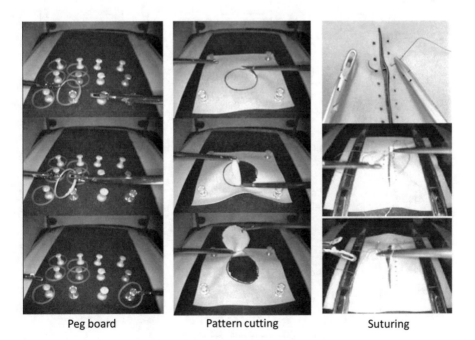

Peg board   Pattern cutting   Suturing

Fig. 2. Typical training tasks for endoscopic surgery

3   Results

The pegboard, pattern cutting, and suturing exercises were ontologically described as shown in Figs. 3, 4, and 5, respectively. Basic surgical maneuvers for endoscopic surgery performed in the training box were successfully classified and described by referencing ontological concept.

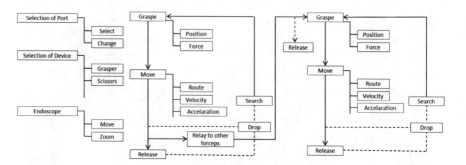

Fig. 3. Ontological description of "Pegboard" task

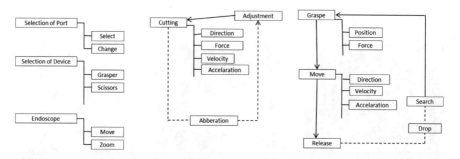

Fig. 4. Ontological description of "pattern cutting" task

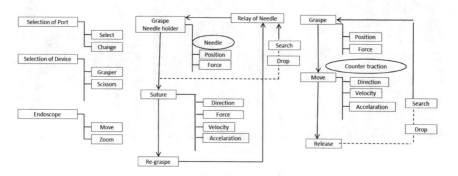

Fig. 5. Ontological description of "Suture" task

4 Discussion

Modern endoscopic surgery began with endoscopic cholecystectomy by Mouret et al. in 1987. In Japan, the first endoscopic cholecystectomy was performed in 1990 and the procedure rapidly spread throughout the country (Fig. 6). The great advantage of endoscopic surgery is its minimization of skin incision and post-surgical pain. Endoscopic surgery is performed using small-caliber instruments under endoscopic guidance. These instruments are inserted via incisions in the visceral wall between 5 and 10 mm in diameter. Surgeons must operate on internal organs via small-caliber access port in a limited space with minimal tactile sensation using a two-dimensional image. Consequently, the level of surgical skills required and the ergonomic stresses imposed on surgeons are much higher in endoscopic surgery than in conventional open surgery. To overcome these technical difficulties, research to improve endoscopic surgery began in the field of surgical robotics, imaging, and navigation technology in the 1990s. A number of training programs and tools were developed during that time. Before more hands-on training, novice surgeons honed their basic skills "*in vitro*" using training boxes, and "*in vivo*" by performing surgical procedures on animals.

The three "*in vitro*" training box exercises—pegboard, pattern cutting, and suturing —had been widely used to train surgeons for a long time. These exercises are suitable for the technical and cognitive training of coordinative movements required for operating surgical devices under the conditions of restricted sensation, visual cues, and operating space. Surgical training using training boxes has generally been evaluated by measuring the speed and precision of the work performed during the exercises. However, AI will be required for more rapid and objective evaluations.

The application of AI in medicine has been attempted since the 1960s. Automated diagnosis based on laboratory data and X-ray images has been achieved in internal medicine and radiology. However, there have so far been few successful applications of AI in surgery. Ontological description and analysis of surgical procedures will be indispensable if AI is to be successfully implemented in surgical training and evaluation. An ontological approach has been applied to disease terminology, diagnosis, and clinical guidelines. Several ontologies, such as OGEM and POID, which focus on ontological definitions of diseases, have been proposed for medical information systems [1–4]. Despite these advances, ontological analysis has not yet been applied to surgery.

This study revealed that ontological descriptions of surgical procedures are possible. Ontological description of endoscopic surgical procedures can be applied in many ways, including computerized evaluation of medical training and certification, automated robotic surgical systems, and medical safety check and alert systems. The need for simple and practical methods for the ontological description of clinical medicine was also clarified in this study [5, 6].

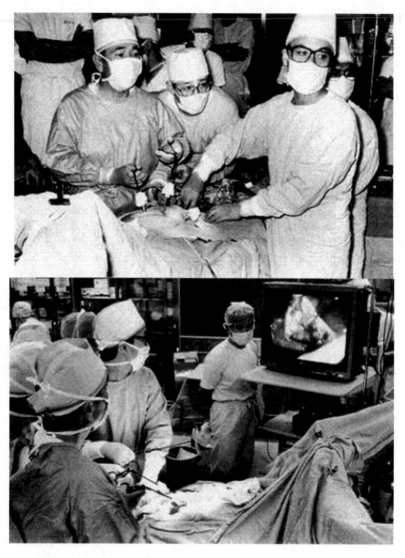

Fig. 6. A scene of preparation for laparoscopic surgery, Department of surgery, University of Tokyo, 1990

5   Conclusion

This study applied ontological analysis to basic endoscopic surgery training. The training box exercises used to train novice surgeons were successfully described with reference to ontological concepts. Such ontological descriptions can be applied in many ways, including robotic surgical systems, medical safety, and the evaluation of medical training. The need for simple and practical methods for the ontological description of clinical medicine was also clarified in this study.

References

1. Kozaki, K., et al.: Browsing Causal Chains in a Disease Ontology, Poster & Demo Notes of 11th International Semantic Web Conference, Boston, USA, November 11–15 (2012)
2. Scheuermann, R., et al.: Toward on ontological treatment of disease and diagnosis. In: Proceedings of the 2009 AMIA Summit on Translational Bioinfomatics, San Francisco, pp. 116–120 (2009)
3. Osborne, J., et al.: Annotating the human genome with Disease Ontology. BMC Genomics **10** (1), 56 (2009)
4. Kumar, A., et al.: An Ontological framework for the implementation of clinical guidelines in healthcare organization. In: Ontologies in Medicine. IOS Press, Amsterdam (2004)
5. Shinohara, K.: Preliminary study of ontological process analysis of surgical endoscopy. In: Advances in Human Factors and Ergonomics in Healthcare and Medical Devices, pp. 455–461. Springer, Cham (2017)
6. Shinohara, K.: The feasibility of ontological description of medical device connectivity for laparoscopic surgery. In: Advances in Human Factors and Ergonomics in Healthcare and Medical Devices, pp. 88–95. Springer, Cham (2018)

An Alternative Graphic and Mathematical Method of Dimensional Analysis: Its Application on 71 Constructive Models of Social Housing in Guayaquil

Byron Sebastian Almeida Chicaiza[1,2](✉) (iD), Jesús Anaya Díaz[2],
Pamela Bermeo Rodríguez[1], Jesús Hechavarria Hernández[1],
and Boris Forero[1]

[1] Facultad de Arquitectura y Urbanismo, Universidad de Guayaquil,
Guayaquil, Ecuador
{byron.almeidac,pamela.bermeor,jesus.hechavarriah,
boris.forerof}@ug.edu.ec
[2] Departamento de Construcción y Tecnología Arquitectónicas,
Universidad Politécnica de Madrid, Madrid, Spain
jesus.anaya@upm.es

Abstract. In this study, a method of alternative graphic and mathematical analysis is presented to determine important information such as the ratio between length and width of the analyzed orthogonal spaces and their relation with the useful surface, in order to determine the most suitable relationships for the user, also allowing to carry out an analysis of the spaces that are part of the architectural program among the studied models, with the aim of establishing the constructive tendencies and the dimensional characteristics framed in specific periods of time.

The methodology has been applied to 71 single-family low-income housing solutions that represent the buildings where 41.3% of the population of Guayaquil inhabits; of which 50 models belong to public and NGO's promotions and 21 to self-built models by users. It shows the materiality and the used techniques, classifying the analyzed models according to the type of promotion, number of levels, the adopted grouping typology and the architectural program that predominates in most of the cases.

Keywords: Social housing · Graphical analysis · Guayaquil ·
Dimensional analysis

1 Introduction

The objective of this study is to determine the most common dimensional characteristics or frequent dimensions, as well as predominant materials; highlighting the relationship between the two dimensions that define the architectural plant (without considering the height for this time). In this study, it has been considered the relationship between comfort and the surface of a space, although it is not the only parameter that defines a "good" home. However, even though a surface by itself is not a value that indicates the

This is a U.S. government work and not under copyright protection in the U.S.;
foreign copyright protection may apply 2020
T. Ahram (Ed.): AHFE 2019, AISC 965, pp. 598–607, 2020.
https://doi.org/10.1007/978-3-030-20454-9_59

functional suitability, it has been considered that if we have the aspect ratio (ratio between its dimensions), two spaces can be compared more objectively; for example, we can have a room of 12.00 m^2, but with 6.00 m long and 2.00 m wide, and this would not be a functional space for a bedroom.

Social interest housing[1] [1] in Ecuador can be classified chronologically according to the main economic activities that generated income to the country, which is intimately related to the production of housing. This is how the following productive periods are defined [2–4]: Cocoa Era (1880–1950), Banana Era (1940–1970), Oil Era (1970–1980), Globalization Era (1990–2010), and the last stage has been defined as the Bonanza-indebtedness-pre-sales period (2011–2018) [5].

In Guayaquil, in a period of analysis of 76 years, it has been determined that the number of built homes would reach 350 thousand units of public promotion and Non-Governmental Organizations (NGOs). These have been divided into 2 large groups: formal housing and informal housing [6]. The formal housing have been production of Institutions of the National Government and the Local Government; while the homes of the informal ones, in certain cases, have been produced by NGOs; In addition, self-construction housing, which is why this study has analyzed cases of self-construction by the owners who are in vulnerable sectors.

According to the analysis of the population data of the INEC (National Institute of Statistics and Census), the population of the urban area of Guayaquil has gone from 258 966 inhabitants (1950) to 2 563 636 inhabitants (2017) in 60 years; that is to say, it has increased 10 times its size [7–12] (Table 1).

Table 1. Evolution of the population of Guayaquil.

	Years								
	1950	1962	1974	1982	1990	2001	2010	2017	2030
Years gone by	0	12	12	8	8	11	9	7	13
Population	258,966	510,804	823,219	1,199,344	1,513,437	1,985,379	2,278,691	2,563,636	3,191,749
Annual population growth rate	–	5.82	4.06	4.82	2.95	2.50	1.54	1.70	1.70

*Rates of annual housing growth have been calculated by intercensal growth formula. Urban population in 2017 have been taken from INEN projection and percentage of urban population with respect to the total (2010 Census).
Projection of urban population to 2030 have been calculated with the growth rate 2010–2017. Sources: [4, 8, 9, 13]

This study contemplates the analysis of the architectural plans^2 of 71 single-family housing models that belong to different social housing programs of Guayaquil that have been designed/built in Guayaquil between 1942 and 2017; in total they reach approximately 289 thousand units (see Table 2), 236 thousand built units and 53 thousand, that were not built.

---

[1] *Ley Orgánica de Ordenamiento Territorial Uso y Gestión de Suelo (LOOTUGS)* (Organic Law of Land Use and Land Management): "The housing of social interest is that destined to the population of low income and groups of priority attention".

[2] Obtained from institutions responsible for housing development, historical archives and field surveys.

The average number of inhabitants per dwelling in Guayaquil is 3.8 inhabitants/unit [13], so the population that live in these dwellings would be close to 1.1 million inhabitants. Considering that the population of Guayaquil in 2017 was 2 563 636 inhabitants, this would represent 675 thousand homes [13]; Therefore, the studied models represent the homes where 43% of the population of Guayaquil live. On the other hand, if we consider that 72% of the population of Guayaquil lives in these popular neighborhoods [14, 15], these models represent approximately 60% of Guayaquil's social housing.

1.1   The Conceptual Basis

In the first steps of conceiving the grouping of housing units, there are fundamental elements as indicated by Sabeté [16] the blocks and streets; On the one hand, the streets define circulation, while the dimensions of the blocks are related to the size of the lots and consequently to the buildable depth and housing typologies.

The search of spatial relations between architectural spaces, dates back to the theories of the russian architect Alexander Klein [17], who carries out studies on the biological minimal of air, light and spaces essential for living, "*existenzminimun*" [18]. Klein concentrated his attention on the logical-mathematical individualization of the "objective" parameters of assessment of the sizing and of the optimal distribution of the dwellings type [19], drawing upon the use of three methods: the graphical method, the method of points and the method of successive increments.

Klein's work was mainly focused on typological and economic problems and in his work, he uses three coefficients: the so-called *Betteffekt* (bed-effect, surface built by bed), which is the dimensioning of the spaces in relation to those who will inhabit it. It is followed by the *Nutzeffekt* (utility effect, quotient between useful surface and built surface), and the *Wohneffekt* (habitability effect, quotient between living space, bedrooms and living room, and the constructed surface) [18].

If we consider that the dimensions of each space that make up the spatial arrangement will define the total dimension of the house (in the architectural plant) and consequently, these dimensions will define the dimensions of the land and its proportions [16, 18], we can define certain numerical relationships between the dimensions and areas, which will facilitate the comparison and interpretation of data. Prior to the application of the proposed method, different scales of analysis have been used, such as those proposed by Bamba in his multiscale approach [20] which is based on the use of graphic and numerical codes that are applied to a series of "zoom levels": 100 x 100 m, 30 x 30 m and 15 x 15 m, stablishing a smaller grid of 4 x 4 m (which in certain cases had to be extended to 6 x 6 m, because of the space dimensions) for the direct comparison between spaces with the same functions but from different models (bedrooms with bedrooms, kitchens with kitchens, etc.)

2   Methodology

71 single-family housing models have been classified into 4 groups: A (Villa type - Public Promotion), B (Villa type - NGO), C (Department type - Public Promotion) and D (Villa type - Self-construction); in each group in turn, it has been separated into two

sub groups: Solutions of 1 plant and 2 plants or more; obtaining 7 possible classifications with a code for each one (See Table 2).

Table 2. Classification and quantity of analysed housing models.

Analyzed housing models						
No.	Code	Housing type	Type of promotion	Number of floors	Number of analyzed models	Projected units
1	A1	House	Public	1	13	66,376
2	A2			2+	14	
3	B1	House	NGO	1	5	200,000
4	C1	Flat	Public	1	17	22,974
5	C2			2+	1	
6	D1	House	Selfbuilt	1	19	21
7	D2			2+	2	
Total					71	289,371

In order to standardize the procedure of data digitization, the architectural floor plans has been aligned with an XY Cartesian plan. The alignment with the axes assign the location in the data as it is shown in Table 3.

If we look at Fig. 1, we analyze Bedroom 1 (D1) of an "A1" type of house (house-public-one floor), one of the housing models from the Socio Vivienda project [21, 22]. The value of the "y" axis corresponds to 2.64 m, while the value of the "x" axis is 2.68 m; Subsequently, by calculation formulas, the lower value of both (DIM_{minor}) and the higher value (DIM_{major}) is automatically obtained (Table 3).

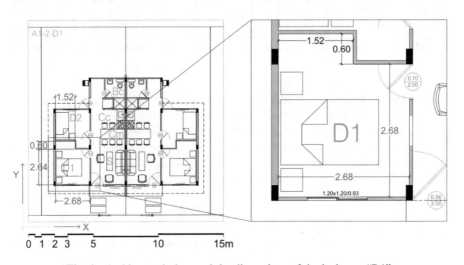

Fig. 1. Architectural plant and the dimensions of the bedroom "D1".

The schematic simplified plants (polygons) are represented in graphs and, as a general rule, the minor measurement is oriented on the "y" axis and the major measurement, on the "X" axis. Any additional space such as the closets, or additional area to the polygon that defines the architectural space, have been represented as a shaded area (Fig. 2).

Table 3. Classification and quantity of analysed housing models and detail of dimensions according to referential orientation; determination of major *measurement (DIM$_{major}$)* and minor measurement *(DIM$_{minor}$)* of the analyzed space.

A1	Symbology	Space/Room (F	X	Y	Partial area	Total area	DIM minor	DIM major	Ratio (DIM minor/ DIM major	Resulting degree Dminor / Dmajor
Type: House	S	Living room	2.61	2.64	6.89	6.89	2.61	2.64	0.99	44.67
	Cm	Dining room	2.61	1.49	3.89	3.89	1.49	2.61	0.57	29.72
	Cc	kitchen	2.61	1.44	3.76	3.76	1.44	2.61	0.55	28.89
	BC1	Bathroom 1	1.51	1.41	2.13	2.13	1.41	1.51	0.93	43.04
	BC2	Bathroom 2								
	BC3	Bathroom 3								
Country: Ecuador	BV	Washroom								
	D1p1	Bedroom 1 Part 1 (masterl)	2.68	2.64	7.08	7.08	2.64	2.68	0.99	44.57
Promoter: Public	D1-p2	Bedroom 1 Part 2 (master)	1.52	0.60			0.60	1.52	0.39	21.54
	D2-p1	Bedroom 2 Part 1	2.68	2.27	6.08	6.08	2.27	2.68	0.85	40.27
	D2-p2	Bedroom 2 Part 2	1.09	0.60			0.60	1.09	0.55	28.83
Floors: 1	D3-p1	Bedroom 3 Part 1								
	D3-p2	Bedroom 3 Part 2								
	D4-p1	Bedroom 4 Part 1								
	D4-p2	Bedroom 4 Part 2								

Table header: A1-1 — "Socio Vivienda 1" y "Socio Vivienda 2" House (Not built) — 1 floor

In order to translate the numerical values into an equivalent graphic representation, we have chosen to draw a diagonal from the lower left corner of the schematic simplified plants, to the upper right corner in the polygons (regularly square or rectangular) also determining the angle that the diagonal forms with the horizontal or abscissa axis. This angle can also be determined with the ratio (DIM$_{minor}$/DIM$_{major}$) which in other words would be the tangent of this angle (see formulas below and Fig. 2).

$$\text{Tan } \alpha = \text{DIM}_{\text{minor}}/\text{DIM}_{\text{major}} \tag{1}$$

$$\alpha = \text{Tan}^{-1} \text{DIM}_{\text{minor}}/\text{DIM}_{\text{major}} \tag{2}$$

Ratio: In previous studies, the aspect ratio or ratio between space dimensions are calculated by dividing DIM$_{major}$/DIM$_{minor}$, which generated results that were very dispersed and difficult to interpret when represented graphically. For this reason, it was decided to invert the calculation formula, dividing DIM_{minor}/DIM_{major}, obtaining values that would always keep below 1.00; being 1.00 a square space, and any value below 1.00, a rectangular space; the smaller the coefficient, the longer the space would be.

Resulting degrees: If we draw the diagonal of a square, the angle that it forms with the horizontal would be 45°. If the angle is calculated by the application of trigonometry, this would be the inverse tangent of the ratio (DIM_{minor}/DIM_{major}) (formula 2). Considering that the larger dimensions are plotted on the "x" axis, the angle formed by the diagonal with the horizontal will never be greater than 45°.

As an example, two different bedrooms have been drawn, each one from a different Type: A1 (Public promotion- one storey –villa type) and D1 (Self built- one storey – villa type). Additional information been calculated, such as the average area, the average of the ratios (DIM_{minor}/DIM_{major}) and the proportion. The polygon that represents the media have been drawn (Fig. 3) with its diagonal, also specifying the angle it forms with the horizontal obtaining its sides by using trigonometry.

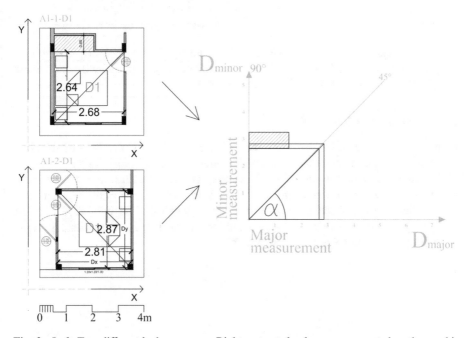

Fig. 2. Left: Two different bedroom cases. Right: extracted polygons represented on the graphic according to the stablished rules.

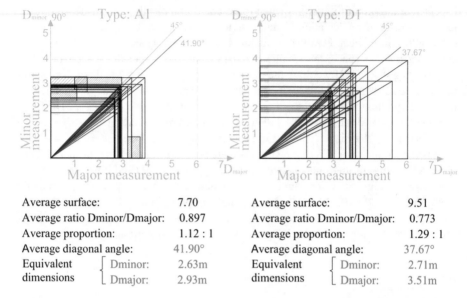

Average surface:	7.70	Average surface:	9.51
Average ratio Dminor/Dmajor:	0.897	Average ratio Dminor/Dmajor:	0.773
Average proportion:	1.12 : 1	Average proportion:	1.29 : 1
Average diagonal angle:	41.90°	Average diagonal angle:	37.67°
Equivalent [Dminor:	2.63m	Equivalent [Dminor:	2.71m
dimensions [Dmajor:	2.93m	dimensions [Dmajor:	3.51m

Fig. 3. Representation of polygons that define the bedroom spaces of model types A1 (*Public promotion- one storey –villa type on left*) and D1 (*Self built- one storey –villa type*); diagonals, important data and representation of average dimensions.

2.1   Predominant Materials

From the analyzed projects data, 69% corresponds to dwellings promoted by the NGO Hogar de Cristo, dwellings that correspond to houses built mostly of wood and bamboo [23]. When analyzing the rest of the solutions, without counting those of the afore-mentioned NGO, concrete is the predominant material with 87.6% of the cases, making it clear that the constructive systems that are included can be reinforced concrete columns, reinforced concrete walls, and reinforced concrete systems with lost form-work (Royal Building System RBS) (Fig. 4).

Used materials in structures (Including Hogar de Cristo project)				
Color	Main Structure	Housing units	%	
	Reinforced concrete columns.	23,603	8.16%	
	Reinforced concrete walls	17,298	5.98%	
	Unweighted due to lack of data. Columns Reinforced Concrete, Simple Concrete Walls, Reinforced Concrete Walls (RBS and others).	37,319	12.90%	
	Steel structure.	11,148	3.85%	
	Wood-bamboo structure.	200,003	69.12%	
Total		289,371	100.00%	

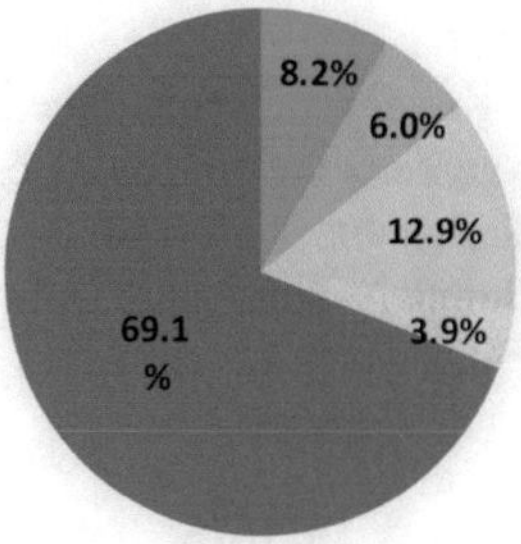

(Pie chart: 69.1%, 12.9%, 8.2%, 6.0%, 3.9%)

Used materials in structures (Not including Hogar de Cristo project)				
Color	Main Structure	Housing units	%	
	Reinforced concrete columns.	23,603	26.41%	
	Reinforced concrete walls	17,298	19.36%	
	Unweighted due to lack of data. Columns Reinforced Concrete, Simple Concrete Walls, Reinforced Concrete Walls (RBS and others).	37,319	41.76%	
	Steel structure.	11,148	12.47%	
	Wood-bamboo structure.	3	0.00%	
Total		89,371	100.00%	

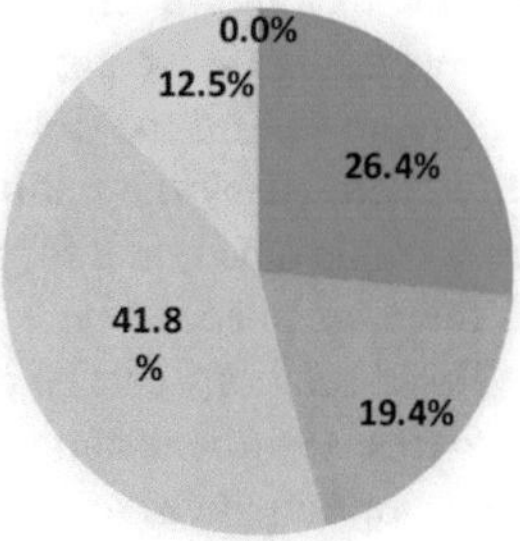

(Pie chart: 41.8%, 26.4%, 19.4%, 12.5%, 0.0%)

Fig. 4. Predominant materials, housing units per category, graph of statistical distribution.

3 Results and Discussion

From the analyzed data of the 71 cases specified in Table 2, it has been determined that the space allocated to the dormitory with the largest area has 23.75 m^2, with 3.93 m in the short side (DIM$_{minor}$) and 6.05 m in the large side (DIM$_{major}$), with a ratio (DIM$_{minor}$/DIM$_{major}$) of 0.65 and proportion of 1.5:1. The space with the smallest area for this same use has an area of 3.55 m^2, 1.87 m by 1.90 m, a ratio (DIM$_{minor}$/DIM$_{major}$) of 0.98 and a proportion 1:1. As for the aspect ratio, for the highest case, the ratio is 1.00 (several models) and the average of the dimensions is 2.74 m per side; while the more elongated one has an aspect ratio of 0.47 with a smaller side of 1.63 m and 3.5 m of larger side.

If the use destined to each space is considered, statistically the bedrooms analyzed are the spaces that have a larger area; so if this is related to the predominant material, we could establish a material-spatial dimension relationship.

In the spaces that have been self-constructed, the information obtained is very varied, having rooms that can be very small and where practically only the bed fits, such as very large rooms where there is a lot of space and that are usually used for other activities.

Klein's studies and the proposals (saving distances) that seek to quantify the spaces through coefficients obtained from dimensional relationships, should not be analyzed in isolation and should not be the only data taken into account since the concept of housing or home is dehumanized; however, these data can provide important information about the interpretation of spaces trying to bring the qualitative to the quantitative.

4 Conclusions

The statistical data obtained about most common dimensions, surfaces and the relationships between both values combined with its graphic representation, provide information that can be interpreted in a simpler and practical way, especially considering the volume of information handled.

Depending on the analyzed space (kitchen, bathroom, toilet, living room, bedroom) when representing the polygons in the proposed graph, there will be lower limits that will be defined by the minimum dimensions that are required to fit a piece of furniture (for example a bed) and for it to be functional (circulations); although it has been shown that there are spaces where functional comfort has been sacrificed due to lack of space.

These relations can be taken to a next level, by relating the interior dimensions of the spaces with the distances to the structural elements and their materiality, allowing to identify options to give an added value to the spaces, such as provide flexibility to the spaces [24, 25] through the definition of growth zones and constructive components that could make this possible.

Another aspect that can be taken into account in future research is the relationship of the dimensions of an architectural floor plan and its relationship with heights, since the unevenness or height affects (positively or negatively) the functionality of a space.

Finally, by conducting this study has identified the need to conduct a thorough study of the furniture used in social interest housing, as well as all the belongings (clothing, pots, bicycles, etc.) that mean a spatial need that regularly do not appear in an architectural plan at the design stage.

References

1. Registro Oficial No.790, Ley Orgánica de Ordenamiento Territorial, Uso y Gestión de Suelo, p. 20 (2016)
2. Solano, A., Villacres, C.: Categorización y Definición Socio espacial del Barrio Como unidad Básica territorial. AUC Revista de Arquitectura 31, 49–60 (2011)
3. Guerra, F.: Análisis del modelo económico y social Ecuatoriano. De los años: 70s y 80s, en el marco de la globalización. Facultad Latinoamericana de Ciencias Sociales FLACSO, pp. 5, 55, 56, 62, 63, 194 (2001)

4. Almeida, S., Anaya, J., Hechavarría, J.: Caracterización dimensional de modelos constructivos de Vivienda de Interés Social construídos en Guayaquil, Ecuador. In: Libro de Memorias Científicas del V Conrgreso Internacional de investigación y actualización en ingenierías, pp. 234–249, 227–239. CIDE, Guayaquil (2017)

5. Ekonegocios (ed.): Ecuador tuvo cuatro principales fuentes de financiación en 2016, pp. 1–3 (2016)

6. Peek, O., Hordijk, M., d'Auria, V.: User-based design for inclusive urban transformation: learning from 'informal' and 'formal' dwelling practices in Guayaquil, Ecuador. Int. J. Hous. Policy, 9, 14, 15 (2017)

7. INEC: Cantón Guayaquil: fascículo Guayaquil, población según el censo (2001)

8. INEC: Resultados del Censo 2010 de la población y vivienda del Ecuador: Fascículo Nacional (2010)

9. Sánchez, B.: Mercado de suelo informal y políticas de hábitat urbano en la ciudad de Guayaquil, pp. 37–43. FLACSO Sede Ecuador, Quito (2014)

10. El Telégrafo (ed.): El presupuesto de Guayaquil se ha incrementado el 48% en 7 años, pp. 1–5 (2014)

11. Municipio de Guayaquil: Demografía de Guayaquil (2010)

12. INEC-SENPLADES, Proyecciones referenciales de población a nivel cantonal 2010–2030 (2017)

13. Almeida, V.: Promedio de Personas por Hogar a Nivel Nacional (2017)

14. Forero, B., Hechavarría, J.: TUS 015. Análisis de las condiciones de confort térmico en el interior de las viviendas del complejo habitacional Socio Vivienda 2, Etapa 1, en la ciudad de Guayaquil, Ecuador. In: 3er Congreso Científico Internacional Tecnología, Universidad y Sociedad TUS, pp. 128, 129, 133, 134 (2015)

15. Hechavarría, J., Forero, B.: TUS 052. Aplicación de la metodología de análisis y síntesis de sistemas de ingeniería en la búsqueda de soluciones a problemas de la sociedad. In: 3er Congreso Científico Internacional Tecnología, Universidad y Sociedad TUS, pp. 479, 480 (2015)

16. Sabaté, J., Pesoa, M.: Consideraciones sobre las 'buenas medidas' y la composición de los crecimientos residenciales en las ciudades intermedias españolas (1980–2010). Urban, pp. 17, 20, 22, 23 (2016)

17. González, L.: Socialismo y vivienda obrera en España (1926–1939), vol. 120, p. 56. Universidad de Salamanca, Salamanca (2003)

18. Klein, A.: Vivienda mínima: 1906–1957, no. 728.3, pp. 32, 90–96. Gustavo Gili, Barcelona (1980)

19. Gravagnuolo, B.: Historia del urbanismo en Europa 1750–1960, no. 14, pp. 370, 371. Ediciones Akal, Madrid (1998)

20. Bamba, J.C.: Vivienda Colectiva Pública: Guayaquil (1970–1990). Universidad Católica de Santiago de Guayaquil, pp. 31, 35 (2018)

21. El Universo (ed.): Gobierno entrega 204 casas de Socio Vivienda (2010)

22. El Universo (ed.): Proyecto Habitacional Socio Vivienda 2 en Guayaquil (2011)

23. Almeida, B., Muscio, E., Iparreño, L., Anaya, J.: Panel prefabricado de guadua-acero-mortero micro vibrado con ce-niza de cáscara de arroz para vivienda de interés social. In: Congreso Internacional de Innovación Tecnológica en la Edificación (2018)

24. Reyes González, J.M.: D21 System: un juego para ser habitado. Mairea Libros, Madrid (2007)

25. Reyes, J.M.: Tendencias actuales de los proyectos de arquitectura doméstica fabricados con sistemas de construcción por componentes compatibles, Universidad Politécnica de Madrid, (1997)

Improving the Blockchain User Experience - An Approach to Address Blockchain Mass Adoption Issues from a Human-Centred Perspective

Leonhard Glomann[✉], Maximilian Schmid, and Nika Kitajewa

LINC Interactionarchitects GmbH, 80469 Munich, Germany
{leo.glomann,maximilian.schmid,
nika.kitajewa}@linc-interaction.de

Abstract. Blockchain is seen as one of the most promising technologies that appeared in recent years. However, blockchain-based decentralised apps, DApps, have failed to receive mainstream attraction so far. The objective of this paper is to outline potential solutions for the most severe adoption problems from a human-centred perspective. Existing adoption issues will get analysed via unstructured quantitative as well as qualitative user research, in order to discover causes and consequences of these issues. The identified four major issues for mass adoption are: The motivation to change, the onboarding challenge, the usability problem and the feature problem. The insights gained from this research form the basis for solution approaches, which can be summarized in the following statements: DApps need to offer a distinct user benefit, focus on learnability, be conveniently accessible and its constraints need to be turned into assets.

Keywords: Blockchain · DApps · Cryptocurrencies · Technology adoption · Learnability · Human-Centered design

1 Introduction and Objective

The blockchain technology is seen as one of the most significant innovations of our time: Many highlight its high potential for future societies, while some even go as far as claiming it already marks a breaking point for the entire internet [1]. This comes as no big surprise, considering the potential benefits this technology offers to its users. It grants a greater transparency than conventional methods, as all users of one blockchain share the same information. This ties into higher security of the data as changes require a high number of users to agree for changes, as opposed to single user authentications. Additionally, all transactions are linked, merged and encrypted which increases the difficulty for hackers to change the data in the blockchain. For the users however, the technology offers an easy way to view all data points by offering ways to track the origin of an object in addition to all its stopovers [2]. These are just a few examples of advantages this new technology offers to its users. However, a widespread acceptance of the blockchain technology and decentralized applications, so-called DApps, is yet to

© Springer Nature Switzerland AG 2020
T. Ahram (Ed.): AHFE 2019, AISC 965, pp. 608–616, 2020.
https://doi.org/10.1007/978-3-030-20454-9_60

be seen due to a number of reasons. The overarching objective of this paper is to identify the most hindering issues blockchain faces, especially regarding usability and user experience. In addition to that, solution approaches are provided from a human-centred point of view.

2 Definitions of Blockchain and DApps

The first big issue when dealing with analysing the usage of blockchain as a technology is the lack of a clear definition of the term blockchain itself. As stated by The Verge in their report "blockchain is meaningless" [3], there is no definition of the term that is universally agreed upon. The term has gone a long way since it was only used to describe the technology of a shared linear transaction-logging database behind Bitcoin. Nowadays, the term has become a buzzword, used for a synonym for everything from inter-bank transactions to a supply-chain database used by Walmart [3]. The Oxford Living Dictionaries take the approach of defining blockchain as "a system in which a record of transactions made in Vitcoin or another cryptocurrency are maintained across several computers that are linked in a peer-to-peer network" [4]. While the "digital ledger" characteristic is quite commonly agreed on, the mentioned dependency to cryptocurrencies, however, ignores the existence of blockchain-based services, without an associated cryptocurrency. Other definitions include the statement that a blockchain has to be public and is always decentralized [5]. However, there are many blockchains which are not generally publicly accessible or based on a decentralized infrastructure [3]. So again, the definition is unable to do the term justice. A good representation of the lacking definition for the term can be found in an article published in the Harvard Business Review called "Blockchain will help us prove our identities in a digital world". The article states, that Estonia operates a blockchain based digital identity scheme since 2007 already. The system in place uses a distributed ledger, where every individual can control the data and pass it along to others [6]. According to David Birch, who works as a financial technology consultant and is the author of the book "Before Babylon, Beyond Bitcoin", the system used in Estonia is nothing but an urban legend, only getting so much attention due to a lack of knowledge of many people. Speaking from a technical level, this happened due to a misunderstanding of the use of hashes [7]. On the other hand, Estonia can pretty well claim it actually operates on blockchain due to the lack of a commonly agreed on definition [3]. When referring to blockchain in this paper, the definition of it being an open decentralized database where any kind of information can be stored, shared and accounted, while the individual blocks of data are connected in a cryptographic way, will be used [8]. However, the aim of this paper is not to define terminologies. For this paper, the definition only serves the purpose of giving a general sense of direction. Its impact on the identification of the problems and the solution attempts is limited.

Additionally, the terminology of DApp is important: A DApp is "an application that runs on a network in a distributed fashion with participant information securely (and possibly pseudonymously) protected and operation execution decentralized across network nodes" [1]. There are four important points they have to fulfil: First of all, the app has to have an open source code, accessible and changeable for and by every user.

Secondly, this source code has to be stored within a blockchain, together with every other piece of data and information in the app. The use of cryptographically encrypted tokens is the third important detail of DApps. The last point is the possibility to create tokens as a reward for miners. Only apps fulfilling these four points can be considered a DApp [9].

3   Research Method

This paper focuses on the mass adoption of the blockchain technology and DApps. Various models exist that describe the adoption or acceptance of technology. Hillmer groups them by purpose into diffusion theories, user acceptance theories, decision making theories, personality theories and organisation structure theories [10]. Within the user acceptance theories, the Technology Acceptance Models TAM 1 and TAM 2 [11] and the Unified Model of User Acceptance of Information Technology (UTAUT) [12] are commonly used for academic reference. Despite a lot of criticism of these models [13], the adoption of a (new) technology is supposed to be founded on various user-dependent factors: The use behaviour is considered to be influenced by expectations (performance, effort), conditions (social, facilitating) and perception (perceived usefulness and perceived ease of use). The underlying dependency of the aforementioned technology adoption models on the specific context of use and specific user needs, justifies to address the adoption of DApps via an user-centred approach like Human-Centred Design [14].

Following Human-Centred Design, the initial unstructured information gathering process uses both quantitative and qualitative user research. The quantitative research consists of a survey, which was conducted in various online communities (reddit forums, discord channels, telegram groups and twitter feeds), all having a certain connection to the blockchain technology. This was done to reach users in diverse fields dealing with the blockchain technology. Additionally, it allowed to get in touch with both experts and beginners in the field, to gather wide-ranging knowledge. This was necessary to understand potential reasons for the lack of mainstream adoption, as experts tend to see other issues than beginners. Therefore, the survey revolves around the existing knowledge of the users, their current and potential future use of blockchain-based services, how easily they access DApps and their general problems faced when using them and when having gotten into using them in the first place. For the qualitative interviews, the questions were modified to fit the knowledge level of the expert interview partners. The focus was the elaboration of the problems hindering the technology from receiving mainstream adoption so far and the various problems the interview participants faced at the time they got in touch with the blockchain technology. The experts were asked about the problems they see regarding current and future blockchain-based services and possible ways to address them from a user's perspective. Besides that, expert reviews have been conducted with a number of DApps.

In addition to the conducted survey and interviews, further sources like specialist community forums and blogs were consulted. Most of the data gathered here matched the one from the user research. Nevertheless, this additional information collection was necessary to validate the data.

4   Research Results

Over the course of the whole initial unstructured research activity, a number of problems have already been identified. Obviously, the degree of how much these problems affect a certain user type varies – depending on the background, knowledge, experience and the will and determination to overcome said issues. If applications and services using this technology aim to receive mainstream appeal and get adopted by a larger audience, it is vital to offer an easy starting point for newcomers, as well as a pleasant experience for everyone over the whole process of use. The following paragraphs will outline the identified four major problems and corresponding solution approaches from a user experience perspective. The described solution approaches aim to be generically applicable in virtually any blockchain-based scenario.

4.1   The Motivation to Change

Problem outline. According to Jason Bloomberg, an IT industry analyst, the "blockchain is a solution looking for a problem" [15]. However, the starting point of the technology was the invention of the peer-to-peer electronic cash system Bitcoin [16], addressing the capability of transacting money from one person to another without the need of a banking institution. The system became known, adopted and used primarily by users with interest or background in computer programming or information technology [17]. Within that specific user group, the factors to influence the adoption of the underlying blockchain technology have been positively addressed, especially when it comes to the expectations of performance effort: Due to the perceived innovation potential of the technology, combined with a high level of IT knowledge, this user group had a relatively low entry barrier to experiment and use the technology. However, for the average internet user with an average technology affinity, this entry barrier is placed higher. The average internet user feels satisfied with common web-based services for areas like communication, banking or entertainment. Why should this kind of mainstream user switch to Steemit* from Facebook, use Keybase* instead of Slack, Storj* instead of Dropbox or Essentia.one* instead of Android?[1] What might be a motivation to change from a well-established, commercial service to a DApp?

Solution approach: DApps need to offer a distinct user benefit. First and foremost, what is the motivation to change from a current web service to a DApp? The mission of the Internet of Blockchain Foundation IOBF is to "foster the adoption of and transition to […] a user-centric web where users fully own their data, identity and digital assets" [18]. As part of this mission, the IOBF supports a decentralized framework called

---

[1] www.steemit.com, www.keybase.io, www.storj.io, www.essentia.one

Essentia.one. It provides a service "where cryptography rules and data remain in the control of the user" [19], natively integrating DApps and enable interoperability between multiple blockchains. The aim is to provide an easy-to-use interface for anyone to access, own and manage their digital lives in a decentralized web. The decentralized web is defined by the following standards: "Persistent, Not-Corruptible, Privacy-Focused, Anti-Censorship, Language Agnostic, Host Agnostic, Fault-Proof and Future-Proof" [18]. In conclusion, it's required to give mainstream users a clear motivation to change: By making people aware about the lack of these standards in traditional centralized internet services, the characteristics of DApps are likely to be understood as personally valuable.

4.2   The Onboarding Challenge

Problem outline. A key issue for many users is the knowledge gathering process, someone has to undergo before even being able to use a blockchain-based service or product properly. The biggest issue here is the accessibility of the information required to learn the basics. It is very difficult for mainstream users to find a starting point, as most of the available information aims at people with a higher degree of technology affinity, software developers or cryptocurrency experts. For someone just getting started, the amount of time and frustration before finding what they are looking for is a major pain point and might result in them losing interest in the topic.

Solution approach: DApps need to focus on learnability. The problems with the knowledge gathering process of the users could be addressed with various solutions. But first of all, it needs to be stated that the situation already drastically improved over the past years. During an interview with an expert who got into the blockchain topic back in 2013, it has become easier to figure out where to get the information you are looking for and who the information is aimed at since then. It is likely that this problem gets less imminent in the future as DApps grow in maturity. However, in order to get there, the information beginners are looking for needs to be easily accessible. Thus, a solution for this problem is to create general as well as platform-specific tutorials for users on various levels of knowledge. These trainings should be separated not only by knowledge level, but also by specific topic. For example, there should be courses about using cryptocurrency exchanges for beginners, advanced users and experts. Everybody could access the information they are looking for at the right level. It will probably be quite hard for someone just getting into the field of blockchain-based services to understand and assimilate an information aimed at an expert who also happens to be a DApp programmer. Additionally, trained support staff or specific bots should be considered to teach beginners and help them with their problems, give them more information about what to do next or even point them to literature, blogs, etc. – widening their knowledge in the field. The idea here is to have this medium in place, specifically for beginners, on top of regular tool support. This may be advertised as a unique selling point (USP) for certain services. Another idea is to view the onboarding challenge holistically, gathering common pain points of first-time users of cryptocurrency exchanges, platforms and DApps. The aim would be to create a beginner's guide or tool, materializing in the shape of an overarching DApp operating system. Its objective should be to give people the information they need, to increase the knowledge about the technology, its advantages, but also its disadvantages and risks on a broad level. This could even lead to more curiosity amongst individuals to try out these services, which would lead to a bigger possible audience for products in this field.

4.3   The Usability Problem

Problem outline. The next issue identified revolves specifically around the user experience and user interface design. The problem boils down to the matter of fact that there are just very few usability engineers, user researchers and interaction designers working in the field up until now. The whole conceptual and design part is mostly covered as a secondary activity by developers. The majority of developers of blockchain services focus on functionality, maintenance and stability. Usability, efficiency and accessibility are less in focus. In practice, this leads to potentially confusing interfaces. The consequence is that users have to learn their way around with different services and applications almost from scratch, as similar things are handled completely different each time, which sometimes leads to frustration and irritation amongst users.

Solution approach: DApps need to be conveniently accessible. For issues regarding the user interface and its interaction, there are multiple possible solutions. A simple one is to hire a specialized counterpart focusing on usability and user experience. However, for most development teams, this is a funding question. For others this is a question of finding the right specialists, as the usability and design field is relatively underrepresented in the area of DApps. A right balance has to be found between the cost of development and interaction design resources. When aiming for convenient accessibility for users, the amount of features in applications are not corresponding to the quality in user experience. In order to achieve a good user experience step-by-step, it is seen as a good practice to first release a minimum viable product (MVP), considering all aspects including acceptable usability and improve upon it iteratively. Another way to improve application usability is to stick as far as possible to common standards in interaction patterns, that the user group has learnt by using traditional centralized apps and platforms.

4.4   The Feature Problem

Problem outline. Another problem revolves around certain features which are specific to blockchain, like account management, the management of passwords and the need for multiple system accounts in some cases. This part is very application-specific, however it is a pain point for many nonetheless. Generally speaking, there are multiple perceived issues with password management. Due to the underlying logic of truly decentralized technologies, resetting passwords is absolutely impossible as there is no central authority to allow that technically – which means losing a password results in being unable to access the user account ever again. All the information and value storage, like holdings of cryptocurrencies, will be inaccessible for its owner. Therefore, all services and apps have to be very transparent about this issue as it can be a huge risk, especially for newcomers in the field. The problem also includes the fact that passwords tend to be a long string of random small and capital letters and numbers cobbled together, which makes it easy to have misspellings when writing them down. Another issue is the need to register and use multiple platforms and respective accounts, just to be able to make use of one single DApp. A key issue here, in many cases, is the need for owning cryptocurrencies, such as Ethereum as a prerequisite to use certain applications based on the Ethereum ecosystem. This means, people who are interested in using a certain DApp, first need to familiarize themselves with the ecosystems' prerequisites, including all the problems of gathering information, setting up accounts, storing passwords securely, getting used to the way the application works, before even being able to create the account for the service you wanted to use in the first place; which again comes with all the issues mentioned previously. This can be incredible time demanding and infuriating, especially for beginners.

Solution approach: DApps' constraints need to be turned into assets. Regarding the password problem, there are already some apps out there, utilising the so-called mnemonic seed. Mnemonic means "something such as a very short poem or a special word used to help a person remember something" as stated in the Cambridge dictionary [20]. An example for such a thing would be "never eat sea-weed" for remembering the directions north, east, south, west. In the context of passwords, this technique can be used to make passwords easier to read, manage, store, and memorize [21]. This might make it more convenient to users, as managing a string of words is simpler than managing a long chain of random characters. The fact that passwords cannot be restored is another major issue for mainstream users, as stated above. This is especially problematic as setting up an account and thus handling a password is the first thing to do when using a service or product. Focussing on the unexperienced user, it has to be clearly communicated, that passwords once lost cannot be restored. The experts in the interviews agreed, that this cannot be stressed enough: A simple "don't lose your password" is not sufficient. Therefore, during a registration process, the user needs to be educated via precise and repetitive dialogues about the risk of losing the password. It is important to add a short description why the technology won't allow password restorations by pointing out the technical limitations and specialities. Here, not stressing the limitations themselves, but communicating the connected benefit – the user is the one and only owner of his personal data – is key to turn DApps' constraints into USPs. Another aspect of the feature problems is about the need for holdings of cryptocurrencies for using most DApps. A simple solution would be to detach at least the initial use of the service from the requirement to own cryptocurrencies. This lowers the entry barrier, similar as it is known for successful traditional centralized services, which are free to use and don't require for example a bank account. Additionally, registering on one specific traditional centralized platform, for example Facebook, lets the user allow to re-use the login credentials for associated services. Shared accounts and single sign-on fosters convenience for mainstream users as well. Learning from existing patterns like these, which are seen as common practice amongst traditional internet users, is key to allow for mass adoption of the technology. Only then, technology restrictions or limitations may be turned into useful assets.

5 Conclusion

In conclusion, it has been shown that there are important challenges for providers of blockchain-based products and services to undergo. Today, the complexity of the underlying blockchain technology, the performance issues, the impossibility to revert mistakes, coupled together with the mainstream adaption problems make the blockchain a difficult technology to deal with [15]. As a provider, it is important to address the aforementioned issues to achieve a pleasant experience for every user, including an easy start for newcomers. In addition, technological advantages like increased transparency, higher privacy and security levels for personal data or better information tracking abilities [2] need to be strongly emphasized via easy ways of communication. In the end, the question whether a service profits from using the blockchain technology should always depend on a simple question: "Which user needs are supposed to be addressed by the service?" Based on the answer to this question, there should be a very clear reasoning for using blockchain as a technology.

6 Further Research

The initial unstructured research and solution approaches, as described in this paper, open up various opportunities. First of all, in-depth research of clearly split sub target groups of current and potential users, structurally identifying user needs is an interesting research object. Further analysis should consider specific categories of DApps, domain-specific use cases and potential future business cases. Secondly, adding to and detailing out the identified problems, starting from the above-mentioned key areas: The motivation to change, the onboarding challenge, the usability problem and the feature problem. This might be done by deriving potentials from the identified user needs. Lastly, the solution approaches are subject to be more concrete and precisely matched to problems and potentials with further details.

References

1. Swan, M., Blockchain: Blueprint for a New Economy. O'Reilly and Associates, Sebastopol (2015)
2. Hooper, M.: Top five blockchain benefits transforming your industry. IBM. https://www.ibm.com/blogs/blockchain/2018/02/top-five-blockchain-benefits-transforming-your-industry/ (2018). Retrieved 31 January 2019
3. Jeffries, A.: 'Blockchain' is meaningless. The Verge. https://www.theverge.com/2018/3/7/17091766/blockchain-bitcoin-ethereum-cryptocurrency-meaning (2018). Retrieved 31 January 2019
4. The Oxford Living Dictionaries: https://en.oxforddictionaries.com/definition/blockchain (2019). Retrieved 31 January 2019
5. Fortney, L.: Blockchain, Explained, Investopedia. https://www.investopedia.com/terms/b/blockchain.asp (2019). Retrieved 31 January 2019

6. Mainelli, M.: Blockchain will help us prove our identities in a digital world. Harvard Business Review. https://hbr.org/2017/03/blockchain-will-help-us-prove-our-identities-in-a-digital-world (2017). Retrieved 31 January 2019
7. Birch, D.: Before Babylon, Beyond Bitcoin: From Money That We Understand to Money That Understands Us, London Publishing Partnership, London (2017)
8. Schiller, K.: Was ist Blockchain?, Blockchainwelt. https://blockchainwelt.de/blockchain-was-ist-das/#Blockchain_Definition_und_Erklaerung (2018). Retrieved 31 January 2019
9. Schiller, K.: Was ist eine DApp (dezentralisierte App)?, Blockchainwelt. https://blockchainwelt.de/dapp-dezentralisierte-app-dapps/ (2018). Retrieved 31 January 2019
10. Hillmer, U.: Existing theories considering technology adoption. In: Technology Acceptance in Mechatronics. Gabler (2009)
11. Davis, F.D., Bagozzi, R.P., Warshaw, P.R.: User acceptance of computer technology: a comparison of two theoretical models. Manage. Sci. 35(8), 982–1003 (1989)
12. Venkatesh, V., Morris, M.G., Davis, G.B., Davis, F.D.: User acceptance of information technology: toward a unified view. MIS Q. 27(3), 425–478 (2003)
13. Chuttur M.Y.: Overview of the technology acceptance model: origins, developments and future directions, Indiana University, USA. Sprouts: Working Papers on Information Systems, 9(37) (2009)
14. International Organization for Standardization: DIN EN ISO 9241-210 Human-centred design for interactive systems (2010)
15. Bloomberg, J.: Eight reasons to be skeptical about blockchain. Forbes. https://www.forbes.com/sites/jasonbloomberg/2017/05/31/eight-reasons-to-be-skeptical-about-blockchain/ (2017). Retrieved 31 January 2019
16. Nakamoto, S.: Bitcoin: a peer-to-peer electronic cash system. Bitcoin Foundation. https://bitcoin.org/bitcoin.pdf (2008). Retrieved 31 January 2019
17. Yelowitz, A., Wilson, M.: Characteristics of Bitcoin users: an analysis of Google search data. Appl. Econ. Lett. 22, 1–7 (2015)
18. Internet of Blockchain Foundation: https://iobf.co (2019). Retrieved 31 January 2019
19. Essentia.One: https://essentia.one (2019). Retrieved 31 January 2019
20. Cambridge Dictionary: https://dictionary.cambridge.org/dictionary/english/mnemonic (2019). Retrieved 31 January 2019
21. Khatwani, S.: What Is Mnemonic Phrase & Mnemonic Passphrase? Coinsutra. https://coinsutra.com/mnemonic-passphrase/ (2018). Retrieved 31 January 2019

Human Factors in Energy

Beyond COSS: Human Factors for Whole Plant Management

Roger Lew[1(✉)], Thomas A. Ulrich[2], and Ronald L. Boring[2]

[1] University of Idaho, Moscow, ID, USA
rogerlew@uidaho.edu
[2] Idaho National Laboratory, Idaho Falls, ID, USA
{thomas.ulrich, ronald.boring}@inl.gov

Abstract. A Computerized Operator Support System (COSS) has been envisioned and prototyped for nuclear power plant (NPP) operations. The COSS supports operator decision making during abnormal events by making use of an online fault-diagnostic system known as PRO-AID. PRO-AID uses existing sensors and confluence equations to detect system faults before plant threshold alarms would alert operators. In a full-scope, full-scale nuclear power plant simulator we demonstrated that early diagnosis in conjunction with computer-based-procedures can assist operators in mitigating faults in circumstances that would normally lead to shutting down the plant. Here we conceive of a computerized plant management system (CPMS) that would complement COSS. COSS is intended to aid in the short-time scale operations of a plant. The CPMS is aimed at supporting longer time-scale activities involving maintenance, calibration, and early warning fault detection. Digitization allows staff to manage and monitor assets remotely without being physically present locally at the equipment under surveillance. This allows for specialization by equipment and systems rather than location. Cyber-physical systems and the industrial internet of things have no shortage of possibilities and promises, but often the benefits come with conflicting considerations: availability may tradeoff with increased cyber risk, increased redundancy may come at the cost of increased complexity, outsourcing engineering requires more communication across organizations and the associated administrative burden. In too many instances, humans tend to fill the gaps between what automation cannot reliably or cost-effectively accomplish. Human-centered design using human factors considers the role of the human as part of a human-machine-system. Our COSS concept examined how a control system, a diagnostic system, expert guidance system, and computer-based procedure system could work in concert to aid operator actions. Here we examine the potential for human factors and human-centered design for whole plant management using a CPMS. A CPMS could serve several valuable functions from providing computerized procedures for field operations to using prognostics for early failure detection, or managing information flows from sensors to business networks, system engineers, cyber-security personnel, and third-party vendors.

Keywords: Nuclear power · Nuclear human factors · Control room · Computerized Operator Support Systems (COSS) · Process control

© Springer Nature Switzerland AG 2020
T. Ahram (Ed.): AHFE 2019, AISC 965, pp. 619–630, 2020.
https://doi.org/10.1007/978-3-030-20454-9_61

1 Introduction

1.1 Modernization vs. Advanced Reactors

The U.S. has 99 operating nuclear power plants (NPPs) that account for roughly 20% of total electricity production. NPPs have high initial capital expenses, but fuel is inexpensive in contrast to gas turbine plants that have high fuel costs and relatively low initial capital expenditures. For this and other reasons it is worthwhile to keep plants in operation. These plants are being modernized to enable continued operation beyond their current 40-year licensing periods. In contrast to other electricity generation sources, nuclear power has high labor costs with the industry average at one person per 2 MWe. Nuclear energy is also comprised of a highly skilled and knowledgeable but aging workforce.

Previous efforts have conceived, designed, and prototyped a Computerized Operator Support System (COSS) for existing Generation II NPPs [1]. COSS is envisioned to be deployed alongside control room modernization upgrades in hybrid digital/analog control rooms. Our COSS concept contains a conventional distributed control system (DCS) human-machine interface (HMI) and enhances it with a computer-based procedure system and diagnostic system. The diagnostic system is capable of detecting many faults ahead of plant alarms, and an expert guidance system can then convey the fault diagnosis to operators as well as suggest mitigations for the fault. As envisioned, COSS would not fundamentally change how NPPs are operated, but would provide insight, information, and review capabilities to supplement operators during normal as well as abnormal and highly time critical events. A key principle with COSS is that it does not displace the autonomy of the operators. It is also important to note that COSS does not alter how the plant is operated. COSS was envisioned as a system that would operate in a heterogeneous environment of analog interlocks, conventional control systems, and distributed control systems. The underlying operation of the plant remains unchanged, relying primarily on single point controllers as envisioned by the original plant engineers many decades ago.

Here we turn our attention to advanced reactors. Generation IIIb reactors iterate on existing Generation II designs with increased passive safety and a substantial increase in the use of digital instrumentation and control (I&C) technology, but their control systems maintain traditional control strategies relying primarily on single-input, single-output classical control. Passive safety, digitization, autonomous control systems, and artificial intelligence provide new possibilities for reactor control and operations. Of particular interest to the authors are small modular reactors (SMRs). This class of reactors are defined as producing less than 300 MWe and manufactured prior to transportation to the site of operations. An SMR plant could operate as a single unit, but it is most cost effective to operate several as a single plant. The NuScale Power reactor concept, for example, features 12 reactors at each site. SMRs have many attractive features including competitive costs, increased flexibility with load generation, high capacity factor, and co-generation of combined heat and power generation. However, new reactor designs and in particular SMRs have a critical challenge in reducing the high personnel costs associated with nuclear power. U.S. Nuclear Regulatory Commission (NRC) regulations for Generation II plants suggest three operators per reactor;

twelve reactors would therefore require 36 operators staffed around the clock. Such a requirement could make operating a plant cost prohibitive. The NRC allows for plants with advanced automation (i.e., GE's ESBWR) to have only one licensed operator at the controls during normal, at power operations, and in future Generation IV reactors, two to four operators may manage up to a dozen modular units [2]. This is a move in the right direction toward reducing operational labor costs.

Historically plants have used periodic maintenance schedules that would require personnel to service equipment at established intervals or after operating durations. Such an approach is not only costly, but unnecessary maintenance always introduces the possibility for human error[1]. In contrast, one could use predictive maintenance to determine when fluids or components should be serviced. Predictive maintenance could reduce labor needed for maintenance while reducing the risk to the plant and human workers. The catch is that predictive maintenance requires understanding components, their fault modes, and having available data to predict faults ahead of actual fault events. In addition to having an appropriate sensor set, data from across installations needs to be examined by engineering and data science experts.

1.2 Autonomous Control vs. Automatic Control

Autonomous control systems would move beyond the capabilities of automatic control systems. Wood, Upadhyaya, and Floyd remind us that in Greek *automatos* means self-acting, whereas *autonomos* means independent [3]. A fully automatic control system would have the ability to perform normal operational control activities like startup and load changing as well as automated responses to alarm conditions. However, an automatic system could fall short if component or system malfunctions alter how the control actions impact the state of the plant. Nuclear plant operations also have unique and sometimes conflicting objectives that would need to be managed under autonomous control including temperature performance, load-following operation, alarm processing and diagnosis, optimal fuel loading, fault detection and identification, and sensor signal validation [4]. Under current operating conditions, human operators become responsible for identifying, diagnosing, and mitigating a fault.

Humans are adept problem solvers in novel situations and have demonstrated remarkable performance in extreme situations. Chesley Burnett "Sully" Sullenberger landing U.S. Airways Flight 1549 on the Hudson River is a prime example. There are also numerous cases of human actions being at least partially responsible for accidents, and humans are often regarded as the least reliable components of human machine systems. Thus, humans have unique strengths and weaknesses and their reliability can suffer when under time pressure and complex operating environments. An autonomous plant on the other hand could operate with much less variability. In a recent publication, researchers developed and validated an autonomous safety control system

---

[1] Anecdotally, people tend to service their vehicles right before a road trip. While well-intentioned it would be less risky to have the vehicle serviced days before the service to make sure the vehicle is operating normally before embarking. I once drove 300 miles with my oilplug merely hand tight. Had I taken my own advice, I would have seen a tell tail oil stain on my driveway and not risked the plug coming all the way out while on the highway.

designed for a Westinghouse 930 MWe, three-loop pressurized water reactor. The system was trained with 206 scenarios and over 244,000 training sets. When compared to automation and human operators the autonomous system responded faster and better controlled critical plant parameters like reactor water level [5].

While there is much speculation as to how autonomous control would be accomplished, control engineers envision conventional digital control systems layered with intelligent agents for fault detection, planning, supervision, cyber security and so forth. High fidelity physical models and simulation technologies could be used to vet autonomous systems under a much wider set of error conditions than would be possible with human operators. Autonomous control hinges on the ability to estimate the result of control actions and the performance of machine learning to optimize those actions [6]. Intelligent agents could use expert systems (i.e., hard-coded information), fuzzy logic, neural networks, and genetic algorithms. The intelligent agents would need to form a fault tolerant system that would be capable of safely controlling the system under prospective novel fault conditions with a high degree of certainty. The agents would also need the ability to retain the controllability and observability following component failure. The plant does not have to recover (i.e., be resilient) but it should have the ability to safely shutdown. A component failure could invalidate the normal operation of the control system, and the control systems needs to be able to determine an alternate method of retaining controllability. Controllability could be assured through analytical redundancy in combination with examining controller strategies and parameters for all conceivable failed states to develop alternate control schemes for component failures off-line and store them for eventual use. Alternatively, artificial intelligence could be used to learn on-line how to control the plant, or an expert system could be devised with hard-coded strategies [7]. Some are naturally skeptical of the feasibility of such a system. However, much of this work is actively being pursued in nuclear or related industries. Here we continue under the assumption that such an autonomous plant could be designed and examine the repercussions it could have on human operators and plant management.

1.3   Plant Operation and Monitoring

To maximize cost efficiency, reactor vendors could contract maintenance and ongoing engineering support for utilities. This would allow reactors to be maintained as a fleet across sites. Keeping plants similar to one another at the component level simplifies the engineering workload and regulatory burden compared to managing plants independently. While the reactors may be identical or versioned across multiple sites, the plants could be tailored to the needs of the operator. For example, plants in areas with lots of solar power might need co-firing gas capabilities. Plants in cold climates might make use of excess heat for municipal or industrial heating purposes. Autonomous plant control capable of remaining vigilant 24/7 would significantly reduce the necessity for human operators. As envisioned, plants would be able to detect abnormalities and safely shutdown the plant, and isolate any remaining issues without intervention. Then humans could be dispatched to resolve any remaining problems and get the plant back up and running. In the aftermath, engineers would need to examine the root cause of the failure, identify whether other plants or units are susceptible, refine plant models, and push updated fault detection and remediation protocols.

Humans would still need to interact with the plant and maintain a deep understanding of how the plant functions, but the autonomous systems would supplant many of the responsibilities that currently befall human operators. Thus, the HMI requirements would substantially change and new cyberinfrastructure systems would need to be developed to support operations. With current HMIs the ability to understand the state of the plant in real-time and rapidly respond to abnormal circumstances is the primary consideration. With autonomous plants and autonomous smart grids, real-time monitoring of the minutia of a plant may not be necessary or even desirable. Future energy grids and energy producers will form an ever increasingly complex and dynamic system intertwined with weather, transportation, and resource commodities. Human working memory and information bandwidth capacities would quickly become overwhelmed compared to an autonomous machine counterpart. For instance, consider telephony automation, one might find it remarkable that telephone switchboard operators were employed well into the 1960s with the last operator in the U.S. retiring in 1983. At present day, providing human supervisory oversight of telephone operations seems absurd. Electro-mechanical automatic exchanges replaced human operators and digital systems eventually replaced those. The operation of telephone switching is now almost entirely automated and fault tolerant to losing exchange nodes. Much of the human capital can then be transitioned to higher level tasks like examining network topology for vulnerabilities and ensuring exchange capacities are sufficient. Autonomous control could result in a similar abstraction of humans from the moment-to-moment operation of plants in order to guarantee safety and enhance efficiency.

2   Humans in the Loop

2.1   Maintaining Plant Health

More capable autonomous systems would reduce the actions required by humans similar to how autopilot reduces the workload of pilots. Autonomous systems would be able to carry out routine and abnormal operations. However, humans may still serve vital roles as supervisors and decision makers.

Instrumentation and controls would be self-diagnosing. So, for example, if a sensor stopped responding or even started drifting, the system would be able to detect and report the abnormality. In such a hypothetical event the plant would be able to continue to operate by enlisting redundant sensor values, virtual sensors, intelligent process-based control, etc. Humans could leisurely investigate the issue and correct it to maintain plant health. Some of this type of monitoring could even be accomplished remotely and coordinated with on-site personnel who would still be needed for field-inspections and potentially coordinated during major plant operations like startup, shutdown, etc. The company Ormat operates geothermal plants and recovered energy plants in 22 countries. They had remote monitoring capabilities for several years. At the time IT infrastructure was unreliable and slow. They have now transitioned to incorporating increased automation into their control systems and remote monitoring. This allows for fewer personnel to oversee more assets. From the perspective of the operators the experience is the same regardless of whether their control room is situated at the plant, a few hundred feet from the plant, or several thousand miles from the plant. This strategy by Ormat also allows for specialization by equipment or system instead of physical location. This allows the most capable personnel to be tasked with problem-solving when problems do arise [8]. SunPower has remote operations for controlling approximately 1.3 gigawatts of solar power as of 2016 [9]. The same business logic for justifying remote operations would apply to nuclear power. Another possibility is a hybrid arrangement where reactor vendors or turbine vendors are responsible for some of the monitoring and diagnostics.

2.2   Outage Control

Outages with Generation II NPPs require a great deal of human labor to safely and quickly get the units back up and running. With large reactors, the utility loses potentially $1 million or more in revenue everyday they are down. Many of the maintenance activities are planned for months or even years in advance. Refueling for SMRs should be simpler and since they have multiple units could implement a rolling outage schedule so that the plant does not have to be taken completely offline for refueling and maintenance. Refueling would still require a good deal of human diligence and coordination both internally and with supporting vendors.

2.3   Load Control, Strategic Planning

Current NPPs shut down for refueling every 18 or 24 months for 30 days or more, and in the interim the more or less operate continuously. Increased proliferation of renewables in combination with other factors results in this strategy becoming less viable. Energy markets are becoming more dynamic with requirements for utilities down to even 5-minute intervals. Texas electric grid operators, the Electric Reliability Council of Texas (ERCOT), have to contend with 23 GW of wind power, about 4 times that of California. Texas operates market rules on 5-minute intervals with severe penalties for non-compliance [10].

Grid operations and utilities need the ability to easily and autonomously curtail or ramp assets in a coordinated fashion to meet demand, contend with cascade disruptions or bringing power back online. Utilities must also manage conflicting objectives. For example, plants located next to rivers or lakes might use that water as a tertiary cooling loop. Plants may have to reduce usage or be restricted on the amount of heat that is returned to the body of water. Ultimately, humans may need to decide what strategies to employ. Also, their strategies may shift depending on environmental circumstances and changing risk tolerances. For instance, California wildfires have demonstrated that operating transmission lines during hot and dry periods can be hazardous. With a smarter grid and plants risk of fire could be augmented at the cost of less efficient production or utilization of stored energy.

2.4   Incidence Response

The CPMS would also need to support timely actions and decision making if an incident where to occur that the autonomous system could not handle independently. For example, quickly initiating lockout-tagout of equipment might be needed to safely send personnel into otherwise unsafe areas of the plant.

2.5   Security

As with any cyber-physical infrastructure security is paramount. Security includes monitoring physical and cyber activities at the plant. Security includes detection and response of unauthorized access of resources as well as monitoring and auditing for suspicious activity.

2.6   Oversight, Regulations, and Reporting

Operating a plant falls under numerous jurisdictions and regulations. Plants would need to ensure they are meeting those requirements and have record-keeping to substantiate compliance. If a minor or major event were to occur investigations and reporting would need access to plant data.

2.7   Innovation

Over the past decade autonomous systems have proliferated in the cyber realm as computational power in conjunction with machine learning has advanced. Autonomous systems for better or worse are responsible for filtering our inbox, cataloging the internet, assessing our media preferences, deciding what content is shown on our social media feeds, and so forth. In coming decades, autonomous systems will increasingly make inroads into our physical world through manufacturing, unmanned and manned vehicular automation, and cyber-physical systems. From 2012 to 2014 the number of sensors shipped has increased more than 4 fold to 23.6 billion [10]. These sensors serve a variety of purposes and industries, resulting in an evolving trend of interconnected ecosystems that were segregated by traditional industry boundaries (e.g., manufacturing, transportation, power generation, power distribution, etc.). When combined with the proliferation of digital control systems there exists vast potential for innovations that improve productivity, efficiency, safety, and reliability. Some of this work might take place in industry and or as academic or government research. However, software and data analytics are needed to access and process the volumes of data being collected, securely connect disparate systems, and, perhaps most importantly, provide meaningful insights to key stakeholders.

3   Computerized Plant Management System Considerations

We have reviewed some of the possibilities of autonomous control systems for nuclear power as well as postulating where humans would interface with the plant. From that discussion we can see that the scope extends beyond what is typically thought of as

NPP control. A plant's production, health, and maintenance might be guided by several different departmental units scattered across organizations. These humans would have varying roles from varying disciplines. They would need to coordinate with one another as well as communicate with the plant. For some, read-only access would be sufficient. Others might need to initiate actions or create work-orders for onsite personnel. Some might need to monitor the plant on a moment-by-moment basis, while others might want to query and aggregate plant data over its entire life. Some users will need real-time dashboards for monitoring data, others might need to link in engineering tools, and some users may need to generate reports. The cyber-infrastructure would need to support all of these use cases as well as communicating with plant control systems. In essence middleware or a software layer between the plant and autonomous agents, services, and end-user interfaces. Other domains provide some guidance on how this might be accomplished and in particular what features might be useful.

3.1 Industrial Internet of Things

Smart devices are capable of sensing and acting on their environment. However in isolation they are not particularly useful. It is in the aggregation of data and control that the versatility and flexibility of the internet of things is meant to shine. Research on decentralized platforms exist that would make use of blockchain technology, but for our purposes platforms that utilize centralized network technologies are ahead of the curve. Platforms like ThingsBoard provides a middle layer for collecting, analyzing, and visualizing data, as well as managing devices through rules or microservices. Things-Board can be cloud deployed with redundancy and load-balancing for operations. The ThingsBoard is Open Source with a System as a Service (SaaS). The microservices could be autonomous agents for intrusion detection or asset optimization. The architecture provides functionality for building real-time web-based dashboards (i.e., HMIs) for humans as well as integrating with other services through a variety of protocols.

3.2 Smart Home Automation

Due to the proliferation of commodity smart home devices (e.g., lights, locks, thermostats, switches, cameras, etc.) and adoption by major technology companies like Google, Amazon, and Apple home automation is coming of age. A major headache for adopters has been integrating components across platforms. One would need to choose from Google Home, Amazon's Alexa, or Apple's HomeKit as the controller for their smart home and inevitable use third party mobile phone apps and use other additional services like IFTTT (If This Then That) to add automation functionality not otherwise natively possible. However, an alternative approach is to use the Home Assistant, an open source smart home integration platform, to link devices from incompatible platforms. Automations can be configured through a web-interface, in YAML, scripted in Python, or configured in Node-RED's graphical node based interface. Home Assistant has a standard set of entities (e.g., lights, switches, thermostats, etc.) and is platform agnostic, allowing devices from most all smart-home vendors to be integrated, automated, and controlled under a standard platform. Home Assistant also has support for Google Home, Alexa, and HomeKit, which provides wide flexibility for end-user control.

3.3   Robot Operating System

The Robot Operating System or ROS is a bit of a misnormer in that it is not an operating system for a computer but rather a toolkit for simulating, programming, and controlling robots. ROS is self-described as a robotics middleware first publicly released in 2007 by the Stanford Artificial Intelligence Library. Under the hood, ROS implements a subscriber/publisher model for message passing. Independent nodes (similar to services) control or sense a robot. The nodes could be simple procedural scripts, or more sophisticated autonomous agents. ROS has a three-dimensional simulation environment for developing robots in a virtual environment before deploying to physical environments.

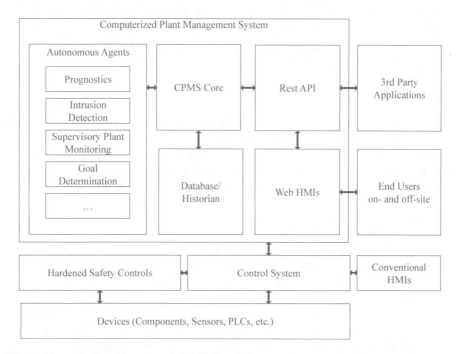

Fig. 1. Conceptual architecture of CPMS in relation to plant control systems and end-user HMIs. The CPMS is envisioned as a centralized system that would serve numerous functions related to autonomous plant control, monitoring, diagnosis, security, and ongoing research and development.

Our CPMS concept is distinct because of the distinct feature set we envision it would provide to advanced reactors. However, the CPMS would share commonalities with the platforms highlighted above (See Fig. 1). The CPMS could be implemented as a middleware (software) layer on more traditional controller architecture. The CPMS would tie in autonomous agents as services. The CPMS would also need functionality to implement human machine interfaces for viewing plant data, performing actions, and communicating with other personnel. The CPMS would also need the ability to interface with 3rd party applications for diagnostics, engineering, business analytics.

4 Human Factors Challenges

4.1 New Challenges

Autonomous control systems may reduce the responsibilities of human operators and make conventional HMIs obsolete, but they also present new use cases and new challenges across several domains: engineering, control, cyber-security, and human factors. Sensor technology is improving and becoming cheaper to deploy. Advances in modelling result in virtual sensors and more generally the ability to extract much more detailed spatial-temporal data. There is great potential for this data if insights can be gained from it. However, gaining insights and understanding the complex data requires the right combination of data analytics, statistics, and visualization.

4.2 Collaboration

Optimizing human resources is vitally important to controlling cost. New plants will need to share experts across units, sites, and even utilities. A given problem might require a team of experts with varied backgrounds from project managers, to engineers, to field technicians, to network technicians. All of these experts would need to acquire detailed technical information and communicate with other team members.

4.3 Visualization Ergonomics

There is no shortage of novel methods for translating data to graphical representations for human representation. Modern web frameworks like d3 and plotty provide powerful techniques for generating useful interactive plots with minimal effort. Often the challenge with visualization is not the lack of possibilities but determining an appropriate representation for the data. Perceptual psychology and human factors could aid in this arena by developing guidelines for specific datasets and translating those to standard but versatile widgets that could be configured in visualization frameworks.

4.4 Cybersecurity Monitoring

Cybersecurity is highlighted as a concern as digital control systems become interconnected [10]. The ability to visualize network topologies and network traffic can aid in the identification of suspicious activities.

4.5 Augmented Reality/Virtual Reality

Augmented reality is the enhancement of natural vision with virtual overlays. Augmented could be implemented using projection mapping, with see through head mounted displays, or utilizing a forward facing camera and a mobile device. Augmented reality is an emerging technology with potential for aiding diagnostics, troubleshooting, and engineering of mechanical systems. Augmented reality could overlay thermal imagery over imagery of components or represent microscopic vibrations detected in rotating equipment.

Virtual reality could likewise serve as a valuable tool for designing and optimizing systems in virtual environments. Virtual reality could also be used for remote inspections is plants were equipped with tele-operated unmanned robots that could move about equipment in plants. The robots could have thermal imaging cameras, telescopic lenses and sound processing that could make remote inspections from a first person view with better than normal human perception possible.

5 Conclusions and Discussion

There is great potential for advancement of energy infrastructure and advanced reactors in the coming decades. Reactors designed from the ground up will incorporate passive safety and robust safety systems. Autonomous control technologies could reduce operational and maintenance personnel costs to make nuclear power cost competitive with competing technologies. The proliferation of renewables provides clean energy but also complicates managing a stable electrical grid. Digitization of critical infrastructure provides opportunities for innovation to increase the productivity and efficiency of systems but also introduces new challenges related to security and interdependency that will need to be solved. Implementation of autonomous control systems and other systems needed to support humans in the loop will require computerized plant managements system technology with capabilities not yet available.

6 Disclaimer

This work of authorship was prepared as an account of work sponsored by an agency of the United States Government. Neither the United States Government, nor any agency thereof, nor any of their employees makes any warranty, express or implied, or assumes any legal liability or responsibility for the accuracy, completeness, or usefulness of any information, apparatus, product, or process disclosed, or represents that its use would not infringe privately-owned rights. Idaho National Laboratory is a multi-program laboratory operated by Battelle Energy Alliance LLC, for the United States Department of Energy.

References

1. Lew, R., Ulrich, T., Boring, R.: Nuclear reactor crew evaluation of a computerized operator support system HMI for chemical and volume control system. In: International Conference on Augmented Cognition (2017)
2. O'Hara, J.M., Higgins, J.C.: Human-System Interfaces to Automatic Systems: Review Guidance and Technical Basis. Brookhaven National Laboratory, Upton, New York (2010)

3. Wood, R.T., Upadhyaya B.R., Floyd D.C.: An autonomous control framework for advanced reactors. Nucl. Eng. Technol. **49**(5) (2017)
4. Basher H., Neal. J. S.: Autonomous control of nuclear power plants. ORNL/TM-2003/252, Oak Ridge National Lab (ORNL), Oak Ridge, TN (United States) (2003)
5. Lee, D., Seong, P.H., Kim, J.: Autonomous operation algorithm for safety systems of nuclear power plants by using long-short term memory and function-based hierarchical framework. Ann. Nucl. Eng. **119** (2018)
6. Pachter, M., Chandler, P.R.: Challenges of autonomous control. IEEE Control Syst. Mag. **18** (4), 92–97 (1998)
7. Stengel, R. F.: Intelligent failure-tolerant control. IEEE Control Syst. Mag. 14–23 (1991)
8. Reshef, R., Fever, T.: Ormat plants: from remote monitoring to remote operation. GRC Transactions (2012)
9. Pickerel, K.: Sunpower launches remote operations control center for monitoring solar plants. Solar Power World (2015)
10. Wolrd Economic Forum. 020315, industrial internet of things: unleashing the potential of connected products and services (2015)

Imagine 2025: Prosumer and Consumer Requirements for Distributed Energy Resource Systems Business Models

Susen Döbelt[(⊠)] and Maria Kreußlein

Professorship of Cognitive and Engineering Psychology,
Chemnitz University of Technology,
Wilhelm-Raabe-Str. 43, 09120 Chemnitz, Germany
{susen.doebelt,
maria.kreusslein}@psycholgogie.tu-chemnitz.de

Abstract. The user-centered design and the acceptance of smart grid technologies is one key factor for their success. To identify user requirements, barriers and underlying variables of acceptance for future business models (DSO controlled, Voltage-Tariff, Peer-to-Peer) a partly-standardized interview study with $N = 21$ pro- and consumers was conducted. The results of quantitative and qualitative data demonstrate that the acceptance of each future energy business model is relatively high. The overall usefulness was rated higher for future business models than the current business model. Prosumers had a more positive attitude towards the Peer-to-Peer model, whereas consumers preferred models in which the effort is low (DSO controlled) or an incentive is offered (Voltage-Tariff). The DSO controlled model is not attractive for prosumers, who criticize the increased dependency and external control. From the results it can be concluded that tariffs should be adapted to the user type.

Keywords: Acceptance · Consumer and prosumer requirements ·
Distributed energy resource · Energy tariffs · Energy business models

1 Introduction

The energy revolution and smart grids (SG) are two of the most important energy topics today. Recently, the demand for energy is growing while traditional resources of energy supply (coal, natural gas, and oil) will not meet the increasing energy demand any longer [1]. In order to face ecological challenges a flexible power grid enabling the integration of renewables - temporal dynamic energy sources - is required [2]. As a result, distributed energy resources (DERs) have become increasingly important because of their advantages for the grid. DERs enable demand response, grid stabilization and reduce the distribution of transmission costs. Consequently, integration of DERs is more energy-efficient [9] and saves money for the customers.

However, the increase of DER integration probably results in a reduction of energy sales of the distribution system operators (DSOs). In consequence, the DSOs are predicted to increase their grid tariffs in order to counterbalance their sales loss, which in turn might attract more consumers to become prosumers, who probably increase self-consumption

© Springer Nature Switzerland AG 2020
T. Ahram (Ed.): AHFE 2019, AISC 965, pp. 631–643, 2020.
https://doi.org/10.1007/978-3-030-20454-9_62

and invest in local storage technologies or even disconnect from the grid - called the spiral of death [3]. Consequently, in order to maintain their market role, DSOs have either to adjust their business models or create new business models that integrate DERs into the SG. There are some studies [4, 5], which examine financing topics of renewable infrastructure barriers for future business models. Renewable Energy Cooperatives were found to facilitate the market uptake of renewable energies by applying community-based marketing initiatives [6], but business models of DER integration have not been in the focus of research yet.

New energy business models will involve new market players and give the opportunity of active con- and prosumer involvement in terms of optimal production and consumption of energy. Their acceptance is a main contributor to the adoption [7] of these models. Hence, integration of user preferences and barriers is of fundamental importance in the design process, but research on acceptance of DERs and business models is scarce until now [7, 8]. To enlarge the body of user research, this study focuses on the empirical assessment of underlying variables for acceptance.

2 Related Work

2.1 Acceptance

There is sparsely empirical research on technology acceptance of DER business model so far due to the novelty of the topic. Recently, Von Wirth, Gislason and Seidl [10] investigated drivers and barriers for the social acceptance of DERs. Results of the literature research and semi-structured interviews with representatives of pilot regions indicate that the awareness of local advantages could be a decisive argument promoting these systems. Furthermore, the authors conclude that the ownership of infrastructure fosters the acceptance of such systems. There is a plethora of research focusing on the concept of acceptance of new technologies. As DER systems, combining renewable energy generation, energy conversion, and energy storage on different local scales [10] technology acceptance is one crucial construct when it comes to user adoption.

Technology Acceptance Model
A well-known theoretical framework to describe technology acceptance is the Technology Acceptance Model (TAM) proposed by Davis [11]. According to the author, acceptance of new technologies depends on the perceived usefulness and the perceived ease of use of the technology. The first is defined as "[…] *the degree to which a person believes that using a particular system would enhance his or her performance […]*", whereas the latter refers to "[…] *the degree to which a person believes that using a particular system would be free of effort.*" [11]. The combination of these concepts determines a person's attitude towards the usage of a technology, which influences the intention to use and finally the actual use of a technology. With respect to SGs, beside all the merits of DERs in a SG, the integration of multiple features in a business model might hamper the ease of use even though usefulness ratings are high and lead to reduced acceptance consequently.

Norm Activation Model
Another main motivator for the acceptance of DERs in general might be their positive outcomes. Beside monetary savings, DERs allow an environmentally friendly energy production and grid stabilization. Therefore, mere technology acceptance falls too short describing user adoption. According to Schwartz [12, 13] the behavior of using technologies that benefit others or the environment is motivated by personal norms or self-expectations: *"Personal norms focus exclusively on the evaluation of acts in terms of their moral worth to the self."* [14, p. 245]. Therefore, personal norms might contribute to the accepance of future energy business models with a higher amount of DERs.

Responsible Technology Acceptance Model
The Responsible Technology Acceptance Model (RTAM), [15] combines aspects of the TAM [11] with the Norm Activation Model (NAM) [12, 13]. Here acceptance of new technologies depends on the rational cost-benefit assumptions and personal moral deliberations. A survey with 950 subjects from Denmark, Switzerland, and Norway showed that the RTAM successfully predicted the acceptance of the SG application [15]. Rational assessments as well as personal norms triggered by *"[…] feeling of moral obligation or responsibility towards the environment and a positive contribution to the society […]"* [15, p. 398] are essential for a positive evaluation. Individual benefits like monetary saving or incentives are not the only motivator for usage [16]. The authors discuss, that societal as well as environmental benefits should be stressed when promoting new technologies, which is in line with research by Bolderdijk et al. [17], who found, that pointing out societal and environmental benefits in the communication induces more positive feelings than solely mentioning private ones.

2.2 Hypotheses

As present research [10] showed, SG actors who own parts of the energy infrastructure - such as prosumers - should have a higher acceptance of future energy business models than consumers. Furthermore, the acceptance should be higher the more local advantages an energy model provides. Accordingly, it can be derived from the NAM [12, 13], that the acceptance of business models, that integrate personal norms like pro-environmental and pro-social behavior, is higher. This is confirmed by the research of Toft et al. [15], and Bolderdijk et al. [17], who integrated environmental and societal benefits. The Peer-to-Peer (P2P) model described above represents the most complicated model in terms of (technical infra-) structure, but also the most locally anchored one in terms of social involvement and energy infrastructure. Therefore, the following hypotheses are derived from the literature for underlying varibles of acceptance:

H1.1:   The current business model (BaU) is perceived more easy to use than any future business models (DSO, Volt, P2P).

H1.2:   With increasing complexity, due to more infrastructure and interaction with the user needed, ratings on perceived ease of use diminish. Hence lowest ratings are expected with P2P, followed by Volt and DSO.

H2:   Perceived usefulness is rated higher for future energy business models (DSO, Volt, P2P) than the current model (BaU).

H3.1:   The influence of personal norms is highest for future business models (DSO, Volt, P2P) compared to the currently used business model (BaU).

H3.2:   Ratings on personal norms will be higher for models with a high level of DERs (P2P, Volt) compared to the other presented energy business models (DSO, BaU).

H4.1:   The attitude towards future energy business models (DSO, Volt, P2P) is more positive than towards the current model (BaU).

H4.2:   Compared to consumers, prosumers` attitude rating is higher with models with a high level of DERs (P2P, Volt).

2.3   Method

Participants

Overall, $N = 21$ persons had been interviewed (9 prosumer, 18 male). The sample consisted of three Swedes and 18 Germans. On average they were $M = 43$ years old ($SD = 13.75$; min = 21; max = 71). The most interviewees ($n = 18$) indicate holding a university degree or higher. The average household size was 3.6 residents. The income was indicated by 20 interviewees with a most frequently ($n = 7$) income category of "3000-4500€", followed by "more than 6000€" ($n = 6$).

Material

Three future business models were investigated: 1.) DSO controlled (DSO); 2.) Voltage-Tariff (Volt) and 3.) Peer-2-Peer (P2P). The business models incorporated different degrees of DER ranging from large to small-scale distributed energy generation and were developed within the research project "NEMoGrid - New Energy Business Models in the Distribution Grid". Business as Usual (BaU) represents a baseline measurement. In order to facilitate comparability, the descriptions of the models comprised following characteristics: (1) Energy source (e.g., PV plant or DSO), (2) Basic tariff structure (e.g., static depending on the consumption), (3) Installed infrastructure (soft-/hardware; e.g., algorithms and storage), (4) Possible effect for daily life (e.g., shifting of energy consuming activities), (5) Composition of the energy bill (e.g., quota of network service usage, time-specific energy costs), and (6) Financial benefits. Detailed business model descriptions can be found in the project deliverable "D2.3" [19].

Interviews with Swedish participants were conducted in English and with English materials. For German interviewees material and interview questions were translated.

Procedure

Demographic variables like user type (pro-/consumer) were assessed in a pre-questionnaire, which was used to preselect subjects for the interviews. Subjects were provided with the energy business models descriptions before the interview. These descriptions varied depending on the pro-/consumer classification. The interviews were conducted via Skype or phone call. After the introduction on the project subject and interview procedure, consent was obtained. Subsequently interviewees were asked to evaluate each of the four energy business models, which were presented in randomized order, by answering closed- and open-ended questions. The questions on the variables presented in this paper (1) the perceived ease of use, (2) the perceived usefulness, (3) personal norms, and (4) the attitude, remained the same for each model. Interviewees were asked to give their evaluation on a scale ranging from 1 to 6 for the respective items of each variable and were asked to explain their ratings subsequently. The interview lasts for about one hour and the interviewees received a remuneration of 40€ for their participation.

Data Analysis

Quantitative data was analyzed descriptively. After verifying distribution of normality parametric inferential statistics were applied. Since the research question was to identify differences between the user group and/or the business models, a repeated measurement analysis of variance (RMANOVA) was calculated with model as within-subject factor and user type as a between subject factor.

The open-ended questions of the interview have been transcribed using the software easytranscript (Version 2.50.7) [20]. Answers of each interviewee were split up into single statements and categorized into a bottom-up built category system (example statements for the reported categories are listed in Annex A). The category system for the variables (except personal norms) distinguished between positive and negative statements on an overall level. On a sub-level, the categories contain detailed information. A second coder was included to ensure reliability of codings. Intercoder-reliability (unweighted Kappa) on the sub-level varied between $\kappa = .72$ (ease of use) and $\kappa = .86$ (personal norms). According to literature this is a "substantial"/"almost perfect agreement" [18]. Discrepancies of codings were identified and eliminated. The frequencies of the sub-level categories of this consensus solution are reported relatively to the overall amount of either positive or negative statements.

2.4   Results

Perceived Ease of Use

Interviewees rated BaU as the most easy to use, followed by DSO (Table 1). Perceived ease of use was rated the lowest with the P2P. RMANOVA showed, that models significantly differed from each other ($F(3, 57) = 29.24$, $p < .001$, $\eta^2 = .60$). Post-hoc pairwise comparisons became significant ($p < .001$) for all except DSO and the Volt.

Table 1. Mean evaluation of perceived ease of use ($N = 21$).

Statement	BaU	DSO	Volt	P2P
It is easy to use this business model.	5.43	4.24	4.05	2.67
Using this business model does not require any effort from me.	5.43	3.71	3.33	2.24
It is easy to learn how to operate in this business model.	5.76	4.19	3.38	2.43
Score perceived ease of use	**5.54**	**4.05**	**3.59**	**2.48**

In total 162 single statements explaining the ease of use ratings had been categorized. With exception of BaU - evaluated mainly positively ($n = 40$; 85%), the future energy business models received more critique than positive statements (P2P: 37; 82%; DSO: 30; 65%; Volt: 24; 61% negative statements).

Positive Statements. BaU was appreciated for the "Low Cognitive Effort" (50%), the "Predictability of Costs" (23%) and its "Accessibility" (20%). DSO and Volt were appreciated for the "Low Cognitive Effort" (DSO: 56%; Volt: 33%) and the "Accessibility" (DSO: 25%; Volt: 13%) as well. For P2P various other aspects were pointed out positively, e.g., "Financial Incentives" (38%).

Negative Statements. BaU received critique for the "Initial Installation Effort" (57%). In particular, prosumer criticized the initial bureaucratic effort for an energy plant installation. The future models especially P2P (54%) and Volt (46%) were criticized for the "Increased Cognitive Effort" the interviewees assumed. The "Uncertainty of Costs" (27%) was mentioned negatively during the P2P evaluation. Furthermore, the interviewees named that DSO will probably lead to a "Discrepancy with their own Habits" (27%) and criticized the "External Control" applied within this model (23%).

Perceived Usefulness

Usefulness was rated the highest for the Volt, followed by the P2P and the DSO model (Table 2). BaU was rated the least useful. Future models were all perceived significantly more useful than BaU ($F(3, 57) = 31.75$, $p < .001$, $\eta^2 = .62$, pairwise comparisons: $p < .001$). However, there were no significant differences between future models. RMANOVA revealed a significant interaction ($F(3, 57) = 5.82$, $p = .002$, $\eta^2 = .23$) for user type and model. Whereas prosumers rated BaU more useful than consumers, the DSO was rated better by consumers.

Table 2. Mean results for perceived usefulness ($N = 21$).

Statement	BaU	DSO	Volt	P2P
The current business model uses electricity efficiently.	2.48	4.81	4.62	4.90
This business model enables me to adjust my electricity consumption so that I can benefit from fluctuations in electricity prices.	1.62	3.71	4.57	4.57
This business model contributes to a reliable electricity supply.	2.95	4.57	4.81	4.43
Score perceived usefulness	**2.35**	**4.37**	**4.67**	**4.63**

The open-ended explanations resulted in 155 single statements. Overall, all future business models were evaluated positively (positive statements: DSO: 21; 60%; Volt: 33; 75%; P2P: 32; 73%). Underlining the quantitative ratings, the majority of statements (32; 76%) about the BaU model was negative.

Positive Statements. BaU was especially valued because of its "Grid Stabilization and the Secure Supply" (40%). This advantage was also mentioned for DSO (43%) and Volt (36%). Furthermore, diverse other categories were mentioned during BaU evaluation: "Increased Level of Freedom on Influence and Consumption Costs" (20%), "Predictability of Costs" (10%), and "Integration of Renewables" (10%). In general, participants prefer the "Increased Level of Freedom" with future models, especially on consumption costs within the Volt (33%) and the P2P (31%). The interviewees liked the "Effective Usage of Energy" with DSO (24%) and P2P (25%).

Negative Statements. Considering BaU the interviewees mainly fear the "Destabilization of the Grid" (38%) and complain about the "Few Incentives to Adjust Behavior" (25%). DSO was criticized even more frequently for "Few Incentives to Adjust Behavior" (43%) compared to BaU. Again, the "External Control" was evalauted disadvantageously (36%). Regarding P2P interviewees worried about "Destabilization of the Grid" (50%) and again the "Increased Cognitive Effort" (17%). Least was also the main critique on the Volt (27%).

Personal Norms

Interviewees' personal norms towards P2P were the highest, followed by the Volt and DSO model (Table 3). There was a significant effect of the model ($F(3, 57) = 3.96$, $p = .012$, $\eta^2 = .17$). However, pairwise comparisons only became significant for BaU and P2P ($p = .045$). Again, prosumers and consumers differed in their rating between models ($F(3, 57) = 4.06$, $p = .011$, $\eta^2 = .17$; see Fig. 1). Whereas Volt and P2P result in similar ratings, consumers see DSO more positive than BaU, whereas prosumers are in favor for BaU (Fig. 1).

Table 3. Descriptive results on personal norms ($N = 21$).

Statement	BaU	DSO	Volt	P2P
It is my duty to participate in this business model if it is necessary for an environment friendly and well-functioning electricity supply.	3.90	3.95	4.38	4.48
I feel a moral obligation to participate in this business model regardless of what others do.	2.38	3.33	3.52	3.95
To participate in this business model is the right thing to do.	2.81	3.90	4.29	4.14
Score personal norms	**3.03**	**3.73**	**4.06**	**4.19**

Overall, 150 open-ended question statements had been analyzed. Especially with BaU, the influence of personal norms seems low (30; 71% rejection). Contrary to that, this is the case for the majority of DSO (22; 63%) and Volt statements (24; 63%, P2P: 18; 51%).

Social Norms Present. For BaU the interviewees most frequently named "Ecological reasons" (42%). This agreement was caused by prosumers, e.g. by stating that the integration of renewables is feasible already. For DSO (27%) and Volt "Grid Stabilization and Secure Supply" (21%) was mentioned. "Financial Incentives" (Volt: 25%; P2P: 22%) and "Ecological Reasons" (P2P: 22%; DSO: 18%; Volt: 17%) were pointed out for the future models. "Innovativeness" was particularly pointed out with P2P (22%) and Volt (17%).

Social Norms Not Present. In contrast, some participants denied the influence of personal norms. For BaU "Non-Ecological Production" of energy was mentioned (23%) as a reason. Furthermore, there was criticism to the extent, that there is "Room for Improvement" (20%) and the model leads to a "Destabilization of the grid" (17%). Especially for the DSO evaluation, interviewees criticize the "Dependence on Others" (46%). In general the interviewees see "Room for Improvement" (36%) and explicitly mention that "Norms are not Applicable in this Context" (29%) for Volt. Least was also

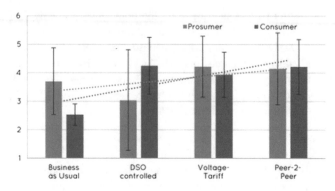

Fig. 1. Personal norms (1 = strongly disagree to 6 = strongly agree) by user type with 95% confidence interval (*N* = 21).

stated with regard to P2P (29%). Again, and strictly speaking for P2P the interviewees fear the "Increased Cognitive Effort" (24%).

Attitude

Overall, the attitude towards the business models is rather neutral to slight positive. BaU is rated the safest and P2P the least safely. However, a vice versa pattern was found for the evaluation on the scale from very bad to very good, where P2P and Volt received the highest ratings. Feelings towards P2P were the most positive (Table 4).

Table 4. Descriptive results on attitude (*N* = 21).

On a scale from 1 to 6 how do you feel about being a consumer/prosumer in this business model when 1 means … and 6 means …?"	BaU	DSO	Volt	P2P
very negative…very positive	4.00	4.19	4.24	4.52
very bad… very good	3.81	4.24	4.43	4.43
very unsafe… very safe	4.86	4.38	4.05	3.57
Score attitude	**4.22**	**4.26**	**4.23**	**4.17**

RMANOVA of overall score revealed no significant differences between models or user group. However, an interaction between both factors occurred ($F(3, 57) = 5.65$, $p = .002$, $\eta^2 = .22$). Whereas prosumers were in favor for BaU and P2P, consumers prefered the DSO and the Volt model.

The open-ended question, gathering reasons for the attitude ratings, lead to 168 single statements. Overall, the evaluation of BaU indicated a positive attitude (37; 71%). There were less positive statements for DSO (25; 58%), Volt (24; 51%) and P2P (18; 46%).

Positive Statements. "Low Cognitive Effort" (27%) and the "Effectiveness and Reliable Functionality" (27%) were appreciated most frequently for BaU. Interestingly, the "Effectiveness and Reliable Functionality" were mentioned also for all future models (DSO: 28%; P2P: 22%; Volt: 21%). Additionally, future models (esp. Volt:

46%), were appreciated because of their "Increased Level of Freedom". Statements belonging to this category refer mainly to an increased influence on the consumption costs. Furthermore, the interviewees appreciated the "Innovativeness", especially for P2P (28%). Additionally, P2P was valued because of its potential of "Ecological Integration of Renewables" (28%).

Negative Statements. BaU received critique because the interviewees broadly see 1) "Room for Improvement" (67%) and think that; 2) energy is produced "Non-Ecological" (20%). A negative attitude towards the future models was explained by the "External Control and Dependence" on other actors. This was the case for DSO (61%) and Volt (30%). In contrast, P2P was criticized for "Increased Cognitive Effort" (43%). Furthermore, the interviewees complained about the "Uncertainty of Costs" with P2P (33%).

2.5   Discussion and Conclusion

Our interview study aimed on the identification of factors fostering or preventing the acceptance of future energy business models. In general, the quantitative ratings of the interviewees showed that acceptance of future models is relatively high.

In accordance with the hypotheses (H1.1, H1.2) we found that perceived ease of use diminished with increasing complexity of the business model. Not surprisingly, it was rated the best for the current business model, possibly because user already interact with it. However, perceived usefulness was rated significantly higher for future business models, even though characterized by higher complexity, which is in line with our hypothesis (H2). Toft et al. [17] stated personal norms should be taken into consideration for the evaluation of smart grids. Assumptions on the influence of personal norms can only be confirmed partly. The only statistically significant difference from business as usual is to be found in comparison to P2P. Therefore, H3.1 must be rejected. However, prosumers and consumers do rate some models differently. Wherereas, the Voltage-Tariff and the P2P model result in similar ratings, consumers see the DSO controlled model more positive than BaU, whereas prosumers are in favor for BaU. Therefore, H3.2 can only be confirmed partly. Acceptance in terms of a positive, good and safe attitude towards the model is higher for all future models but not significantly higher. Hence H4.1 must be rejected. However, taking into account the different user groups we see that prosumers were in favor for the current business model and the P2P model, whereas consumers seemed to prefer the DSO and the Volt model, which speaks for a partial confirmation of H4.2. Therefore, we can confirm and extend the results of Von Wirth, Gislason and Seidl [10], who underlined the importance of infrastructure ownership for the acceptance of distributed energy systems. Qualitative data helps to better understand these results.

The analysis of qualitative data showed that future energy business models received critique due to their specific characteristics, such as the increased external control within the DSO controlled model. Probably this affected specifically prosumer ratings, as they value independence of energy supply and therefore acceptance is diminished for this model. In contrast, consumer acceptance is remarkably higher for this model. Probably they are more concerned about the destabilization of the grid, as they are more depending on a secure supply. Least could be a reason for the different perspectives on the DSO controlled model. In turn, the appreciated "Increased Level of Freedom" (resp. influence on consumption costs) especially for the Voltage-Tariff is probably more decisive for consumers than for prosumers, as they do not have any possibilities to influence their costs at the moment. The P2P model was criticized for its complexity and the uncertainty of costs, which could be more discouraging for consumers than prosumers, who are used to a certain degree of complexity, high investments for energy generation or the – also financial – security of self-consumption.

It can be concluded that as energy business models acceptance differ with regard to user group, they should be adapted to the specific pro- and consumer requirements. Generally speaking our interviewees valued the innovativeness of future energy business models, especially the P2P model. We can conclude that they see need for change as they criticized the current model for its backwardness.

2.6   Limitations and Future Work

As DER systems imply the installation of technical infrastructure the theoretical foundation of this study built up on research of technology acceptance. Even though the RTAM model [15] focuses on the technology acceptance of smart grid infrastructure and incorporates personal norms, aspects of social acceptance are neglected. Moreover, some interviewees denied influence of personal norms in this study. Interviewees mentioned economical thinking mostly drives acceptance. Therefore, future work should enlarge the scope of variables including acceptance on a social and financial level.

The present findings should be considered with care because sample size is relatively small and self-selection bias can lead to increased acceptance and interest in the topic. Due to regional differences in energy market, national differences should be considered more intense. Findings in other countries might differ from the ones generated here.

Further, in order to identify tendency more clearly, we used 6-point scales instead of 7-point scales suggested by Toft et al. [15]. Hence, comparisons with other technologies ratings are not possible with the present results.

Regarding the qualitative data, the categories identified for each variable were very similar. Positive and negative category statements frequently represent opposite opinions. Future work could investigate main motives and barriers in a quantitative manner, avoiding contrary queries of motives and barriers.

For our interview study different business models descriptions and user group have been prepared. The descriptions were phrased in colloquial language to ensure comprehension of the models, but could be improved with regard to implementation aspects to improve con- and prosumer ideas of the future energy business models.

Acknowledgments. The current research is part of the "NEMoGrid" project and has received funding in the framework of the joint programming initiative ERA-Net SES focus initiative Smart Grids Plus, with support from the EU's Horizon 2020 research and innovation programme under grant agreement No. 646039. The content and views expressed in this study are those of the authors and do not necessarily reflect the views or opinion of the ERA-Net SG+ initiative. Any reference given does not necessarily imply the endorsement by ERA-Net SG+. We appreciate the support of our student scientists, who supported data collection and analysis.

Annex A: Example Statements for the Sub-level Categories of Qualitative Data Analysis

Sub- category (positive or negative)	Example Statement (source: interview no.; model; acceptance variable)
Low Cognitive Effort (+)	*"It is a quite simple business model. It is easy to understand [...]"* (4; BaU; attitude)
Effectiveness & Reliable Functionality (+)	*"That's a business model which had worked out in the past."* (13; BaU; attitude)
Increased Level of Freedom (+)	*"Well, I see some opportunities for personal influence here, e.g., with the own behavior for both, the balancing of energy production and consumption as well as to save some money."* (14; Volt; attitude)
Innovativeness (+)	*"But I think it is a positive model, because it is new and because it sound interesting."* (15; P2P; attitude)
Ecological Integration of Renewables (+)	*"That's the way which leads to 100% renewables in the German electricity grid."* (16; P2P; attitude)
Room for Improvement (−)	*"Actually this model is outdated, as the future is the decentralized energy supply."* (11; BaU; attitude)
Non-Ecological Production (−)	*"The energy suppliers reputation is very bad [...] as they heavily produce coal-based and nuclear powe."* (11; BaU; attitude)
External Control and Dependence (−)	*"For me the DSO influence is too big."* (19; DSO; attitude)
Increased Cognitive Effort (−)	*"[...] I don't want to participate in auctions for my energy permanently and I don't want to think about my energy price. Well for me it sounds very complicated and it causes a lot more effort than energy usage is worth for me."* (20; P2P; attitude)
Uncertainty of Costs (−)	*"[...] not really knowing what things are going to cost [...] trying to understand the bill at the end of month will be a nightmare."* (14; P2P; attitude)
Predictability of Costs (+)	*"[...] there is one price, there is one amount of consumption, and depending on what one consumes you just have to pay."* (7; BaU; ease of use)
Accessibility (+)	*"[...] there's no sort of requirement, there's no plenty of requirement. You just use when you want."* (3; BaU; ease of use)
Financial Incentives (+)	*"Probably it would motivate the people, if the earnings are good, to refinance the PV plant."* (7; P2P; ease of use)
Initial Installation Effort (−)	*"[...] at the beginning before you are connected to the grid, there are a lot of bureaucratic things at the beginning."* (8; BaU; ease of use)
Discrepancy with their own Habits (−)	*"you need to [...] just go out of your normal routine of electricity consumption."* (1; DSO; ease of use)

(continued)

(continued)

Sub- category (positive or negative)	Example Statement (source: interview no.; model; acceptance variable)
Grid Stabilization & Secure Supply (+)	"[…] one have made positive experiences with this business model. Energy was always available." (21; BaU; usefulness)
Increased Level of Freedom (+)	"Within this model the price is varying. Thus one could probably profit, if it runs well." (6; Volt; usefulness)
Effective Usage of Energy (+)	"[…] and it [the P2P model] will lead to a more effective utilization of the energy grid, the energy consumption […]." (13; P2P; usefulness)
Destabilization of the Grid (−)	"To realize an even grid load […] it [the BaU model] is not very beneficial to do so." (17; BaU; usefulness)
Few Incentives to Adjust Behavior (−)	Well however I don't have any incentive if the energy price stays always the same" (9; BaU; usefulness)
Ecological reasons (+)	"Me myself I favor environment friendly energy production." (6; BaU; personal norms)
Norms are not Applicable (−)	"[…] for such approaches moral-ethical aspects are not of importance, or they are ranked very low." (13; P2P; personal norms)

References

1. Bhatti, H.J.: The future of sustainable society—the state of the art of renewable energy and distribution systems. Thesis Industrial Management and Innovation, Halmstad University (2018)
2. Döbelt, S., Jung, M., Busch, M., Tscheligi, M.: Consumers' privacy concerns and implications for a privacy preserving Smart Grid architecture—results of an Austrian study. Energy Res. Soc. Sci. **9**, 137–145 (2015)
3. Severance, C.A.: A practical, affordable (and least business risk) plan to achieve "80% clean electricity" by 2035. Electr. J. **24**(6), 8–26 (2011)
4. Yildiz, Ö.: Financing renewable energy infrastructures via financial citizen participation—the case of Germany. Renew. Energy **68**, 677–685 (2014)
5. Richter, M.: Business model innovation for sustainable energy: German utilities and renewable energy. Energy Policy **62**, 1226–1237 (2013)
6. Viardot, E.: The role of cooperatives in overcoming the barriers to adoption of renewable energy. Energy Policy **63**, 756–764 (2013)
7. Wolsink, M.: The research agenda on social acceptance of distributed generation in smart grids: renewable as common pool resources. Renew. Sustain. Energy Rev. **16**(1), 822–835 (2012)
8. Kubli, M., Loock, M., Wüstenhagen, R.: The flexible prosumer: measuring the willingness to co-create distributed flexibility. Energy Policy **114**, 540–548 (2018)
9. Akorede, M.F., Hizam, H., Pouresmaeil, E.: Distributed energy resources and benefits to the environment. Renew. Sustain. Energy Rev. **14**(2), 724–734 (2010)

10. Von Wirth, T., Gislason, L., Seidl, R.: Distributed energy systems on a neighborhood scale: Reviewing drivers of and barriers to social acceptance. Renew. Sustain. Energy Rev. **82**, 2618–2628 (2018)

11. Davis, F.D.: Perceived usefulness, perceived ease of use, and user acceptance. MIS Quart **13**, 319–340 (1989)

12. Schwartz, S.H.: Normative influences on altruism. In: Leonard, B. (ed.). Advances in Experimental Social Psychology, 10, pp. 221—279, Academic Press (1977)

13. Schwartz, S.H., Howard, J.A.: A normative decision-making model of altruism. In: Rushton, J.P., Sorrentine, R.M. (eds.) Altruism and Helping Behaviour: Social, Personality, and Developmental Perspectives, pp. 189–211. Lawrence Erlbaum, Hillsdale, NJ (1981)

14. Schwartz, S.H., Howard, J.A.: Internalized values as moderators of altruism. In: Staub, E., Karylowski, B.-T., Reykowski, J. (eds.) Development and Maintenance of Prosocial Behavior: International Perspectives on Positive Morality, pp. 229–255. Plenum Press, New York (1984)

15. Toft, M.B., Schuitema, G., Thøgersen, J.: Responsible technology acceptance: model development and application to consumer acceptance of Smart Grid technology. Appl. Energy **134**, 392–400 (2014)

16. Energy, D.: Demand Response—The eFlex Project. Virum, Denmark (2012)

17. Bolderdijk, J.W., Steg L., Geller, E.S., Lehman, P.K., Postmes T.: Comparing the effectiveness of monetary versus moral motives in environmental campaigning. Nat. Clim. Chang. 3(4), 413–416 (2013)

18. Landis, J.R., Koch, G.G.: The measurement of observer agreement for categorical data. Biometrics, pp. 159–174 (1977)

19. Döbelt, S., Kreußlein, M.: Results regarding consumer/prosumer requirements. Public project delivearable NEMoGrid (2018). http://nemogrid.eu/wp-content/uploads/D2.3-Results-regarding-consumer-prosumer-requirements_TUC.pdf

20. Easytranscript (version 2.50.7) [Transcription Software]. E-Werkzeug. https://www.e-werkzeug.eu/index.php/de/produkte/easytranscript

Nuclear Power Plant Accident Diagnosis Algorithm Including Novelty Detection Function Using LSTM

Jaemin Yang, Subong Lee, and Jonghyun Kim$^{(\boxtimes)}$

Department of Nuclear Engineering, Chosun University, 309 Pilmun-Daero,
Dong-Gu, Gwangju 501-709, Republic of Korea
jonghyun.kim@chosun.ac.kr

Abstract. Diagnosis of the accident or transient at the nuclear power plants is performed under the judgment of operators based on the procedures. Although procedures given to operators, numerous and rapidly changing parameters are generated by measurements from a variety of indicators and alarms, thus, there can be difficulties or delays to interpret a situation. In order to deal with this problem, many approaches have suggested based on computerized algorithms or networks. Although those studies suggested methods to diagnose accidents, if an unknown (or untrained) accident is given, they cannot respond as they do not know about it. In this light, this study aims at developing an algorithm to diagnose the accidents including "don't know" response. Long short term memory recurrent neural network and the auto encoder are applied for implementing the algorithm including novelty detection function. The algorithm is validated with various examples regarding untrained cases to demonstrate its feasibility.

Keywords: LSTM · Auto encoder · Accident diagnosis · Novelty detection

1 Introduction

Current operated nuclear power plants (NPPs) have the goal of electricity production ensuring safety [1]. However, recent Korean denuclearization argument has shown considerable social anxiety. Above all, it is certain that such NPP accidents recorded in history (e.g., Chernobyl, TMI, Fukushima) cause distrust to the public [2, 3]. Although utilities and related organizations have shown careful efforts to improve accident prevention, the distress and concern has still not been fully resolved, as damage predictions cannot be predicted. Therefore, securing safety from the consequences of an accident or potential danger is not only a prerequisite for the NPP operation, but it is also a great challenge for the general public to accept without worry.

All actions at nuclear power plants (NPPs) are carried out at the discretion of the operator in accordance with procedures. However, in the case of an accident or a transient, there are several factors that can cause human error, even if the procedure provided to the operator is the result of numerous research and studies. First of all, many parameters measured from various instruments (e.g., indicators and alarms) are

© Springer Nature Switzerland AG 2020
T. Ahram (Ed.): AHFE 2019, AISC 965, pp. 644–655, 2020.
https://doi.org/10.1007/978-3-030-20454-9_63

generated, which can cause difficulties or delays in interpreting the situation. In addition, even if emergency response procedures are provided to operators to respond appropriately to the situation, diagnostics are still considered to be the most challenging task for operators, due to intense conditions such as time pressure or rapid change of parameters. Thus, while the operator is diagnosing the accident, a person's mistake (i.e., improper behavior or judgment that could worsen safety and integration of the facility) can occur. On account of these features, not only a delay in effective response but also more severe consequences can happen from wrong procedure selection [4–7].

In response to these emergency situations, there are various approaches such as operator assistance systems and algorithms to lessen burdens of operators. Above all, with the remarkable advancement of computer technology, the artificial intelligence approach that is attracting attention as the fourth industrial revolution is increasing significantly. Some of them based on such techniques (e.g., fuzzy inference [8], artificial neural networks (ANNs) [9], hidden Markov model [10], support vector machine [11]) demonstrate applicability of those techniques. Representatively, ANN is considered to be one of the best approaches because it has demonstrated outstanding performance against pattern recognition problems and nonlinear problems. In this sense, several studies have been proposed ANNs to develop algorithms for accident response in NPPs and some of them give potential applicability of ANNs [12–15].

Although a number of ANN based methods have been proposed, this study suggests applying recurrent neural networks (RNN) to develop accident response algorithms due to NPP data characteristics (i.e., nonlinear, time-sequential and multivariate). Because the RNN is considered as a neural network that can handle time-sequential data, it must be appropriate to reflect the time-dominant dynamic feature of the NPP [16]. Despite the appropriate characteristics to be applied, two limitations due to back propagation errors are well known in relation to weight control of the network. Those are 'vanishing gradients' and 'weights blowing up', the latter one means a rapid oscillation of the weights, whereas in the former case weights can be zero. Therefore, it may take a lot of time to learn, or the perform to be good even if the algorithm learning is done. In order to solve the problem of the RNN, a long-term memory (LSTM) has been proposed to improve the intrinsic defect [17]. Having the same RNN architecture base, but improvement of it can make it cover long temporal sequence of data as well as varying length sequential data. In addition, recently several studies using LSTM have shown clear applicability and satisfactory performance in various fields (e.g., image captioning, natural language processing, genome analysis, handwriting recognition) [18–21].

Several studies have proposed a method based on LSTM to take time-sequential characteristics into account, but given unknown (or untrained) inputs, they can not respond as if they do not. This problem is called novelty detection and is used to identify abnormal or unexpected phenomena contained in a large amount of normal datasets [22]. In addition, if this algorithm is implemented as an operator-assisted system in an actual NPP, this feature is important because it can define the limits of the system or the scope of an accident. Therefore, several studies have been proposed to solve this problem [23, 24]. However, since there is no approach compatible with LSTM, this study applies AE (Auto Encoder) to identify unknown data. AE is an unsupervised learning method that learns functions that approximate inputs and outputs. By using the characteristics of AE, if an unknown incident is given as inputs it can be identified whether it has been trained. Furthermore, some studies show the compatibility of AE with LSTM [25–27]. Therefore, this study will apply LSTM-AE (i.e., network structure that uses both LSTM and AE) so as not to give up the characteristics of LSTM.

From this point of view, this study aims to develop an accident response algorithm including novelty detection function. The first part of this work deals with a methodologies for developing algorithm. After describing the methodology, the implementation of the accident diagnostic algorithm is described and the results are presented individually. The network for this algorithm has been trained and validated using compact nuclear simulator (CNS) based on the Westinghouse 930MWe three-loops pressurized water reactor (PWR) as the reference plant.

2   Methodology

This chapter describes LSTM and AE for developing the accident diagnosis algorithm. LSTM is applied to implement the network modeling for the algorithm. In addition, AE is used in conjunction with LSTM to design the novelty detection function.

2.1   Long Short-Term Memory (LSTM)

LSTM is a widely used deep learning method that improves RNNs. Unlike other ANNs, RNNs show excellent functionality that represents time-sequential features in data. The basic characteristic of RNNs is that it can be cycled within the loop as shown in Fig. 1, since at least one feedback connection is included in the network.

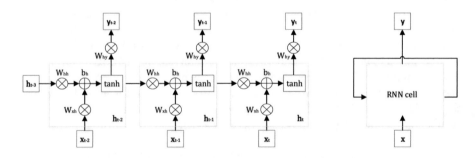

Fig. 1. The process of RNN cells working

The current cell state (h_t) is updated by the previous cell state (h_{t-1}) and the current input vector (x_t) considering bias (b_h) and activation function (tanh), then, output (y_t) is updated according to Eqs. (1) and (2). Based on this architectural feature, networks can learn sequences and process time series data.

$$h_t = \tanh(W_{hh}h_{t-1} + W_{xh}x_t + b_h) \tag{1}$$

$$y_t = W_{hy}h_t \tag{2}$$

However, the larger the time difference between past and present information, the lower the ability to reflect past information. In the case of conventional back propagation used in other ANNs, it goes back to the network and updates the weights and biases taking into account the error portion of the output. However, too much back propagation through the time spent in the RNN can cause vanishing gradients or blowing up due to repeated weight multiplication in the course of the past. In this case, since the recursive multiplication operation is configured many times over the network during backtracking of the RNN of time, even very small values can be multiplied, ultimately resulting in too large a value. In other words, existing RNNs have problems handling long sequences [28].

To solve this problem, several improved networks based on RNNs have been proposed, such as LSTM and gated recurrent unit (GRU) [29]. Both of them utilize gating structure to overcome the problem of RNNs. Although GRU was introduced after LSTM, it has a simpler structure, but the performance is almost identical [30]. In addition, with the long history of LSTM has various research results, so this study develops the algorithm by applying LSTM.

In case of LSTM, the most distinctive feature is the gate structure, which appears in the LSTM cell architecture as shown in Fig. 2. In the figure, the cell state (i.e., the line passing through the center) is an essential part of the LSTM. It passes through the whole like a conveyor belt, and the information can continue to pass to the next level without change. Gates are used to update or exclude information based on this cell state. Through the input modulation (g_t) and the input gate (i_t), the LSTM regulates the degree to which the input is updated to the cell state. Equation (3) represents the input conditioning node and has tanh (ϕ) activation function. Equation (4) represents the input gate and has sigmoid (σ) Activation function. Through the sigmoid activation function, it outputs a value of 0 or 1, which determines whether each component will be affected. The forget gate (f_t) and output gate (o_t) are represented by Eq. (5) and (6). This gating structure allows the cell state to control the influence of previous state information on the current state, update the information associated with the current input, and determine the influence level on the output through gate modulation.

$$g_t = \phi\left(W_g \cdot [h_{t-1}, x_t] + b_g\right) \tag{3}$$

$$i_t = \sigma(W_i \cdot [h_{t-1}, x_t] + b_i) \tag{4}$$

$$f_t = \sigma\left(W_f \cdot [h_{t-1}, x_t] + b_f\right) \tag{5}$$

$$o_t = \sigma(W_o \cdot [h_{t-1}, x_t] + b_o) \tag{6}$$

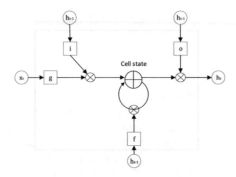

Fig. 2. The Architecture of the LSTM cell

2.2   Auto Encoder (AE)

In general, machine learning algorithms fall into three categories: supervised learning, unsupervised learning, and reinforcement learning. In this study, the accident diagnosis algorithm developed on the basis of LSTM can be classified as supervised learning, but AE is classified as uncontrolled learning. Unsupervised learning, unlike supervised learning, does not require labeled data given in both the input and target values of the sample data for training. Since the function of the data can be found in the state where only the input values of the data are provided, an approach based on unsupervised learning can handle new data that is unknown or untrained.

$$h(x) = f(W_e x + b_e) \tag{7}$$

$$y = f(W_d h(x) + b_d) \tag{8}$$

For the purpose of responding to the untrained or unknown accident, this study applies AE that is one of the most prominent unsupervised learning methods. AE is a way to learn functions that approximate input values to output values. Figure 3 shows the structure of AE. As shown in this figure, the AE is composed of an encoder that encodes the input to the hidden layer and a decoder that decodes the encoded hidden unit and outputs the same size as the input. Depending on the input $x \in R^n$ and the hidden representation $h(x) \in R^m$, these are described in Eq. (7), in which the nonlinear activation function is denoted by $f(z)$. Typically, the logistic sigmoid function $f(z) = \frac{1}{1+\exp(-z)}$ is applied. [36] $W_e \in m \times n$ and $b_e \in R^m$ mean a weight matrix and a bias vector, respectively. The hidden representation of the network output and reconstruction $y \in R^n$ are described in Eq. (8), where $W_d \in n \times m$ and $b_d \in R^n$ mean a weight matrix and bias vector, respectively.

After performing the process of extracting and reconstructing feature expressions from the input data via the encoder and decoder, the parameters $\theta(W_e, b_e)$, $\theta'(W_e, b_e)$ are optimized to minimize the loss function (L) as in Eq. (9). As a loss function, the square of the error between the input and the output is generally used as described in Eq. (10).

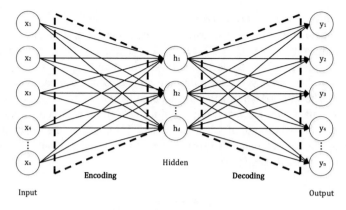

Fig. 3. A structure of auto encoder

$$(\theta, \theta') = argmin_{\theta, \theta'} \frac{1}{n} \sum_{i=1}^{n} L(x, y) \tag{9}$$

$$L(x, y) = \|x - y\|^2 \tag{10}$$

3 Accident Diagnosis Algorithm

This chapter describes both the accident diagnosis algorithm and the untrained accident identification algorithm based on methodologies introduced in Chapter 2. Then, it is verified with test scenarios after the training and optimization process.

3.1 Algorithm Modeling

The framework of accident diagnosis algorithm is developed considering complex NPP characteristics and dynamic states. Accident diagnosis is performed with the NPP datasets through the trained classifier. The model for accident diagnosis is designed for multiple labeling classification, because diagnostics are not mutually exclusive [31]. To predict an accident, a sequence of such variables is needed as inputs. Thus, a many-to-one structure is applied to design the model.

The figure on the shows the outlie of the accident diagnosis algorithm. According to a certain number of NPP input sequences, the designed model is capable of diagnosing an accident by capturing a pattern (i.e., NPP trend). Input variables are selected based on procedures and their importance that can affect the plant states and system availability. Through the network consisting of three LSTM layers, the output is diagnosed and the final diagnostic value is outputted via the output layer for transformation.

Input preprocessing is performed to convert the values to get better performance and to prevent slowing down of the learning speed. Because of different scale of inputs (e.g., RCS temperature: 300°C, SG level: 50%, Valve State: 0 or 1), normalizing the value can help prevent convergence at the local minimum (i.e., not a global minimum, but a minimum point among several minimum points), which can cause degradation of performance. The min-max normalization method is applied to scale parameters. The minimum and maximum are determined within the values of collected data (i.e., not real minimum or maximum of plant variables) [32]. Using the method, a linear transformation of the raw data is performed, and the datasets are calibrated within the range of 0 to 1 through Eq. (11).

$$X_{norm} = \frac{(X - X_{min})}{(X_{max} - X_{min})} \qquad (11)$$

In addition, the softmax function is mainly used to normalize the output value. Softmax is a function that exponentially increases the importance by an exponential function; it also raises the deviation among the values and then normalizes [33]. The function compensates the output value from the 0 to 1 via the Eq. (12), and the sum of the output values is always 1.

$$S(y_i) = \frac{e^{y_i}}{\sum e^{y_i}} \qquad (12)$$

In the case of general algorithms based on neural networks, the output is printed even if an untrained input is given. Therefore, since it is not possible to learn about all accidents, it requires an algorithm that can respond against untrained accidents.

The overview process of the untrained accident identification algorithm is shown in the Fig. 5 on the right. To model the algorithm, AE is applied to design the network. Through the encoder, inputs are compressed as latent variables, and then these are represented via a decoder based on latent variables. The reconstruction errors due to differences between represented outputs and original inputs are calculated, and then it can be identified whether it is trained or not considering the threshold. Like the accident diagnosis algorithm, the same input variables are used that are normalized with the LSTM cell.

In this study, the threshold is fixed at 1.0, that is, if an output (i.e., reconstruction error) is higher than the maximal trained one, then it decides as untrained data. When adjusting the threshold, it is possible to observe the output and determine that the threshold is indicative of high accuracy. However, if it were determined in this way, the result would be too optimistic about the data only at the time. That is, if new data is given, i its optimal performance about normal data cannot be guaranteed. However, there is no golden rule to determine thresholds in other areas. In this study, most of the reconstruction errors based on 40 s, which is the latest of the time when the algorithm diagnosis is completed (i.e., when the oscillation range is less than 0.02) are higher than 1.0, so it is fixed as the threshold in this study.

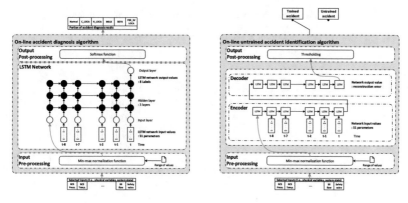

Fig. 5. Overview process of the accident diagnosis and the novelty detection (left: accident diagnosis, right: novelty detection function)

3.2 Training and Optimization

The algorithm is trained and implemented using the CNS developed by Korea Atomic Energy Research Institute. The network for incident diagnosis algorithms consists of three hidden layers of 64 batch sizes and 100 epochs are trained based on 82 scenarios (i.e., 11,571 datasets). 41 scenarios without MSLB (i.e., 9549 datasets) are used. Table 1 describes the training scenarios used for the accident diagnosis algorithm.

Table 1. Training scenarios used for accident diagnosis algorithm

Initiating events	Trained	Untrained
LOCA	40	32
MSLB inside the containment	15	0
MSLB outside the containment	15	0
Steam generator tube rupture (SGTR)	12	9
Total	82	41

Manual search methods are applied by changing hyper parameters (e.g., batch size, input sequence length, number of layers) one by one because there is no golden rule for hyperparameter determination to optimize the model [34, 35]. This study only considers accuracy for optimization. This is because training and verification data cannot be false positive or false negative if they are not artificially created. It is defined in Eq. (10). Taking into account the accuracy of the diagnostic results, the network consists of the optimum values (i.e., input sequence length: 30, batch size: 64, number of hidden layers: 3) as shown in Table 2.

$$Accuracy = \frac{Correct\ results}{Diagnosis\ results} \qquad (13)$$

Table 2. Accuracy comparison results between configured networks

No.	Sequence	Batch sizes	Layers	Accuracy
1	10	32	2	0.9314
2	10	32	3	0.9297
3	10	64	2	0.9211
4	10	64	3	0.9453
5	30	32	2	0.9980
6	30	32	3	0.9954
7	30	64	2	0.9973
8	30	64	3	0.9984

3.3 Test Results

The proposed algorithm is verified by considering the status of each plant as well as the training. Figure 6 shows an example of accident diagnosis result. The figure on the left shows diagnosis results for 40 cm^2 cold-leg LOCA in loop1 and the figure on the right shows diagnosis results for 800 cm^2 MSLB in loop3. The dotted line indicates the diagnostic result and the solid line indicates the answer. The malfunctions are injected at 10 s for all accident scenarios. Time and accident diagnosis results from the network outputs are represented on the X-axis and Y-axis, respectively. The results on these graphs mean that the accidents are diagnosed correctly and continuously (i.e., oscillation range of the diagnosis result is under 0.02) after several seconds.

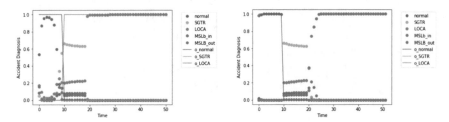

Fig. 6. An example of accident diagnosis result (Left: 40 cm^2 cold-leg LOCA in loop1, Right: 800 cm^2 MSLB in loop3)

In case of validation about the untrained accident identification algorithm, Figs. 7 and 8 illustrate how to deal with cases in which that are not trained. Figure 7 shows an example of the diagnostic result based on the accident diagnosis algorithm about untrained accident (i.e., MSLB). Although the accident is MSLB, the algorithm only diagnoses with SGTR or LOCA depending on the knowledge-based similarity. There is no ability to answer "untrained accident". As shown in this example, if untrained datasets (i.e., 700cm2 MSLB in loop2) are given, the algorithm can only diagnose based on accumulated knowledge. In order to prevent this problem, we propose an algorithm with an identification function for untrained accident.

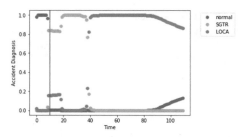

Fig. 7. An example of accident diagnosis result (Left: 40 cm^2 cold-leg LOCA in loop1, Right: 800 cm^2 MSLB in loop3)

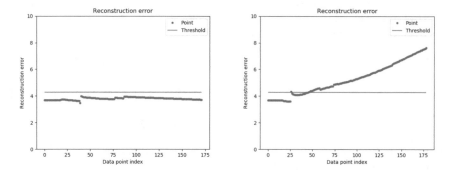

Fig. 8. An example of untrained accident identification (MSLB untrained)

Figure 8 shows an example of untrained accident identification using AE. In this figure, the blue dotted line indicates a reconstruction error between input and output. Thus, if the threshold value indicating the red line (i.e., the maximum value of the reconstruction error during training) is exceeded, it means that the current input is untrained or unknown data. The figure on the left shows the results for the trained accident (i.e., 40cm2 LOCA in loop2) and the figure on the right shows the result for the untrained accident (i.e., 700cm2 MSLB in loop2). Since the MSLB is not trained, the reconstruction error in the figure on the right is risen over time, whereas the reconstruction error of LOCA scenario is maintained under the threshold in the figure on the left. Therefore, if an untrained accident is given as inputs for the accident diagnosis algorithm, it can be identified by this algorithm.

4   Conclusion

This study suggests the algorithms for accident diagnosis and untrained accident identification based on LSTM and AE. The feasibility and applicability of the proposed algorithms are demonstrated by using CNS, which is used for producing accident data. The validation results show that the algorithms improve the previous study, and these can contribute to make the plant safety better.

Acknowledgement. This research was supported by Basic Science Research Program through the National Research Foundation of Korea(NRF) funded by the Ministry of Science, ICT & Future Planning (N01190021-06). Also, it was supported by the NRF grant funded by the Korean government (Ministry of Science and ICT) (2018M2B2B1065651).

References

1. O'Hara, J., et al.: Human Factors Engineering Program Review Model. United States Nuclear Regulatory Commission, Office of Nuclear Regulatory, Wasington, DC (2012)
2. Loganovsky, K.N., Zdanevich, N.A.: Cerebral basis of posttraumatic stress disorder following the Chernobyl disaster. CNS Spectr. **18**(2), 95–102 (2013)
3. Ohtsuru, A., et al.: Nuclear disasters and health: lessons learned, challenges, and proposals. The Lancet **386**(9992), 489–497 (2015)
4. Woods, D.D.: Coping with complexity: the psychology of human behaviour in complex systems. In: Tasks, Errors, and Mental Models. Taylor & Francis, London (1988)
5. Meister, D.: Cognitive behavior of nuclear reactor operators. Int. J. Ind. Ergon. **16**(2), 109–122 (1995)
6. Vaurio, J.K.: Safety-related decision making at a nuclear power plant. Nucl. Eng. Des. **185**(2–3), 335–345 (1998)
7. de Oliveira, M.V., et al.: HSI for monitoring the critical safety functions status tree of a NPP. (2013)
8. Lee, M.r.: Expert system for nuclear power plant accident diagnosis using a fuzzy inference method. Expert Systems **19**(4), 201–207 (2002)
9. Santosh, T., et al.: Application of artificial neural networks to nuclear power plant transient diagnosis. Reliab. Eng. Syst. Saf. **92**(10), 1468–1472 (2007)
10. Kwon, K.-C., Kim, J.-H.: Accident identification in nuclear power plants using hidden Markov models. Eng. Appl. Artif. Intell. **12**(4), 491–501 (1999)
11. Zio, E.: A support vector machine integrated system for the classification of operation anomalies in nuclear components and systems. Reliab. Eng. Syst. Saf. **92**(5), 593–600 (2007)
12. Fantoni, P., Mazzola, A.: A pattern recognition-artificial neural networks based model for signal validation in nuclear power plants. Ann. Nucl. Energy **23**(13), 1069–1076 (1996)
13. Embrechts, M.J., Benedek, S.: Hybrid identification of nuclear power plant transients with artificial neural networks. IEEE Trans. Industr. Electron. **51**(3), 686–693 (2004)
14. Uhrig, R.E., Hines, J.: Computational intelligence in nuclear engineering. Nucl. Eng. Technol. **37**(2), 127–138 (2005)
15. Moshkbar-Bakhshayesh, K., Ghofrani, M.B.: Transient identification in nuclear power plants: A review. Prog. Nucl. Energy **67**, 23–32 (2013)
16. Nabeshima, K., et al.: On-line neuro-expert monitoring system for borssele nuclear power plant. Prog. Nucl. Energy **43**(1–4), 397–404 (2003)
17. Hochreiter, S., Schmidhuber, J.: Long short-term memory. Neural Comput. **9**(8), 1735–1780 (1997)
18. Vinyals, O., et al.: Show and tell: A neural image caption generator. In: Proceedings of the IEEE Conference on Computer Vision and Pattern Recognition (2015)
19. Karpathy, A., L. Fei-Fei.: Deep visual-semantic alignments for generating image descriptions. In: Proceedings of the IEEE Conference on Computer Vision and Pattern Recognition (2015)

20. Auli, M., et al.: Joint language and translation modeling with recurrent neural networks (2013)
21. Liwicki, M., et al.: A novel approach to on-line handwriting recognition based on bidirectional long short-term memory networks. In: Proceedings of the 9th International Conference on Document Analysis and Recognition, ICDAR 2007 (2007)
22. Ma, J., Perkins, S.: Online novelty detection on temporal sequences. In Proceedings of the Ninth ACM SIGKDD International Conference on Knowledge Discovery and Data Mining (2003)
23. Bartal, Y., Lin, J., Uhrig, R.E.: Nuclear power plant transient diagnostics using artificial neural networks that allow "don't-know" classifications. Nucl. Technol. **110**(3), 436–449 (1995)
24. Antoˆnio, C.d.A., Martinez, A.S., Schirru, R.: A neural model for transient identification in dynamic processes with "don't know" response. Ann. Nucl. Energy, **30**(13), 1365–1381. (2003)
25. Marchi, E., et al.: A novel approach for automatic acoustic novelty detection using a denoising autoencoder with bidirectional LSTM neural networks. In: Acoustics, Speech and Signal Processing (ICASSP), 2015 IEEE International Conference on (2015)
26. Srivastava, N., Mansimov, E., Salakhudinov, R.: Unsupervised learning of video representations using lstms. In: International Conference on Machine Learning (2015)
27. Gensler, A., et al.: Deep Learning for solar power forecasting—An approach using autoencoder and LSTM neural networks. In: Systems, Man, and Cybernetics (SMC), 2016 IEEE International Conference on (2016)
28. Hochreiter, S., Schmidhuber, J.: Bridging long time lags by weight guessing and "Long Short-Term Memory". Spatiotemporal Models in Biol. Artif. Syst. **37**, 65–72 (1996)
29. Cho, K., et al.: Learning phrase representations using RNN encoder-decoder for statistical machine translation. arXiv preprint arXiv:1406.1078 (2014)
30. Chung, J., et al.: Empirical evaluation of gated recurrent neural networks on sequence modeling. arXiv preprint arXiv:1412.3555, (2014)
31. Yang, J., Kim, J.: An accident diagnosis algorithm using long short-term memory. Nucl. Eng. Technol. **50**(4), 582–588 (2018)
32. Jain, Y.K., Bhandare, S.K.: Min max normalization based data perturbation method for privacy protection. Int. J. Comput. Commun. Technol. **2**(8), 45–50 (2011)
33. Duan, K., et al.: Multi-category classification by soft-max combination of binary classifiers. In: International Workshop on Multiple Classifier Systems. Springer. (2003)
34. Bergstra, J., Bengio, Y.: Random search for hyper-parameter optimization. J. Mach. Learn. Res. **13**(Feb), 281–305 (2012)
35. Snoek, J., Larochelle, H., Adams, R.P. Practical bayesian optimization of machine learning algorithms. In: Advances in Neural Information Processing Systems. (2012)

Application of Economic and Legal Instruments at the Stage of Transition to Bioeconomy

Valentyna Yakubiv[1], Olena Panukhnyk[2], Svitlana Shults[3],
Yuliia Maksymiv[1], Iryna Hryhoruk[1], Nazariy Popadynets[3(✉)],
Rostyslav Bilyk[4], Yana Fedotova[5], and Iryna Bilyk[6]

[1] Vasyl Stefanyk Precarpathian National University, Shevchenko, Str. 57,
76018 Ivano-Frankivsk, Ukraine
yakubiv.valentyna@gmail.com, j.maksumiv@gmail.com,
ira.hryhoruk@gmail.com
[2] Ternopil Ivan Puluj National Technical University, Ruska Str. 56,
46001 Ternopil, Ukraine
panukhnyk@gmail.com
[3] SI "Institute of Regional Research Named After M. I. Dolishniy of the NAS
of Ukraine", Kozelnytska Str. 4, 79026 Lviv, Ukraine
swetshul@i.ua, popadynets.n@gmail.com
[4] Yuriy Fedkovych Chernivtsi National University, Kotsjubynskyi Str. 2,
58012 Chernivtsi, Ukraine
n.filipchuk@chnu.edu.ua
[5] Military Institute of Taras Shevchenko National University of Kyiv,
Lomonosov Str. 81, 03189 Kiev, Ukraine
fed.yana25@gmail.com
[6] Lviv Polytechnic National University, Bandera Str. 12, 79010 Lviv, Ukraine
ibilyk193513@gmail.com

Abstract. The article presents the results of the analysis of official data on the energy produced from biomass in Ukraine on the basis of application of a linear regression model. The comparative analysis of general trends in bioenergy was conducted for Ukraine and Poland. The analysis showed that while preserving existing trends, there is a high probability of not achieving the strategic goals of a country that has a significant potential for sustainable biomass. The necessity of application of economic and legal instruments that can stimulate the production and consumption of biofuels made on the basis of waste from agricultural enterprises and wood waste is substantiated for the stage of transition to bioeconomy. The bioeconomic approach to the development of bioenergy can only be realized if there is an interest in the rational use of biomass resources as of its founders, as well as the state and end users of biofuels.

Keywords: Bioeconomy · Biomass · Bioenergy · Biofuel ·
Economic and legal instruments · Interested parties

© Springer Nature Switzerland AG 2020
T. Ahram (Ed.): AHFE 2019, AISC 965, pp. 656–666, 2020.
https://doi.org/10.1007/978-3-030-20454-9_64

1 Introduction

Overcoming the growing energy dependence, economic, environmental and social problems facing the world is possible if all groups of society are attached to the process of rational use of resources and if confidence between them is increasing. Therefore an essential priority is the reorientation of linear model of economy into bioeconomy (closed loop economy), whose theoretical substantiation has not been clearly formed yet, though its specific role in achieving the world's goals for resource-saving is obvious. There are different views and definitions of the concept bioeconomy. We will point out the position of the European Commission [1], which defined bioeconomy as the production of renewable biological resources and the transformation of these resources and waste into value added products such as food, feed, biological products, as well as bioenergy. The bioeconomy is the production of biomass and the conversion of biomass into value added products, such as food, feed, bio-based products and bioenergy. It includes such sectors as agriculture, forestry, fisheries, food, pulp and paper production, as well as parts of chemical, biotechnological and energy industries [1]. That is, a bioeconomic approach can be applied to the development of various industries, including the energy sector, an important part of which is bioenergy based on the use of biomass resources (waste from agricultural enterprises and wood waste) for biofuel processing.

The application of economic and legal instruments at the stage of transition to bioeconomy will enhance using rationally sustainable biomass as a source of energy and creating value for consumers and society. The multifaceted nature of the problems that prevent it from being implemented requires the study of the concept of bioeconomy and the tools that will stimulate recycling waste (biomass) into valuable products for society, in particular biofuel.

Scientists and specialists from research institutions, international organizations (in particular, the US Agency for International Development (USAID), the International Renewable Energy Agency (IRENA), the European Commission, etc.) are working on bioenergy issues. The concept of bioeconomy is mentioned in the European Union's development strategies.

European Commission's Strategy and Action Plan on Bioeconomy state that in order to achieve more innovative and low-emission economy, reconciling demands for sustainable agriculture and fisheries, food security, and the sustainable use of renewable biological resources for industrial purposes, while ensuring biodiversity and environmental protection, we should focus on the following key areas: developing new technologies and processes for the bioeconomy; market development and competitiveness in bioeconomy sectors; encouraging politicians and stakeholders to work more closely together [2].

The authors [3] affirm that the bioeconomy development is related with the evolution of agriculture (especially the role of biomass) and of R&D sector. Their research proves that biotechnology is an essential component of the bioeconomy and that biomass remains the basis feedstock along with bio-based products obtained through biotechnological processes. The research provides some insights regarding the future importance of bioeconomy, especially regarding the opportunities and challenges rising from biomass usage and food supply. The authors [4] show how scientific research in different countries after the practical implementation of biotechnology affects their economic development.

At the same time, we emphasize that the bioeconomic approach to the development of bioenergy can only be realized only if its founders (agricultural and woodworking enterprises), the state, and end users of biofuels are interested in the rational use of biomass resources.

As reported in [5] there is a need for inclusion of citizens in the development of bioeconomy, because it requires an interactive common approach to empower various institutions and people to meet and discuss the development of their own living environment and environmental capability (i.e. those bioeconomy opportunities to achieve outcomes people value). Citizens may not be able to find and create new solutions and innovations which the bioeconomy strategy require, but it is citizens who live under varying access to opportunities and entitlements including environmental services.

In our opinion, citizens are the interested party in the development of bioeconomy, which, in case of its implementation, will receive economic, ecological and social benefits at the same level as business entities and the state represented by the relevant bodies.

The authors [6] identified the priority directions of the development of the market for solid biofuels as a factor that will positively influence the strengthening of the energy and economic components of the national security of Ukraine. Attention is drawn to the need to promote domestic production of solid biofuels from wood and agricultural biomass, aimed at the development of the domestic market.

Scientists [7] describe the key concepts, factors and limitations that keep down the growth of the bio-economy in the world and propose constructive measures of government support. The transition toward a bioeconomy will rely on the advancement in technology of a range of processes, on the achievement of a breakthrough in terms of technical performances and cost effectiveness and will depend on the availability of sustainable biomass [8]. This statement is valid for all regions of the world, but some countries, with significant biomass potential, do not achieve the planned strategic goals of the share of bioenergy in the energy balance of the country due to lack of proper state regulation, stimulation of its development and effective economic instruments.

An economic instrument is any instrument that changes the behavior of economic agents by influencing their motivation (as opposed to establishing a standard or technology) [9]. The International Working Group on Economic Instruments has devoted considerable attention to the disclosure of the role and mechanism of securing economic instruments in the world's environmental policy (International Working Group on Economic Instruments). The authors [10] report that economic instruments work by redistributing the rights and obligations of firms, groups or individuals so that they have the incentive and the right to act in a more environmentally responsible manner. The common element of all economic instruments is that they work at a decentralized level because of their impact on market signals, and not through government guidance. At the same time, the International Working Group on Economic Instruments states: "... in fact, legal and economic instruments often work in tandem. For example, governments can set permissible pollution levels for a region or country with market-based approaches such as sales permits that are used to allocate allowable emissions in an efficient way" [10]. The experts point out that determining which economic instrument to use and how to apply it to solve a particular problem can be a daunting task. The central elements of this process are the assessment of existing information and basic conditions; and working with stakeholders to better understand and support individual approaches [10].

According to Francis X. Johnson [11] "energy can be seen as the "glue" in the bioeconomy due to the high energy requirements of modern economies and the physical interrelationships between energy and non-energy products and processes. The bioeconomy will ultimately replace the fossil economy that is characterized by high reliance on non-renewable resources". In addition, social and economic sustainability needs to be addressed along with environmental sustainability. Although it is expected that the bioeconomy will be more equitable than the fossil economy, there is no guarantee that biological resources will not be monopolized as fossil resources have been. Policies and institutions are especially needed to promote the interests of developing and emerging economies where biomass resources are concentrated.

2   Materials and Methods

The purpose of the study is to substantiate the need to apply economic and legal instruments at the stage of transition to bioeconomy as a factor in achieving the goals set by the Energy Strategy of Ukraine. To achieve this purpose, the scientific literature, documents on strategic development in the EU and Ukraine and statistical data were analyzed. Such methods as calculation and construction, grouping, comparison, modeling, prognosis and algorithm were used in research. A comparative analysis of the dynamics of the increase in the use of bioenergy for Ukraine and Poland has been made taking into account the current trends in bioenergy. The linearization of bioenergy production forecasts for these countries for the next 15 years has been applied using the Levenberg-Markuat algorithm

3   Results and Discussion

According to Anja Karliczek, Federal Minister of Education and Research of Germany, one of the principles of bioeconomy is that it is local, but it must also think globally. At the same time, different regions may have different biological resources or be good in different fields of research and technology depending on local conditions [12]. In the context of bioenergy and bioeconomy, which are interconnected concepts, most countries in the world think globally, since they define ambitious plans for achieving some level of bioenergy development. But each country, given its economic development, geographical location, etc., has a different potential for generating bioenergy and needs a different approach to the application of economic and legal instruments.

The bioeconomy covers a wide range of established and emerging policy areas at the global level. EU, national and regional level which share and pursue their objectives, yet result in a complex and sometimes fragmented, political environment. The bioeconomic strategy requires a more substantive dialogue, in particular on the role of scientific advancement and the better interaction between existing bioeconomy-supporting policies at EU and Member States level [2], as well as those countries that want to become full members of the EU, in particular Ukraine.

The sectoral structure of the Ukrainian economy enables to rely on the great potential of biomass available for the production of biofuel and the need for its use in view of the problems of dependence on non-renewable energy sources. However, the pace of development of Ukrainian bioenergy, in comparison with global trends, is insufficient from the point of view of production and consumption, and it requires the use of certain economic and legal instruments to stimulate this process.

Before proposing the application of certain types of economic and legal instruments at the stage of transition to bioeconomy and development of the bioenergy sector, it is necessary to analyze the current state of this sphere and the proximity to the plans approved by the state and the possibility of achieving these goals in the long term. To do this, we will show schematically (Fig. 1) the percentage ratio of renewable energy resources (RER) and bioenergy resources (BER) to gross final energy consumption (GFEC) in Ukraine.

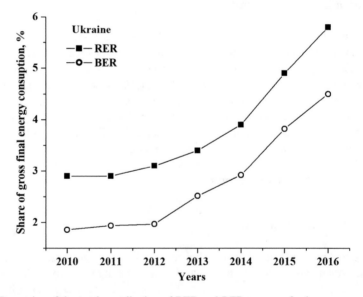

Fig. 1. Dynamics of the total contribution of RER and BER to gross final energy consumption in Ukraine. Source: developed by the author based on data [13]

Percentage of RER and BER in gross final consumption is presented for the period from 2010 to 2016. The share of RER in total final consumption was determined by the data of the State Agency for Energy Efficiency and Energy Conservation of Ukraine, and the share of BER was calculated on the basis of energy balances for 2010–2016, guided by the EU Directive 2009/28/EC [3]. As we see, the contribution of renewable energy resources, in particular bioenergy, continues to grow. Similar tendencies are observed in the countries of the European Union (Fig. 2).

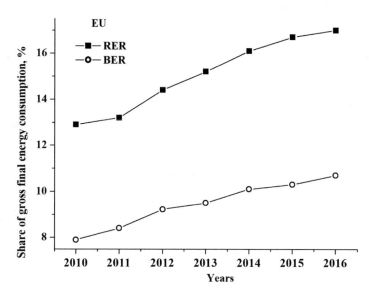

Fig. 2. Dynamics of the total contribution of RER and BER to GFEC in EU countries. Source: developed by the author based on data [15]

Comparing the dynamics of the total contribution of RER and BER to GFEC in Ukraine and the EU, we see that Ukraine is characterized by faster growth of RER and BER. This is primarily due to the political and economic factors affecting the realization of existing potential, which was almost out of focus 5 years ago.

According to the State Statistics Service of Ukraine [13], bioenergy resources in the structure of Total Primary Energy Supply (TPES) are characterized by an annual increase. The ambitious plans of the Ukrainian government predict further growth. Thus, according to the Energy Strategy of Ukraine for the period up to 2035 "Security, Energy Efficiency, Competitiveness" [14], by 2035 the contribution of BER in the TPES should be 11.5%.

The comparative analysis of total energy production from biomass Poland and Ukraine with the comparison of planned indicators in the period from 2015 to 2025, 2035 was made (Fig. 3). Poland was chosen as a comparative base because it is characterized by generally similar initial conditions of the economy, a similar structure of agriculture and similar geographic condition. We can conclude that the time dynamics of the primary energy obtained from renewable sources of biological origin for Ukraine in the 2010–2016 periods is close to the behavior of corresponding indicators for Poland in the 2005–2010. The stabilization and some slight downtrend in 2014–2016 is mainly caused to global price decrease of traditional energy sources. At the same time the predicted growth rate of planned indicative values for Poland accordingly [16] is set at the level of about 220 ths. TOE per year when for Ukraine this

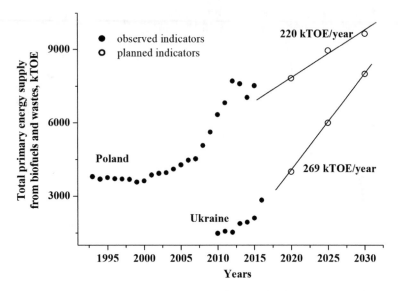

Fig. 3. Dynamics of BER for Ukraine and Poland. Source: developed by the author based on data [14–16]

parameter is about 18 percent bigger – 269 ths. TOE per year. It is clear that these targets for Ukraine are unrealistic and will not be realized in practice without stimulus politics.

The analysis shows that while preserving existing trends, there is a high probability that the goals of the Energy Strategy of Ukraine will not be achieved. Therefore, we consider the economic and legal instruments of influence, which should be implemented in response to existing problems at the stage of transition to bioeconomy. Their task is to create conditions for the economic, ecological and social interest of the participants in the bioenergy market on the basis of the fulfillment the requirements of legislation and demands of society.

It can be realized only if there is commitment of all participants of the process, that is, the participants of the bioenergy market, such as business entities, state bodies and scientific institutions (which are interested parties in establishing cooperation on the production, sale and consumption of biofuels, namely: wood processing and agricultural enterprises – waste generators, suitable for further processing on solid biofuels); producers of solid biofuels (often the same companies, within the main activities of which waste suitable for further processing is formed); machine-building enterprises specializing in the manufacture of solid fuel boilers, equipment for the production of briquettes and granules; research institutions engaged in the design of specialized plants, as well as training personnel with required qualifications; local authorities; financial and credit institutions, etc. whose activities in interaction can achieve a synergetic effect and create favorable conditions for attracting investment [17].

We emphasize that the state incentive policy in the form of certain instruments should be used only for the development of production and consumption of biofuels made from so-called "sustainable resources", that is, taking into account the bioeconomic aspect. According to [18] biomass produced in a sustainable way – the so-called modern biomass – excludes traditional uses of biomass as fuelwood and includes electricity generation and heat production, as well as transportation fuels, from agricultural and forest residues and solid waste. The authors [18] note that "traditional biomass" often produced in an unsustainable way and it is still used as a non-commercial source, usually with very low efficiencies for cooking in many countries. In most statistics, renewables include traditional biomass, despite the fact that in many countries the widespread use of trees as fuelwood is not considered sustainable.

The need to isolate the biomass produced by the sustainable way involves the need to avoid the use of land for the sowing food crops in favor of energy crops, given that "their production is more cost-effective" [19]. Concerns also arise internationally: the report of the Food and Agriculture Organization of the United Nations "The state of affairs in the field of food and agriculture" points to the need for in-depth studies on the role of biofuels in food and energy security [20]. However, it is unequivocal to assert that the threat of using land for cultivating crops suitable for processing into biofuels instead of food crops, cutting down forests for processing into biofuels is not appropriate, as Ukraine has significant areas of untapped land. It should be noted that in the study [21] a model and system algorithm for ensuring balanced development of agriculture in terms of using the potential of biomass energy has been developed.

But in order to prevent the use of bioenergy exclusively for the economic "profit" of a particular economic entity on the basis of the lack of "sustainable use" of biomass and to prevent negative environmental impact of bioenergy, the state incentive policy with economic and legal instruments should be used. First of all the secondary raw materials of agriculture and the wood industry as the basis for the production of biofuels should be encouraged because they meet the requirements for sustainability under Art. 17 (3) -17 (5) of the Directive of the European Parliament and of the Council of Europe 2009/28/EC [22]. Thus, according to the document, in order to meet the requirements of sustainability, biofuels can not be produced in the areas of high biodiversity value and carbon stocks [22].

We believe that the most effective are economic instruments supported in some cases by the state regulation, that is, economic and legal instruments. In the Ukrainian economy, an element of regulation is necessary, for example, when it comes to the obligation to use vegetable waste for processing into biofuels, as opposed to burning it "near the road". This is due to the fact that the mechanism of application of the economic instruments in legal acts gives them practical. At the same time, economic instruments can operate without state intervention, for example, it is a classic market approach like transparent competition, which in the context of the biofuel market manifests itself in free pricing based on the relationship of supply and demand. Although justice on the biofuels market can be achieved, for example, through the existence of a legally-established auctions. A purely market-based approach does not always give the desired result.

"Some false state policies are a real threat to the sustainable use of natural resources and ecosystems, such as subsidies for goods, services or practices that cause serious environmental degradation. Such subsidies are so widespread that many economists (especially those working with developing countries) call the "exclusion of subsidies" one of the main instruments of environmental policy" [23]. Thus, this situation is also relevant for Ukraine in the context of gas subsidies, when one part of society establishes RER in the households, while the other does not make any effort to save heat and electricity, because it uses subsidies. Therefore, the "withdrawal of subsidies" and their orientation towards the promotion of energy saving measures can be considered as an economic and legal instrument of state incentive policy. In Ukraine, only since 2018, an important step has been taken in this direction, since an unified state register of recipients of subsidies has been created, the introduction of which will contribute to attracting a certain part of citizens to the process of energy saving, and as a result to the transition to solid biofuels as a source of renewable energy.

Among the main measures for realization strategic goals in the RER sector, the increase of biomass usage in the generation of electricity and heat by informing about the possibilities of using biomass as a fuel in individual heat supply has been mentioned in the Strategy [14]. We believe that such information may be provided by local authorities, which should have information about the producers of biofuels in their region and the possibilities for their implementation to the local population. The most economically feasible way is to publish such information (with contact details of potential suppliers) on the sites of Utilities or other specially created platforms. The possibility of realization of this tool of social interaction will be possible in practice when it is legally fixed.

4   Conclusions

The transition to bioeconomy is an irreversible process, but developing countries are confronted with a number of problems that need to be addressed. It is impossible to apply the same management decisions to stimulate bioenergy in different countries, taking into account different political, economic, social and environmental conditions. Having analyzed the potential of bioenergy development in Ukraine, we conclude that in order to achieve the strategic goals of Ukraine it is necessary to apply economic and legal instruments for the development of production and consumption of biofuels made from so-called "sustainable resources" that is, taking into account the bioeconomic aspect. Determining the specificity of economic and legal instruments and guiding the motivational components of implementing each of them by different groups of stakeholders is the direction of our further research.

References

1. European Commission. Communication from the commission to the European parliament, the council, the European economic and social committee and the committee of the regions. In: Innovating for Sustainable Growth: A Bioeconomy for Europe (2012). https://ec.europa. eu/research/bioeconomy/pdf/official-strategy_en.pdf

2. European Commission. Innovating for sustainable growth. A bioeconomy for Europe EU publications. Brussels (2012). https://publications.europa.eu/en/publication-detail/-/publication/1f0d8515-8dc0-4435-ba53-9570e47dbd51
3. Toma, E., Stelica, C., Dobre, C., Dona, I.: From bio-based products to bio-based industries development in an emerging BioEconomy. Rom. Biotechnol. Lett. **22**. University of Bucharest (2016). https://www.rombio.eu/docs/Toma-Dona_final.docx
4. Wisz, G., Nykyruy, L., Yakubiv, V., Hryhoruk, I., Yavorsky, R.: Impact of advanced research on development of renewable energy policy: case of Ukraine. Int. J. Renew. Energy Res. **8**(4) (2018). https://www.ijrer.org/ijrer/index.php/ijrer/article/view/8688
5. Mustalahti, I.: The responsive bioeconomy: the need for inclusion of citizens and environmental capability in the forest based bioeconomy. J. Clean. Prod. **172**, 3781–3790 (2018)
6. Popadynets, N.M., Maksymiv, Yu.V: Developemnt of domestic solid biofuel market in Ukraine under current conditions. Economic Annals-XXI **159**(5–6), 93–96 (2016)
7. Bobyliv, S.N., Mikhailova, S.Yu., Kiryushin, P.A.: Bioeconomics: problems of formation. Econ. Taxes Right **6**, 21–25 (2014)
8. Scarlat, N., Dallemand, J.-F., Monforti-Ferrario, F., Nita, V.: The role of biomass and bioenergy in a future bioeconomy: policies and facts. Environ. Dev. **15**, 3–34 (2015)
9. Panayotou, T.: Instruments of Change: Motivating and Financing Sustainable Development. Earthscan Publications, UNEP (1998)
10. International Working Group on Economic Instruments. Opportunities, prospects, and challenges for the use of economic instruments in environmental policy making United Nations Environment Programme (2002). https://unep.ch/etu/etp/events/Economic_Instruments/Opportunities.pdf
11. Johnson, F.X.: Biofuels, bioenergy and the bioeconomy in north and south. Ind. Biotechnol. **13**(6), 289–291 (2017)
12. Karliczek, A.: Federal Minister of Education and Research of Germany. Global Bioeconomy Summit Conference Report. Innovation in the Global Bioeconomy for Sustainable and Inclusive Transformation and Wellbeing (2018). http://gbs2018.com/fileadmin/gbs2018/GBS_2018_Report_web.pdf
13. State Statistical Service of Ukraine. Energy balance for 2010–2016 (2018). http://www.ukrstat.gov.ua/
14. Cabinet of Ministers of Ukraine. 2035 energy strategy of Ukraine "Safety, energy efficiency, competitive ability" (2017). http://zakon2.rada.gov.ua/laws/show/605-2017-D180/paran2#n2
15. European Commission. The energy balance 2010–2016 (Energy balance sheets 2010–2016 edition) (2016). https://ec.europa.eu/eurostat/web/products-statistical-books/-/KS-EN-18-001
16. European Commission Directorate-General for Energy. EU energy trends to 2030 (2009). https://ec.europa.eu/energy/sites/ener/files/documents/trends_to_2030_update_2009.pdf
17. Maksymiv, Y.: Reporting as an important tool in ensuring interaction between stakeholders. Actual Probl. Econ. **4**(178), 304–310 (2016)
18. Goldemberg, J., Teixeira Coelhob, S.: Viewpoint Renewable Energy—Traditional Biomass vs. Modern Biomass Energy Policy, vol. 32, pp. 711–714 (2014)
19. Sinchenko, V.M., Gomentik, M.Ya., Bondar, V.S.: Classification of biofuels and their production prospects in Ukraine. Bioenergetics **1**, 5–6 (2014)
20. Committee on World Food Security. Biofuels and Food Security. 164 p. Rome (2013)
21. Yakubiv, V., Zhuk, O., Prodanova I.: Model of region's balanced agricultural development using the biomass energy potential. Economic Annals-XXI, **3–4**, pp. 86–90 (2014)

22. European Parliament. Directive 2009/28/EC of the European Parliament and of the Council of the promotion of the use of energy from renewable sources and amending and subsequently repealing Directives 2001/77/EC and 2003/30/EC (2009). http://eur-lex.europa.eu/legal-content/EN/TXT/?uri=CELEX:02009L0028-20151005
23. Sterner, T., Coria, J.: Policy Instruments for Environmental and Natural Resource Management, 613 p. New York and London (2015)

Author Index

© Springer Nature Switzerland AG 2020
T. Ahram (Ed.): AHFE 2019, AISC 965, pp. 667–669, 2020.
https://doi.org/10.1007/978-3-030-20454-9